Vehicle Thermal Management Systems Conference (VTMS 12)

Nottingham, United Kingdom
10 - 13 May 2015

ISBN: 978-1-5108-5611-0

Printed from e-media with permission by:

Curran Associates, Inc.
57 Morehouse Lane
Red Hook, NY 12571

Some format issues inherent in the e-media version may also appear in this print version.

Printed by Curran Associates, Inc. (2018)

For permission requests, please contact Institution of Mechanical Engineers (IMechE) at the address below.

Institution of Mechanical Engineers (IMechE)
One Birdcage Walk
London, SW1H 9JJ
United Kingdom

library@Imeche.org
www.imeche.org

Additional copies of this publication are available from:

Curran Associates, Inc.
57 Morehouse Lane
Red Hook, NY 12571 USA
Phone: 845-758-0400
Fax: 845-758-2633
Email: curran@proceedings.com
Web: www.proceedings.com

Contents

Heat Exchange

Alternative Powertrain

Engines

Simulation / Energy Management

Waste Heat Recovery

Heat, AC & Energy Management

Underhood & Simulation

Heat & AC

VTMS 12 ORGANISING COMMITTEE

Heat Exchange

Optimizing the cost of thermal management components through the use of high performance engineering thermoplastics

M WRIGHT
Solvay Specialty Polymers, UK

ABSTRACT

Water/Glycol coolant flow is needed in many more parts of the modern powertrain systems to help drive operating efficiency improvements. Advanced and integrated coolant flow in combination with down-sized engines has increased the design complexity and performance requirements of thermal management components. Weight reduction and productivity improvements are also being sought in order to provide suitable cost effective solutions. High performance engineering thermoplastics are leading candidates for specification in the next generation of thermal management components. This paper will highlight the suitability of one family of thermoplastics, Polyphthalamides, matching the complex needs of thermal management components including mechanical property retention and design features that can enable productivity gains.

1 INTRODUCTION

Thermal Management is defined as any technology that uses a medium (air or liquid) to cool the heat that is produced by the engine [1]. The growth of highly down-sized turbo-charged diesel and gasoline engines with direct fuel injection plus the introduction of new exhaust gas recirculation cooler concepts require more advanced thermal management technology. The market penetration of the new turbo-charged family of engines is mainly driven by the need to comply with new regulations surrounding CO_2 and tailpipe emissions and fuel efficiency targets [2,3]. The consequence to the components within the thermal management of the engine is to drive increased architectural complexity, increased temperature and pressure and fuel the need for new functional components that overall allow for a more efficient use of energy and a tighter temperature control.

Historically thermal management was related to engine block heating and cooling. Here the 3 primary functions were (1) to dissipate the heat created from the engine thereby preventing the engine from over-heating (2) enabling faster heat up periods to prevent engine component wear and (3) maintain engine temperature. The same basic principles still hold true given that vehicle drive uses barely one-third of the fuel energy however modern water/glycol circuits have become much more complicated with high and low temperature circuits, electrically controlled thermostats and primary and secondary water pumps [4].

As thermal management becomes more demanding and more complex, other features such as weight reduction, more efficient packaging and productivity benefits are also a very important dynamics to consider. This is where the use of high performance engineering plastics (HPEP) becomes important [5]. Thermoplastics inherently allow for more part or feature integration compared to metals or thermosets thus enabling part cost optimization and weight reduction and now there is a family of thermoplastics that allow metal replacement in some of the most demanding applications. Key performance attributes of these thermoplastics are thermal stability, mechanical strength, chemical resistance, toughness.

This paper will highlight the suitability of HPEP's, in particular Polyphthalamides (PPA), to next generation thermal management components and also describes the practical design steps that need to be taken to ensure that a metal replacement project is successful. PPA's are a family of high performance polyamides that are one of the fastest growing thermoplastics within thermal management applications. One of the main reasons is the chemical resistance behaviour to anti-freeze even too modern long life coolants such that extended warranty periods can be achieved.

2 POLYPHTHALAMIDES

Polymers are made of repeating units of monomers to form a long chain. Thermosets and Thermoplastics are the main classifications. PPA's are thermoplastic so can be re-melted and are semi-crystalline in nature; contain both crystalline and amorphous regions. PPA materials are polyamides and as such contain the –CH-NO- repeat unit in their structure. However they differ from their more well-known aliphatic counterparts, polyamide 66, by incorporating aromatic or carbon ring structures into the polymer back bone [6]. This imparts the temperature, mechanical, chemical and dimensional characteristics of PPA compounds. PPA's were first introduced into the automotive market more than 20 years ago. Thermal management was quickly established as one of the most important application areas however other major automotive applications can be located within Fuel systems, Electrical / Electronic parts, Transmission and Chassis. PPA's provide a bridge between the lower performance of standard engineering thermoplastics (Polyamide 66, Polybutylene Terephthalate (PBT), PolyOxyMethylene (POM)) and the high cost of thermoplastics like Liquid Crystal Polymer (LCP) and Polyetheretherketone (PEEK). PPA's typically exhibit better practical toughness compared to another high performing thermoplastic, Polyphenylene Sulfide (PPS).

3 MECHANICAL PERFORMANCES FOR THERMAL MANAGEMENT APPLICATIONS

One of the critical drivers to the adoption of PPA's is their mechanical property retention performance in long term exposure to water/glycol based coolants. This has become more critical through the increased adoption of highly turbocharged down-sized engines. The consequence being that the coolant test temperature stated in OEM specifications have increased, some requests now specify thermoplastic materials to be tested above 130°C. When water glycol specifications require testing at 140°C and above, PPS would be the most suitable thermoplastic choice wrt mechanical retention of properties and dimensional stability.

The chemistry of automotive coolants uses ethylene glycol / water as a base. Corrosion inhibitors are added such as silicates but they tend to degrade quickly over time. Most of the current corrosion inhibitor technology is based on organic acid technology (OAT) or hybrid organic acid technology (H-OAT). The organic acids used have better stability, allowing for longer time between changing of engine coolant hence the term "long life coolants".

Solvay completed a 5000hr water/glycol study for a major automotive OEM comparing the performance of four materials in Halvoline Dexcool XLC organic acid technology (OAT) long life coolant. Coolant tested as a 50/50 mix with water. The materials tested were 35% glass reinforced polyamide 66, 33% glass reinforced grade of PPA, 45% glass reinforced grade of PPA and 40% glass reinforced grade of PPS. All polyamide grades have modified formulations for improved hydrolysis resistance.

During the testing protocol both tensile (strength, modulus and elongation), flexural (strength, modulus and elongation) and Izod notched impact strength were measured. Injection moulded samples were immersed in the water/glycol mixture at 125°C with mechanical retention of properties being measured on samples every 500hrs up to 5000hrs. Mechanical property retention was measured at 23°C. Figures 1, 2 and 3 provide mechanical retention of properties for tensile strength, tensile elongation and tensile modulus with figure 4 providing retention of Izod notched impact strength.

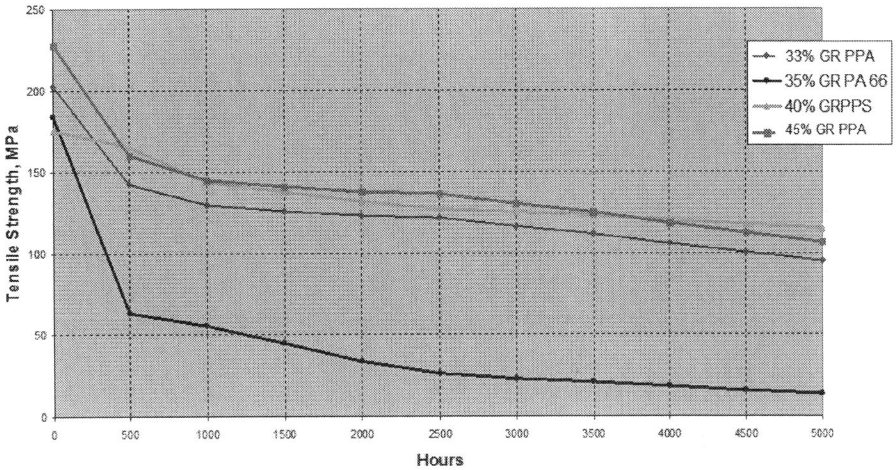

Figure 1: Tensile strength retention (ISO178) of 35%GF PA66, 33% GF PPA (Amodel® AS-1933HS), 45%GF PPA (Amodel® AS-1945HS) and 40% GF PPS exposed to Dexcool XLC at 125°C

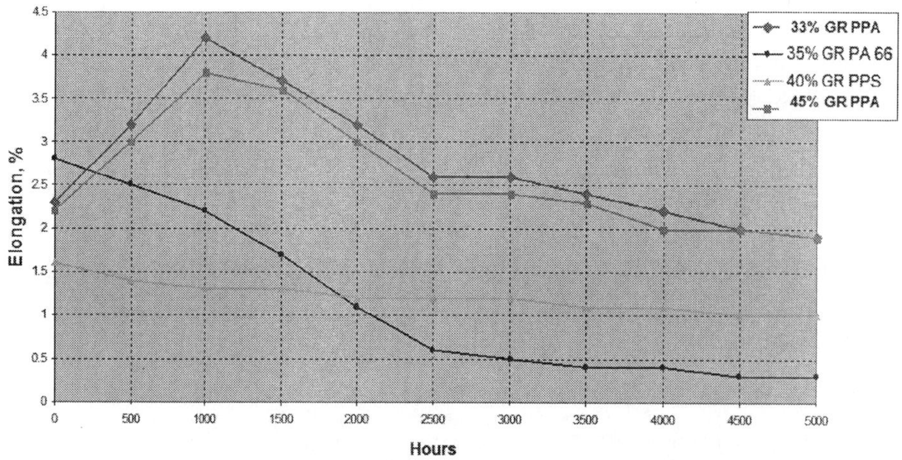

Figure 2: Tensile strain at break retention (ISO178) of 35%GF PA66, 33% GF PPA (Amodel®AS-1933HS), 45%GF PPA (Amodel®AS-1945HS) and 40% GF PPS exposed to Dexcool XLC at 125°C

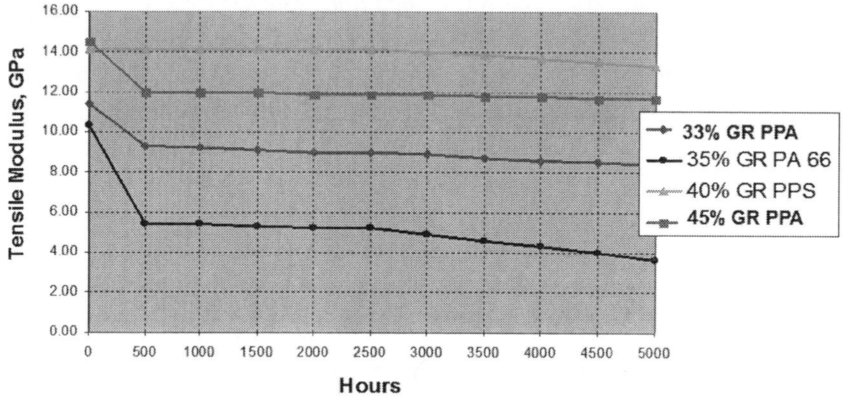

Figure 3: Tensile modulus retention (ISO178) of 35%GF PA66, 33% GF PPA (Amodel® AS-1933HS), 45%GF PPA (Amodel® AS-1945HS) and 40% GF PPS exposed to Dexcool XLC at 125°C

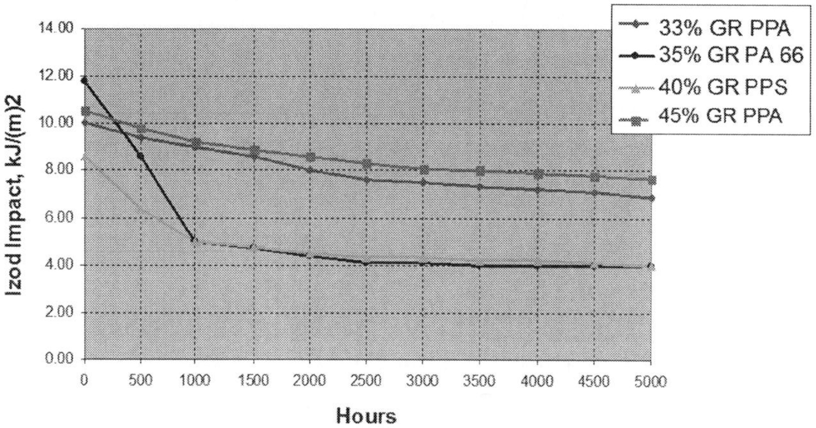

Figure 4: Izod notched impact strength retention (ISO179) of 35%GF PA66, 33% GF PPA (Amodel® AS-1933HS), 45%GF PPA (Amodel® AS-1945HS) and 40% GF PPS exposed to Dexcool XLC at 125°C

PA66 material shows the largest drop in properties, losing most of its performance within the first 500 hours of the testing protocol. As the temperatures and pressures of the modern down-sized engine increase more applications will need to look towards thermoplastic materials that can provide an improvement in mechanical property retention, which is one of the main drivers toward the increased specification of higher performing thermoplastics like either PPA or PPS. The tensile strength retention of PPA and PPS grades, as shown in figure 1, are very similar over the testing period. The clear differentiation between PPA and PPS is apparent in elongation at break, see figure 2 and impact properties, see figure 4, and were PPA materials show superior property retention. PPS grades provide a benefit in tensile modulus, see figure 3. Therefore when considering material selection for components that are in direct contact with water/glycol environment, PPA's provide the more practical durability with respect to impact and elongation which are attributes that provide most benefit is applications like thermostat housings, water inlets/outlets and water pump housings.

3.1 Creep

Many thermoplastic components within the engine compartment are bolted directly to the engine block or have clip features that are used in assembly. Examples of these are provided in Figure 5.

Additional to retention of long term mechanical properties, creep is another important mechanical property to characterize the long-term performance of a thermoplastic under constant load. Figure 6 illustrates the influence of glass fibre loading on the tensile creep properties of PPA when exposed to a temperature of 150°C and a constant stress of 41.4MPa. Figure 6 also compares the relative creep performance of PA66, PPA and PPS materials under the same conditions.

Figure 5: Typical automotive water outlet applications

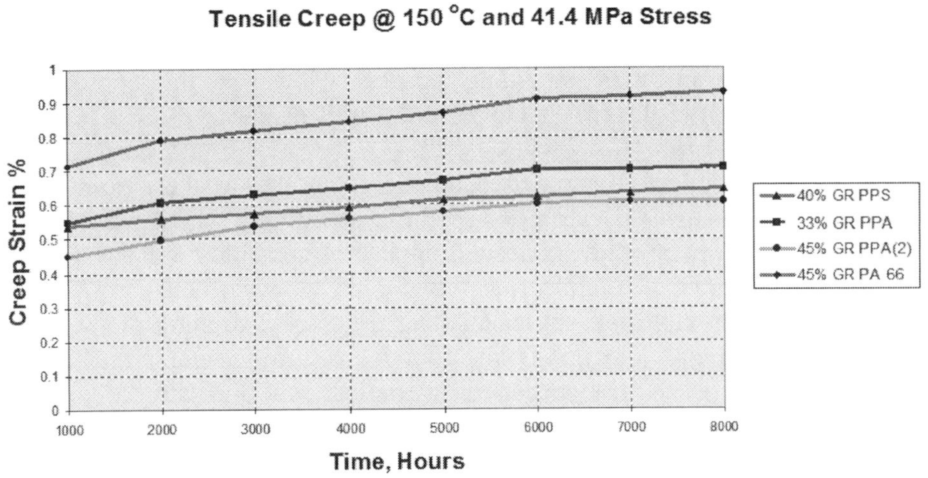

Figure 6: Tensile Creep Performance over time of 35%GF PA66, 33% GF PPA, 45%GF PPA and 40% GF PPS when exposed to 150°C and 41.4MPa Stress

In Figure 5, the first application has a long span between the individual bolt holes therefore making a 45% glass reinforced grade of PPA more suitable. In the second example the housing is clipped into position and the constant spring load of the clip exerts a creep load on the material. Here again a higher glass loading would be desirable although this does increase component weight slightly due to higher specific gravity of a thermoplastic compound with increased glass content (1.57 to 1.45 for 45% GR compared to 33% GR PPA respectively).

Figure 7: Typical automotive thermostat housing containing (1) press fit insert, (2) moulded-in insert and (3) ultrasonically inserted

Another design consideration related to creep is the use of metal inserts which are typically used to maintain bolt torque when thermoplastic components are secured to engine blocks etc. There are several different methods for using the metal insert which are highlighted in Figure 7.

Considerations to take into action when choosing a particular metal insertion method will greatly depend on the pull-out force and torque requirement of the application and also if there are any other design restrictions. Moulded in inserts, typically made from Brass, provide the highest performance with respect to pull-out force, vibration resistance and shock and also allow frequent disassembly. The disadvantages are that they need pre-heating before moulding and the cycle time of the moulding process can be extended. When considering press fit inserts it is vital to maintain tight dimensional tolerance as variations could potentially increase the risk of cracking. To further optimize the design, the insert should be 0.05-0.1mm taller than the plastic part. This design feature enables the control of the different thermal expansion coefficients of the materials used and stress relaxation behaviour.

3.2 Weldline Strength

Thermal management component complexity is increasing therefore it is mandatory that computer simulation is utilized to establish the most optimum design and most suitable gate location for the injection moulding process. However, the consequence of increased part complexity is the inevitable presence of weld-lines.

Weld-lines within an injection molded component are the weakest point therefore computer simulation analysis is extremely useful in identifying whether the location of the weld-lines will be in a critically loaded area. PPA materials are one of the best performing high performance engineering thermoplastics when considering weld-line strength. This has been confirmed by calculating the tensile strength of ISO test bars that have been molded with injection points at both ends of the sample thus inducing a weld-line in the center of the test specimen. Internal Solvay Specialty Polymer testing showed the weld-line tensile strength retention of various PPA grades to be 35-42% of the datasheet value for tensile strength. The study included competitor PPA grades and the glass fiber loading was restricted to 33% or 35% glass reinforced grades. In comparison, in the same study PPS tensile strength retention ranged from 14-20%. The PPS grades used in the analysis ranged from 40-65% reinforcement levels.

Part design and processing can also be used to help increase weld-line strength. Figure 8 illustrates how a typical weld-line is formed when two thermoplastic flow fronts meet. One method for optimizing weld-line strength is to reduce the flow length which allows for more effective packing. If the processing window allows it, slightly higher melt temperatures can also be utilized during the molding process to keep the melt front fluid for longer. Figure 8 illustrates how a flow front is formed when two melt fronts meet at 180° to each other. More effective weld-line strength can be achieved by enabling flow fronts to meet at angles of approximately 130° which is known as a Meld-Line.

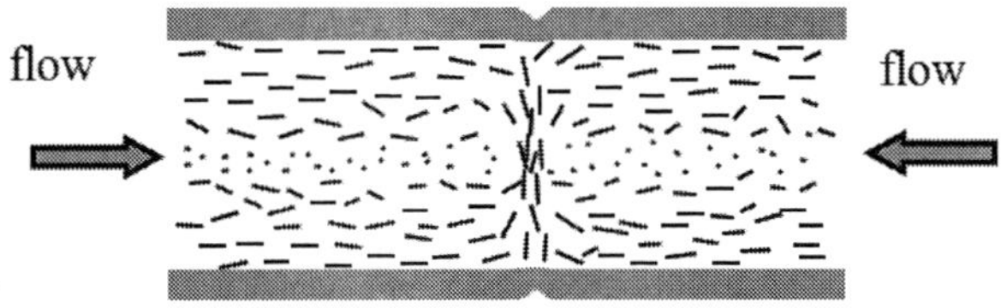

Figure 8: Thermoplastic flow that causes weld-lines

4 "PULL-PASS" BARB CAPABILITY

So far we have investigated mechanical retention performance in water/glycol environments which lead to the increased adoption of HPEP for modern thermal management components. Also some basic but critical design features have been explained. There is another feature that also needs to be taken into account when looking at thermal management components which is the key driver to productivity gains in these complex components. Thermostat housings, water outlets, water pump housings all have inlets and outlets that allow the water/glycol to flow in and out of the component. The influence of the design of the inlet/outlet feature on component manufacturability is often under-estimated. Elastomeric hoses are normally pushed on over the inlets and outlets. Specifications indicate the required minimum pull off force. Retention of the hoses is achieved through the use of an under-cut or barb feature. In traditional metal designs these features needed to be machined and therefore incur an additional secondary operation cost. As components move towards use of thermoplastic the most conventional method for producing the bead feature on the inlet and outlet is to use an injection mould tool with a parting line on the inlet/outlet feature as indicated in figure 9.

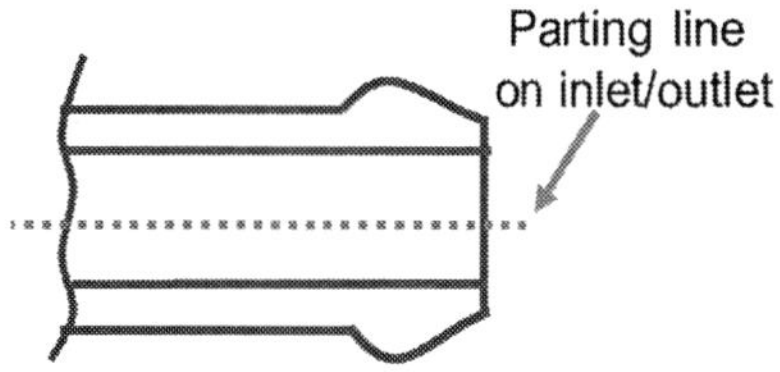

Figure 9: Standard injection moulding tool design that enables "pull-pass" features to be moulded in thermoplastics

The split in the tool enables the formation of this feature but results in a witness line on the part which over time can become more severe due to tool wear. Flash can start to appear on the part and tool maintenance is necessary to prevent this becoming an issue or flash needs to be manually removed, adding cost and reducing cost effectiveness. The presence of the flash in the parting line area increases the risk of potential leak paths. Some OEM's are limiting the maximum level of witness line height that can be tolerated ranging from 0.01mm to 0.1mm. Some OEM's are even not allowing any parting lines on inlets and outlets of thermal management components.

This is an issue when component performance is increasingly directing specification of higher performance engineering thermoplastics. These materials are normally structural but not very elastic in performance. The production process that is required to produce inlets and outlet features with no parting line would require the material to be able to be what is called "Pull-Pass" Barb capable. This feature is shown in Figure 10.

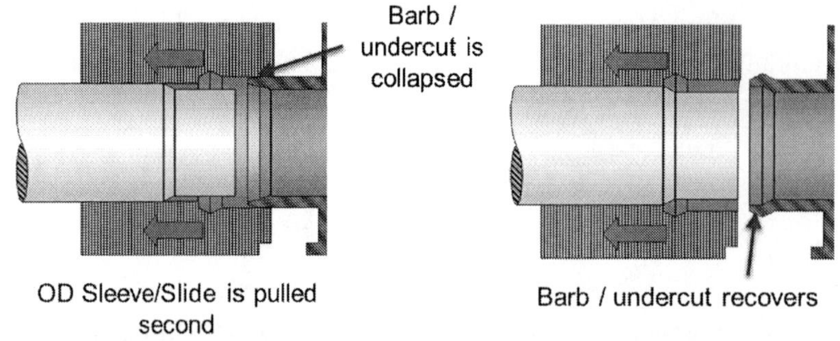

Figure 10: Injection moulding process illustrating "pull-pass" barb capability

The plastic component is represented in green colour. To de-mould the under-cut or "Pull-Pass" feature without the use of a parting line requires the thermoplastic material to have an elastic deformation that would allow the barb or undercut to momentarily collapse or deform during the ejection process which would then be recovered with no permeant deformation.

The most important mechanical property that would enable this feature is elongation at break at the tool temperatures that are needed to process HPEP. Typically for water/glycol resistant grades of PPA and PPS the tool temperatures are 130-160°C. Therefore if elongation at break was considered at these temperatures the most suitable materials could be identified. Figure 11 shows elongation at break at 130°C.

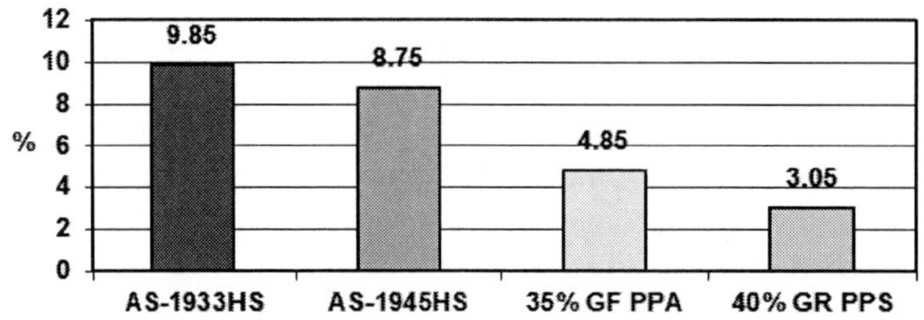

Figure 11: Elongation at break for Amodel® AS-1933HS and AS-1945HS, competitive 35%GF PPA grade and 40% GF PPS measured at 130°C

The figure highlights that the most suitable HPEP materials that can provide the combined mechanical performance with respect to retention of properties and creep in long term exposure to water/glycol and pull-pass barb capability to enable manufacturing productivity benefits are Amodel® grades AS-1933HS and AS-1945HS.

Solvay investigated the "Pull-Pass" Barb capability of Amodel® PPA and other competitive PPA's by investing in an injection moulding tool that had inter-changeable inserts. 7 different designs of "Pull-Pass" barb were studied with Undercut Ratio's and Ejection Angles varying from 0 to 16.7% and 130° to 160° respectively. Figure 12 should be used to understand Undercut Ratio and Ejection angle.

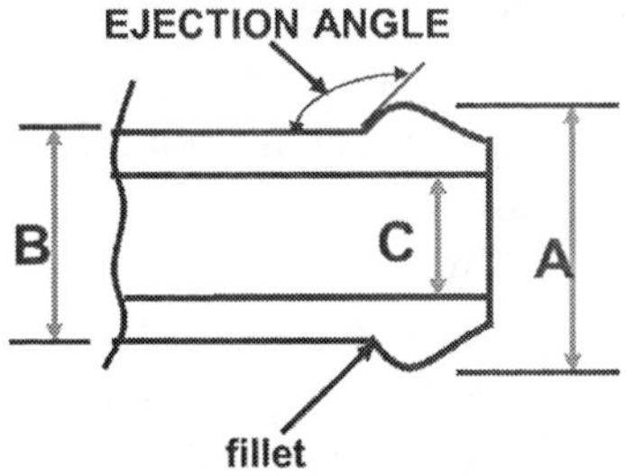

Figure 12: Diagram illustrating a typical inlet/outlet design with main dimensions used to calculate under-cut ratio

Undercut ratio of a given design is calculated through the following equation (A-B)/C *100. During the injection moulding trials material performance was assessed through several functions: - (1) could the part be de-moulded, (2) was there permanent deformation during the de-moulding, (3) the surface of the de-moulded part was smooth or damaged and (4) what ejection pressure was needed for the de-moulding. All materials were moulded according to their published recommended conditions.

Some designs (i.e A=17mm, Undercut Ratio = 10.8%, ejection angle = 160°) did not differentiate between any of the PPA materials. However as the smaller diameter designs with more severe undercut ratios with lower ejection angles were investigated it become clear that the superior elongation behaviour of Amodel AS-1933HS and AS-1945HS differentiates these grades performance versus other PPA grades. Quality issues associated with competitive PPA grades are shown in Figure 13.

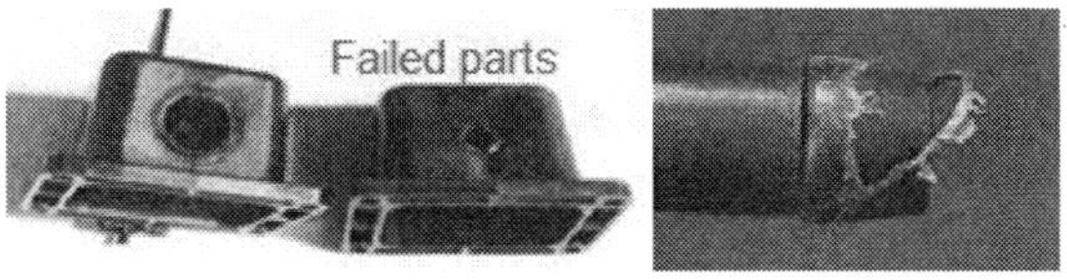

Figure 13: Typical failure modes observed during "Pull-Pass" Moulding trials on competitive PPA Materials

The first image illustrates a failure mode in which the part could not be de-moulded at all with the inlet/outlet feature breaking off and being retained in the tool rather than be ejected with the part. This is the most severe of the failure modes. The second image provides an example of a material that could be ejected from the tool but the act of ejection has severely damaged the outer surface of the part.

The benefit of "Pull-Pass" Barb capability is best illustrated in a commercial example. Figure 14 shows a 10 port water outlet that is used on a Japanese OEM vehicle platform manufactured for Asian and North American markets. Material specified is Amodel® AS-1933HS BK324.

Figure 14: "Pull-Pass" Barb capability of Amodel® AS-1933HS BK324 in 10 Port water outlet

In the previous vehicle platform this moulding was actually two separate metal parts. Thermoplastics have allowed the integration of 2 separate parts to be achieved in one single moulding thus enabling productivity gains through injection moulding and eliminating the mounting hardware for the second part which is a key benefit when current under-hood packaging space is at a premium. In addition each of the 10 ports was moulded without any split lines on the tooling in accordance with OEM requirements. The performance requirements of 5000hrs at 130°C in water/OAT glycol coolant resulted in HPEP specification. This meant that only Amodel® PPA resins were able to meet the performance requirements and the design complexity constraints imposed on this part. The design won an SPE innovation award in 2012 for automotive part design.

5 CONCLUSION

Thermal Management components are becoming more complex in design and performance requirements are increasing. HPEP are excellent candidates to meet the challenge set by these new set of requirements. One family of thermoplastics, PPA's, are particularly suited to provide the long term mechanical property retention required for extended warranties and also the design flexibility such that productivity gains can continue to be made. However, Amodel® PPA seems to be the best suited due its differentiating "Pull-Pass" barb capability.

REFERENCE LIST

1. E.Barrieu, D.Gafforelli, N.Becker "Powertrain Efficiency : a System Approach to Speed up Engine Warm Up during Cold Starts thanks to Exhaust Heat Recovery Combined with Suitable Engine Thermal Management", SIA The Spark Ignition Engine of The Future, Dec 4th, 2013
2. T.Gissom, Automotive World Seminar : Supercharging vs Turbocharging, Dec 2012
3. The Automotive Turbocharger and Supercharger Report 2013, HIS
4. A.Carman, "Electric Water Pumps in Vehicles Current and Future", ITB Automotive Air and Cooling Systems Conference, June 5th, 2014
5. A.Carvalho, "Engine Thermal Management and High Performance Polymer Applications", SPE Auto EPcon, April 30th 2013
6. ASTM D5336 "Standard Specification for Polyphthalamide (PPA) Injection Molding Materials

High temperature polyamides: the latest developments in high temperature applications

E SPINI
Radici Plastics, Italy

Using a novel approach, RadiciGroup scientists have developed two specialty polyamide families that are stable at high temperatures greater than 170°C and up to 230°C. This technology has been developed mainly in response to newer automotive performance demands for under-bonnet applications, including charge air cooler end tanks, turbo air ducts, etc. The first product family, Radilon® HHR[1] (High Heat Resistance), can be used at long-term service temperatures in air, from 180°C to 210°C. To protect against thermal degradation in such applications, a blend of special additives has been incorporated to inhibit the thermal oxidative process as much as possible. This process is a surface phenomenon which, in polyamides with conventional stabilization, leads to the degradation and carbonization of an increasingly deeper polymer layer, with a consequent dramatic reduction in the resistant section [1].The second family, Radilon® XTreme HHR, has exceptional retention of properties after heat ageing and has been designed to resist thermal degradation at continuous use temperatures of up to 230°C. Radilon® XTreme HHR contains a semi-aromatic component that, in combination with the additives used against thermo-oxidation, further raises the continuous service temperature of the product.

1 INTRODUCTION

During recent years, the automotive industry has seen a trend towards smaller, more efficient engines installed in less space but nonetheless capable of delivering high performance and cutting emissions of harmful gasses such as CO_2. All these factors have contributed to a general increase in under-bonnet temperatures and a harsher operating environment for components in engine cooling and air intake systems [2].

In this paper we are going to focus our attention on new materials developed by RadiciGroup Plastics, particularly materials designed for applications where the component is in direct contact with hot air.

In the last few years, continuous operating temperatures for turbo engine components, such as intercooler end tanks and turbo ducts (Figure 1), have jumped from values between160 and 170°C to the 220 to 230°C range. Pressure pulsations and induced vibrations cause higher mechanical fatigue on components, and thus the related test

[1] Radilon® is Radici Plastics registered trademark

conditions for material approval have generally become harsher. The test duration has become longer and, in the case of pressure pulsations, test pressure values have been increased from 1.5 to around 3.7 bars, with a corresponding increase in stress on the material.

Figure 1. Turbo duct prototype made with Radilon® HHR successfully tested at temperatures of up to 210°C.

Lastly, of crucial importance is the chemical resistance of the material against the aggressive action of nitric and sulphuric acid condensates with pH values of between 2 and 4 that form in turbo diesel engines. The formation of these chemical substances in this type of engine has posed a serious engineering challenge in designing materials capable of withstanding prolonged direct contact at elevated temperatures.

In designing the two classes of materials mentioned above, we especially focused on defining the requirements of the final applications through the continual exchange of information with some of our partners who are world leaders in the development and manufacture of under-bonnet components. After completing what we call a "functional analysis", we attempted to implement the application requirements using material properties that could be measured on test specimens.

2 CHANGES IN MECHANICAL PROPERTIES AFTER AGEING IN HOT AIR

A relevant criterion for assessing the resistance of a material to thermal oxidation is to measure the residual values of mechanical properties after prolonged exposure to high temperatures.

In Figure 2 the recommended continuous use temperature (CUT) ranges are shown for the two product families. For the Radilon A HHR family (highly thermally stabilized PA66), the maximum recommended temperature is 210°C for exposures of up to 3000 hours. In contrast, for the Radilon XTreme-HHR family (speciality polyamide with a semi-aromatic component for high temperature applications), the recommended threshold for exposures of up to 3000 hours increases to 230°C.

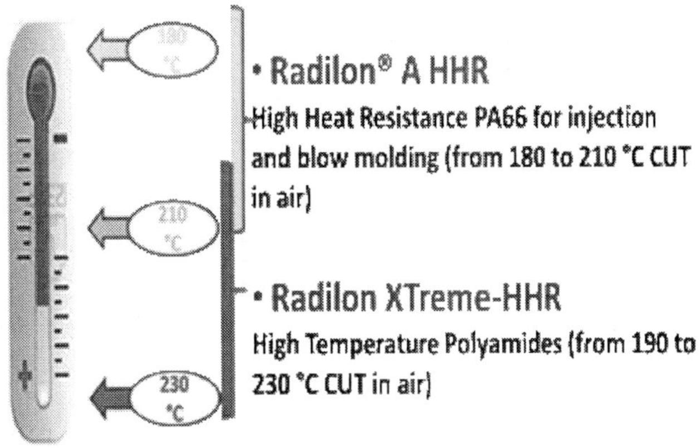

Figure 2. Recommended continuous use temperature (CUT) ranges for the Radilon® HHR and Radilon® XTreme HHR polyamide families.

Figure 3 shows the tensile strength at break (measured according to ISO527) of Radilon® XTreme RV350HHR at temperatures of 190°C, 210°C and 230°C. Even at 230°C, the harshest condition, the residual tensile strength at break does not decrease below 50% of its initial value.

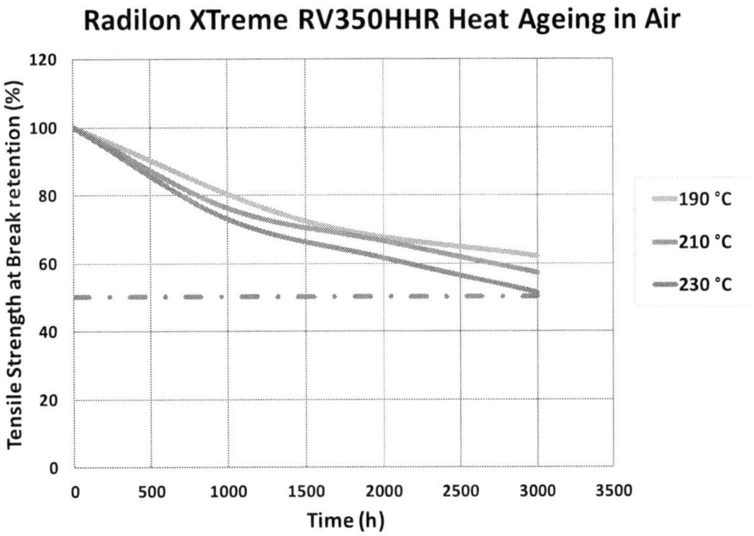

Figure 3. After 3000 hours of exposure, the residual tensile strength at break stays above 50% of its initial value at all test temperatures.

Comparative tests at 220°C were also conducted on high temperature polyamides (PPA, PA46, etc.) commonly used for under-bonnet components, and the results are shown in Figures 4, 5 and 6 below. Radilon XTreme exhibits a behaviour similar to PA46 but better than PPA and PA*-GF35 (speciality polyamide for high temperature applications).

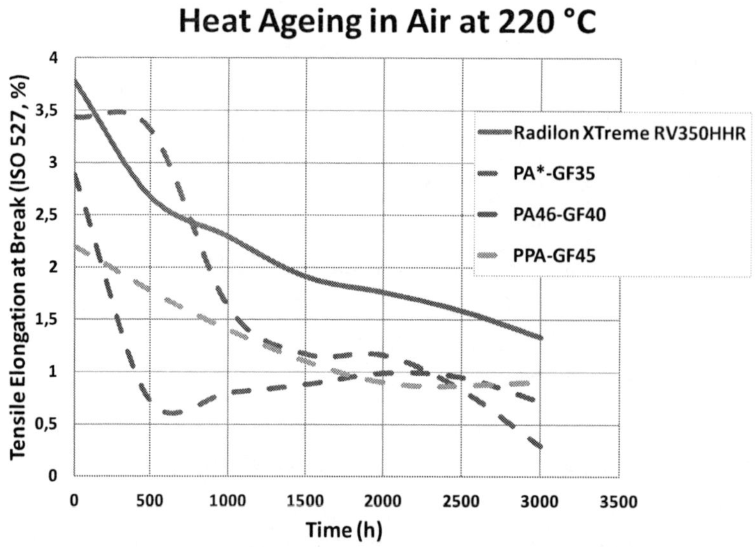

Figure 4. After 3000 hours of exposure at 220°C, Radilon® XTreme RV350HHR has a residual tensile elongation at break of 1.35%, higher than the benchmark PAs.

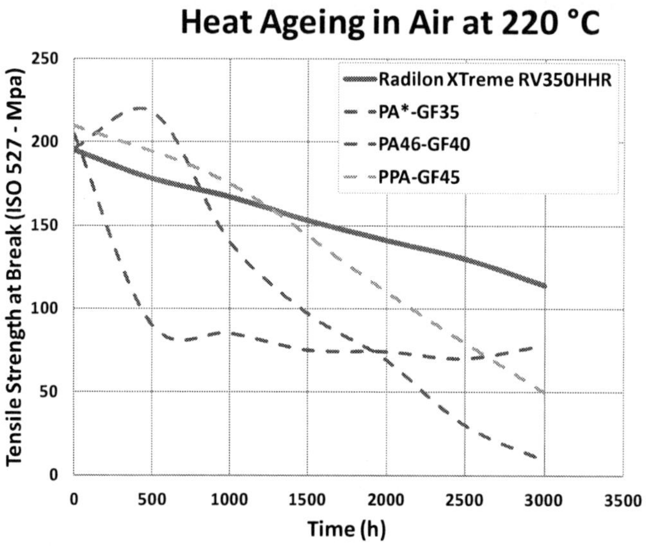

Figure 5. After 3000 hours of exposure at 220°C, Radilon® XTreme RV350HHR has a tensile strength at break of 115 MPa, higher than all the benchmark PAs.

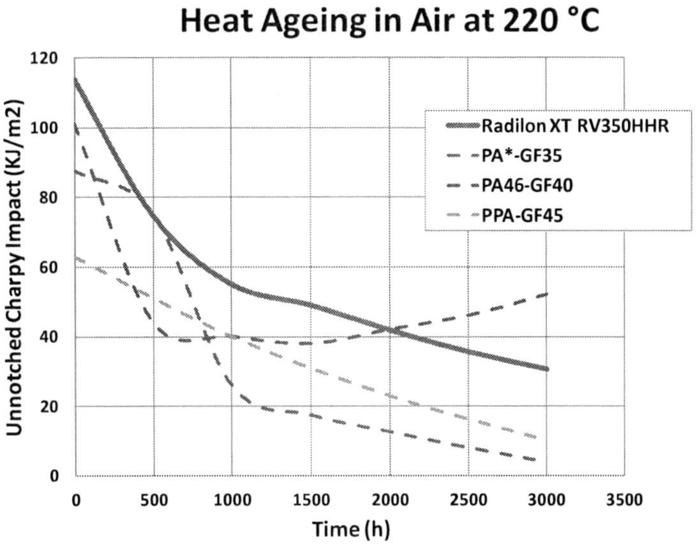

Figure 6. After 3000 hours of exposure at 220°C, Radilon® XTreme RV350HHR has an unnotched Charpy impact strength above 30 KJ/m², lower than PA46 but higher than PPA and speciality PA-GF35.

3 MECHANICAL PROPERTIES AT HIGH TEMPERATURES

As already mentioned above, Radilon® XTreme is particularly suited to the manufacture of parts with high service temperatures that are typical of vehicle under-bonnet components. Therefore, it is important to know how the mechanical properties of the material change as a function of service temperature. For proper component sizing, one needs to know characteristics such as tensile modulus (Figure 7) and tensile stress at break (Figure 8) under extreme service conditions. Indeed, the properties of this kind of material are known to change noticeably with temperature. Despite this drawback, it is generally possible to meet a given set of specifications by optimizing the shape and thickness, and by adding stiffening ribs where needed. Ultimately, one must always follow the best design practices typically used for engineering plastics[3]. When it comes to designing high service temperature components, the fact that plastics allow for creating more complicated and complex shapes compared to metals is an extremely important feature. Figures 7 and 8 show how the tensile modulus and tensile stress at break change with temperature for Radilon® XTreme RV350HHR and Radilon® XTreme RV500 HHR, with 35%- and 50%-glass-fibre fill, respectively. The 50%-glass-fibre fill version exhibits higher stiffness and mechanical resistance at all temperatures. At 230°C, the stiffness and mechanical resistance are40% and 20% higher, respectively, compared to the 35%-glass-fibre fill specimens. The variation of the tensile modulus and tensile stress at break with temperature does not differ significantly with respect to other high use temperature polyamides (Figure 9). Thus, Radilon XTreme can be considered equivalent to the latter when used in components with the same geometry and thickness.

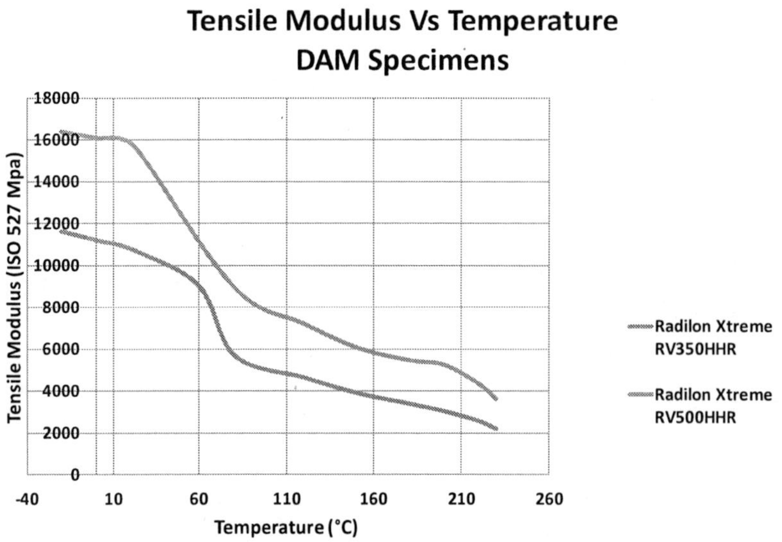

Figure 7. Variation of tensile modulus with temperature for Radilon® XTreme, 35%- and 50%-glass-fibre fill versions.

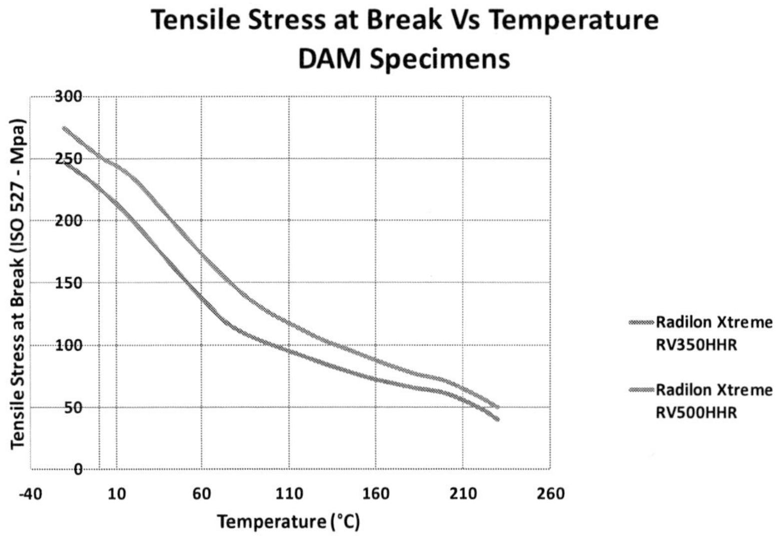

Figure 8. Variation of the tensile stress at break with temperature for Radilon® XTreme, 35%- and 50%-glass-fibre fill versions.

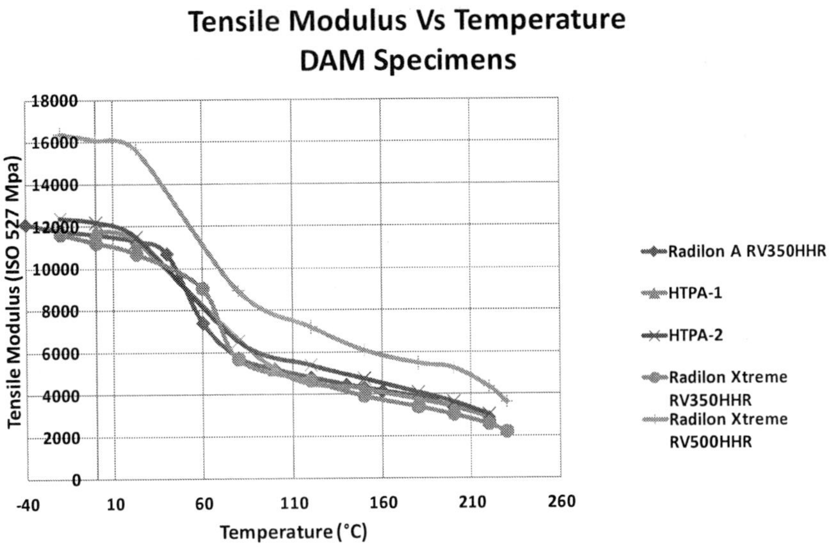

Figure 9. Comparison of Radilon® XTreme to other speciality polyamides for high temperature applications. The graph shows a nearly equivalent change in the tensile modulus values as a function of temperature for the materials tested.

4 FATIGUE BEHAVIOUR

As pointed out above, in the presence of cyclic stress (vibrations and pressure pulsations) the values of validation test parameters such as duration, temperature and pressure pulsation are greater than in the past. Therefore, for proper component sizing, it is also necessary to take into account the fatigue resistance of the material and, from the beginning of the design stage, not to exceed stress limits that could be dangerous for the selected material. For that purpose, the information on fatigue tensile stress, as shown in Figure 10 for Radilon A RV350HHR at 150 and 190°C, becomes very useful.

Similar curves are also available for ISO527 test specimens moulded using a simple two-sided injection mould. These graphs provide reliable data on stress in highly critical areas such as weld lines formed during the moulding process.

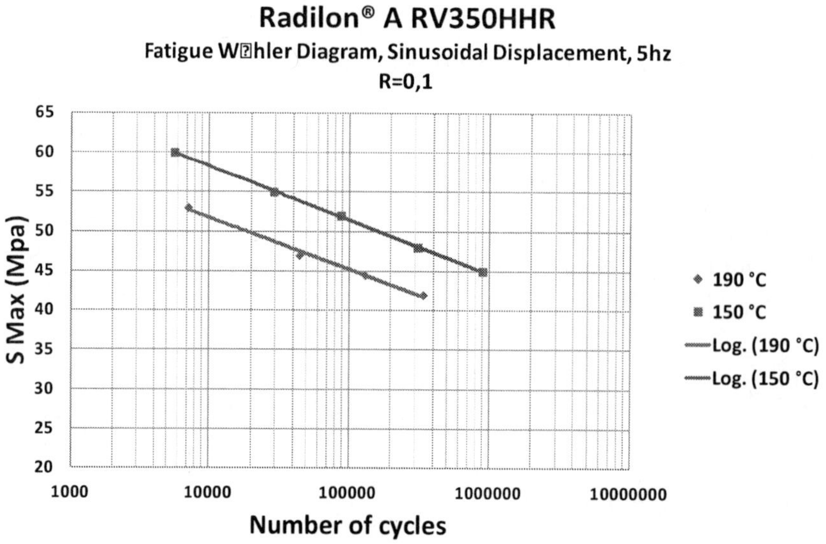

Figure 10. Whöler curves for Radilon® A RV350HHR at 150 and 190°C. Specimens were subjected to cyclic tensile stress with a frequency of 5 Hz and R=0.1.

5 CHANGES IN MECHANICAL PROPERTIES DUE TO THE PRESENCE OF JOINT LINES OR WELD LINES BETWEEN VIBRATION-WELDED PIECES

For almost all plastic moulded parts for engineering applications, one must take into account the joint lines that form during the mould filling process and the exact location of such lines. The mechanical resistance of the material along these lines is, as a rule, much lower than the nominal values (obtained, for example, by measuring the tensile strength on ISO 527 standard specimens). For this reason, the formation of joint lines, particularly in high stress areas, should be avoided wherever possible through proper design and choice of injection points. Additionally, information such as that provided by the graphs in Figures 11 and 12 can also be very useful. In fact, stress fractures occur rather frequently in areas of the component that are considered to be low stress areas. The differences in material behaviour are often quite significant and may depend on the chemical nature of the materials. The graphs show that the mechanical properties at weld lines for Radilon® A RV350HHR and Radilon® XTreme RV350HHR are higher than for the PPAs.

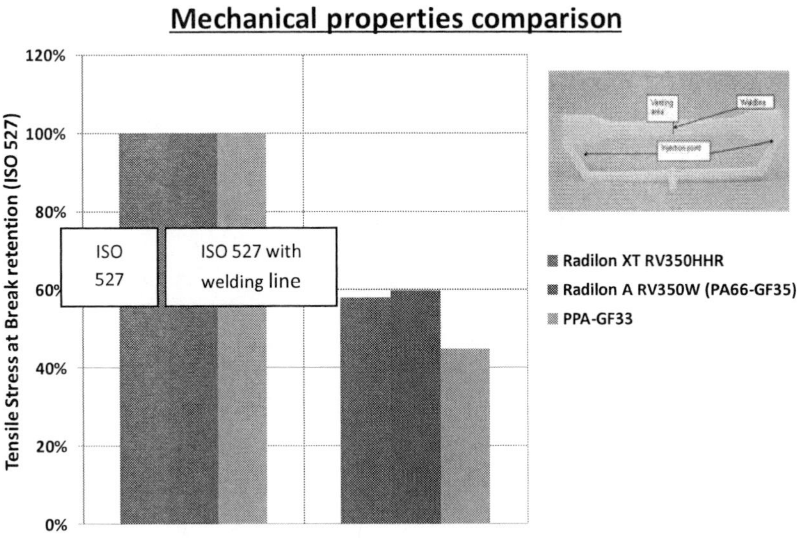

Figure 11. The drop in tensile stress at break due to the presence of a weld line is 40% for Radilon® A RV350HHR and Radilon® XTreme RV350HHR, compared to 55% for the 33%-glass-fibre filled PPA.

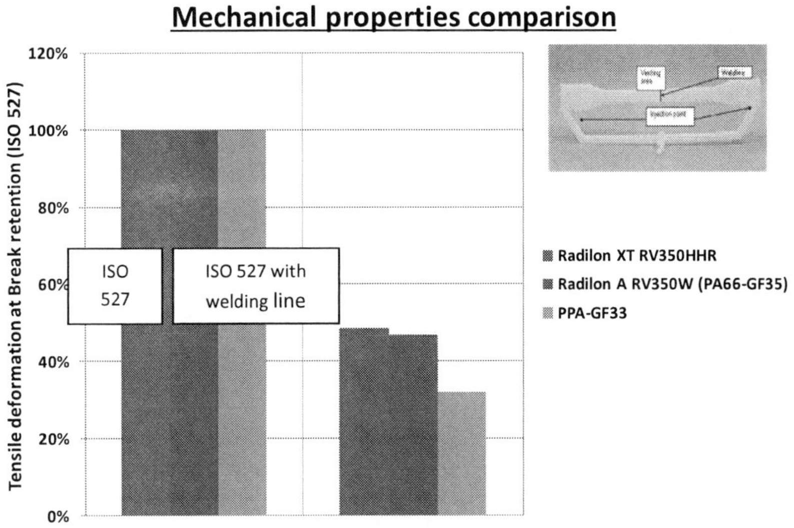

Figure 12. The drop in tensile deformation at break due to the presence of a joint line is 50-55% for Radilon® A RV350HHR and Radilon® XTreme RV350HHR, compared to 65-70% for the 33%-glass-fibre filled PPA.

Many under-bonnet components, such as the turbo resonator shown in Figure 13, are moulded in two or more separate parts, which are then joined together after the moulding process. A frequently used method for joining the parts is linear vibration welding, a very efficient and economical technique. As in the case of joint lines, proper component design must take into consideration the stress along weld lines and compare the properties of the material with and without the weld line. Figure 14 shows comparative graphs for various materials, whose values were obtained by measuring the tensile strength of rectangular cross-section specimens welded together using linear vibration welding. For all the materials tested, the values were obtained by testing specimens welded together under optimal conditions for the welding parameters, such as amplitude, depth, frequency and time.

Figure 13. Turbo resonator made of Radilon® A RV350HHR. This mass-produced component passed the tensile strength test, among others, at the weld line (formed by vibration welding) under the most critical thermal and mechanical stress conditions.

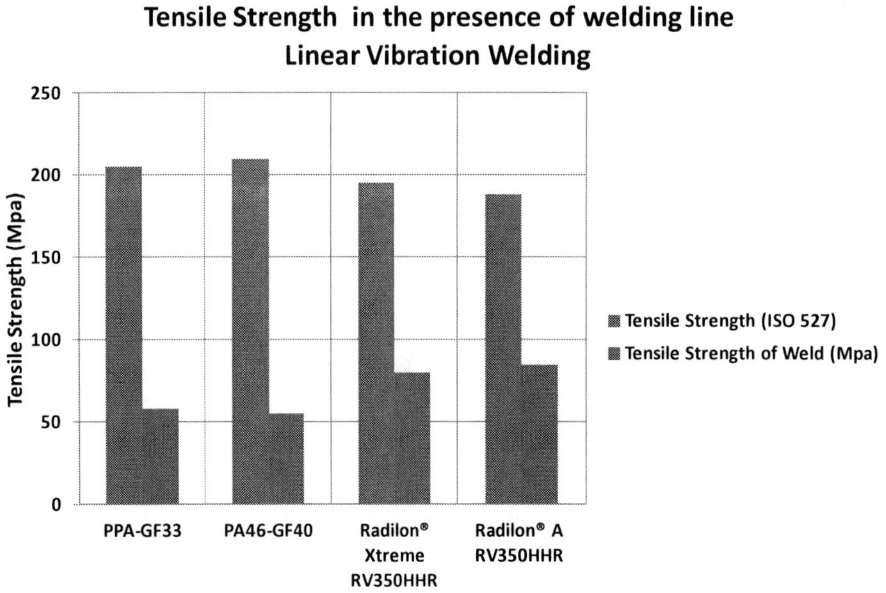

Figure 14. Radilon® A RV350HHR and Radilon® XTreme RV350HHR exhibit 30 to 40% higher tensile strength at weld lines formed by vibration welding, compared to PPA and PA46.

6 MECHANICAL PROPERTIES AFTER IMMERSION IN ACID SOLUTIONS

Chemical resistance to acid substances at high temperatures plays a very important role in the correct choice of a material. Generally, pH values of between 2 and 4 are considered. Acid solution immersion tests at pH 2 were conducted at the RadiciGroup Plastics laboratories. Figure 15 shows the changes in mechanical properties as a result of 1000 hours of immersion at 90°C.

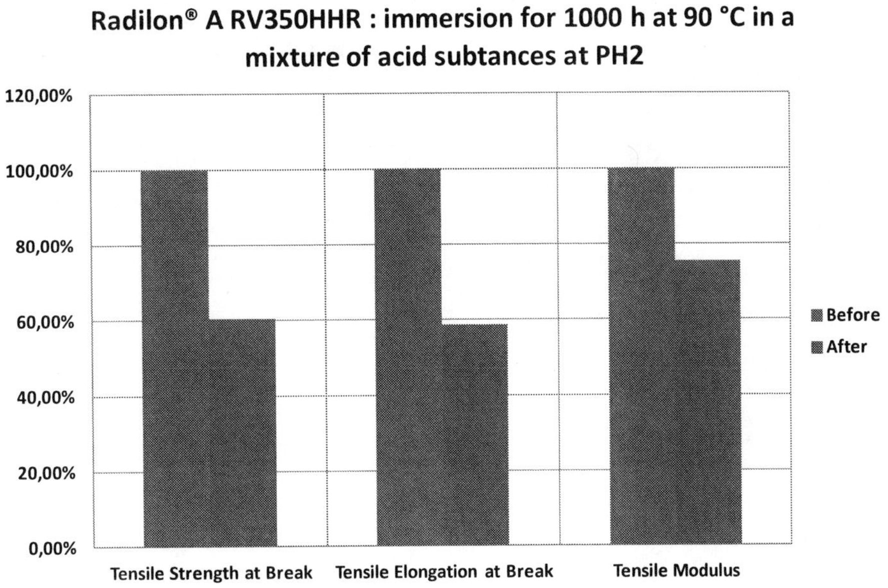

Figure 15. Radilon® A RV350HHR before and after immersion in pH 2 acid solutions. Tensile strength and tensile elongation retain 60% of their initial values while the tensile modulus drops by about 25%.

7 ADVANCED CALCULATION AID FOR CRITICAL COMPONENT DESIGN

The proper design of critical components, such as those used in automotive cooling systems, can be a decisive factor in both ensuring the functionality and reducing the cost of plastic alternatives to metal. The great majority of applications require the use of fibre-filled polymers (most often, glass-fibre-filled). The traditional design approach is usually based on considering the material as isotropic and introducing appropriate correction factors. This technique limits both their liability and precision of calculations. To overcome this limitation, RadiciGroup Plastics has implemented a new design approach that treats the material as anisotropic and requires the characterization of the polymeric matrix, the fibre fill, and the bond between the fill and matrix. In the case of glass-, carbon-, or other fibre-filled materials, the characterization is performed on specimens cut out of plates in various directions as shown in Figure 16.

Using appropriate calculation tools, the mould filling simulation results, including fibre orientation information, are used in both linear and non-linear structural calculations [4,5,6]. Using this approach, the model adopted for the structural calculations will necessarily change, at times radically, as will the mould injection points, marking a departure from the isotropic model approach.

Figure 17 schematically illustrates this new approach, which can offer truly substantial advantages in calculation reliability and, ultimately, lead to significant economic gains.

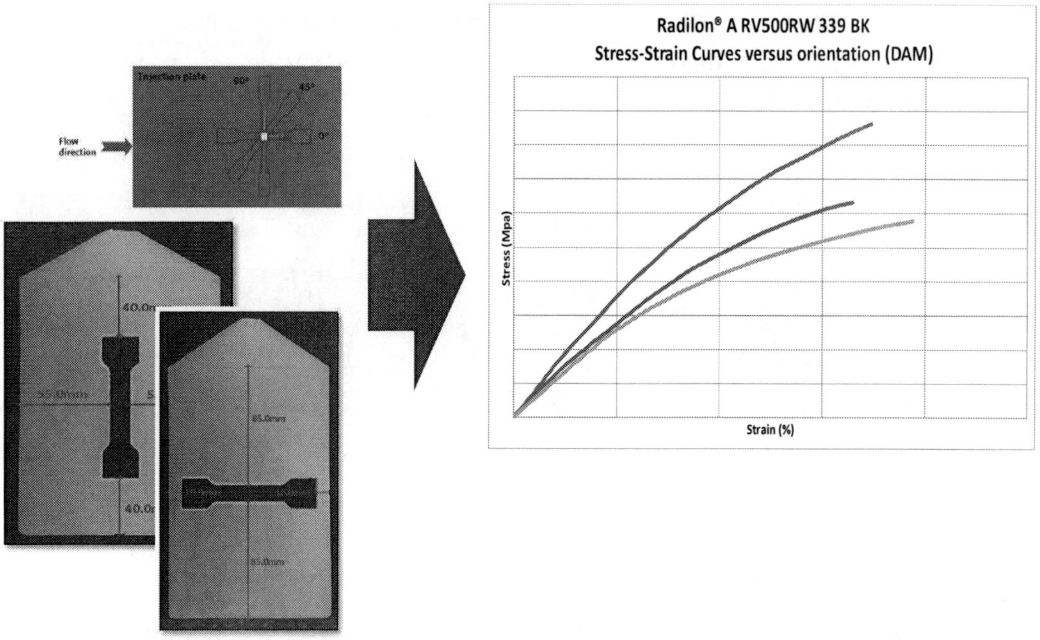

Figure 16. Characterization of the material in the anisotropic model approach. Specimens for tensile strength tests were cut out in directions other than the direction of the melt flow during the injection moulding process.

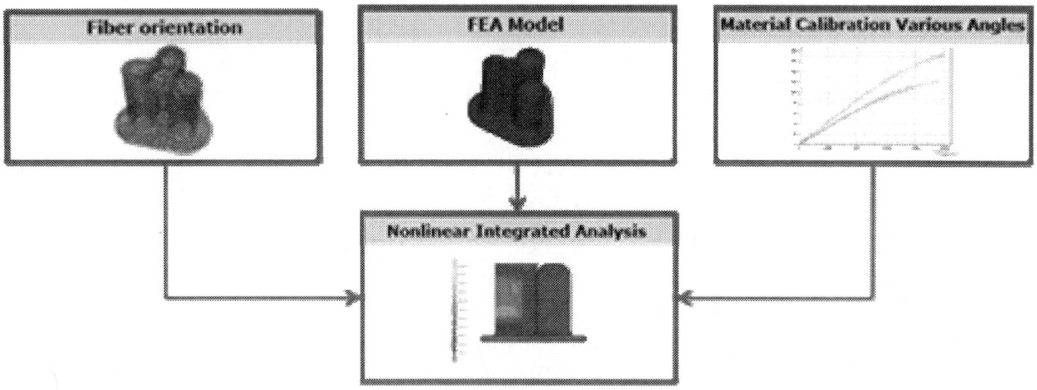

Figure 17. For the structural analysis, material characterization data taking anisotropy into account were used. Furthermore, the structural analysis was preceded by a mould fill analysis that, among other data, provides important information on the orientation of the glass fibres inside the part wall.

8 CONCLUSION

The introduction of two distinct families of products for high temperature applications allows for adapting the material to the actual component use conditions, thus avoiding oversizing the components and potentially increasing the cost. Research efforts were directed at designing materials that would have the right mechanical characteristics, i.e., mechanical, thermal and chemical resistance, while keeping a keen eye on the material processing properties during injection moulding. The Radilon® HHR and Radilon® XTreme families do not have the negative aspects that are typical of highly crystalline, semi-aromatic polyamides, such as weak joint lines formed during moulding and decreased resistance at vibration weld lines.

Specifically, the moulding conditions and joint/weld line properties for Radilon® A RVHHR products, which are recommended for service temperatures of up to 210°C, can be considered equivalent to those of standard PA66. Similar considerations concerning ease of processing and resistance along weld lines formed by vibration welding apply to the products of the Radilon® XTreme family, which are recommended for service temperatures of up to 230°C and have a higher polymer melting point compared to PA66.

A comparison of Radilon® XTreme and the more popular polymers for high temperature applications is given in the table below, where the strengths and weaknesses of each product are graded.

Table 1. Radilon® XTreme comparison with other polymers for high temperature applications. (Glass filled version 30 to 40%)

Polymer	Ease of processing	Surface aspect	Heat ageing properties retention at 230 °C	Stiffness	Tensile strength	Impact resistance
PPA	*	*	**	***	***	**
PPS	*	**	****	***	***	*
PA46	*	*	***	**** DAM *Cond	**** DAM *Cond	***
Rad® XTreme	***	****	***	***	***	***
* Poor ** Fair *** Good **** Excellent						

Ease of processing (by injection moulding): Radilon® XTreme is extremely easy to mould. The temperature of the polymer melt ranges from 190 to 310°C and the mould temperature must be around 90°C. Unlike PPA, PPS and PA46, mould temperatures greater than 100°C are not required.

Surface aspect of moulded parts: surface quality is excellent, partly because of the high fluidity of the material, and is clearly superior to the quality achieved with PPA and PA46.

Property retention after heat ageing: comparable to PA46 and PPA; slightly less than PPS.

Mechanical properties: Radilon® XTreme features values comparable to those of PPA, PA46 and PPS for both stiffness and tensile strength within a temperature range from -20 to + 230°C. Water absorption is comparable to PPA, higher than PPS and lower than PA46.

Impact resistance: better than PPS and PPA; comparable to PA46.

In conclusion, we can say that Radilon® XTreme meets the demand of the automotive industry for polymers capable of withstanding higher service temperatures to be used in new-generation engines. Ease of processing, with all the related benefits, is a further advantage, because it simplifies moulding process control and decreases the risk of lowering the finished product quality.

REFERENCE LIST

1. Gijsmann, P., Tummers, D., & Janssen, K., Polymer Degradation and Stability, 1995
2. A.Weber, Plastics under the hood, 1993, Assembly Magazine article
3. G.Erhard, Designing with Plastics, Hanser, 2006
4. Digimat, Nonlinear Multi-Scale Material Modeling Presentation
5. Digimat, Composite and Plastics Simulation presentation
6. Converse, Process Structure Interaction for injection molding plastics parts presentation

Plastic material development for lightweight fan systems

P HAVET and **A TANGHE**
VALEO Thermal Systems – Powertrain Thermal Sytems, France

ABSTRACT

Issues
European regulations on automotive pollution are becoming more and more severe. One of the innovative solutions proposed is to limit emissions by downsizing the weight of plastic parts through foaming technology.

Content
Global overview of foaming technologies and highlight the key factors of interaction between plastic material, part and process.

Synthesis of tests performed for process and part validation, for fan system support.

Analysis of mechanical property evolutions linked to weight reduction.

Conclusion
VALEO expects to get positive results thanks to these solutions and to integrate them in serial production with thermal applications in the very near future.

Keywords: Lightweight, plastic, under hood parts, climate control parts, foaming

1 CONTEXT OF THE STUDY

The European regulations for automotive pollution levels are getting more and more restrictive. Euro 6 regulation became effective in September 2014 for new launches and will be effective in September 2015 for all vehicles, with an objective to reach 130 g of CO_2 /km. It will evolve again around 2018 with a second component (Euro 6.C) [1]. Forecast estimate that the polluting emissions should not exceed 95 g CO_2/km by 2020 (cf. Figure 1). Therefore, the automotive industry must innovate to find technical solutions that meet these requirements.

Based on assumption that 10 kg less on vehicle means 1 g CO2 not emitted, one of the technical solutions proposed by VALEO Thermal Systems, as Tier1 for automotive, is to limit polluting emissions by downsizing the weight of parts in the current systems.

This paper presents the evaluation of different lightweight solutions with our current plastic materials through the addition of physical or chemical foaming agents, mixed with glass fibers or mineral fillers and without reducing our technical requirements.

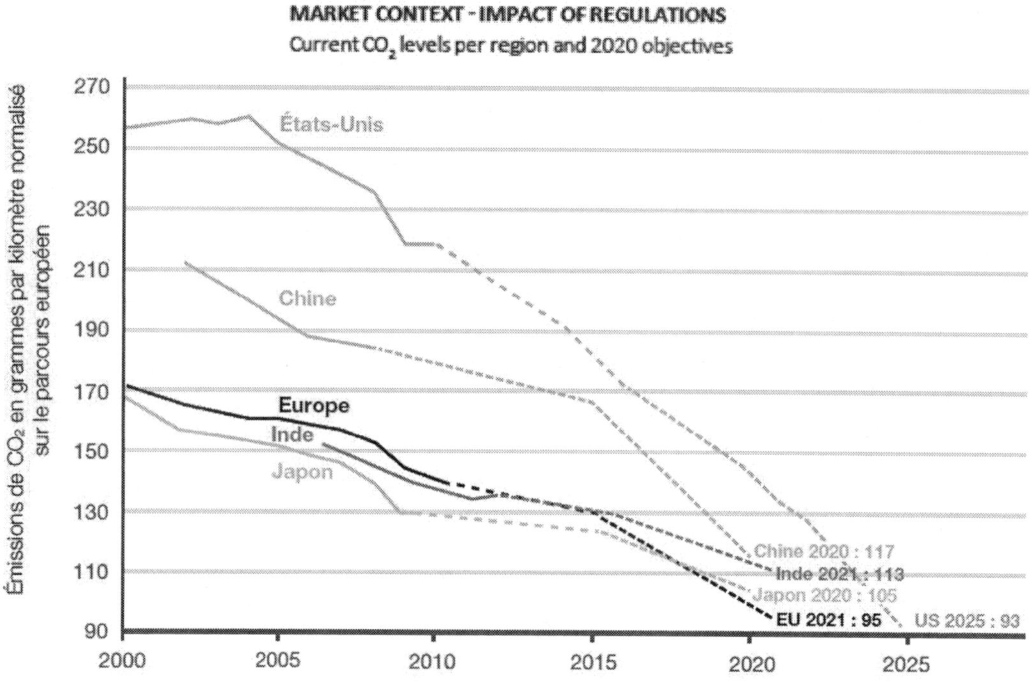

Figure 1: European regulation [2]

2 INTRODUCTION

In polymer engineering and injection moulding processing, there are two main orientations to downsize the CO_2 emission by the weight of components: the density and the thickness through the flow. Currently, based on theses orientations, lightweight is led by the wall thickness and process according to the following technology:

Lightweight by downsizing the wall thickness involves: melt flow, flow length, injection velocity capacity, interface behaviour improving the filling of thin wall.

Lightweight by downsizing the density involves: formulation by additive or polymer structure providing melt flow and stiffness features, composite concept bringing up the addition of short or long and continuous fibre reinforced thermoplastic, process by creating micro cellular structure with foaming technology.

Foaming technology comes up with wide advantages not only to reduce the weight of the final parts but also to provide capacity and flexibility of our injection process plant, quality of our products (dimensional stability): these improvements are confirmed for VALEO Thermal Systems parts, injected in mineral filled polypropylene for Heating Ventilation and Air-Conditioning system (Figure 2a) and glass fibres filled polyamide polymers for Fan systems (cf. Figure 2b).

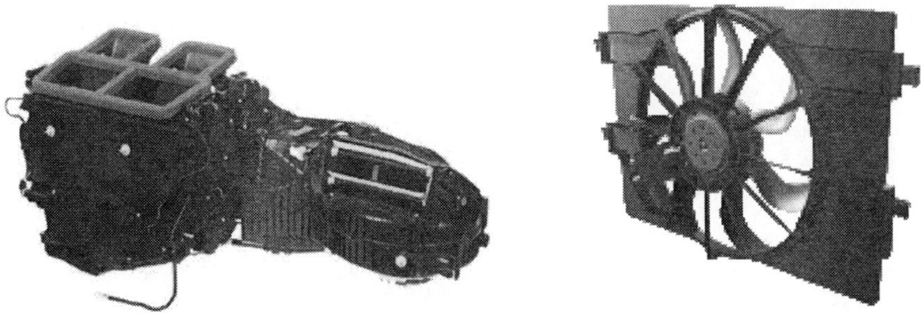

(a) Heating ventilation & air-conditioning system (b) Fan systems

Figure 2: Examples of foamed applications

3 THEORETICAL BACKGROUND

Different processes can be used to create a micro cellular structure in thermoplastics. These different kinds of process use specific substances such as:

- glass hollow microspheres capable of resisting to injection moulding pressure
- supercritical fluids for physical foaming represented by the MuCell® process
- chemical foaming agents introduced in the melt material and decomposing to release gas when heated to their activation temperature

In this paper, we will present different solutions but we will mainly focus on the results obtained with chemical foaming agents, which provide an industrial flexibility for VALEO Thermal Systems within all standard injection moulding machines with a lean approach in order to roll out widely through the current applications.

3.1 Hollow glass microspheres [3]

Hollow glass microspheres offer a variety of advantages over conventional irregularly-shaped mineral fillers or glass fibers. Their size is between 40 to 80 μm, with a density in g/cm³ between 0,46 and 0,6. Main benefits are expected on thin walls until 0,5 mm.

Unlike irregularly-shaped fillers, glass microspheres roll easily over one another. This can help increasing volume loading capacity (cf. Figure 3a&b), improving dimensional stability (isotropic shrinkage), reducing warpage and decreasing the resin consumption.

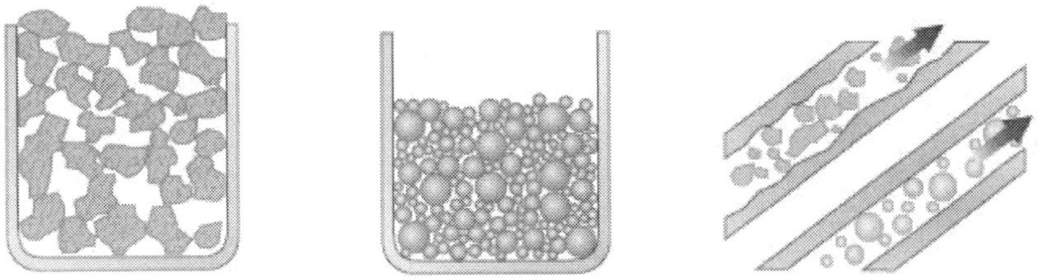

(a) (b) (c)

Figure 3: Advantages of glass microspheres [3]

As hollow glass microspheres roll easily due to their spherical shape (cf. Figure 3c), it comes out to an easy-flow process, which can help to reduce cycle time.

Glass microspheres can improve elastic modulus due to their isotropic reinforcing role.

Figure 4 below highlights expected weight reduction when adding glass microspheres in an initially 33% glass fibres polyamide 66 material.

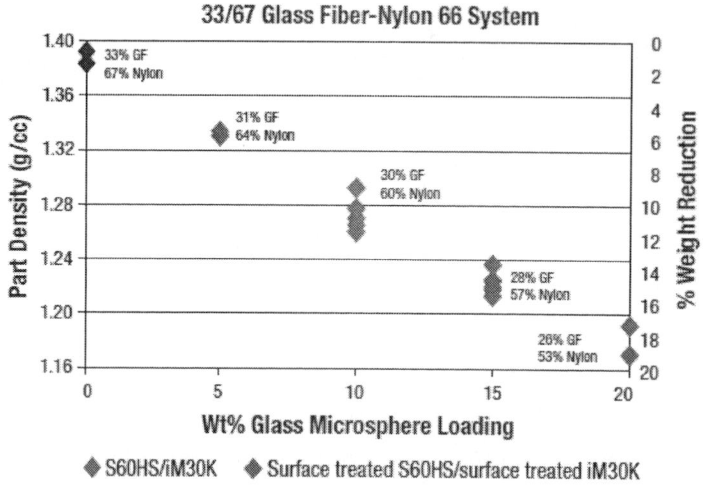

Figure 4: Lightweight regarding amount of glass microspheres [3]

Nevertheless, working under pressure, the resistance to crushing of microspheres due to the process can be a limiting factor.

3.2 Physical foaming by Mucell [4] [5]

The MuCell® Injection Molding Process involves use of gas (commonly nitrogen, but occasionally carbon dioxide) in its supercritical state to create micron-sized voids in molded parts (cf. Figure 5). The voids are created when a single-phase solution of polymer and gas passes through the injection gate into the mold, resulting in a homogeneous nucleation.

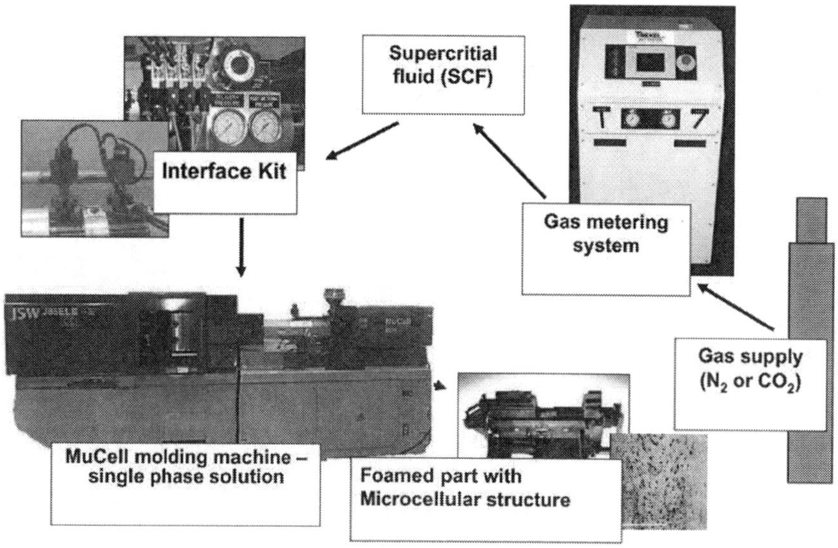

Figure 5: MuCell® process [5]

In addition to lightweight advantage given by foamed structure to the part, we can put in emphasis the dimensional stability, the reduction of deformations and sink marks. Improvements concerning process are cycle time reduction and downsizing of clamping force (cf. Figure 6).

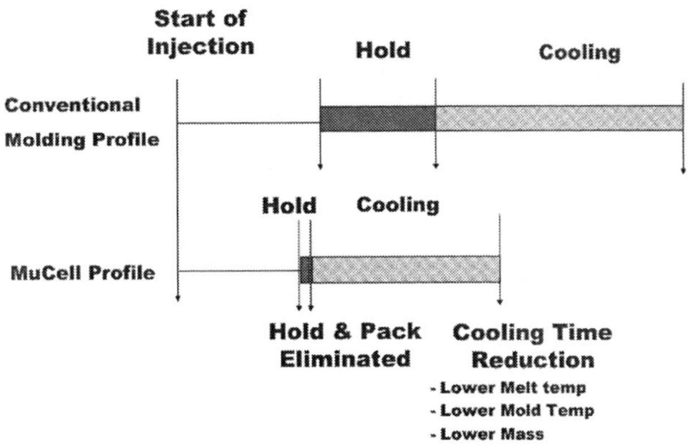

Figure 6: Process modification when using MuCell® [5]

Among principle limits of using physical foaming by supercritical fluids, we can highlight the needs of specific investments and by consequence the decrease of flexibility of equipments.

3.3 Chemical foaming agents

3.3.1 Kind of agents

The chemical foaming agents (CFA) can be divided into two groups (cf. Table 1 for their specific characteristics):

- Endothermic chemical foaming agents are inorganic chemicals such as sodium bi carbonate and citric acids. They absorb heat during the decomposition reaction and decompose themselves into carbon dioxide.
- Exothermic chemical foaming agents are organic in nature. They release N2 gas and heat during the decomposition reaction. Exothermic agents have a high gas yield per gram of CFA and are useful to obtain maximum density reduction.

Table 1: Chemical foaming agents and key properties

Chemical foaming agent	Molecular formula	Reaction	Decomposition temperature	Gas volume
Sodium bicarbonate	$NaHCO_3$	Endothermic	135 °C – 190 °C	120 ml/g
Citric acid	$C_6H_8O_7$	Endothermic	190 °C – 220 °C	120 ml/g
Azo (ADC)	$C_2H_4N_4O_2$	Exothermic	220 °C	220 ml/g
PTSS	$C_8H_{11}N_3O_3S$	Exothermic	220 °C	120 ml/g

3.3.2 Chemical foaming process

The mechanism of the cell nucleation of the chemical foaming is a multi step process which influences the cell size. In both cases, endothermic or exothermic chemical foaming agent, the gas has to be homogeneously dispersed and mixed with the molten polymer at an appropriate decomposition temperature and under pressure.

When the chemical foaming agent is dispersed in the melt, the pressure has to be maintained at sufficient level to avoid that the material gets through to foam prematurely, leading to very inhomogeneous foams with very big cells and very bad mechanical properties. [6]

Then, the gas polymer melt is decompressed when it goes through the barrel or the hot runner system. When the pressure drop occurs in the cavity mould, the nucleation process starts forming the nuclei which will later become the cells of the foam. The number of nuclei formed is depending on the rate of pressure drop.

Then, the gas dissolved in the polymer diffuses into the nuclei which expand to form cells, giving a micro-cellular structure to the part. In this phase, the interaction flow length / wall thickness ratio is critical according to the viscosity of the polymer. If the viscosity is too high, the pressure in the cells is not sufficient to make them expand. On the contrary, if the viscosity is too low, the cell walls will not be stable enough and the cells will start to collapse (coalescence).

In the last step of foaming the polymer cools down and the cell walls freeze: they are able to support the foam structure (cf. Figure 7).

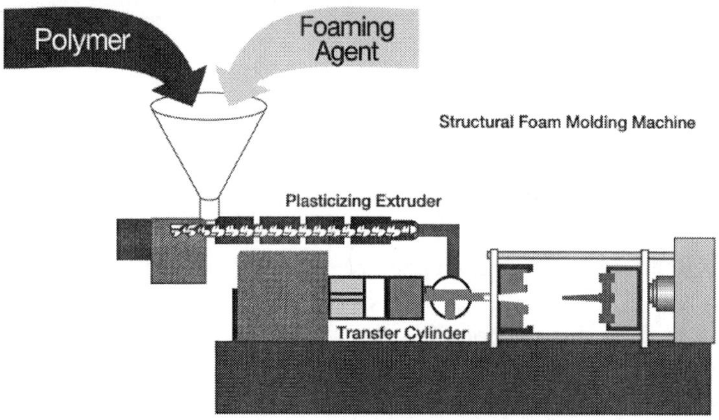

Figure 7: Chemical foaming process [7]

Endothermic chemical foaming agents absorb heat during the decomposition reaction, this involves a reduction of the cooling time.

Exothermic chemical foaming agents release gas and heat during the decomposition reaction. They need intense cooling during injection, for removing the energy excess due to the decomposition reaction.

Furthermore, as the gas expands into the parts, it is no more needed to have back pressure during the injection cycle: this back pressure is ensured by the expansion of the gas, in the thickness of the part.

3.3.3 Fundamental mechanism of cell nucleation

According to the nucleation theory, the surface tension at the liquid-gas interface, and the pressure inside a critical size cell, is led by the gas concentration dissolved in the polymer and also by the free energy barrier that needs to be overcome for cell nucleation emergence.

The free energy barrier is lower if a cell is nucleated at the surface of the additives, or at the surface of an impurity, or at the equipment wall surface (heterogeneous nucleation), instead of nucleated in the bulk phase of the polymer-gas mixture (homogeneous nucleation) [8].

Therefore, additives such as talc are often added to the polymer-gas mixture as a cell nucleating agent. The effectiveness of a nucleating agent depends essentially on its surface geometry.

Alternatively, potential nucleating sites can be found in free volumes within the saturated polymer-gas mixture, due to incomplete wetting between the different phases in the mixture. After a fast decrease in system pressure, critical radius of cell decreases beyond the size of the micro-void, which will grow spontaneously to become nucleated cell.

In regions with extensional stresses, like at the injection gate, the cell nucleation rate increases due to an increase of the super-saturation level (the pressure inside a critical size cell is less than the system pressure). An applied extensional stress could significantly enhance cell nucleation, and such effect is much more apparent when filler particles (like talc) are added.

4 METHODOLOGY

In this section, we will focus on the comparison between parts injected with and without chemical foaming agents, for fan system application for VALEO Thermal Systems.

4.1 Fan systems and their constitutive materials

Fan systems are constituted of several components: a dynamic part, called fan, getting in rotation thanks to a motor and a static part, called support which among other functions bears the motor.

Fan system is dedicated to force the air passing through exchangers under the hood when necessary (for example: sudden stop due to traffic or park whereas motor and exchangers are still at a hot temperature). In this study, we will focus on the fan system support (cf. Figure 8).

The most common plastic materials used for the injection of fan system support are polypropylene filled with 30% of glass fibers (PPGF30), polyamide 6 filled with 30% of glass fibers (PA6GF30) and polyamide 66 filled with 30% of glass fibers (PA66GF30). Chosen material depends on the requirements of the project (temperature, structural or non structural part).

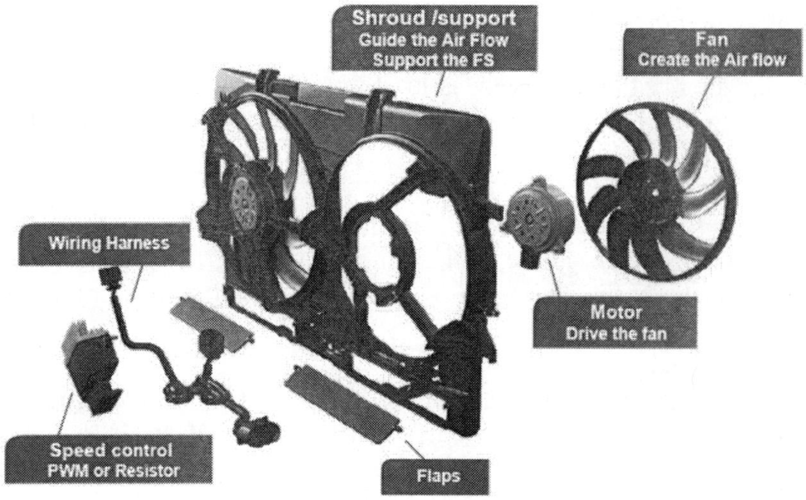

Figure 8: Fan system support

4.2 Parts injection

Foaming technology induces some "voids" in the parts, due to the gas release during the injection process. This gas release is depending on the thickness of the part. During injection, there is no foaming in the two superficial layers of melted material since their contact with the mold makes them directly frozen. The foaming occurs in the core thickness of the part, only if this thickness is high enough in the core melt layer. So, the foaming percentage in the part depends directly on its thickness and is directly linked to the part design.

Typically, our fan systems parts get thickness between 2,5 and 3 mm. Normalized tensile specimens used for material validations are ISO527 1A specimens with a 4 mm thickness.

For this reason, we cannot evaluate the influence of foaming on mechanical properties on classical molded specimens as usual. The foaming would not be representative of what would occur on a thinner part.

Specimens need to be cut directly in flat areas of the real parts. Cutting specimens directly from real parts also means that fibre orientation in specimens will not be as "ideal" as in molded tensile specimens where glass fibres are oriented in the direction of specimen. So, even samples cut from non foamed parts will not exhibit the same levels of properties than classical iso 527 1A molded specimens.

In order to evaluate the impact of foaming on the mechanical properties of the final parts, we firstly injected fan system supports in a production mold, without foaming to get reference parts. In a second step, we molded the same parts with foaming.

4.3 Preparation of samples and tests performed with PA6GF30 material

To perform mechanical characterizations by comparison, samples were cut in the both injected parts, according to ISO norms (ISO527-1BA for tensile tests and ISO179-1fU for impact tests).

To be able to perform accurate mechanical characterizations, specimens have to be cut in flat areas of the fan system support, getting constant thickness. Only a small area of the support can be used to cut the samples (cf. Figure 9). Five samples are prepared for each test configuration.

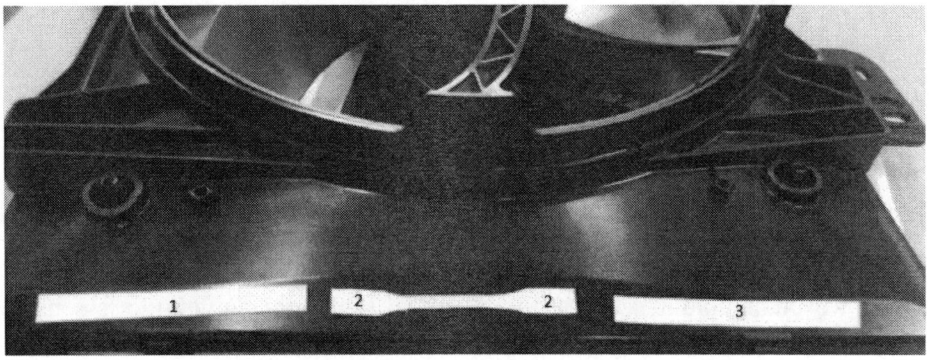

Figure 9: Samples cut in fan system support

First of all, the weight of parts is measured on foamed and non foamed parts in order to evaluate the weight reduction obtained (cf. Table 2).

Table 2: Weight reduction of PA6GF30 injected parts

<table>
<tr><th>Material</th><th>Weight (g)</th><th>Estimated part volumic mass (g/cm³)</th><th>General Density Reduction</th><th>Cutting zone density ISO 1183 (g/cm³)</th><th>Density reduction in cutting zone</th></tr>
<tr><td>Standard</td><td>639 ± 0</td><td>1,337</td><td rowspan="2">7%</td><td>1,337 ± 0,006</td><td rowspan="2">5%</td></tr>
<tr><td>Foamed part</td><td>594 ± 2</td><td>1,244</td><td>1,273 ± 0,003</td></tr>
</table>

Considering that weight reduction by foaming is only on the polymer matrix (and removing the 30% of glass fibers) the weight reduction of 7% on the part is equivalent to 10% on polymer matrix alone.

On Figure 10, we can see the observations of the fan system support cutting zones in foamed parts, with the cellular aspect.

Some density measurements were also performed according to ISO 1183-1 norm, based on Archimedes principle. This methodology consists in weighing the same samples in the air and immersed in water to get their density. These measurements were performed in the area of specimens cutting. The weight reduction by foaming obtained by ISO1183-1 in cutting area is only 5%, compared to the global 7% measured on whole parts (cf. Table 2). These observations confirm that foaming is dependent on thickness and design of the parts and is not homogeneous.

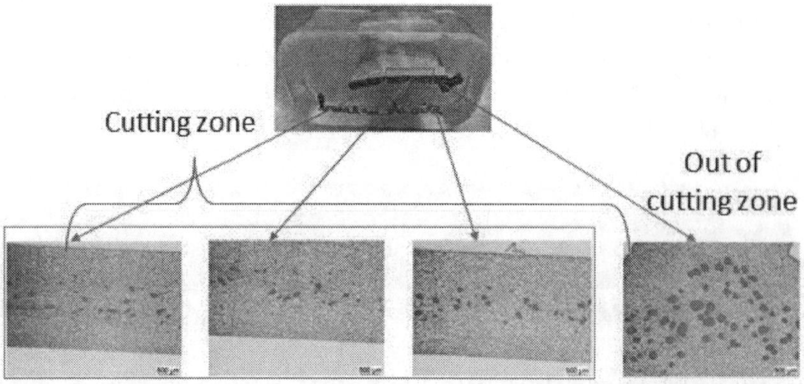

Figure 10: Observation of foam cells in cutting areas

Due to the way it is obtained and to its chemical structure, polyamide is sensitive to humidity absorption. The humidity content in a polyamide sample can greatly influence its mechanical properties. To avoid influence of this parameter on results, all specimens tested in this study were dried in a vacuum oven for 72 hours at 80 °C before characterization to be in dry state.

Mechanical characterizations were performed at initial and after long term heat aging. For such application, we evaluated the impact resistance of the material at 23 °C and the tensile characteristics at 130 °C, which is the highest temperature in use when the fan system is located between the exchangers and the motor, under the hood. And we also performed long term heat aging (cf. Figure 11) at 130 °C during 1000 hours followed by a peak at 150 °C during 20% of the heat aging time (to secure a peak temperature during motor operation). This heat aging is representative of the working conditions that under-hood parts such as fan systems have to undergo.

After this heat aging, we performed the same mechanical characterizations as for initial state, in order to evaluate the retention of properties after long term heat aging.

Figure 11: Heat aging in ventilated oven

We can notice on figure 12 that tensile properties at initial state are equivalent between foamed and non foamed parts: the 5% weight reduction has a very low impact on mechanical properties, even in temperature.

For the 4 next graphs (Figure 12 to 15), Y axis is 100% basis regarding VALEO norm specification on 4 mm injected ISO 527-1A specimens.

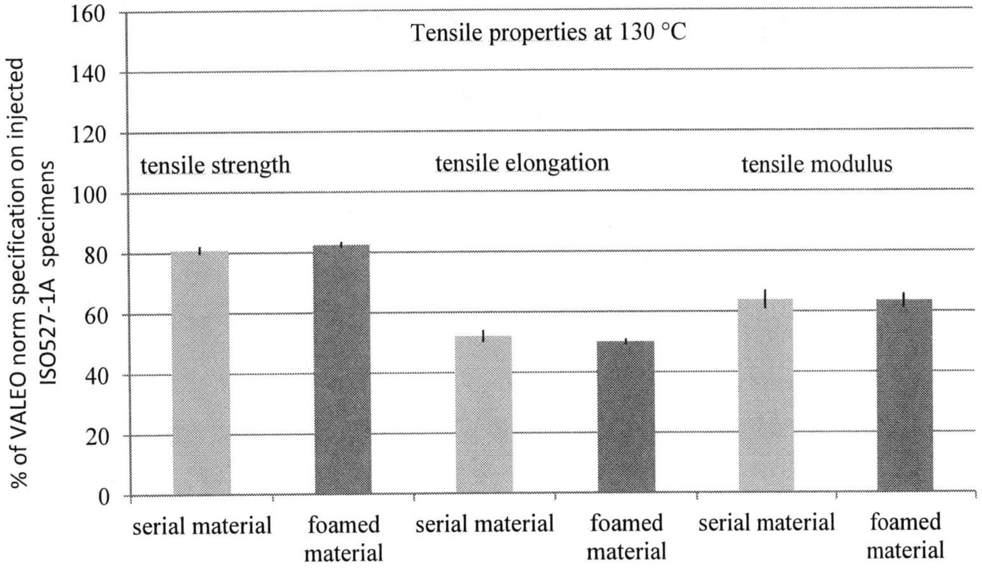

Figure 12: Tensile test, initial state, 130°C

We can notice on Figure 13 that impact resistance is lower (about 25% less) for foamed parts. This can be explained by a preferred path out through the voids in the material, during the crack propagation.

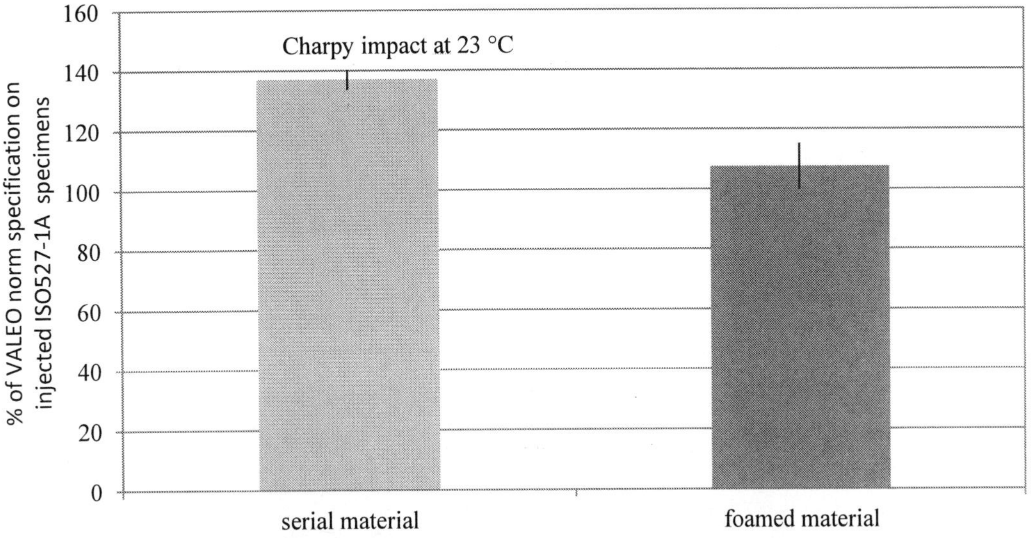

Figure 13: Charpy impact, initial state, 23°C

Tensile test results, after heat aging at 130 °C during 1000hours followed by a peak at 150 °C during 20% of the heat aging time, are given in Figure 14 below.

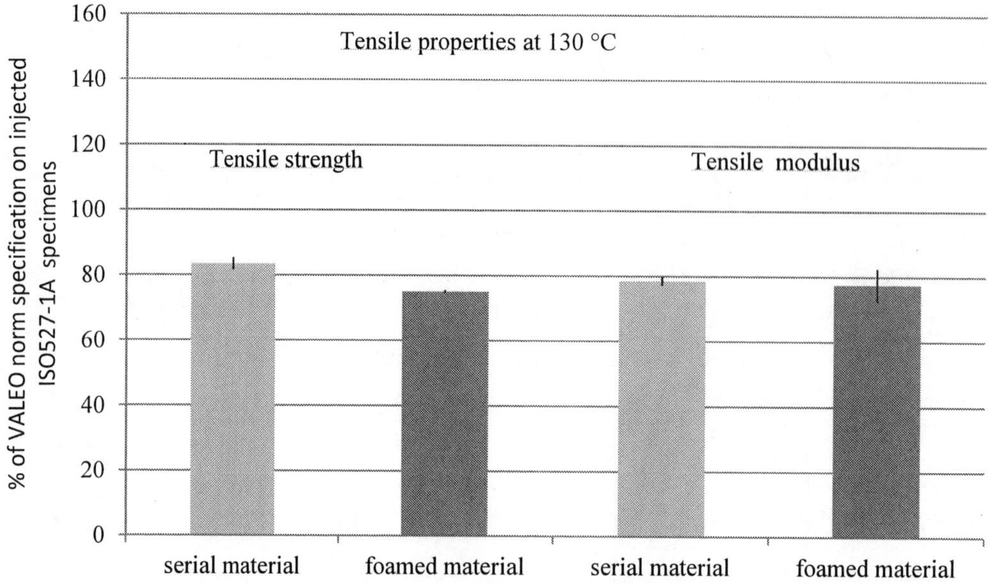

Figure 14: Tensile test after long term heat aging

From Figure 14, we can conclude that both materials (with and without chemical foaming agents) have equivalent tensile properties at 130 °C, even after long term heat aging. A slight decrease of tensile strength on foamed part (about 10 %) is observed. Moreover, the retentions of properties after heat aging are equivalent.

We can notice on Figure 15 that impact resistance is concerned by the addition of chemical foaming agents in the material: the loss of impact resistance can be due to the fact that the "voids" create a weakness during such test at high velocity. We had observed similar behavior at initial state (cf. Figure 13).

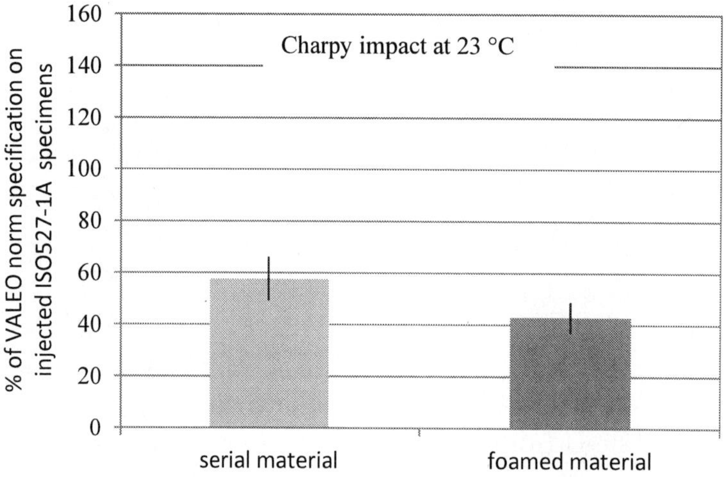

Figure 15: Charpy impact after long term heat aging

For this next graph (Figure 16), the Y axis corresponds to retention of properties after heat aging compared to initial state results.

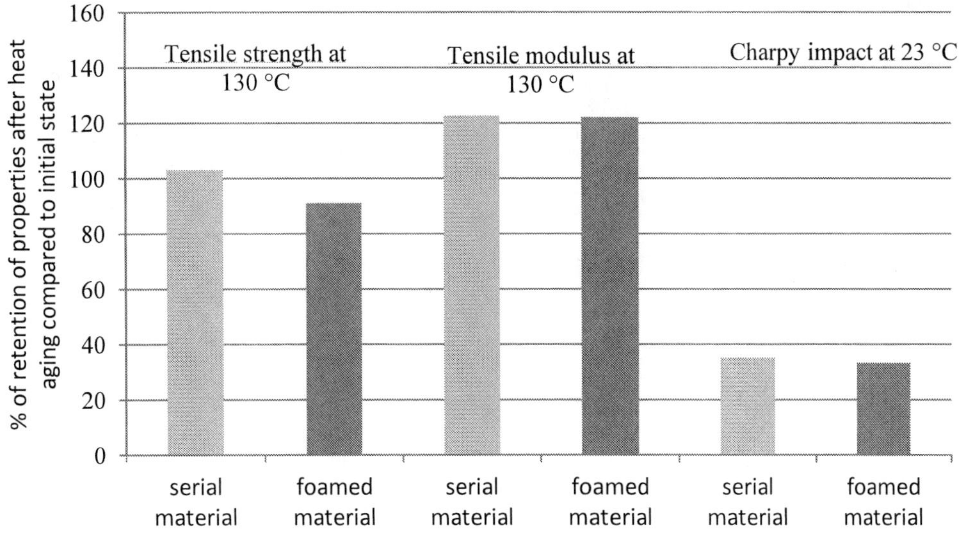

Figure 16: Retention of properties after long term heat aging

By conclusion, we can observe on Figure 16, that the retention of tensile modulus and impact resistance after long term heat aging, by comparison with initial state, is the same for fan system support injected with and without chemical foaming agent. A slight difference could be observed for tensile strength between serial and foamed part but this effect is negligible in the intergrity of the part. So we can consider that for our fan systems part, there is no mechanical impact of the chemical foaming agent added during the injection of the parts, both at initial state and after long term heat aging.

4.4 Tests performed with PA66GF30 material

Similar tests were also performed on parts molded with PA66GF30. Obtained results are similar to the ones with PA6GF30. Globally, 7% of lightweight could be achieved on those parts. The influence of foaming on mechanical performance of specimens cut in parts is similar to what had been found with PA6GF30 parts: tensile characteristics are similar between foamed and unfoamed parts but a decrease of Charpy impact resistance can be observed.

4.5 Tests performed with PPGF30 material

Some tests were also performed on parts molded with PPGF30 material. PPGF30 material is used for fan systems supports in project for which fan systems is a pushing one: fan systems is located in front of the exchangers and push the air towards exchangers. In that configuration, fan system is submitted to lower temperatures than pulling fan systems. For polypropylene material, the foaming agent used is an endothermic one. The weight saving was about 6% and cycle time was reduced by 25% on parts getting thickness between 2 and 2,5 mm. Observations of specimens cut in PPGF30 supports lead to similar observations than the ones performed on PA6GF30 parts: parts molded with chemical foaming agent exhibit a foam structure in the core of their thickness (cf. Figure 18) compared to unfoamed part.

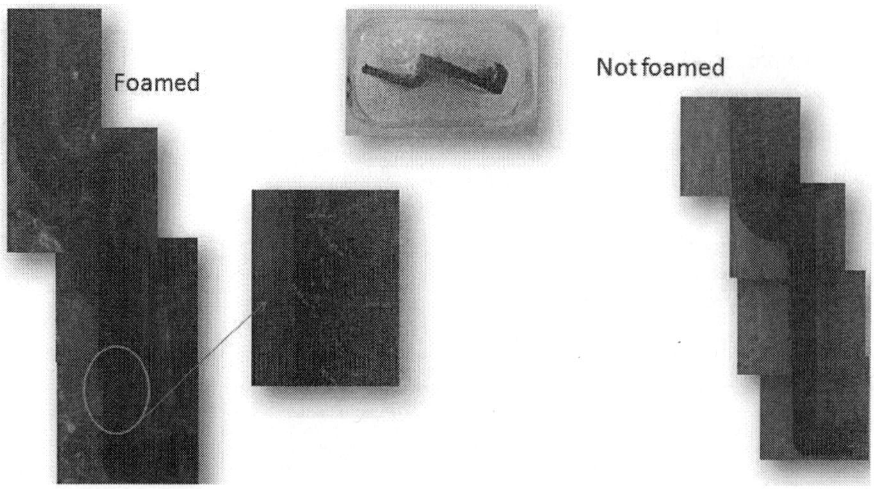

Figure 17: Observation of foam cells in a PPGF30 part

5 LIMITATIONS OF THE STUDY

With reference to the design of our fan system support, we initially expected more than 7% of weight reduction. Injection trials with several percentages of chemical foaming agents, showed that weight reduction higher than 7% was possible but with a degradation of the surface aspect and local deformations of the parts. The best way to improve the weight reduction is to re-design the part, taking into account the presence of foamed structure, depending on the thickness at each area of the part.

The efficiency of foaming is linked to the hot spot in the part due to the non homogeneous wall thickness. As a result, a redesign of the part should be taken into account. In addition, the reduction of the flow length / thickness ratio in foaming injection compared to a solid injection influences the mold design especially the engineering of the filling.

6 CONCLUSION

Chemical foaming agents have allowed achieving a 7% weight reduction on VALEO fan system support.

For mechanical evaluation on parts:
- Results of tensile tests showed that foamed material has, generally, equivalent properties than the serial material
- Long term heat aging as similar effects on the mechanical properties for both serial and foamed materials.
- Foamed material has lower impact properties than serial material

The evaluation of the foamed material must be continued with impact tests on parts directly and all the tests performed during the approval of such parts (acoustic, vibration…).

Beyond the weight saving, the addition of chemical foaming agents induces also a better dimensional stability and reduces the differential shrinkage and deformation of the injected part.

In our case, the use of chemical foaming agent allows also to reduce the cycle time by 20% for VALEO fan system support.

Among other benefits, the clamping force is also reduced involving energy saving and injection machine downsizing in the case where the mould features can be fitted with smaller equipments.

Furthermore, foaming technology comes up with wide advantages improving capacity and flexibility of equipments process in the plants.

REFERENCE LIST

[1] http://www.largus.fr/actualite-automobile/norme-antipollution-euro-6-a-quoi-ca-sert-4931379.html

[2] Valeo 2013 reference document. 3M™ Glass Microspheres for the Automotive Market document

[3] http://multimedia.3m.com/mws/media/431692O/3mtm-glass-microspheres-for-the-automotive-market.pdf?&fn=GlassMicrospheresAuto.pdf

[4] http://www.trexel.com/en/

[5] MuCell®Injection Molding Process Strategic quality improvement & cost savings technology

[6] Dr T.Mergenhagen. Chemical foaming agents in thermoplastics and thermosets. RAPRA Blowing Agents and Foaming Processes, May 2013

[7] Injection_Molding_Processing_Guide document http://www.bergeninternational.com/

[8] A. Wong, Y.Guo, C B.Parck. Fundamental of Cell Nucleation in Polypropylene foaming with supercritical carbon Dioxide effects of extentional stresses and crystals. Journal of supercritical fluids, 2013.

Chemical stress analysis of plastic cooling components from field vehicles: a novel approach for designing plastic components

T LANDTMETERS
Toyota Motors Europe, Belgium
N DELON-ANIK
Solvay Research and Innovation, France
A GUIU
Solvay Engineering Plastics, France

ABSTRACT

When exposed to hot air or hot coolants, plastic components eventually see their properties affected through various chemical reactions. The degradation of glass filled polyamide 66 cooling components was evaluated using a Gel Permeation Chromatography combined with classical mechanical analysis. The effect of exposure to engine coolant at various temperatures and for various durations have first been analysed in a controlled environment to afterwards make the link with the chemical degradation seen on parts recovered from the field. Such an approach allows a better understanding of the chemical reactions taking place on the complete surface of a polyamide cooling component and quantifies the remaining part strength on parts from the field, thereby allowing better optimization of the part design, and eventually reduce manufacturing costs.

1. INTRODUCTION

Under the pressure of increasing environmental constraints, car manufacturers are developing more energy efficient and more environmental friendly engines, while ensuring high quality standards and low manufacturing costs. When it comes to material choices for a specific component, engineering plastics, and more particularly polyamides for powertrain components, have a long and proven track record in a large number of applications such as cam cover, air intake manifolds, radiator end tanks, or turbo charged air tanks or ducts. Low density, high mechanical properties, long lasting performance in temperature or chemical media, low manufacturing costs, high design flexibility, good simulation capabilities, or prototyping possibilities are one of the many advantages offered by glass filled polyamide materials. As a result, polyamide based materials are widely used for automotive and truck powertrains components. Typically, passenger vehicles manufactured in Europe use on average around 13 Kg of polyamide, including 4 Kg for engine components, 2 Kg for cooling circuit components and 2 Kg for air intake circuit components.

Designing a component with plastic in agreement with specific mechanical specifications is a well-known matter. Beyond conventional finite element analysis simulation methods, advanced simulation approaches based on anisotropic mechanical modelling, such as MMI TECHNYL® Design, offers an unmatched degree of predictability. However, predicting accurately the behaviour and eventually the failure mode of a plastic part after thousands of hours of continuous or discontinuous exposure to hot air or aggressive chemical media such as coolants is not possible today. This is mainly due to the fact that a failure can be activated by a chemical transformation of the material, which is dependent of the local environment seen by the material at a specific location in the part. A common approach is therefore to design a part taking into account a certain level of mechanical degradation of the material (say 50%), after long term (say 3000h) exposure to the media at the specified continuous use temperature.

In this paper we are presenting a new approach to better understand the chemical degradation mechanisms of plastic components exposed to engine coolants, that we call chemical stress analysis. Using chemical analysis (Gel Permeation Chromatography) on cooling circuit thermostat housings from field vehicles with different mileages and driving conditions, we could evaluate the level of chemical degradation of the polymer material in different locations of the part, compare it with new and lab aged parts, and finally make the correlation with the part ageing specifications. We think this approach can precisely quantify the remaining part strength on field parts, hence allowing better optimization of the part design, and eventually reduce manufacturing costs.

2. BACKGROUND INFORMATION

2.1. Description of the cooling components used in this study

Figure 1 shows the AD engine (= Toyota 2.0 and 2.2 litre diesel engine) cooling system layout highlighting the thermostat housing considered in this study. As can be seen, 3 different coolant circuits are entering the thermostat housing which then brings the coolant to the water pump on the engine. For the investigation of the field recovered parts, 4 different locations on the part were analysed. Location 1 is on the return line from the radiator, location 2 is on the by pass return line after the thermostat, location 3 is on the return line from the cylinder head and location 4 is on the exit towards the water pump. Location 1 is before the wax thermostat, which opens at 82°C and location 2 → 4 is after the wax thermostat.

The thermostat housing is an assembly of 2 plastic parts and a thermostat. The 2 plastic parts are made out of 35% glass filled, hydrolysis resistance polyamide 66 material (reference TECHNYL® A 218 V35 BLACK 34NG from Solvay Engineering Plastics). Figure 2 shows the thermostat housing in an exploded view together with the branch numbers as used in Figure 1.

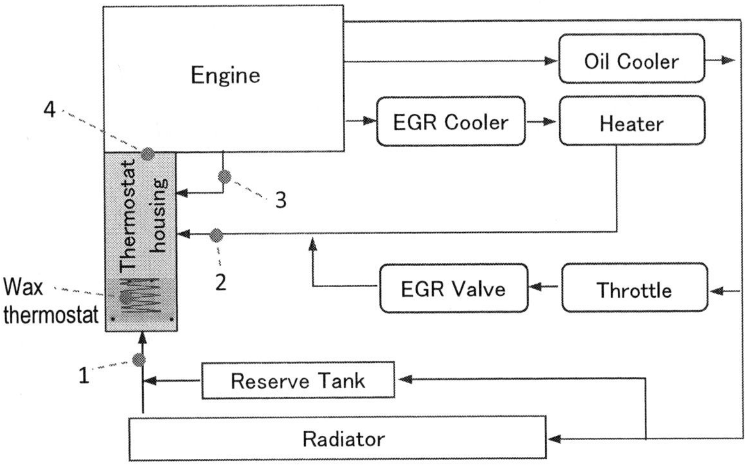

Figure 1 : AD cooling system schematics

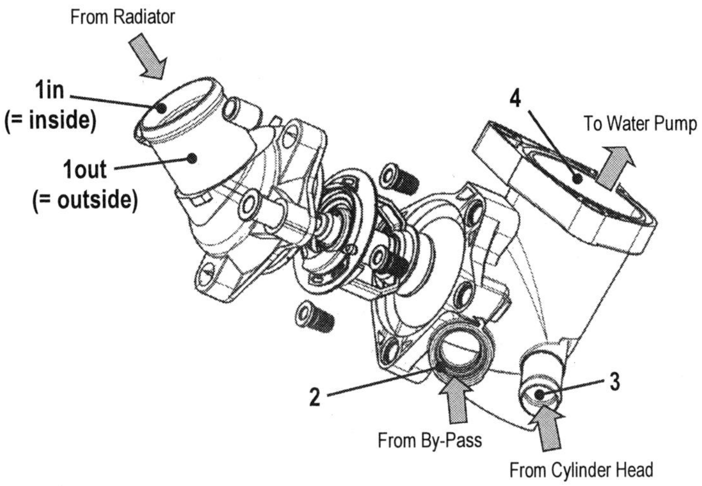

Figure 2 : Exploded view AD thermostat housing

This specific component of the coolant systems is exposed on the inside to a continuous flow of Long Life Coolant (LLC = mixture of water and glycol) coming from 3 different coolant circuits. On the outside, the part will feel the warm air from the engine compartment. Table 1 shows the maximum temperature at each position of the thermostat housing in function of various driving patterns.

Table 1 : Maximum temperature measured on thermostat housing (temperatures in °C / wax thermostat open)

	LLC temperature				Engine compartment temperature
Driving pattern ↓	Position 1	Position 2	Position 3	Position 4	
City	75	90	90	85	70
Congestion	95	100	100	97	95
Highway	85	95	95	90	75
Mountain	100	105	105	103	100

The complex structure of the part together with continuously changing operating temperatures, different from branch to branch, made it an interesting part to study with Gel Permeation Chromatography to understand the actual remaining part performance (based on analyse of field recovered parts) in comparison with the mechanical properties obtained after component durability evaluation.

2.2. Gel Permeation Chromatography

In this study, we performed the chemical analysis by Gel Permeation Chromatography (GPC) or Size Exclusion Chromatography (SEC) for synthetic polymers. GPC is a chromatographic method where a polymer is dissolved in a solvent and the resulting solutes are then separated according to their molecular masses, or more precisely to their hydrodynamic volumes. This method give access to the distribution of the molecular weights of the polymer chains in the sample, which can be visualized in a chromatogram, as shown Figure 3. Note that during the polymer dissolving process, all glass fibres are removed from the polymer matrix. The GPC method in this study will therefore only focus on the degradation of the polymer material.

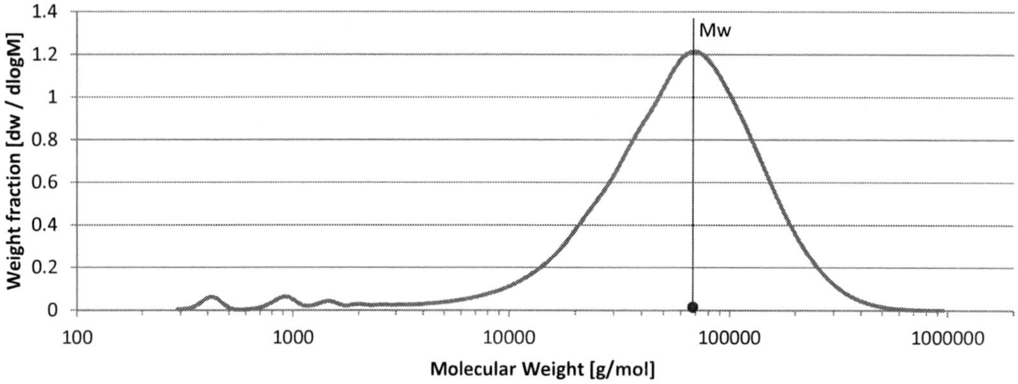

Figure 3 : Example of a GPC chromatogram for polyamide 66 material

The number average molecular mass is noted $<Mn>$ and the weight average molecular mass that we will consider in this article is noted $<Mw>$. Both can be expressed as:

$$< Mn >= \frac{\sum_i (C_i)}{\sum_i \frac{C_i}{M_i}} = \frac{\sum_i (n_i M_i)}{\sum_i n_i} \qquad \text{Equation 1}$$

$$< Mw >= \frac{\sum_i (C_i M_i)}{\sum_i C_i} = \frac{\sum_i (n_i M_i^2)}{\sum_i n_i M_i} \qquad \text{Equation 2}$$

where n_i are the numbers and Ci the concentration of the chains having a molecular mass equal to M_i. The polydispersity index is then defined as $I = \frac{< Mw >}{< Mn >}$ and represents the width of the molecular weight distribution.

The intrinsic viscosity of a polymer can be expressed as

$$[\eta] = \lim_{C\to 0} \frac{\eta_{PM} - \eta_S}{\eta_S \times C}$$ Equation 3

where η_{PM} is the viscosity of the polymer with concentration C, and η_S is the viscosity of the solvent used.

When the polymer chains are not entirely linear but have some level of branching, the apparent intrinsic viscosity of the polymer measured by GPC will appear to be lower than for a linear polymer. This effect can be represented in Figure 4. On this Figure, the two polymer chains (a) and (b) have the same molecular weight Mi, but the structure of (b) is different compared to (a) because of branching. As a result, during the experiment, the intrinsic viscosity of polymer (b) will appear to be lower than polymer (a). The apparent branching ratio g' may be expressed as:

$$g' = \frac{[\eta]_{sample}}{[\eta]_{linear}}$$ Equation 4

g'=1 for a linear polymer and decreases with the formation of branching.

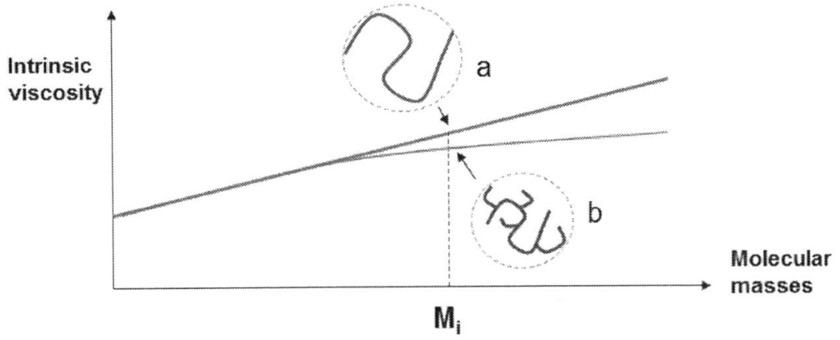

Figure 4 : Effect of branching on the intrinsic viscosity of a polymer

For this study, GPC experiments were performed at Solvay Research and Innovation facilities on a PL-GPC120 Polymer Lab apparatus using four PL-Gel columns, an UV detector and a differential viscosimeter detector (flow rate 1mL/min). Samples were dissolved in organic solvent at room temperature for 6h minimum prior the analysis (PA66 concentration is about 2 mg/mL).

2.3. Degradation of polyamide in water/glycol media

2.3.1. Chemical degradation mechanisms

When exposed to LLC, polyamide 66 chains can undergo two chemical reactions which can eventually occur simultaneously: Hydrolysis (Figure 6) and Glycolysis (Figure 7). The chain scission mechanism is represented in Figure 5. The hydrolysis reaction of polyamides has been studied extensively for decades [1-8]. This phenomenon is statistical [9], which means that the scission of chains is homogeneous in the overall sample at a given temperature and environment. However, it is known [10] that

Glycolysis mechanism only occur at temperatures above 275°C, which is well above LLC temperatures of en engine operating at normal conditions. As a result, Glycolysis is usually not activated during engine operation.

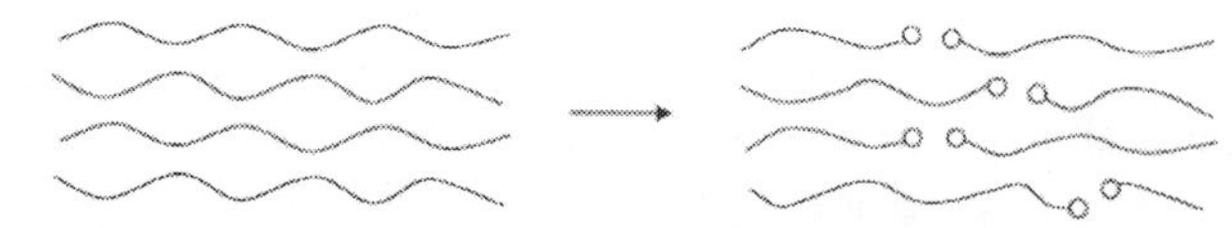

Figure 5 : Schematics of a polyamide chain scission after chemical degradation

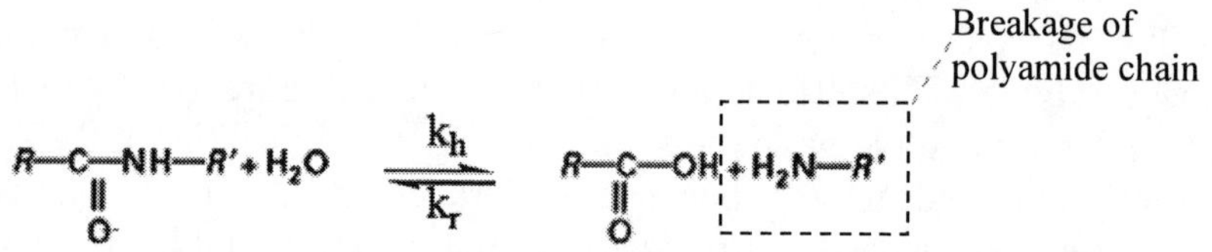

Figure 6 : Hydrolysis reaction of amide function

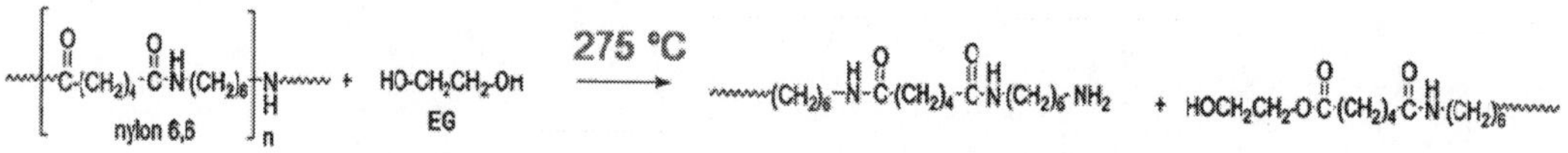

Figure 7 : Glycolysis reaction of polyamide 66

Figure 8 and Figure 9 illustrates the GPC chromatograms of PA66 before and after exposure to hot coolants, and clearly exhibit 3 dominant effects:

① a shift towards lower molecular weights, corresponding to an overall scissions of the molecular chains,

② a deformation of the shape towards non Gaussian distribution, corresponding to a mass redistribution due to a higher population in shorter polyamide chains,

③ the disappearance of small peaks at low molecular weights, corresponding to the extraction during the experiment of oligomers having small chain lengths.

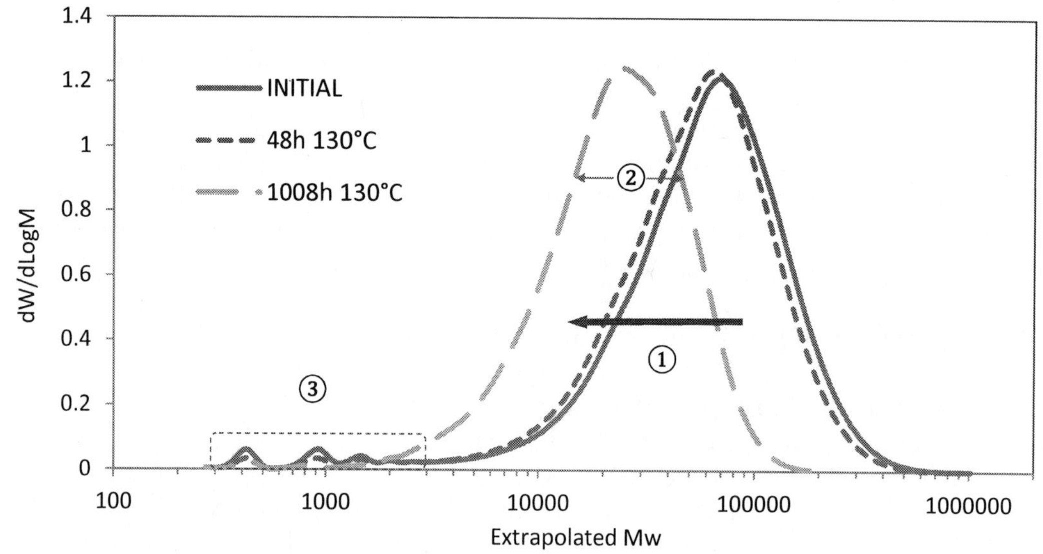

Figure 8 : GPC chromatogram of PA66 after exposure to hot coolant

Those effects are amplified with longer time exposure (Figure 8), higher temperatures (Figure 9), or more aggressive coolants (not studied in this article).

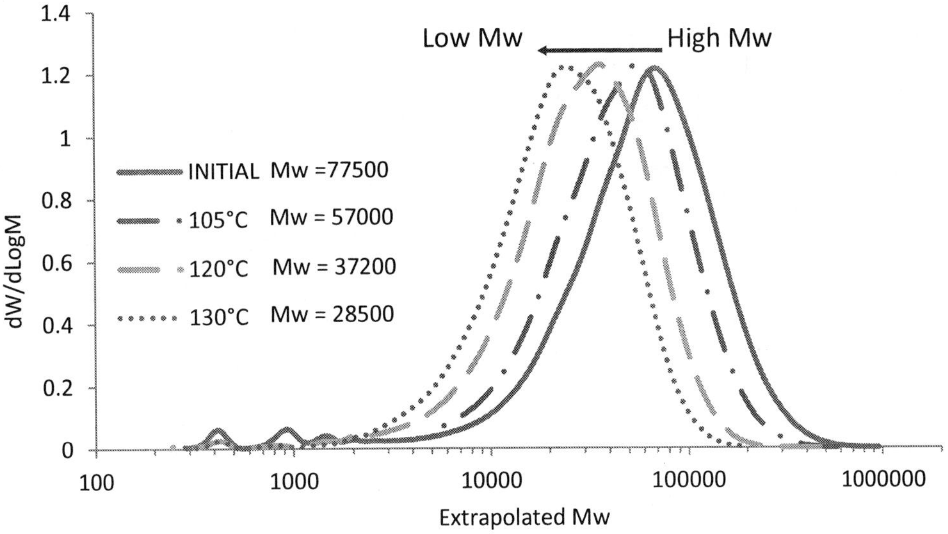

Figure 9 : GPC curves of PA66 after exposure during 1000h to hot coolant at various temperatures

3. RESULTS: GPC MEASUREMENTS IN LABORATORY CONTROLLED ENVIRONMENT

3.1. Analysis of tensile bars and first correlation between chemical degradation and mechanical performance

In a first experiment, we have exposed tensile specimens (ISO 527 tensile bars) of the considered polyamide material to LLC ageing by full immersion of the bars at various temperatures (105°C, 120°C, 130°C). We have then evaluated both molecular weight and tensile properties after ageing, see Figure 10. We can see that the molecular weight has the same behaviour as the tensile strength reduction over time in function of ageing temperature.

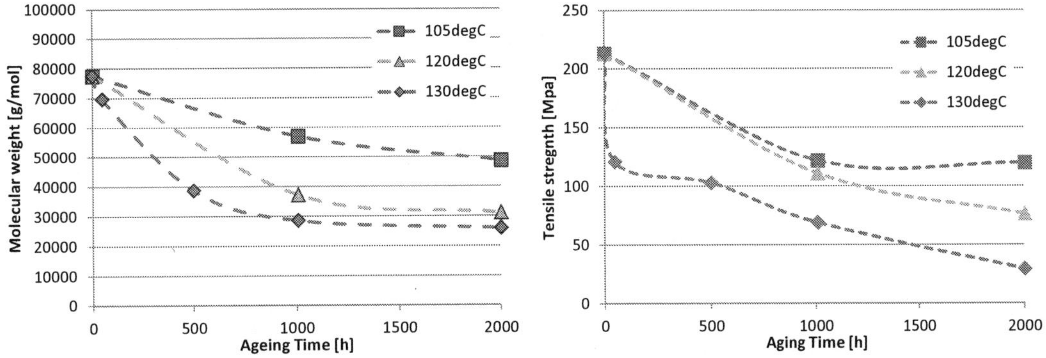

Figure 10 : Molecular weight and mechanical properties on ISO material samples fully immersed in LLC

To quantify the relation between both measurements, a superposition of molecular weight and tensile strength is shown in Figure 11 together with the relative reduction of each property compared to the initial value. For the superposition, the graphs have been adjusted in order to have the initial values of MW and tensile strength at the same scale. From the figure we can see the MW after 2000h ageing at 105°C has been reduced by 37%, the tensile strength after same ageing conditions has been reduced by 43%. This means the MW reduction and tensile strength reduction differ by 6% after such ageing time. Looking at all the measurements points, we have a maximum difference of MW to tensile strength reduction between 4 to 19%, which is very good considering the MW measurements are made on 1 sample only. We can conclude that a clear correlation exists between molecular weight measurements and mechanical strength tests.

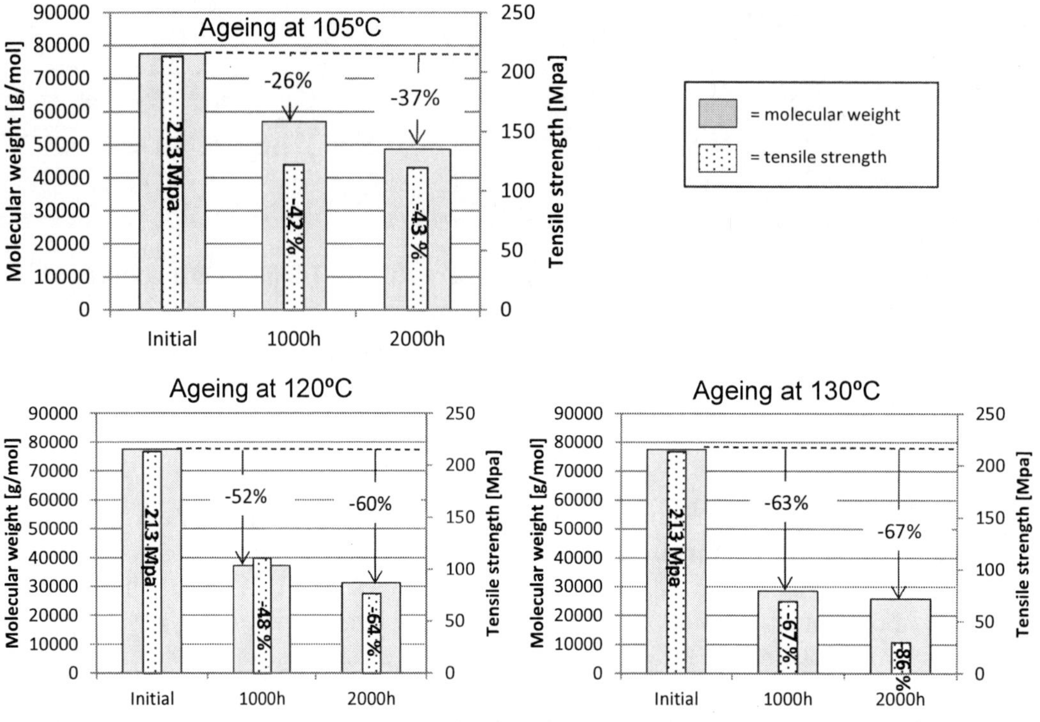

Figure 11 : Comparison Molecular weight and mechanical properties on ISO material samples fully immersed in LLC

3.2. Analysis of thermostat housings aged in the lab by immersion

In a second experiment, we have exposed original thermostat housing to LLC ageing by full immersion in a bath in the laboratory at 130°C. We have then measured molecular weight through GPC method. In this experiment, only 100 mg of material taken from the surface is required for analysis. It then allows evaluating the molecular structure of the material at different specific locations of the part. Here we have chosen to compare the internal face and the external face of the part, corresponding to location 1in and 1out of Figure 2. Figure 12 shows the molecular weight in function of ageing time of the internal and external face of the parts and a superposition of the previous result obtained on tensile bars.

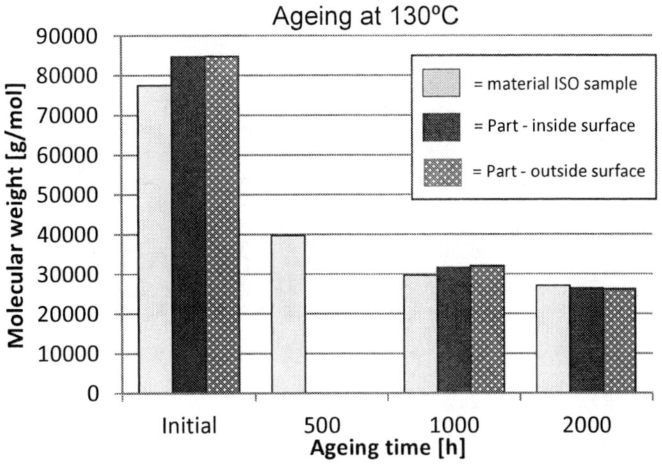

Figure 12 : Molecular weight on actual parts fully immersed in LLC

Here, as expected, we observe that the ageing of the part has been fully homogeneous between the inside and outside surface of the part. As well, we see no significant difference in the kinetics of ageing between a material ISO sample or a part, despite the different geometry. This is due because both geometries have similar thickness and that the time of experiment is much longer than the time needed for diffusion of the coolant into the thickness of the part. With this experiment we again observe that an equilibrium in the hydrolysis reaction has been reached due to the stable molecular weight after 1000h. As a final comment, these is a slight difference in initial molecular weight value since the ISO materials bar and the actual component will not have the same injection parameters which in this case results in slightly longer polyamide chain on the component. However, this difference is not significant when analysing the long term behaviour of material.

3.3. Analysis of thermostat housings aged in the lab on a glycol bench test

In a third experiment, we have submitted some thermostat housing parts to LLC ageing on a coolant circuit bench test designed at Solvay laboratories. On this test bench, six parts were mounted in series and connected together, with a permanent circulation hot LLC (105°C, 120°C and 130°C have been considered in the experiment).

Subsequently, the mechanical resistance of the parts after ageing on the coolant circuit test bench has been evaluated using a specific device mounted on a tensile test machine, as illustrated Figure 13. This test has been performed at a speed of 10 mm/min at room temperature (23°C). The force and displacement are recorded and the force at which the part breaks is noted.

The circle on Figure 13 indicates the breakage location during mechanical test, which was identical for each part. The location was confirmed by simulation, taking the non-isotropic material behaviour due to the glass fibre orientation into account, using MMI TECHNYL® Design technology. The simulated result correlates perfectly with the mechanical tests: the area of the highest stress and where the part breaks is located at the end of the sealing area, as illustrated on Figure 14.

Figure 13 : Picture of the mechanical tests done on thermostat housings

The arrow points at on the location of highest stress concentration, where the part breaks during the mechanical test

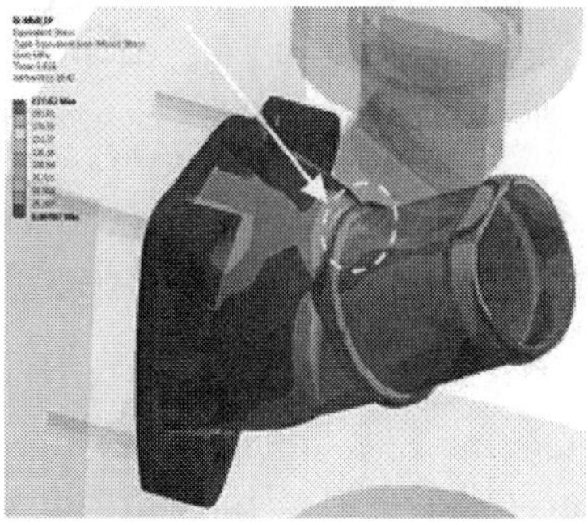

Figure 14 : Stress simulation using MMI TECHNYL® Design anisotropic simulation

Figure 15 presents the results of the molecular weight and the force to break of thermostat housings after ageing on the glycol test bench with a permanent circulation of coolant at 130°C. Here again, one can observe a clear correlation between the drop in molecular weight and the mechanical resistance of the part. The difference in molecular weight reduction and strength reduction is 3% at most.

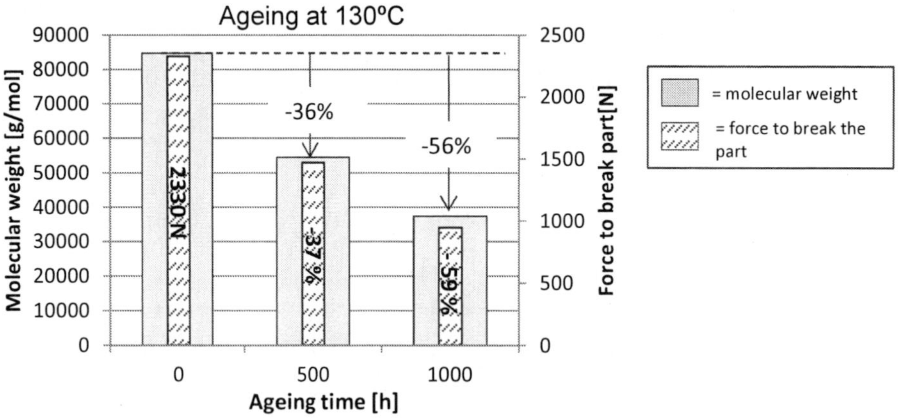

Figure 15 : Molecular weight and mechanical strength on parts installed on coolant circuit bench

Another observation is that, the MW value of a part exposed to LLC on the inside only (as done in the coolant bench) has a slightly higher value of molecular weight compared to when the part is fully immersed in LLC (as can be concluded by comparing the MW values from figure 11 and 14). An immediate conclusion is that the test bench, where there is circulation of coolant, is a less severe test device than the immersion test, where both internal and external surface of the part is in permanent and static contact with the coolant.

4. RESULTS: GPC MEASUREMENTS ON PARTS RECOVERED FROM THE FIELD

4.1. Analysis of parts recovered from the field: overview Mw

In this study, we have analysed thermostat housings taken from 7 used vehicles from the field. The randomly selected vehicles had engines from the same engine family, different mileages (from 102257 to 292116km) and driving country. Table 2 provides the details of each used vehicles from which the thermostat housing have been taken.

Table 2 : Vehicles details of field recovered parts

ID	Vehicle	Engine	Mileage [km]	Owner	Main country of use
A	Auris	1AD	102257	Private	Germany
B	Avensis	2AD	117140	Private	Germany
C	Avensis	2AD	125000	Company	Belgium
D	Avensis	2AD	178548	Company	Germany
E	Verso	2AD	178944	Company	The Netherlands
F	Avensis	2AD	255981	Company	Germany
G	Avensis	1AD	292116	Company	The Netherlands

In order to evaluate the effect of local environment (temperature, chemical media) on the chemical degradation of the material, GPC analyses were performed in the different locations of the thermostat housing assembly, as described in Figure 2 of the previous section.

Figure 16 presents the molecular weight measured in each position of the field recovered parts.

From this figure, we observe 2 different behaviours.

a, For position 1 (in and out), positioned before the wax thermostat, the molecular weight is higher compared to a new sample and fairly constant across all vehicles (= no influence of mileage). We observe that in all cases, location 1out and 1in are barely changed compared to the initial material (reference).

b, For positions 2, 3 and 4, positioned after the wax thermostat, the molecular weight is reducing from vehicle to vehicle (= with mileage increase from 102k to 292k km). The coolant branches exposed to the highest LLC temperatures (branch 2 and 3) have more degradation compared to branch 4.

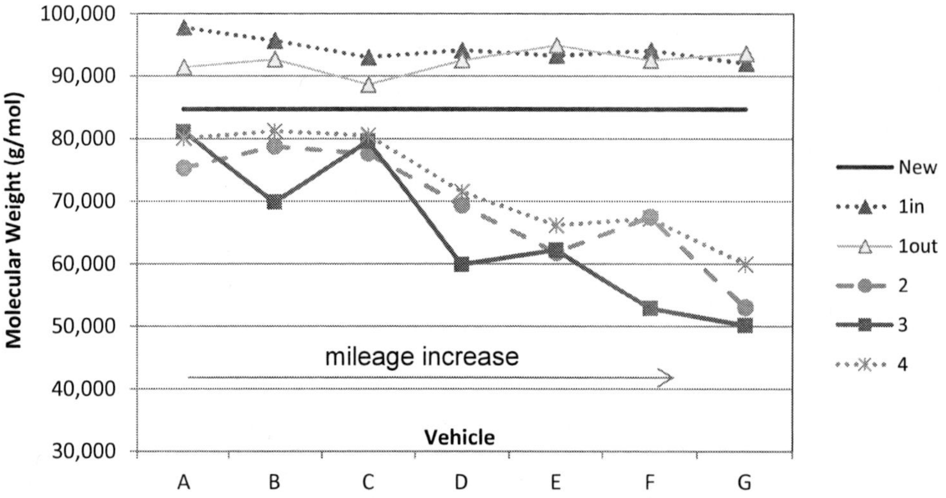

Figure 16 : Molecular weight thermostat housings from field vehicles

4.2. Chemical degradation and mechanical resistance of position 1

Figure 16 shows an increase of around 10% in molecular weight for position 1. To understand the degradation behaviour, the GPC chromatogram has to be investigated, see Figure 17.

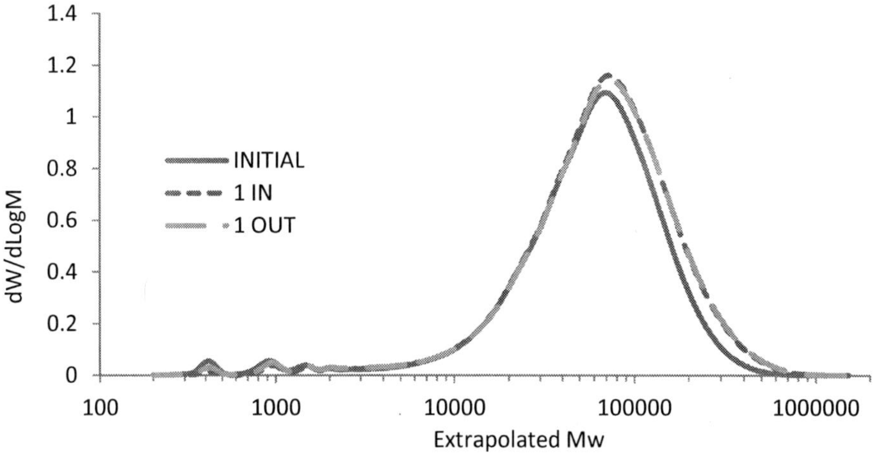

Figure 17 : Chromatograms on position 1 of vehicle G

From this superposition of initial sample against field part sample, we cannot see a shift in molecular weight but a slight change in the profile towards higher molecular weights. This has been confirmed by calculating the viscosity and branching rate and comparing it with the initial sample as done in Table 3 (slight increase of the molecular weight and viscosity but no significant change in branching rate). The increase in molecular weight and viscosity is interpreted by a post-polycondensation of the polymer, occurring at relatively moderate temperatures (75°C to 100°C), and leading to a slight increase of the polymer chain lengths.

Table 3 : Molecular weight, intrinsic viscosity and branching rate of part G

	Mw	η	g'
Initial	84738	0.73	0.99
1in	90185	0.75	0.98
1out	91285	0.77	0.98

Position 1out is located on the outer surface of the part and is in contact with the surrounding air, with temperature ranging from 70°C to 100°C, as shown in Table 1. This temperature is clearly too low to generate observable thermo-oxidation degradation on the polyamide material used for the thermostat housing. This explains why no visible change of the GPC curve is observed on the chromatograms.

Since position 1 is located before the wax thermostat, the coolant inside will only be heated up after the thermostats opens. Due to this transient temperature condition (which can takes up to 30 minutes when driving at low load / high gear) and a relative low temperature during normal use after thermostat opens (city driving = 75°C / highway = 85°C), there is no material degradation on position 1in regardless of vehicle mileage. This can also be demonstrated through the mechanical test results presented in Figure 18. The test procedure is exactly the same as done in paragraph 3.3

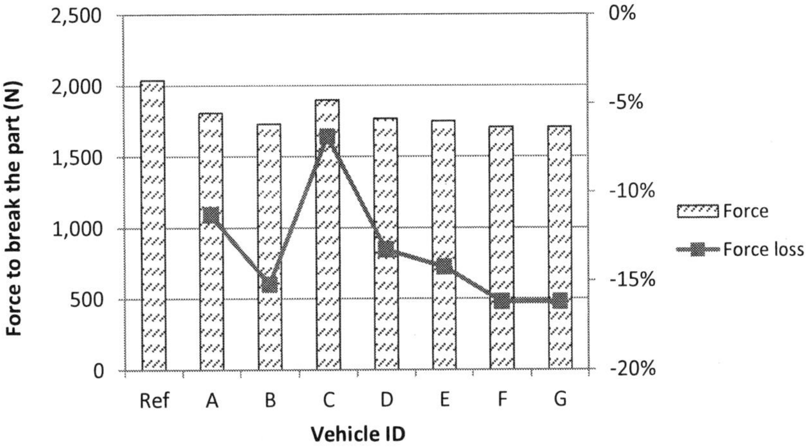

Figure 18 : Mechanical resistance of parts coming from field vehicles

In this graph, we observe a loss in mechanical strength up to 15% but it cannot be related to the molecular weight since the MW increased compared to the new part. We interpret the limited loss of mechanical properties by the physical absorption of LLC by the polymer, and not by a chemical degradation of the polymer. This absorption is a purely physical, and reversible effect and leads to plasticisation of the polymer and, as a result, to a slight and limited softening of the material.

4.3. Chemical degradation seen in position 2, 3 and 4

Figure 19 presents the superposition of GPC chromatograms for position 2, 3 and 4 of the vehicle with the highest mileage (vehicle G). Due to the continuous exposure to hotter coolants compared to branch 1, we can now observe a clear shift of the peaks towards lower molecular value (black array) and a visible widening of the distribution curves.

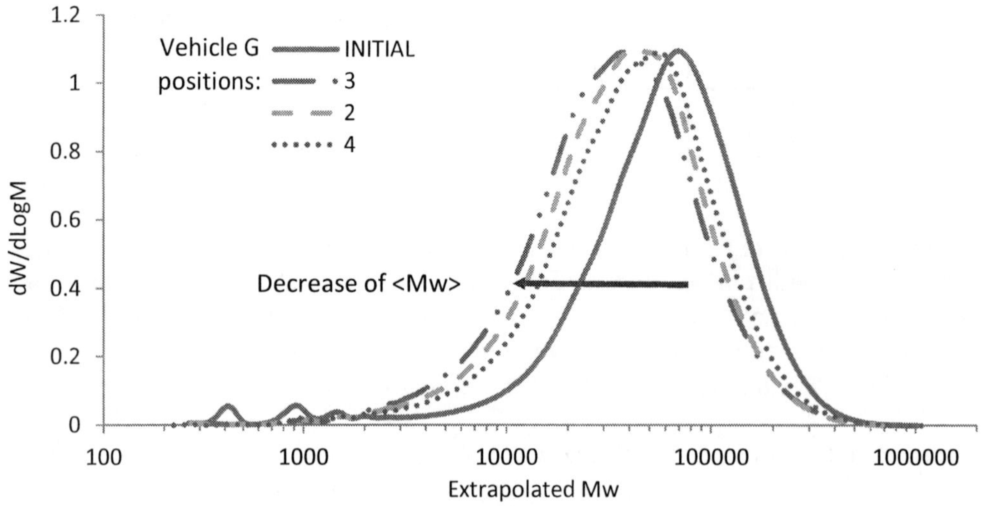

Figure 19 : Superposition of chromatograms of part coming from vehicle G at different positions

This is a clear consequence of a hydrolysis mechanism involving chain scission, modification of distribution length, and to a certain extent, a change in the molecular structure. Table 4 indicates the branching rate (g'), as measured by Equation 4, of the various analysed position. A lower g' value indicates a higher branching effect.

Table 4 : Molecular weight, intrinsic viscosity and branching rate of part G

	Mw	η	g'
Initial	84738	0.73	0.99
2	51495	0.4	0.96
3	48494	0.38	0.95

This result shows that the main chemical degradation mechanism in the material involves chain scission due to hydrolysis reactions and that no significant polymer branching effect is observed.

4.4. Comparison of field aged parts and lab aged parts based on molecular weight

Figure 20 shows the molecular weight for the branch with the worst case degradation (= position 3) in function of vehicle mileage.

We can observe a clear correlation between the molecular weight and vehicle mileage as shown by the dotted trendline. By adding the relative reduction in molecular weight observed on material samples fully immersed in LLC (data from Figure 11), a correlation can be made between vehicle mileage and part ageing conditions. The highest mileage part for example has seen almost the same thermal damage as an ageing of 2000h x 105°C.

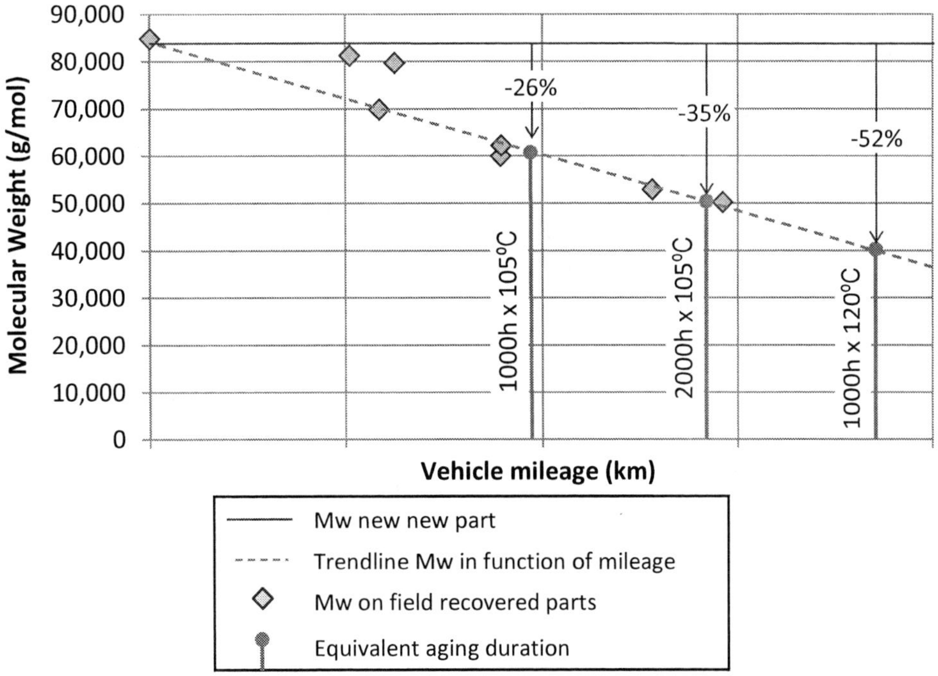

Figure 20 : Molecular weight at position 3 in function of mileage

4.5. Relation between molecular weight and part strength

Finally, by consolidating all data generated in this study, we could build a master plot making the link between the measured reduction in the molecular weight and the reduction in the mechanical strength (all compared to a new reference) obtained on various samples, as shown on Figure 21. We confirm again a strong, almost linear, correlation between both characteristics, independent of test procedure that was used to gather the data.

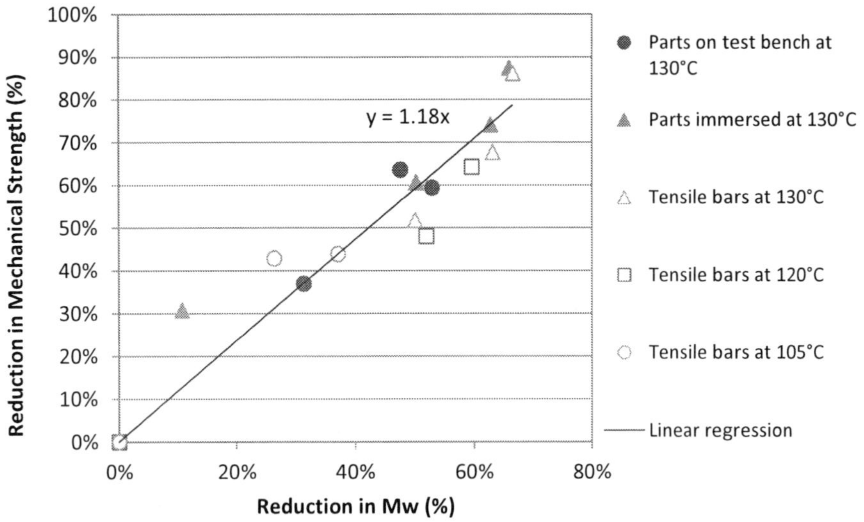

Figure 21 : Link between reduction in mechanical strength in function of the variation of the molecular weight obtained on various tests performed for the study

This relation can be used to determine the remaining strength of branch 3, since no actual measurements were done on this pipe. From the highest mileage field recovered part, Mw is 48000g/mol, which is a reduction of 41% compared to a new material. Using the relation from the masterplot, this would result in a 50% reduction in mechanical strength of the outlet pipe. This is a major difference compared to the 15% reduction in strength seen on pipe from branch 1. As a conclusion, from this data the GPC measurements have highlighted which area to focus on during the part validation process. Furthermore, it provides quantifiable data to judge degradation amount of the part after ageing.

5. CONCLUSIONS AND PERSPECTIVES

This study has brought a better understanding of the chemical degradation of a thermostat housing based on field recovered parts. We showed there is no degradation on the polyamide material located upstream of the wax thermostat. On the surface of the part downstream of the wax thermostat a hydrolysis reaction is taking place as demonstrated by the decrease of molecular weight of the polymer material. Based on investigation of various field parts, this chemical reaction is in function of coolant flow, coolant temperature and exposure duration. These 3 parameters are easy to keep constant during a heat ageing as done for component testing in lab controlled environments. On the vehicle however, these parameters are transient, which makes it difficult to determine the thermal load seen by a component on any given field vehicle. By using the Gel Permeation Chromatography technology, we could determine the thermal load through comparison of molecular weight data of field parts with lab aged parts. Furthermore, since the thermal load is not homogeneous across the parts, this chemical analyse methodology precisely quantify these differences.

In a new approach to design a cooling component, the above observations could be used to optimize the part design, and eventually reduce manufacturing costs. For instance, when designing a cooling plastic component, one should consider the region of part presenting the highest "chemical stress" concentration, which is likely to be where the coolant temperature is the highest. A combination of a high chemical stress with high mechanical stress should be avoided. Regions presenting high mechanical stress concentration but low chemical degradation, for instance not in direct contact with coolant or with coolant at relatively low temperature (below 100°C) are more likely to be safe, because little polymer degradation occur in this region. As a result, even after very long term exposure, the mechanical properties in this type of regions should be close to initial mechanical properties. Combining finite element analysis with chemical stress analysis opens a new area for designing engine plastic components.

REFERENCE LIST

[1] R. Bernstein, D.K. Derzon, K.T. Gillen, Polymer Degradation and Stability, **88**, 480, (2005).

[2] J.H. Hong, D.M. Dhevi, J.S. Lee, K.J. Kim, , Polymer Composites, **28**, 778 (2007).

[3] E.S. Gonçalves, L. Poulsen, P.R. Ogilby, Polymer Degradation and Stability, **92**, 1977 (2007).
[4] M. Ito, K. Nagai, Polymer Degradation and Stability, **93**, 1723 (2008).
[5] A. Bergeret, L. Ferry, P. Ienny, Polymer Degradation and Stability, **94**, 1315 (2009).
[6] R. Bernstein, K.T. Gillen, Polymer Degradation and Stability, **95**, 1471 (2010).
[7] J.L. Thomason, J.Z. Ali, J. Anderson, Composites Part A, **41**, 820 (2010).
[8] C. El-Mazry, O. Correc, X. Colin, Polymer Degradation and Stability, **97**, 1049 (2012).
[9] B. Ledieu, Thèse de l'Ecole Normale Superieure d'Arts et Métiers, collaboration Valéo/Rhodia, 2010.
[10] K.J. Kim, D.M. Dhevi, J.S. Lee, Y.D. Cho, E.K. Choe, Polymer Degradation and Stability, **91**, 1545 (2006) Cw.

High-heat resistant polyamides for the air intake system

O THOMAS, A BAYER, B HOFFMANN and **M HOFFMANN**
EMS-CHEMIE AG, Switzerland

ABSTRACT

By introducing a new technology to our high heat PPA Grivory HT2VS-3HH and Grivory HT2VS-45HH, EMS GRIVORY was able to improve the long term high heat resistance significantly. The new materials retain 50 % of their mechanical properties after heat aging 5000 h / 200 °C or 3000 h / 250 °C respectively. The new technology accelerates the formation of a very thin and narrow defined barrier layer at the surface of the PPA part, which shields the bulk material and inhibits diffusion of oxygen into the PA. Thereby the part is prevented from aging by oxidation over a very long period.

1 LIST OF ABBREVIATIONS

PPA:	Polyphthalamide
PA:	Aliphatic polyamide
GF:	Glass fiber
rt:	Room temperature
FR:	Flame retardant
$[O_2]$:	Concentration of the reactant O_2
t:	Time
x:	Depth of penetration of reactive species
D:	Diffusion coefficient

2 INTRODUCTION

Forced by the ongoing trend of downsizing car engines and the accompanying rise of the operating temperatures, the development of suitable materials is of constant interest. With these requirements polyamide as high performance material, often used in automotive applications, has to be lifted to the next performance level either. [1]

Since heat aging of polyamides means thermal accelerated oxidation – and therefore degradation of the polymer chain – the crucial question while designing a heat resistant polyamide is always how to stop or avoid the radical oxidation process.

This can be realized in different ways:

1. By inorganic radical scavengers (inorganic redox systems). [2]; [3]
2. By organic radical scavengers (mostly based on sterically hindered phenols or aromatic amines). [3]
3. By shielding the material from oxygen.

Whilst organic stabilization systems work well up to a temperature of ~160 °C, the inorganic redox systems can be utilized at even higher temperatures.

The third, not yet common method is well known in the field of intumescent systems in flame retardant polymer formulations, where charring systems are used to protect the bulk material of a part. [4], [5], [6]

The remarkable efficiency in heat stabilization is presented here.

3 RESULTS AND DISCUSSION

3.1 Conventional heat aging at different temperatures

One method to analyze the long term performance of a heat stabilized polyamide is the loss of mechanical properties after heat aging a specimen for a given time.

Heat aging trials were performed at different temperatures. Every data point in the shown graphics represents the average value of stress-strain measurements (conditions: dry, rt) of five specimens.

Figure 1 shows the results of heat aging Grivory HT2VS-3HH in a temperature range of 180 °C to 250 °C for 3000 h.

During the first 500 h the specimen undergoes a temper effect in which a barrier layer is formed (figure 1). As the run of the graphs shows, the speed of this effect depends on the given temperature. The higher the chosen temperature, the faster the formation of the barrier layer takes place. Therefore, the resulting loss of mechanical properties during this phase is low in the case of high temperature (figure 1; 250 °C). When the temper effect is finished, Grivory HT2VS-3HH retains the given level over the whole 3000 h. The run of the elongation at break, shown in figure 3, demonstrates the same effect even more pronounced and confirms this interpretation.

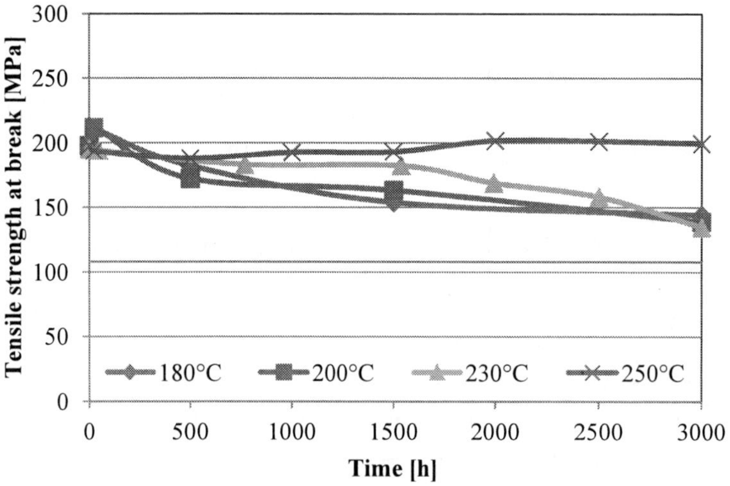

Figure 1: Grivory HT2VS-3HH – Tensile strength at break after heat aging 3000 h (Specimen: EN ISO 3167, type A)

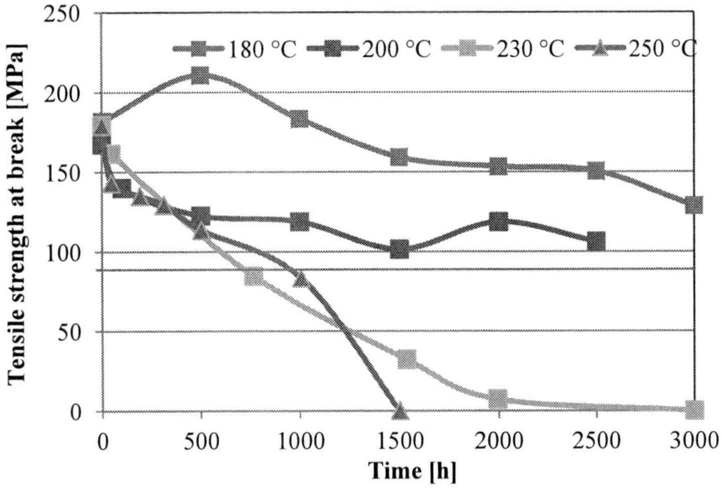

Figure 2: Standard PPA – Tensile strength at break after heat aging 3000 h (Specimen: EN ISO 3167, type A)

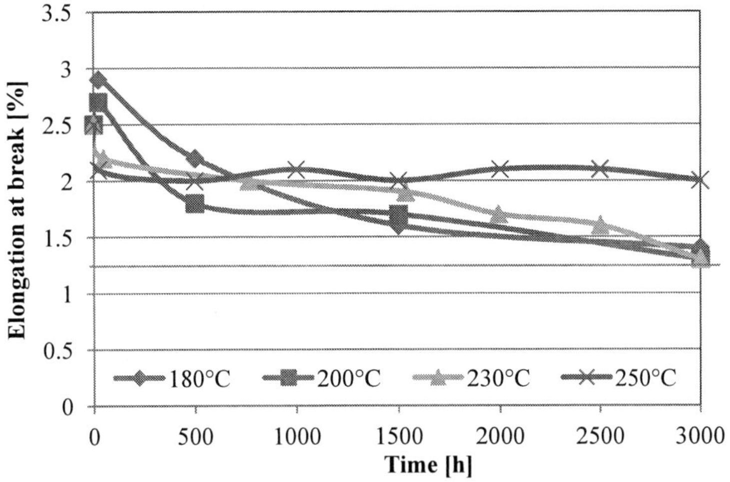

Figure 3: Grivory HT2VS-3HH – Elongation at break after heat aging 3000 h (Specimen: EN ISO 3167, type A)

Here an analogy to the intumescent systems can be drawn – it was shown in literature that by pre-aging FR compounds, the charring in FR tests occurs faster and more selectively.

This leads to a faster formation of the protecting layer. With the quick degradation of the intumescent materials (= low ignition time) the protection of the bulk material can be ensured in a shorter time. [6]

Regarding Grivory HT2VS-3HH, in every case the measured data does not undercut 50% of the starting value after 3000 h. The comparison to a standard PPA product shows the enormous improvement of high heat stability (figure 2). Whilst there is no significant improvement at a temperature of 180 °C detectable, the new stabilizing effect starts at 200 °C. Its full potential becomes visible at 230 °C.

Figure 4 depicts the results of 5000 h heat aging trials at 180 °C and 200 °C respectively.

These results substantiate the long term stability of Grivory HT2VS-3HH. Additionally, figure 5 shows that also the flexural strength after aging at 220 °C does not undercut the 50 % strength bench mark.

First investigations of aging longer than 3000 h indicate the start of undercutting the 50% benchmark beyond 5000 h thermal aging. Since the aging process is still the same, but slowed down extremely by the barrier layer, the loss of properties will appear smooth like it is observed for the standard PPA (figure 2).

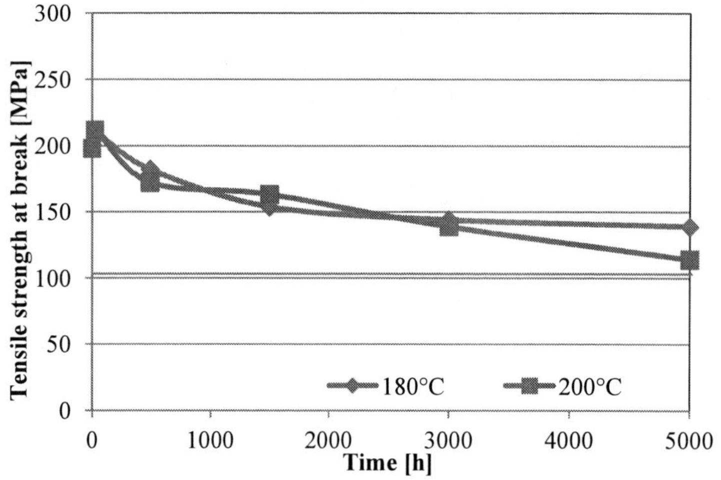

Figure 4: Grivory HT2VS-3HH – Tensile strength at break after heat aging 5000 h (Specimen: EN ISO 3167, type A)

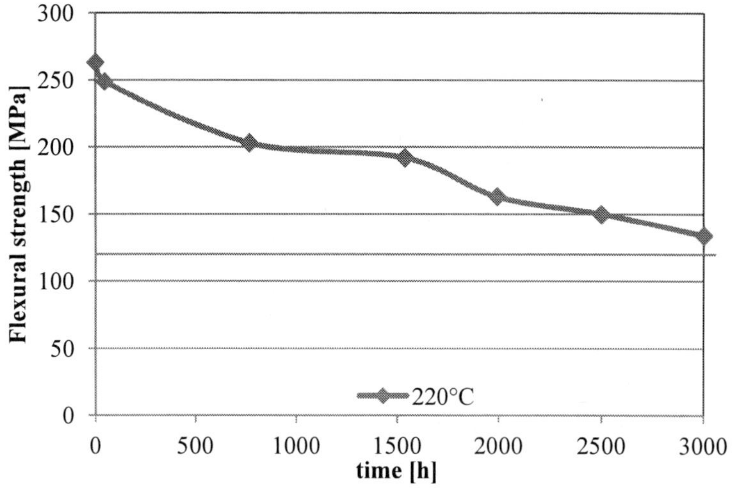

Figure 5: Grivory HT2VS-3HH – Flexural strength after heat aging 3000 h

3.2 Optical assessment *via* microscope analysis

Beside mechanical investigations the heat aging process can be observed optically at the cross section of a specimen. For that reason tensile test bars of Grivory HT2VS-3HH natural were aged at 230 °C and subsequently a sample of the cross section was cut off the test bar and fixed in epoxy resin for further optical investigations.

In figure 6 microscope pictures and pictures of the whole cross section of a heat aged conventional PPA material (30 % GF) are shown. In this case the continuous progress of heat aging can be observed very well with the ongoing formation of carbonized PA from the surface to the core of the tensile bar. The microscope pictures, depicting the surface of the specimen in the cross section view, are showing small cracks growing from the surface to the core of the tensile test bar. These cracks are responsible for a better access of oxygen to the material and accelerated oxidation.

In contrast to the conventional PPA, in figure 7 are shown images of the cross section and the surface of heat aged Grivory HT2VS-3HH.

Within 48 h a well-defined layer of carbonized PA has formed at the surface of the specimen and no cracks, which can act as diffusion channels for oxygen, are visible. This layer protects the core of the specimen from oxidation and retains the mechanical properties over 3000 h and longer. As the image taken at 3000 h shows, even after that time most of the bulk PA is not affected by oxidation.

Investigations regarding the thickness of the formed layer in dependence of the chosen temperature show behavior according to *Arrhenius* (Eq. i). Figure 8 presents the graphical analysis after aging tensile test specimens for 2000 h at a given temperature. Here the qualitative results of the optical assessment are confirmed.

$$k = A \cdot e^{\frac{-E_A}{RT}}$$

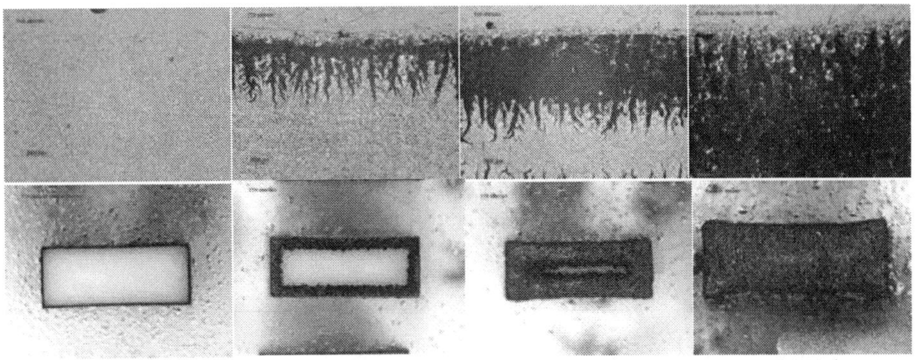

Figure 6: Heat aging (230 °C) of a conventional PPA (30% GF)

Figure 7: Heat aging (230 °C) of Grivory HT2VS-3HH (30% GF)

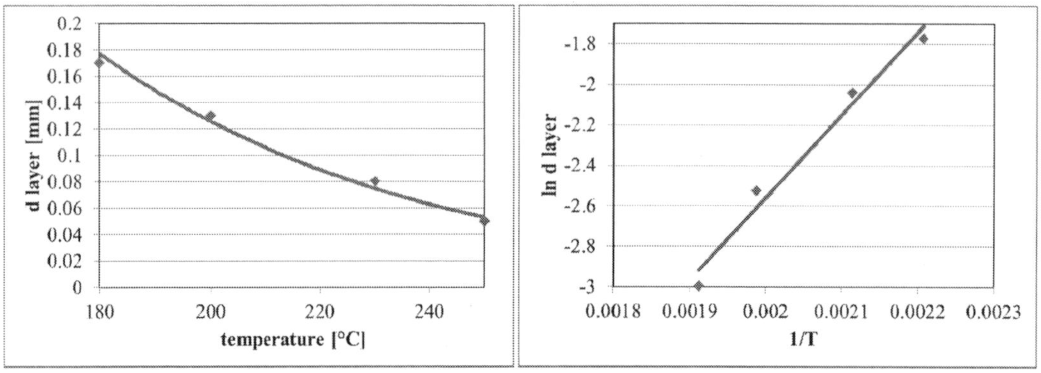

Figure 8: Dependence of the layer thickness on the temperature (after 2000 h aging time)

3.3 Theoretical background

The oxidation of polyamide is based on a radical chain mechanism which follows the three steps:

Initiation:

R-H, R-R → R• **(I)**

Propagation:

R• + O_2 → R-OO• **(II)**

R-OO• + R-H → ROOH + R• **(III)**

ROOH → RO• + •OH **(IV)**

2 ROOH → RO• + ROO• + H_2O **(V)**

Termination:

R• + R• → R-R **(VI)**

R• + RO• → R-O-R **(VII)**

2 RCOO• → RCHO + RCOH + O_2 **(VIII)**

Figure 9: Reaction steps of a radical oxidation reaction

The initiation rate at ambient temperature is extremely low due to the activation energy. Also the reaction of a macro molecule with O_2 can be mostly precluded with regard to kinetic and sterical studies.

At elevated temperatures the initiation reactions are becoming faster and allow the addition of O_2 to active centers (figure 10; phase I). Depending on the application of a stabilization system, the oxidation process reaches a steady state over a certain time (phase II), followed by the raise of radical oxidation reactions after the scavenger system is readily consumed (phase III). The oxidation rate stagnates in phase IV with the decrease of available active reaction centers.

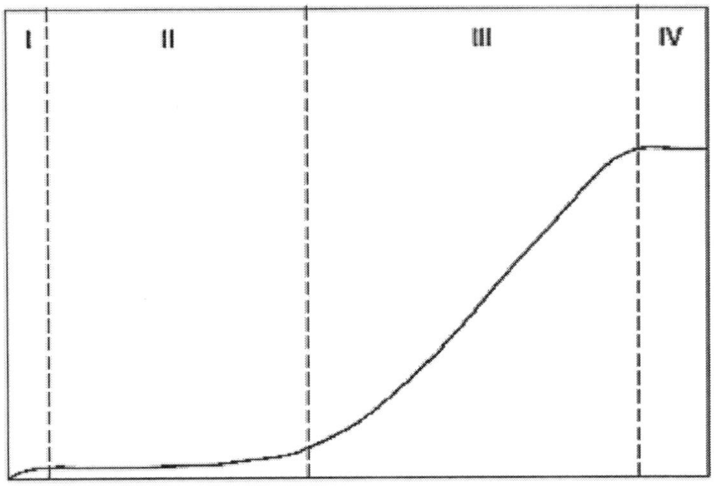

Figure 10: Kinetic sequences of a radical oxidation process with time [7]

Since oxidation of polymers is a reaction between a solid phase and a gaseous one, the oxidation is diffusion controlled. The ability of oxygen diffusion in polymers depends on the ratio of amorphous vs. crystalline phases and on the thermal induced movement of molecule segments. The solubility of oxygen in the given polymer is lowered by the presence of polar groups or methyl groups for example. Activation of a PA, which means formation of an initial radical, preferably occurs at the α-position next to the carbonyl carbon, from where the oxidation process propagates according to figure 9.

3.4 Kinetic considerations

A suitable prediction of the lifetime of a plastic part must involve the thickness distribution of the reaction products. This fact can be described by the *2. Fick's Law.* [8]

$$\frac{\partial[O_2]}{\partial t} = D\frac{\partial^2[O_2]}{\partial x^2} \tag{I}$$

$[O_2]$ = concentration of the reactant O_2, t = time, x = depth of penetration of reactive species, D = diffusion coefficient.

Since parallel to the diffusion occurs a consumption of the reactants to this equation must be added a correction factor:

$$\frac{\partial[O_2]}{\partial t} = D\frac{\partial^2[O_2]}{\partial x^2} - r$$

(II)

r = local reaction rate

The local reaction rate r is a function of the local concentration of the reactants $r \rightarrow r([O_2])$. If $d[O_2]/dt = 0$ thereof results

$$D\frac{\partial^2[O_2]}{\partial x^2} - r([O_2]) = 0$$

(III)

which means that the system has reached a stationary state.

The bimolecular reaction between O_2 and the polymer chain is at first sight a reaction of 2nd order kinetics ($A + B \rightarrow C + D$) with the rate law:

$$-\frac{\partial[A]}{\partial t} = -\frac{\partial[B]}{\partial t} = k \cdot [A] \cdot [B]$$

(IV)

It becomes a reaction of pseudo 1st order kinetics when the system reaches the stationary state of a saturated O_2 concentration. Here the concentration stays constant whilst the polymer chains get degraded. This simplifies the rate law to:

$$-\frac{\partial[A]}{\partial t} = -\frac{\partial[B]}{\partial t} = k \cdot [A]$$

(V)

and results in a linear dependency of [A], in this case the concentration of the undegraded polymer chains.

In most cases the thickness of the degraded layer is of the order of magnitude of D/k (D = diffusion coefficient, k = pseudo first order rate constant of reactant consumption). [2]

If there is formed a barrier against diffusion, as it is the case in Grivory HT2VS-3HH, the rate law converts back into a rate law of 2nd order and the O_2 concentration is no longer negligible. The reaction speed then depends on the concentrations of O_2 and undegraded polymer chains.

These considerations are regarding only one single reaction. In the case of degradation of a polyamide compound the number of possible reactions raises vastly with the number of the components – so these theoretical considerations can only be regarded as a model system.

4 CONCLUSION

Grivory HT2VS-3HH and Grivory HT2VS-45HH are benefitting from the successful application of a charring additive, which leads to the selective formation of a very narrow and well defined barrier layer.

This layer reduces the diffusion of oxygen and the rate of oxidation in a very impressive way as the comparison of figure 1 & 2 shows. With the very low diffusion/oxidation rate, the continuous loss of mechanical properties by time, as it can be still observed for standard PPA (figure 2) is extended over a very long period.

Here two stabilization systems complement each other:

1. The conventional redox heat stabilization scavenges existing active O_2 / radicals in the bulk material.
2. The barrier layer reduces the formation rate of these active centers and controls the oxidation kinetically.

With this synergistic effect the lifetime of the new Grivory HT materials is lifted to the next performance level.

5 AUTOMOTIVE APPLICATIONS

Grivory HT2VS-3HH and Grivory HT2VS-45HH offer properties for demanding components in automotive air intake systems.

Figure 11: Charged air cooler tank examples

REFERENCE LIST

[1] M. Völker, *KUNSTSTOFFE International*, **2013**, *3*, 33-36.

[2] L. Audouin, V. Langlois, J. Verdu, J.C.M. de Bruin, *J. Mat. Sci.*, **1994**, *29*, 569-583.

[3] F.R. Stohler, *Angew. Macromol. Chem.*, **1985**, *136*, 65-74.

[4] F. Dabrowski, M. le Bras, L. Cartier, S. Bourbigot, *J. Fire Sci.*, **2001**, *19*, 219-241.

[5] S. Bourbigot, M. le Bras, S. Duquesne, M. Rochery, *Macromol.Mat.Eng.*, **2004**, *6*, 499-511.

[6] M. le Bras, S. Bourbigot, *Fire and Polymers*, September 15, **2001**, 136-149.

[7] S. Pongratz, *Dissertation*, **2000**, University of Erlangen-Nürnberg.

[8] R.C. Golike, S.W. Lazoski, *J. Phys. Chem.* **1960**, 895; G.C. Furneaux, K.J. Leddury, A. Davis, *Polym. Degrad. Stab.*, **1986**, *3*, 431; S.P. Fairgreve, J.R. McCallum, **1985**, *11*, 251; K.T. Gillen, R.L. Clough, *Handbook of Polymer Science and Technologie*, Vol. 2, edited by N.P. Cheremisinoff (Dekker, New York, **1989**) p. 170.

Domat/Ems, March 2015

Evaluation of plastic materials for new charge air cooler heat exchanger in the low pressure exhaust gas recirculation loop

A TANGHE and **P HAVET**
VALEO Thermal Systems – Powertrain Thermal Systems, France

ABSTRACT

European regulations of automotive pollution are becoming more severe. In 2020, the limit was fixed at 95 g CO_2 /km. Therefore, the automotive industry must innovate to find technical solutions that meet these requirements. One consists in limiting polluting emissions by recirculation of some gases.

Regarding the technical solution for the exhaust gas recirculation, our challenge is to evaluate new plastic materials used for our under-hood heat exchangers.

Validation tests on thermoplastics materials in contact with exhaust gas condensates: define an exhaust gas condensate solution, define ageing conditions, characterize the impact of this environment on the mechanical properties of plastics, evaluate different materials and make a selection of materials for this application.

Keywords: Charge Air Cooler heat exchanger, plastic material, EGR LP loop

LIST OF SYMBOLS AND ABBREVIATIONS

g CO_2 /km: gram of carbon dioxide emitted per kilometer
EGR: Exhaust Gaz Recirculation
HP: High Pressure
LP: Low Pressure
CAC: Charge Air Cooler heat exchanger
ICP: Induced Coupled Plasma
IC: Ionic Chromatography
ISO norm: International Organization for Standardization norm

1 CONTEXT OF THE STUDY

The European regulations for automotive pollution levels are getting progressively more restrictive. By mid 2014, the objective is to reach 130 g of CO_2 /km, and by 2020 the polluting emissions should not exceed 95 g CO_2 /km (cf. Graph 1). Therefore, the automotive industry must innovate to find technical solutions that meet these requirements.

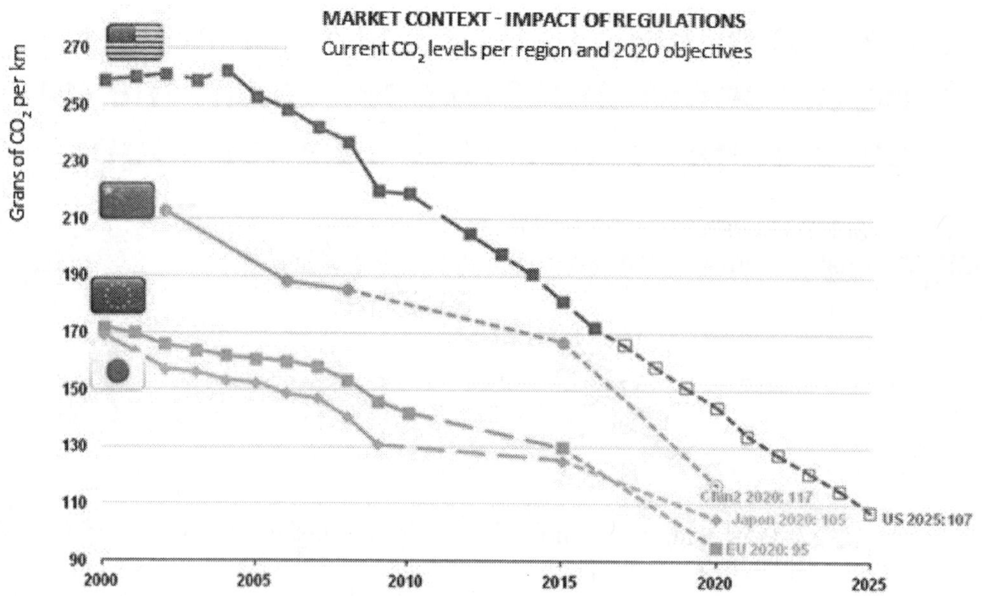

Figure 1: European regulation [1]

One of the technical solutions is to limit polluting emissions by reinjection of a portion of the exhaust gas in the air loop after cooling and pressure decrease.

This paper presents the evaluation of our plastic materials used in our heat exchangers in this new air loop low pressure with exhaust gaz.

2 INTRODUCTION

In our current heat exchanger, for air loop, the aim of a charge air cooler heat exchanger is to cool down the air, in order to re-introduce cooled air in the motor and give it a boost.

Currently, the environment of our charge air cooler heat exchanger (cf. Figure 2) is only hot air charged with engine oil, so our material validation for charge air cooler tank is focused in air aging in hot air and in engine oil.

Figure 2: Valeo Charge Air Cooler

But, with the new system and the re-introduction of exhaust gas in the loop, the environment is changing (cf. Fig.3). The air is charged with exhaust gas after cooling, pressure decrease and clean up.

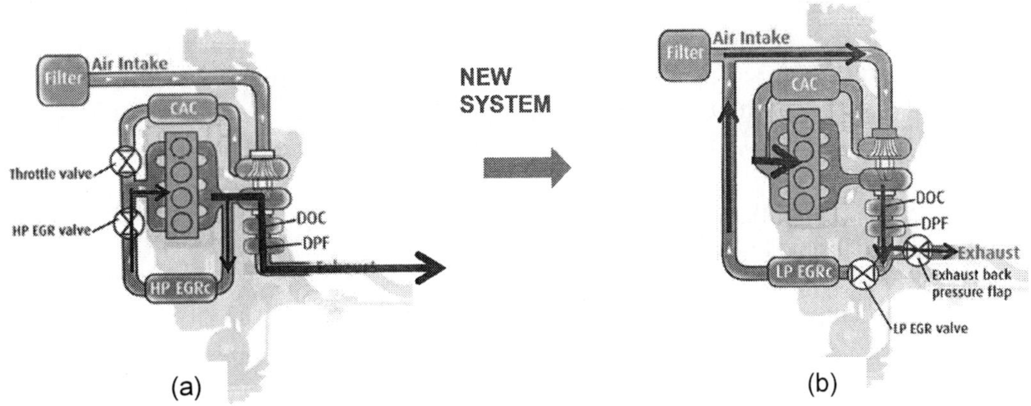

Figure 3: Current and new system - Exhaust Gas Recirculation (EGR) [2]

We can see in Figure 3 (a), that for current air loop, only "new" air is introduced in the Charge Air Cooler heat exchanger (CAC). In this system, exhaust gas are re-circulating but in a high pressure loop, and directly in the motor, through a high pressure EGR valve which regulates the EGR rate, and without passing through the CAC. The drawback of this solution is the limited rate of EGR allowed by the motor, without degrading its efficiency. In this current loop, our plastic material validation for CAC tank is done according to the environment of the part during its lifetime conditions: the validation is done by performing heat aging at the continuous temperature in use and by evaluating the retention of properties after a long term heat aging.

To reduce the rate of exhaust gas at the outlet of the exhaust pipe, and thus the pollution, the strategy is to mix exhaust gas with "new" air, in order to increase the EGR rate (see Figure 3(b)): a new loop needs to be set-up with a low pressure EGR valve. In this case, the EGR can be re-introduced directly upstream of the motor, and can be re-introduced with "new air", directly in the inlet of the CAC and without degrading the motor efficiency even if the rate of EGR is higher, compare to current system. In this case, we need to evaluate our plastic material for CAC tank, according to this environment: air charged with exhaust gas.

3 PLASTIC MATERIAL VALIDATION FOR CAC TANK

In this section, we will focus on the comparison of plastic material validation between the two different loops, and highlight only the differences.

3.1 The part: Charge Air Cooler

The CAC is composed by a core in metallic material, for the air cooling exchange with exterior, two tanks: inlet and outlet, which guide the air to the core and which guide the air outside of the core by the outlet tank. (cf. Figure 2).

This part is located under the hood and near the motor.

3.2 Functional analysis

For each material validation, we perform a functional analysis, for each life step of the part (production step, storage, operating step,…). In our case, the production step analysis is the same for the two loops, contrary to the operating step where environment is changing. We will focus only on this operating step during all the paper.

So, for operating step, the environment is changing from only air charged with engine oil to same air charged but with exhaust gas.

The material validation, done by evaluation of mechanical properties at room temperature and properties at the continuous temperature in use, must implement also chemical compatibility with exhaust gas.

3.3 Material used

The most common plastic material used for the injection of Charge Air Cooler heat exchanger tank is polyamide 66 filled with glass fibers: the percentage of glass fibers can vary from 30% to 50%, depending on the continuous temperature in use, peak temperature and pressure variation inside the part, and thus depending on the mechanical characteristics needed for the application.

3.4 Parts injection

The process used for such part is injection process: the plastic material under pellets is molten in a screw with a temperature up to the glass transition temperature or up to melting temperature, and the molten polymer is pushed in a mold with the design of the CAC.

To perform mechanical characterizations, samples should have been cut in injected parts, according to ISO norms (ISO527-1BA for tensile tests and ISO179-1fU for impact tests), but in this case, specimens have to be cut in flat areas of the CAC tank, getting constant thickness.

Unfortunately, due to high pressure in the air loop, design of the tank needs ribs to re-enforce the part: so, no flat surface could be cut in a part.

Nevertheless, in this study, as we want to compare the behavior of the plastic in two different environments, we have used ISO527 specimen, processed by injection process, as for CAC serial parts.

4 PRELIMINARY TESTS

4.1 Chemical definition of the EGR gas

In order to evaluate the chemical composition of exhaust gas in the EGR Low Pressure (LP) loop, we have built a motor bench test with different driving conditions and different types of fuel.

We have collected the condensates at the CAC location and we have analyzed them by ICP (Induced Coupled Plasma: analysis and quantification of cations) and by Ionic Chromatography (analysis and quantification of anions and organic acids).

These two analyses have highlighted than the composition of the EGR LP condensates is directly linked with the chemical residues issued from the fuel: chemical substances as carbon, hydrogen and also chlorine, sulfur, fluorine, nitrogen. We have quantified sulfate SO_4^{2-} (sulfur oxydation of the fuel directly linked with the sulfur rate of the fuel), nitrate NO^{3-}, nitrite NO^{2-}, chlorine Cl^-, fluoride F^- and organic acids.

During these motor bench tests, we have noticed that below 30 °C, exhaust gas condensate to form what we call EGR LP condensates, which can be found in our heat exchanger, especially in the outlet area.

With these results, we have stated a chemical solution representative of the EGR LP condensates and the temperature to be used around 30 °C. In order to set up an accelerated aging, we have also evaluated the plastic behavior at 60 °C, with same EGR LP condensates.

4.2 Exhaust gas chemical aging

First of all, we want to set-up a comparative aging for different polymers evaluation in order to select the best polymer for our new CAC application in EGR LP loop.

With ICP and Ionic Chromatography analyses, we have defined a solution representative of EGR LP condensates components (see Table1).

To be representative of this new application, we can perform a preliminary test with two different agings in EGR LP condensates: one at 30 °C during 2000 hours in immersion, and with our knowledge in terms of aging, another aging at 60 °C during 504 hours.

Table 1: Exhaust gas chemical aging

EGR LP solution definition	Quantity below 800ppm for each : sulfate SO_4^{2-} nitrate NO^{3-} chlorine Cl^- acetic acid, lactic acid, formic acid and sulfuric acid for pH adjustment
Aging time	504 hours
Aging temperature	60 °C

We have performed the aging in water and in EGR LP condensates in order to evaluate the impact of less than 800 ppm of each element in the exhaust gas condensate in the plastic material, compared to the effect of only water.

4.3 Tests protocol

We have compared plastic behaviors in the two cases, by mechanical characterizations after agings by immersion (cf Table 2) of PA66GF35, our standard material for CAC tank.

Table 2: Test protocol

Material: PA66GF35 used in CAC tank in only air loop	Tensile tests – ISO527 (Tensile strength Tensile Modulus Elongation at break)	Impact resistance ISO179-1fU	Mass evolution of the specimen
Initial state	✓	✓	✓
After 504h immersion at 60°C	✓	✓	✓
After 2000h immersion at 60°C	✓	✓	✓
After 2000h immersion at 30°C	✓	✓	✓

The immersion of ISO527 specimen is done according to ISO 175 for aging in immersion in liquid.

Figure 4: Complete immersion of specimen

Figure 5: Constant stirring of the solution during aging

Due to the way it is obtained and to its chemical structure, polyamide is sensitive to humidity absorption. The humidity content in a polyamide sample can greatly influence its mechanical properties. To avoid influence of this parameter on results, specimens immerged in every solution were dried in a vacuum oven for 72 hours at 80 °C before aging, and specimens for initial characterization were conditioned at 50 % Relative Humidity.

For such application, we evaluated the impact resistance of the material (cf. Figure 6) and the tensile characteristics (cf. Figure 7) at room temperature.

Figure 6: Charpy impact equipment

Figure 7: Tensile test equipment

After the chemical aging, we performed the same mechanical characterizations as for initial state, in order to evaluate the retention of properties after long term chemical aging.

4.4 Results

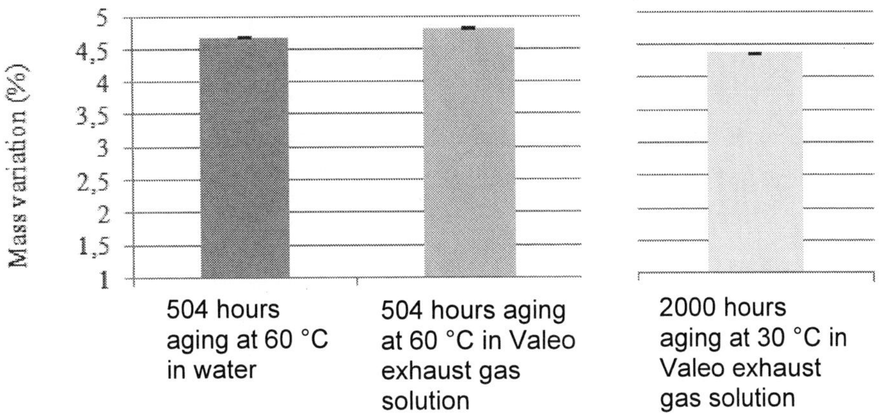

Figure 8: Mass variation of specimens after chemical agings

We can notice on Figure 8 that mass variation between the two short agings (504 hours at 60 °C in water; 504 hours at 60 °C in Valeo exhaust gas condensates) is quite equivalent and is lowest compared to long term 30 °C immersion aging.

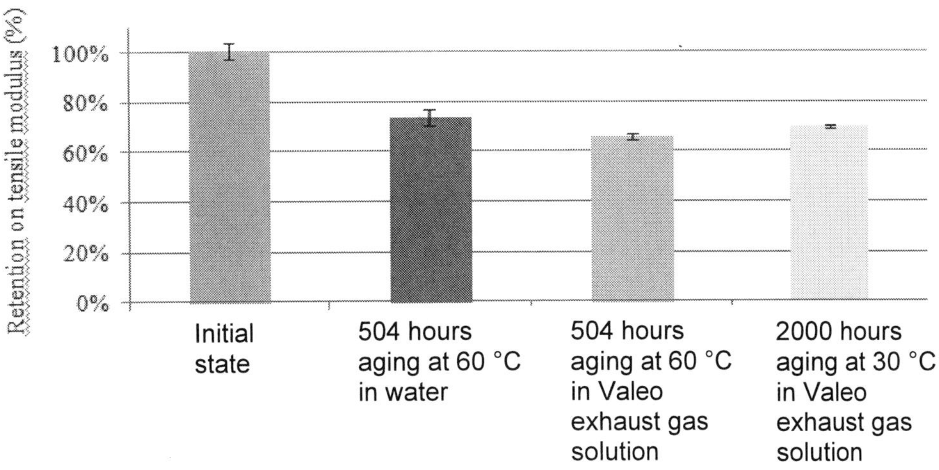

Figure 9: Retention of tensile modulus after chemical agings

We can notice on Figure 9 that tensile modulus is quite equivalent between the three agings, the lowest is after immersion 504 hours at 60 °C EGR LP condensate solution.

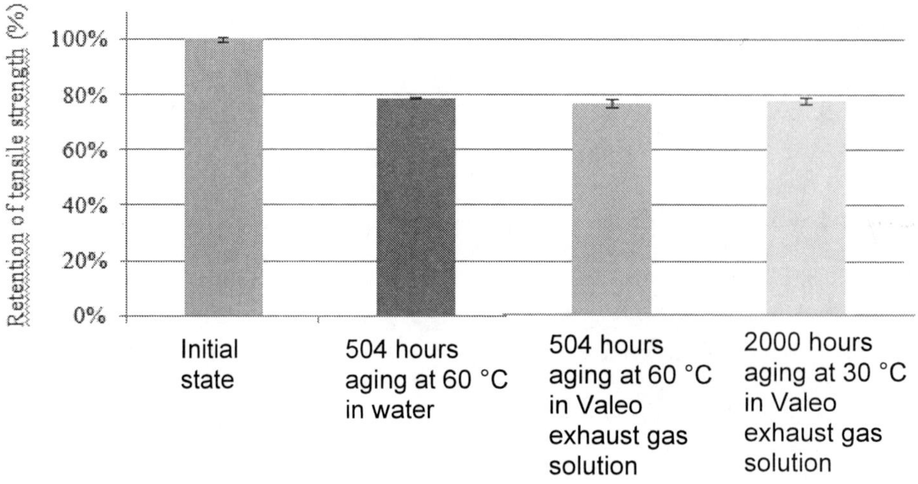

Figure 10: Retention of tensile strength after chemical agings

In Figure 10, we can conclude that the three different agings evaluated have same impact on mechanical characteristics of the material. Moreover, the retentions of properties after chemical agings are quite equivalent.

We can notice on Figure 11 that retention of property for impact resistance is quite equivalent between the three chemical agings. The increase of impact resistance is due to the plasticization of the material, by water absorption. Such mechanical characterization is not relevant for this chemical aging.

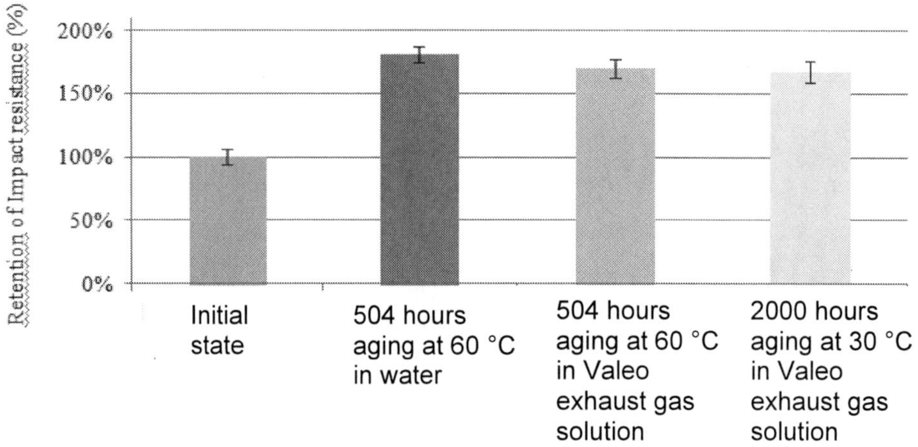

Figure 11: Retention of Charpy impact after chemical agings

By conclusion, we can observe on Figure 12, that the tensile curves are equivalent between the same agings in water and in EGR LP condensates.

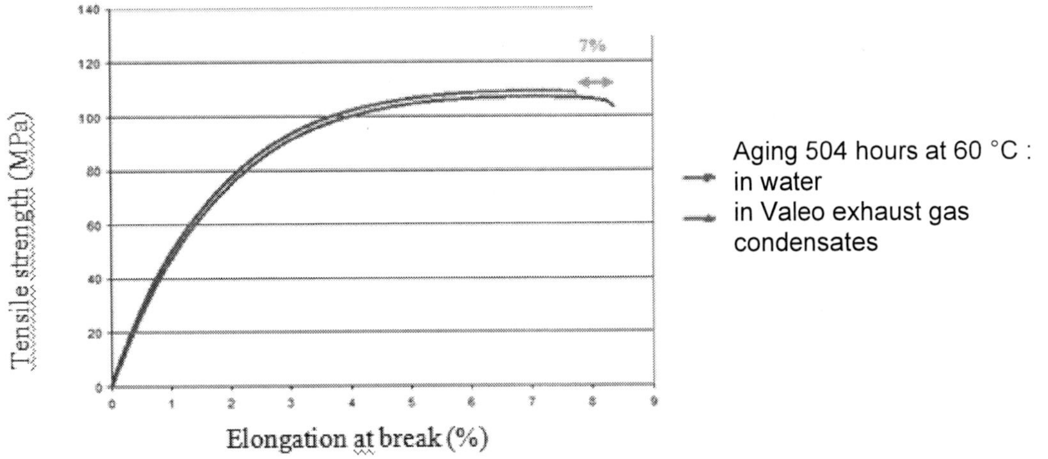

Figure 12: Tensile curves after chemical agings

In order to confirm if the chemical degradation is also the same for these two agings, we have performed viscosity measurements by solvating the polyamide in formic acid and by using a capillary viscosimeter for measurement according to ISO307 (see Figure 13).

We can notice in Figure 14 that as for mechanical properties, chemical degradation of the polymer chain is equivalent between the two agings (504 hours at 60 °C in water and 504 h at 60 °C in Valeo exhaust gas condensates).

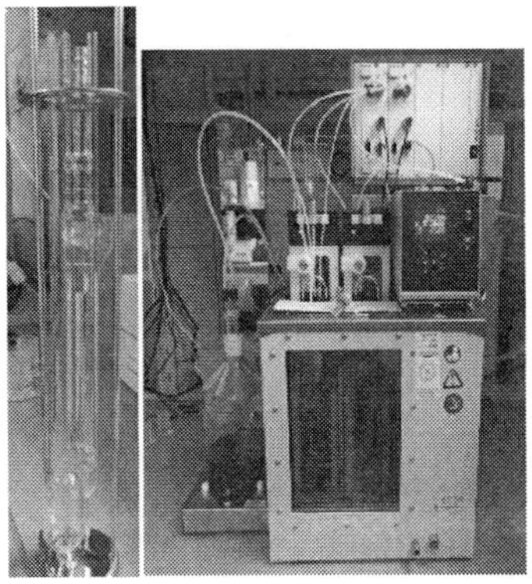

Figure 13: Capillary viscosimeter according to ISO307

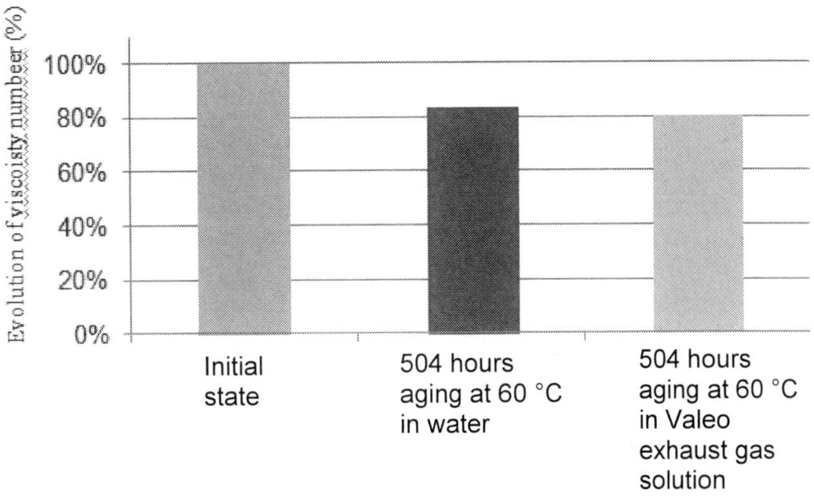

Figure 14: Viscosity Number evolution after chemical agings

In order to status on the parameters of the aging to set-up for plastic material evaluation in case of CAC EGR LP loop, we have also compared long term aging with representative conditions of the CAC application during its life-time (2000 hours at 30 °C in EGR LP condensates) with an accelerated aging at 60 °C during 504 hours.

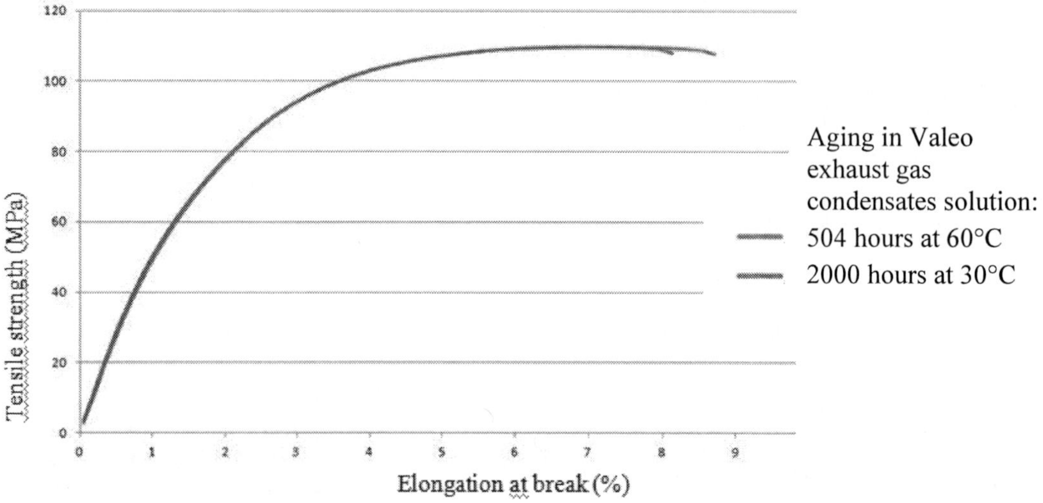

Figure 15: Tensile curves after chemical agings

We can see in Figure 15 that tensile curves are equivalent: we can implement an accelerated aging for our plastic material approval.

4.5 Preliminary tests conclusion

By conclusion of these preliminary tests, we are able to perform comparative evaluation of different polymers for our new CAC tank application under EGR LP loop:

Table 3: Parameters of the test comparison

EGR LP solution definition	Quantity below 800 ppm for each : sulfate SO_4^{2-} nitrate NO^{3-} chlorine Cl^{-} acetic acid, lactic acid, formic acid and sulfuric acid for pH adjustment
Aging time	504 hours
Aging temperature	60 °C
Materials evaluated	PA66GF35 - hydrolysis stabilized PPAGF45 PPSGF40 for its chemical inertia

5 DIFFERENT PLASTIC MATERIALS BEHAVIOR COMPARISON

In summary, here is in Figure 16 and Figure 17, the comparison of the evolution of tensile characteristics of the different plastic materials evaluated in this study.

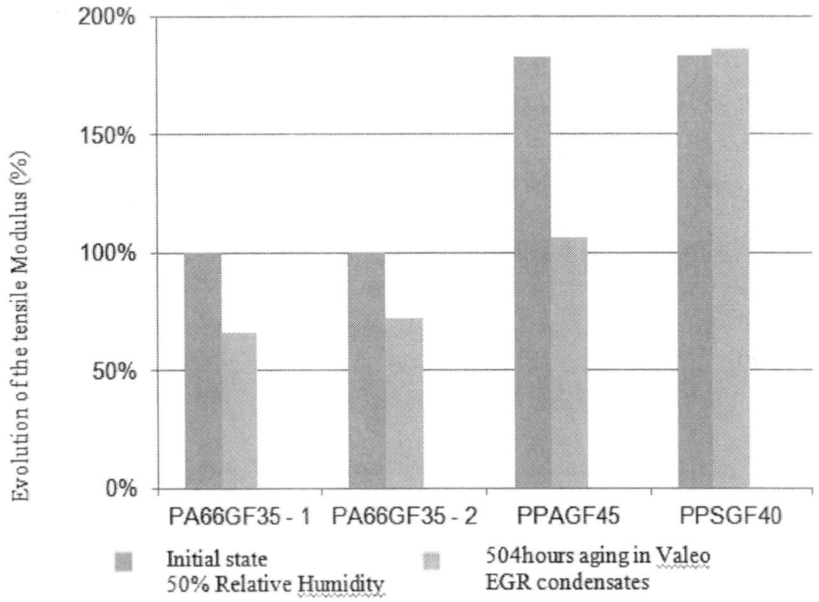

Figure 16: Evolution of tensile modulus of different plastics after chemical aging

We can see on Figure 16 a drop of tensile Modulus for all materials except PPS, already known for its chemical inertia. The tensile modulus after chemical aging is higher for PPAGF45 compared with PA66GF35, even if the drop is higher for PPA.

As for the tensile modulus, we can observe in Figure 17, a drop of tensile strength for all materials except PPS, already known for its chemical inertia. The tensile strength after chemical aging is higher for PPAGF45 compared with PA66GF35, even if the drop is higher for PPA.

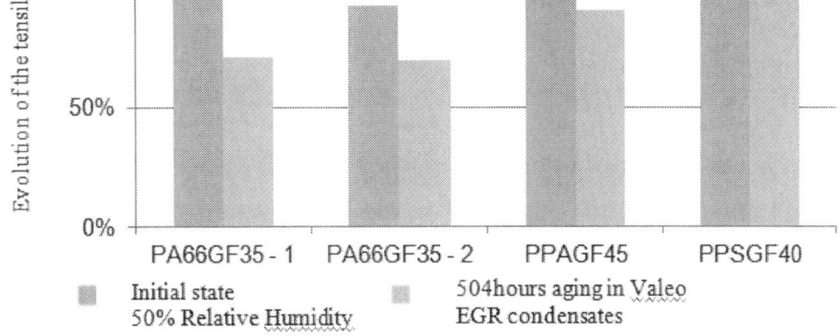

Figure 17: Evolution of tensile strength of different plastics after chemical aging

Nevertheless, knowing the minimum level of mechanical characteristics needed for the CAC tank application, PA66GF35 hydrolysis stabilized keeps a good candidate, if temperature is not so high.

For high temperature, we can keep PPAGF45, as its behavior in the EGR condensates solution is in accordance with the needs for the application.

6 LIMITATIONS OF THE STUDY

This study is limited to the Valeo chemical solution representative of exhaust gas condensates analyzed under motor bench test trials. The limitation concerns not only the chemical solution (species in solution and their concentration) but also the pH of this solution: the pH can be lower, up to minus than 1: in this case, the test of chemical aging should be redone to ensure the compatibility.

7 CONCLUSION

With new EGR low pressure loop, condensates will degrade the plastic material, but in a lower extent compare to hydrolysis degradation, as observed for example, for water tank for a radiator, where the plastic material is always in contact with the coolant.

But hydrolysis degradation is a new aging for CAC plastic material, where currently the environment is only hot air charged with engine oil.

Compare to our current polyamide 66 for CAC tank, and regarding the conclusions of our study, best material for chemical resistance to EGR condensates should be PPS for its chemical inertia, followed by PPA. But regarding the degradation level of the material after exhaust gas chemical degradation, and knowing the limitation of this study in terms of chemical solution representative of exhaust gas condensates, it could be interesting to use a polyamide 66 not only heat but also hydrolysis stabilized, in order to limit the chemical degradation of the polyamide material.

REFERENCES

[1] Valeo 2013 documents
[2] Valeo diagram

Aluminium heat exchanger and brazing – review and outlook of successful symbiosis

M TUERPE and **B GRUENENWALD**
MAHLE Behr GmbH & Co. KG, Germany

ABSTRACT

From the very beginning on, brazing was an important and powerful joining technology for heat exchangers. With the change from copper materials to aluminium about 40 years ago, the importance of brazing increased. Today, aluminium brazing for heat exchangers is well known, well established and state of the art. By adapted technologies as CAB using fluoride fluxes, the strong oxide layer of aluminium is no critical item any longer. Due to the still mandatory use of AlSi filler metals, the low range between working temperature and melting temperature of materials to be brazed is an exception in the world of brazing and has been the starting point for several research activities. In this paper, a number of approaches and developments of the last years are reviewed and assessed concerning the applicability for aluminium heat exchangers brazing.

1 INTRODUCTION

"Automotive heat exchangers have been built of copper and brass with soldered joints since the first water cooled engines were developed. In the early 1970s the U.S. and European automotive industries began replacing copper-brass heat exchangers with mechanically joined aluminium ones. The main reasons for the progressive change to aluminium were lower material cost of aluminum and the demand for lighter cars." [1] "This switch was triggered by the oil crisis in the mid-seventies when increasing fuel economy by light-weighting became a must." [2] "Two main joining technologies have emerged for aluminum radiators: mechanical joining and brazing." The advantages and limitations of mechanically joined radiators are well known. Brazing permits joining of parts of a complex pre-assembled unit in a single step. Brazing of aluminium alloys involves prior coating of the parts with aluminium brazing alloys which have lower melting points than the parent alloys. Typically, the parent alloy (core) is an Al-Mn alloy (3xxx) and the coating one (clad) is near eutectic Al-Si (4xxx). One of the problems is the necessity for the clad alloy to penetrate the oxide film and to wet the surface [1]. "Two solutions have been developed to overcome this: vacuum brazing (fluxless) and controlled atmosphere brazing (with flux).

A disadvantage of the vacuum brazing process is that during brazing, magnesium deposits on the cold furnace walls, which must be regularly cleaned." Compared to other methods, the furnace equipment for vacuum brazing is complex and expensive [1]. "Since the early 1980th, Controlled Atmosphere Brazing (CAB) with non-corrosive fluxes has evolved as the leading technology for manufacturing aluminium heat exchangers in the automotive industry. This technology offers the benefit of a flux for successful oxide removal and working at atmosphere pressure while avoiding the disadvantages of post treatment and corrosion susceptibility. The non-hygroscopic and non-corrosive potassium fluoro-aluminate flux does not react with aluminium in the molten or solid state and the post braze flux residues have a very low water solubility." [3] The main disadvantage of the Nocolok® process (common name for CAB) "is that it can tolerate only a low magnesium content core material (up to 0,3%)." By diffusion of Mg into the molten flux, the oxide dissolution is not effective at higher Mg levels. [1]

It is well known and commonly accepted today that, due to the harsh corrosion requirements, the use of AlSi filler metals in mandatory. Therefore, brazing is done with Al-Si filler metals at a working temperature between 590 and 615 °C. This limits the number of aluminium alloys applicable for brazing [4].

As a consequence, brazers have to handle the narrow range of about 30-40 K between liquidus of filler metal and solidus of parent metal (figure 1), This situation is untypical for brazing. Consequently, aluminium brazing can be seen as a special case of brazing.

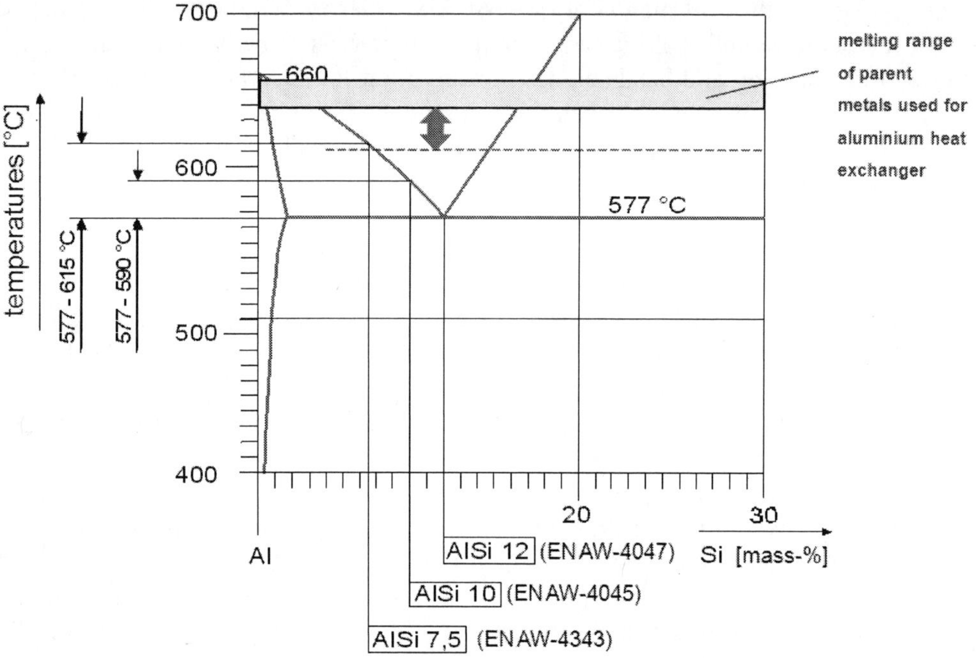

Figure 1: Comparison of melting temperatures filler metals vs. parent metals in schematic equilibrium diagram

2 PROCESS STEPS VS. MECHANISMS OF BRAZING

The pure brazing process consists not only of heating up, holding at brazing temperature and cooling down cycles. During the brazing cycle, different chemical, physical and metallurgical reactions take place and determine the result of the brazing process as well as properties of the brazed products. A schematic correlating the process flow with above named reactions is given in figure 2 [5].

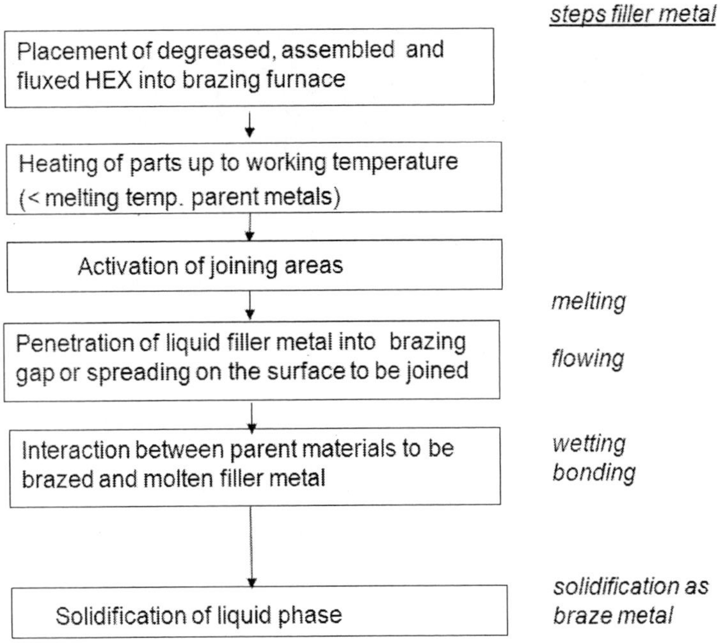

Figure 2: Processes during brazing – principal sequence filler metal [5]

A selected number of developments in aluminium brazing shall be discussed and assessed in the following chapters.

3 REVIEW OF SCIENTIFIC APPROACHES

3.1 Activation – the role of oxide layers

Any soldering or brazing process will run successfully only when the molten filler metals gets in contact with a clean metallic surface of parent metal. It is common knowledge that the oxide layer on aluminium materials is a special challenge for soldering, brazing and welding, as it can be seen in wetting tests in figure 3 [6].

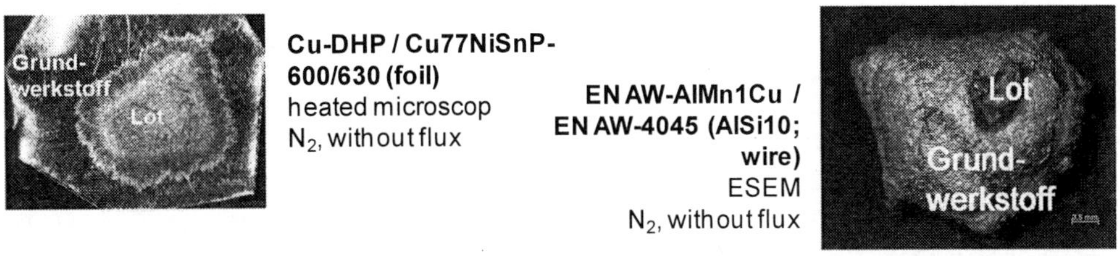

Figure 3: Comparison of good wetting on copper (left) and bad wetting on aluminium (right) [6]

With the development of fluoride bases fluxes (so called "non corrosive fluxes") which can be used in a shielding gas atmosphere, it was possible to eliminate the oxide layer and to achieve good wetting and proper brazed joints (figure 4). With thus development, a milestone in aluminium brazing has been set followed by the leadership of CAB process worldwide.

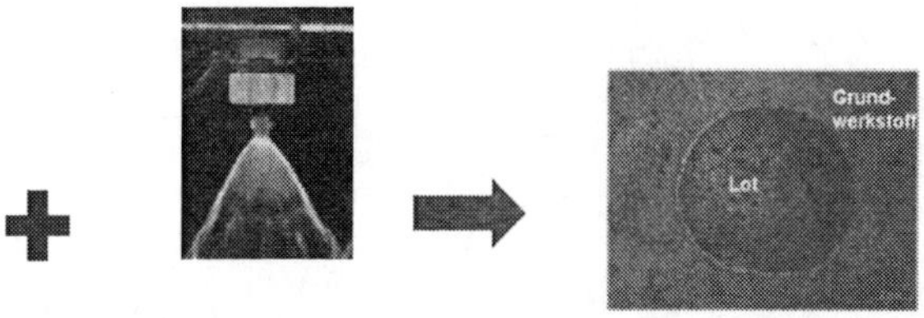

Figure 4: Wetting behaviour of fluxed aluminium surface [6,7]

In general, stable oxide layers on materials as aluminium, titanium or stainless steels have been regarded as existing and unavoidable phenomena for a long time. However, the behaviour of this oxide layer seems to be strange sometimes. Even with brazing parameters inside the allowed range for production, sudden difficulties occur in mass production suddenly without any obvious changes or hints for the reason.

First scientific activities indicate that more new and open questions occurred than answers are found on oxide layers. According to literature, the oxide layer consists of a thin barrier layer and a cover layer, the latter formed due to environmental conditions with pores, further metal oxides etc. Those descriptions are mainly based on postulates; even in a SEM, an aluminium oxide layer with about 30nm thickness can hardly be seen. First with equipment as TEM it is possible to make the oxide layer visible [8].

The measurement and analysis of such layers is a special and important working task [9]. In figure 5, results of an analysis in situ of the development of an aluminium oxide layer during heating up is presented. Whereas the decrease of growth of Al_2O_3 and the increasing growth of MgO parallelly can be explained by Mg diffusion to surface with increasing temperature, the role of the $MgAl_2O_4$-spinel still is not clear.

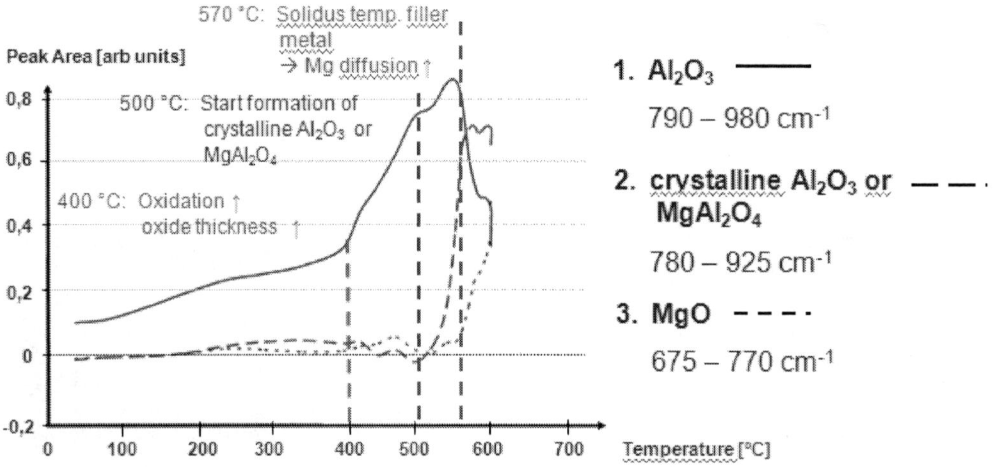

Figure 5: FTIR with heating chamber – summarising the peak areas depending on heating (brazing) temperature; 3003 mod with cladding 4045 [9]

As mentioned in the next paragraph, a lot of research activities concentrate on metallurgical questions for improvement of aluminium brazing. The same situation can be found in other brazing areas where a main focus is set on metallurgy and studies of process parameters. With regard to actual results on nano- and microjoining (see paragraph 3.3), these approaches may turn out to be of secondary importance compared to an understanding of the oxide layer for a better understanding of brazing and soldering mechanisms.

3.2 Melting and flowing– the dream of lower melting points

From the metallurgical point of view, the modification of filler metals for aluminium brazing can be achieved by adding elements as Cu, Ag or Ge in order to lower the brazing temperatures [10]. In automotive environment, it is well known that noble elements do not fit with the corrosion requirements. Therefore, most approaches used or tested e.g. in aviation industry cannot be transferred for other industrial applications.

As Zn is used for a long time in different variants for improvement of corrosion behaviour, its ability as alloying element for filler metals has been subject of similar research projects [11,12]. Although promising results could be achieved, the step of adaption in mass production for automotive heat exchangers could not be reached. A still open item is the question if adapted fluxes for such filler metals with reduced brazing temperatures are available and/ or economic. Another aspect is a possible negative impact on recycling when using high amounts on Zn. In later recent years, developments and field experiences showed the necessity of restriction of Zn use [e.g.13]. Therefore, a further coffin nail for Zn in filler metals has to be added.

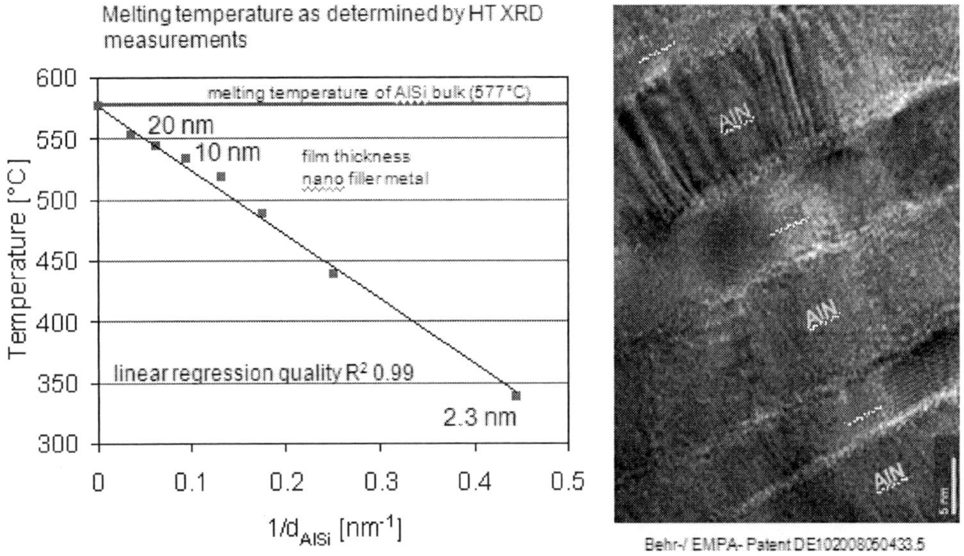

Figure 6: Melting points and structure of AlN-AlSi- nanomultilayers [14]

A much more promising idea is the use of nano filler metals. Depending on the individual position, the intensive research on nano materials and technologies mainly in the recent decade can be seen as hype or as starting point for some revolution in engineering. However, the phenomenon of melting point depression (MPD) by reducing its size can be derived from physical basics. The principal feasibility of this approach for AlSi filler metals has been shown using a multilayer system [14]. Defined numbers of AlSi layers with different thickness (between 2 and 20 nm) were applied by PVD. In order to avoid the loss of nanostructure during heating up, these layers have each been separated by an AlN-nanolayer (figure 6, right side) which had proved as being suitable for this purpose. The melting temperatures of such nanomultilayers were determined by different experiments as XRD and melting trials in a special lab furnace. The results of XRD analysis are shown in the graph on left side of figure 6 confirming that the melting point decreases with reduced thickness of nanolayer filler metal.

But at the same time, a number of new questions arose resulting in a huge need of intensive fundamental research. Actual basic research on AgCu filler metals as model system confirmed that the well known laws of macro and micro brazing are not automatically valid for AlSi nano filler metals [15]. This topic remains an exiting field for research as well as for industry [16].

3.3 Wetting and bonding– the mystery of interfaces

As first step in brazing technology, the ability for brazing of a combination of parent metal and filler metal is tested by a wetting test, described in a number of publications. The filler metal is molten on a flat surface of a sheet specimen, afterwards the spreading of filler metal and in particular the value of wetting angle β is taken as benchmark (figure 7). In brazing literature, it is said in general that β must be lower than 90° and should not exceed 30-37°. However, the angle β is derived from Young's equitation based on the assumption of interfacial energy isotropy [20]. This assumption is valid for an ideal system only, i.e. for the interfaces of fluids and amorphous solids, and the simple transfer for brazing purposes "is overly simplistic" [20]: Technical surfaces are rough, the substrate interface will change shape at elevated temperatures [20], an interlayer (so-called diffusion phase) is formed due to diffusion processes during brazing etc. There is a need for an intensive and very critical discussion concerning the validity of the use of contact angle measurement in soldering and brazing.

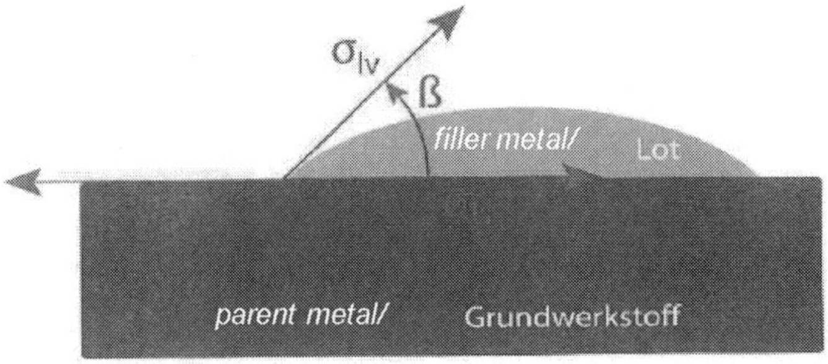

Figure 7: Schematic of wetting angle [17]

Traditionally, an improvement of wetting is achieved by metallurgical means. One example is the addition of Bi to Al filler metals going back to laboratory trials for material development in the 1990s [4]. Based on recent brazing trials in the laboratory as well as in series production, however former conclusion towards an improved wetting of AlSi-filler metals by an addition of Bi could not be confirmed [18]. Additions of Bi did not show a detrimental effect on brazing results but, in the same way, did not offer advantages for brazing. As one conclusion of this work, it should be checked if other mentioned elements predicted to improve wetting really fulfill this expectation or if (up to now) not regarded interactions with other alloying elements may have a negative impact on brazing.

Both approaches, the emphasis of wetting test as well as the metallurgical focus, should be challenged more insistently. Even looking at the wetting behaviour of aluminium, the gap becomes obvious. Taking the above mentioned wetting test for aluminium vacuum brazing, the result is contradictory, see figure 8 (18].

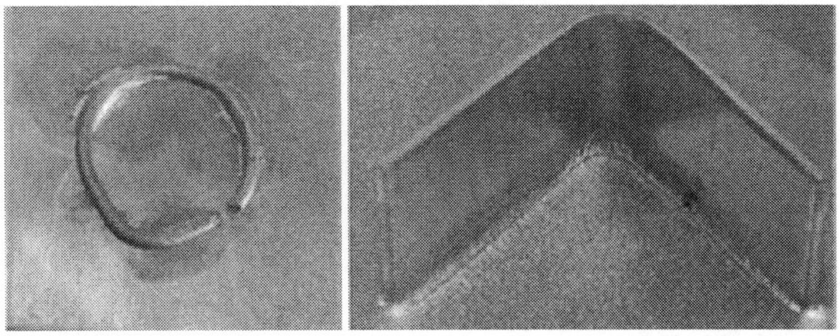

Figure 8: Comparison of wetting test (wire AlSi5, left) and brazing test with AlSi10 cladding (right); brazed in vacuum lab furnace [19]

The filler metal applied as wire on uncladded surface does flow and wet partially whereas a complete fillet can be seen for the angle specimen, i.e. a "brazed joint-specimen". Due to the use of cladded filler metal for aluminium (instead of applied filler metal for other parent metals), a wetting test will neither be possible nor give a significant result compared to a specimen as an angle specimen related to the special application of filler metals for aluminium brazing. This simple example already leads to the conclusion that a single description of phenomena only (as done with measurement of wetting angle or filling of gap in angle specimen) will not be sufficient.

In addition, actual results in the field of nano- and microjoining indicate that atomistic processes and structures in the interface liquid filler metal – solid parent metal may play an important role and determine main parts of joint properties [20]. Those results correlate with an earlier work based on molecular dynamics simulations [21]. In the same way as already mentioned concerning oxide layers in paragraph 3.1, future work should be focused on the much more sophisticated and detailed understanding of wetting and bonding mechanisms and the role of interfaces, their formation as well as their structure – what may lead to an atomistic approach in the end.

4 HOLISTIC APPROACHES FOR FUTURE WORK

Understanding the different mechanisms occurring during brazing is one key for further improvements. However, those mechanisms are not running isolated, but are connected with incidents before and may cause interactions afterwards. Those interactions are not limited to the brazing process only, but may result in the properties of brazed joints or may be the result of chosen boundary conditions. In a first more simple approach, these interactions are described by the brazeability (figure 9), originally defined as" property of a component which allows it to be processed by soldering/ brazing such a way as to comply with specific requirements" [22], but furthermore giving an illustration of interdependency between the items. As (simplified) example for explanation: Changing from CAB to vacuum brazing, i.e. change of process, requires the use Mg-containing filler metals (= change in materials) as well as a more sophisticated design but elimination of the risk of flux inclusions (= change in design/ reliability of brazing).

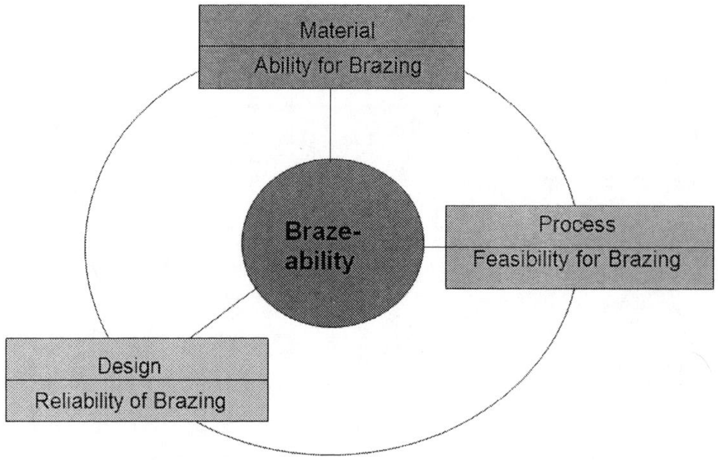

Figure 9: The term of brazeability [22]

A further view on influencing items and their correlations as well as their interactions leads to an extended model (figure 10) [5, 23], in which five main items are defined. A special importance has the non-technical one describing topics as safety in work, environmental protection, availability (of materials, alloying elements etc.) or costs. Engineers tend to neglect those items in early steps of developments imaging that this would be part of work towards the end. In fact, this model allows to find possibly critical items and to assess their importance in an early stage already.

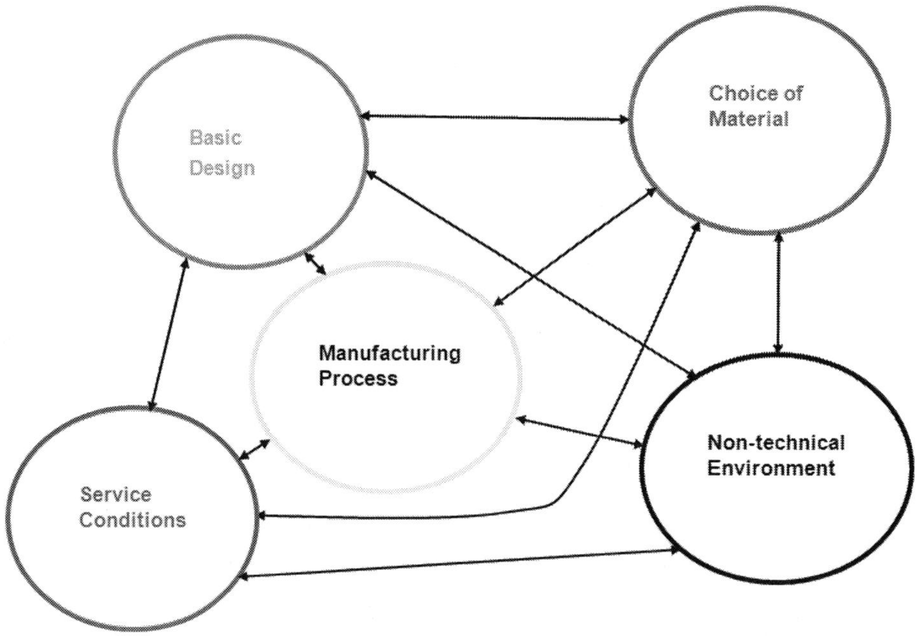

Figure 10: Holistic model for derivation of interactions and connections in technical processes [5]

As (again not completely presented) example: Looking back to the idea of Zn-containing filler metals would mean a lower brazing temperature (= improvement of manufacturing process and parent material properties), but limited possibility of cladding (negative impact by non-technical item availability with following negative impact on manufacturing process as well as on design due to changed filler metal application), furthermore on corrosions resistance, i.e. negative impact on service conditions. More examples will simply be found with further analysis.

5 CONCLUSIONS AND OUTLOOK

Although being state of the art and widely used for aluminium heat exchangers in many kinds and markets, brazing of aluminium still requires further steps of development. Conceived in general terms, understanding of aluminium brazing must be more detailed the more complex the heat exchangers become, including the constantly increasing requirements to be fulfilled. This includes not only the brazing process itself, but also the boundary conditions of application as well as process chains. It is not sufficient to deal with classical brazing parameters only as temperature, time, the analysis of process atmosphere or material choice based on chemical compositions as examples. In fact, the interactions between process chains must be understood more detailed, and the impacts of early manufacturing steps on later ones remain a future work. As example, the history of aluminium surface starting with materials delivery by aluminium producer up to entering the brazing furnace is an ongoing work item.

Furthermore, the processes during brazing as flowing, wetting and bonding of filler metal are often described in general terms and tested by commonly known and/ or standardized methods. It is not clear if this results in a real understanding of soldering and brazing, e.g. of acting mechanisms, or in a simple description of phenomena only. This has a special relevance for aluminium brazing as aluminium is mentioned correctly as a metal with bad wetting properties. Last but not least, it is necessary to challenge existing theories as new methods of analysis as TEM enable a fascinating view into dimensions at atomic scales which may lead to an improved understanding of mechanisms, instead of nowadays description of resulting effects only – surprises included!

REFERENCE LIST

[1] Senaneuch, J.; Nylen, M.; Hutchison, B.: Metallurgy of brazed aluminium heat-exchangers, Part I. Aluminium 77 (2001) 11, 896-899

[2] Gray, A.: The growth of aluminium in automotive heat exchangers. Aluminium 81 (2005) 3, 197-201

[3] Swidersky, H.: Aluminium brazing with non-corrosive fluxes – state of the art and trends in NOCOLOK® flux technology. DVS-Berichte 212 (2001), 164-169

[4] Schoer, H.: Schweißen und Hartlöten von Aluminium-Werkstoffen (welding and brazing of aluminium materials). Fachbuchreihe Schweißtechnik, Band 137, DVS-Verlag Düsseldorf, 1998 (in German)

[5] Türpe, M.: Script lecture "Brazing", TU Dresden, 2014 (translated from German)

[6] Zähr, J et al.: Anwendbarkeit von Benetzungsversuchen für verschiedenen Metalle (Applicability of wetting trials for different metals). METALL 62 (2008) 10, 611

[7] Türpe, M.: Nutzung naturwissenschaftlicher Ansätze für die industrielle Löttechnik (application of natural scientific based approaches for industrial

brazing). Lecture, symposium 60 years manufacturing technology at Dresden, Nov. 2014 (in German)

[8] Sarrafi, R.; Kovacevic, R.: Cathodic Cleaning of Oxides from Aluminium Surface by Variable-Polarity Arc. Welding Journal Supplement 89 (2010), 1s-10s

[9] Hofmann, E.; Füssel, U. et al.: Correlations between the atmosphere conditions on the natural aluminum oxide layer and the brazeability. Lecture IBSC, 20.-25.4.2012, Las Vegas

[10] Kötzing, B.: Entwicklung niedrigschmelzender Aluminiumbasislote zum Fügen höherlegierter Aluminiumwerkstoffe (development of low melting aluminium based filler metals for joining of high alloyed aluminium materials). PhD-Thesis, RWTH Aachen, 1996 (in German)

[11] Wielage, B.; Hoyer. I.; Grund, Th.: Entwicklung eines Controlled Atmosphere Brazing (CAB) Verfahrens zum Fügen von Aluminiumguss- und Aluminiumknet-legierungen (development of a CAB process for joining of cast and wrought aluminium alloys). Final report AiF-project 13.788, TU Chemnitz, 2005 (in German)

[12] Bach, Fr.-W.; Holländer, U.; Langohr, A.: Niedrig schmelzende Aluminiumhart-lote aus dem System Al-Si-Zn (low melting aluminium brazing filler metals derived from system Al-Si-Zn). Final report AiF-project 15.744N, Leibnitz University Hannover, 2010 (in German)

[13] Melander, M.; Woods, R.: A compilation of corrosion studies of brazed aluminium heat exchangers after field service. DVS-Berichte 289 (2012), 115 - 124

[14] Türpe, M.; Janczak-Rusch, J.; Grünenwald, B.; Bissig, V.: Approach for Al brazing with nano filler metals. DVS-Berichte 289 (2012), 125 - 129

[15] Lewandowska, M.; Andrzejzuk, M.; Pigozzi, G.; Jeurgens, L.P.H.; Janczak-Rusch, J.: Microstructural evaluation of nano-structured Ag-Cu/AlN brazing fillers upon heating. Proceedings 2nd Int. Conf. on nanojoining and microjoining, I13, 7.-10.12.2014, Emmetten/ CH

[16] Türpe, M.: Nano brazing technology in industrial environment. Proceedings 2nd Int. Conf. on nanojoining and microjoining, R1, 7.-10.12.2014, Emmetten/ CH

[17] Hartlöten – eine Einführung (introduction to brazing). DVS-Media GmbH, Düsseldorf, 1. edition, 2010 (in German)

[18] Türpe, M.; Grünenwald, B.; Wald, A.: Contribution about Minor Elements in Aluminium Brazing Alloys. DVS-Berichte 305 (2014), 81-83

[19] Hofmann, E.; Füssel, U. et al.: Charakterisierung des Benetzungsverhaltens von Aluminiumlegierungen beim Vakuumlöten (characterisation of wetting behaviour of aluminium alloys at vacuum brazing). DVS-Berichte 284 (2012), 69-74

[20] Kaplan, W.D. et al.: A review of wetting versus adsorption, complexions, and related phenomena: the rosetta stone of wetting. J Mater Sci (2013) 48:5681–5717. DOI 10.1007/s10853-013-7462-y

[21] Hartz-Behrend, K.: Einsatz der klassischen Molekulardynamik (MD) für die quantitative Vorausberechnung des Benetzungsverhaltens beim Löten (application of MD for quantitative projection of wetting behaviour during soldering and brazing). PhD-thesis, Bundeswehr University, Munich, 2012 (in German)

[22] DIN8514: Lötbarkeit (brazeability).edition May 2006

[23] Türpe, M.: Wechselwirkung Löttechnik-Werkstoff, Bedeutung und Potential (interaction brazing – material, importance and potential). Habilitation, TU Dresden, 2004 (in German)

Electrochemical study of aluminium extruded tubes for controlled atmosphere brazed heat exchangers

I PORTAL, C CASENAVE, M WAINER, L AUBANEL, M PHILIPPE, V RENAULT and A-G VILLEMIANE
VALEO Thermal System, France

ABSTRACT

The corrosion behaviour of extruded tubes covered with a coating which contains silica particles and zinc flux and are brazed in controlled atmosphere furnace is studied by electrochemical methods. Tubes are made of AA 1050 aluminium alloys modified either by addition of Manganese (Mn) or Copper (Cu). Improved repassivation ability and different pitting susceptibility are observed for surfaces of tubes brazed with coating depending on aluminium core. The determination of main alloying elements concentrations by EPMA analysis and electrochemical measurements permits to better understand corrosion initiation sites on heat exchangers.

LIST OF ABREVIATIONS

EPMA: Electron Probe Micro-Analysis
SWAAT: Sea Water Acetic Acid Test
TGA: Thermogravimetric Analysis
CAB: Controlled Atmosphere Brazing
SCE: Saturated Calomel Electrode
OCP: Open Circuit Potential
E_{corr}: Corrosion Potential
E_{rp}: Repassivation Potential
E_{pit}: Pitting Potential

INTRODUCTION

Aluminium heat exchangers, located at the front face of the vehicle, are exposed to outside climatic conditions creating variable corrosion conditions. Over the past 10 years, the fuel consumption reduction led to heat exchanger weight reduction and as a consequence, the material thickness of the components has been reduced. The corrosion resistance evaluation of the heat exchanger becomes one of the key topic for its development. Corrosion resistance of heat exchangers is generally evaluated by submitting exchangers to salt spray, and combined hot, wet and dry conditions as defined for SWAAT in ASTM G85 standard. In these test conditions, corrosion proceeds on the exposed tube and fin area until leak of the heat exchanger occurs. The tightness of heat exchangers tubes which contain heat transfer fluid and a good bonding of fin brazed between the tubes are two critical parameters.

The tightness of the heat exchanger in the corrosion test depends mainly on the thickness of heat exchanger tubes which could be reduced as low as 200µm. Hence, material selection has to take into account the corrosion mechanisms to improve exchangers durability. External corrosion resistance of heat exchangers is tested according to ASTM G85 (SWAAT test) which is a reference test and reproduce well field failure. However, this kind of test generally requires several months for leak and corrosion mechanisms information are difficult to draw. Electrochemical measurements can bring supplementary information on material behaviour in different electrolytes.

In this work, a particular focus is brought to extruded tubes coated before exchanger assembly. This coating contains Zinc Flux, potassium aluminium fluoride, silica particles and a polymeric binder. It presents several advantages such as a better control of the flux and zinc load, and avoids a wet fluxing step on assembled exchanger during production. During brazing, the Zinc flux reacts with aluminium and forms Nocolok flux and elemental Zn during brazing $6\ KZnF_3 + 4\ Al \rightarrow 6\ Zn + 3\ KAlF_4 + K_3AlF_6$. The Zn rich layer is sacrificial and permits to delay the tube perforation [1].

Alloying elements have an important impact on corrosion resistance of aluminium. Works from *Shohei Iwao and al* shows that the presence of copper in the core and at least 1% Zn in the filler of brazing sheets ranging from 0,2mm to 0,5mm permit to significantly reduce the corrosion rate after SWAAT test and promote generalized corrosion [2].

During brazing, important modifications of microstructure and alloying elements distribution occur. Copper accumulation in the eutectic phase can cause susceptibility to intergranular corrosion. Formation of α-Al(Fe-Mn)Si particles increases the cathodic reactivity of the surface [3], [4]. Several studies have shown that intermetallics act as pit initiation sites and the anodic reactivity of the matrix is enhanced by heat treatment [5], [6]. Zinc is the element which has the greatest solid solubility in aluminium (82,8 wt% for binary Al-Zn). Copper in solid solution helps to protect aluminium from corrosion. Its solid solubility in binary Al-Cu is equal to 5,67wt% [7]. However, copper intermetallics or any other cathodic intermetallics are sites for pit nucleation by creating a galvanic cell with the aluminium matrix.

The impact of flux on the electrochemical response of an aluminium brazed surface is not clearly established. In works from *Afshar and al*, potentiodynamic experiments were performed in SWAAT solution on fluxed samples of AA3XXX colaminated on both side with an AA4XXX. A slight decrease of the open circuit potential (OCP) measured at the surface of fluxed samples and a change of current density can be seen on anodic curves. In addition, increasing the flux load improved the resistance of the material to SWAAT test [8].

The aim of this work is to determine the electrochemical properties of aluminium extruded tubes brazed with a ZnSilFlux coating in order to have a better understanding of corrosion phenomenon observed on exchangers. By using two different aluminium alloys brazed with the same coating, a particular focus will be brought on the impact of copper in the core material and alloying elements diffusion and concentration in aluminium on the electrochemical behaviour in NaCl solution.

1. EXPERIMENTAL

1.1 Material

Different materials have been used for the experiments. Extruded tubes of type X aluminium or type A covered on both side with coating. The chemical compositions of the tubes have been determined by spark spectrometry and are given in table 1. Type X aluminium corresponds to an AA 1050 series aluminium modified with an addition of Manganese (Mn) and type A aluminium corresponds to a AA 1050 series aluminium modified with an addition of copper (Cu).

1.2 Coating

To ensure brazing, the tube is covered with a coating composed of

- Silica particles (3 g/m^2)
- $KAlF_4$ (6 g/m^2)
- $KZnF_3$ (4,5 g/m^2), giving a Zn load of 1,8 g/m^2
- Acrylic binder (8 g/m^2)

Table 1: Chemical composition of type A and type X aluminium alloys

	Si	Fe	Cu	Mn	Mg	Zn	Cr	Ti	V	Zr
Type A chemical composition	0.05 – 0.10	0.10 – 0.25	0.35 – 0.60	0.05 – 0.15	0.008	<0.0001	0.002	0.024	0.008	<0.0001
Type X chemical composition	0.05 – 0.10	0.10 – 0.25	< 0.05	0.55 – 0.75	<0.0001	0.002	0.017	0.047	0.009	<0.0001

1.3 Brazing

Two sets of experiments were performed

- On the tube without coating
- On the tube with coating

Before brazing, the tube with coating is exposed to a debinding step in order to remove the acrylic binder. This step is performed under nitrogen atmosphere for 10 minutes. A TGA analysis performed under N_2 atmosphere showed that at 400°C, 100% of the organic compound is removed.

The tubes with or without coating were submitted to the conventional controlled atmosphere brazing (CAB), under inert nitrogen gas atmosphere. The brazing cycle consists in brazing at a temperature higher than 577°C for 4 minutes and 1 minute at 605°C.

1.4 SWAAT

SWAAT tests have been performed on entire heat exchangers according to ASTM G85.

1.5 Electrochemical measurements

Electrochemical measurements were performed using 1480 Solartron Analytical potentiostat and a three electrodes system. The reference electrode is a saturated calomel electrode (SCE), the counter electrode is platinum and the working electrode is a 1 square centimetre flat sample cut from the prepared brazed extruded tube. These brazed samples were cleaned with ethanol and then with deionised water in ultrasonic bath (5 seconds US). Aluminium tubes without coatings were polished with 1000 SiC grit paper and then cleaned with deionised water. Working on freshly polished samples permits to work on more reactive materials and improve reproducibility of the measurement. Aluminium tubes with coating were cleaned with ethanol and then with deionised water in ultrasonic bath (5 seconds US).

Corrosion potential have been measured in ASTM G69 solution which consists in 1M NaCl solution to which is added 9mL of 30% hydrogen peroxide (H_2O_2) per liter of solution. Open circuit potential followed by potentiodynamic measurements were performed in 0,5M NaCl solution. NaCl of analytical grade with purity higher than 99,5% and deionised water with resistivity higher than 18MOhms were used for solution preparation.

Anodic and cathodic polarizations were performed separately. Anodic scanning is started 10mV under open circuit potential (OCP) and ends 500mV above open circuit potential. Cathodic scanning is started 10 mV above OCP and ends 500 mV below OCP. The scanning rate is equal to 0,1667mV/s. For cyclic polarization measurements, the scanning is reversed when the current reaches $1mA/cm^2$.

1.6 Material characterization

Material was characterized using spark spectrometer for determination of the bulk chemical composition. Observations were performed with scanning electron microscope Hitachi SU1510. The influence of brazing on elements distribution as a function of the thickness has been determined by electron probe wavelength dispersive spectrometer (WDS) SX100. For each configuration, two profiles are considered, a concentration profile between two brazing joints from the surface of the tube to 125µm in the thickness of the tube and a concentration profile at a brazing joint. Elements concentration is measured every 5 µm on the profile.

2. RESULTS AND DISCUSSION

2.1 SWAAT results

SWAAT tests have been performed on heat exchangers brazed either with type A or type X aluminium alloy. The coating definition for brazing is the same. Then, the only varying parameters come from the core alloys which do not contain the same copper and manganese quantities. Following SWAAT tests, it was noticed that for exchanger brazed with type X aluminium, the corrosion always initiates at the brazing joints whereas for exchangers brazed with type A aluminium, the corrosion always initiates at the center of the space between two fins (figure 1). Furthermore, after 250 cycles SWAAT, almost all brazing joints remain intact with type A aluminium alloy whereas with type X alloy, all the brazing joints are corroded.

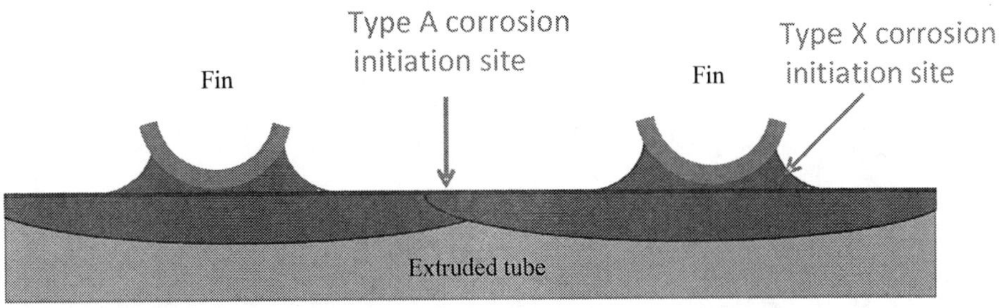

Figure 1: Corrosion initiation sites depending on extruded tube material

2.2 EPMA analysis: difference of element concentration profiles at a brazing joint and between two brazing joints

EPMA analyses have been performed on brazed exchangers. During brazing, the filler metal migrates toward the brazing joints. Therefore, the space between two brazing joints has less filler metal. Thus this can have an impact on elements diffusion and concentration in the tube thickness. The EPMA analysis gives the Silicon, Zinc, Manganese and Copper distribution as a function of the distance compared to the surface of the tube. These measurements are performed at a brazing joint (profile 1) and between two brazing joints (profile 2). An example of the two profile measurements is given on picture 1.

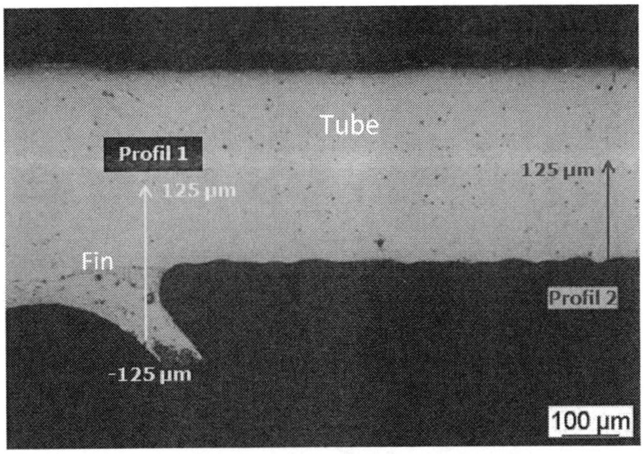

Picture 1: Localisation of the EPMA profiles. Profile 1 at a brazing joint and profile 2 between two brazing joints

2.2.1 EPMA profiles obtained for exchangers brazed with type A aluminium alloy

For type A material, graph 1 shows that copper with a concentration up to 0,7% is present in the brazing joint and there is copper depletion on 25µm between the core alloy and the brazing joint. The only source of copper in our system is the core alloy from the tube. Then, during brazing, copper diffuses from the tube into the brazing joints which can enhance corrosion resistance of the brazing joints. In the same area, the zinc concentration is comprised between 1,5% and 2,4 %. Zinc diffuses in the core alloy over 90 micrometers. A Zinc concentration gradient between the surface of the tube and the core is created during the brazing of the coating. However, this gradient depends on the localization on the tube.

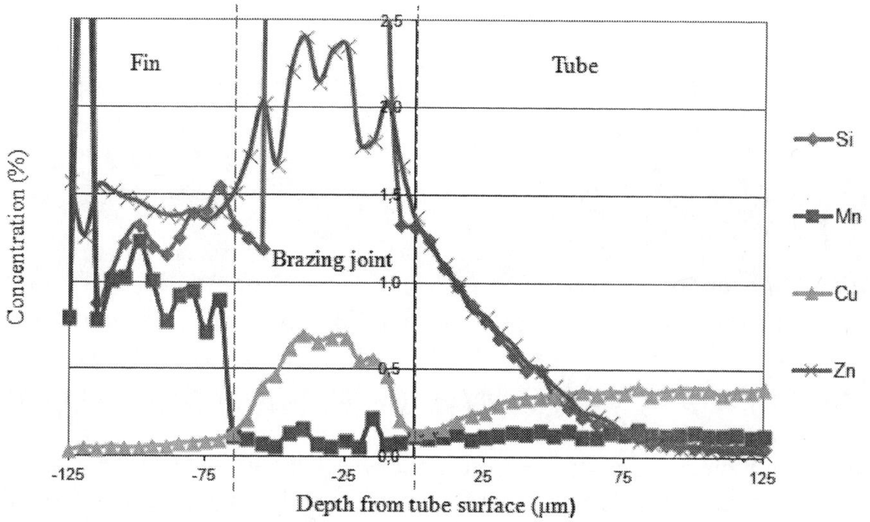

Graph 1: Elements concentration at a brazing joint with Type A aluminium

Graph 2 displays the silicon and zinc concentrations measured at profiles 1 and 2. Zinc is twice more concentrated in the tube under the brazing joint (1,3%) compared to the location between the two brazing joints (0,6%). Si also diffuses from the coating into the core material of the tube. Zn and Si concentrations are the same in the tube under the brazing joint whereas there is more Si than Zn between two brazing joints. This may decrease the protection of the tube against corrosion.

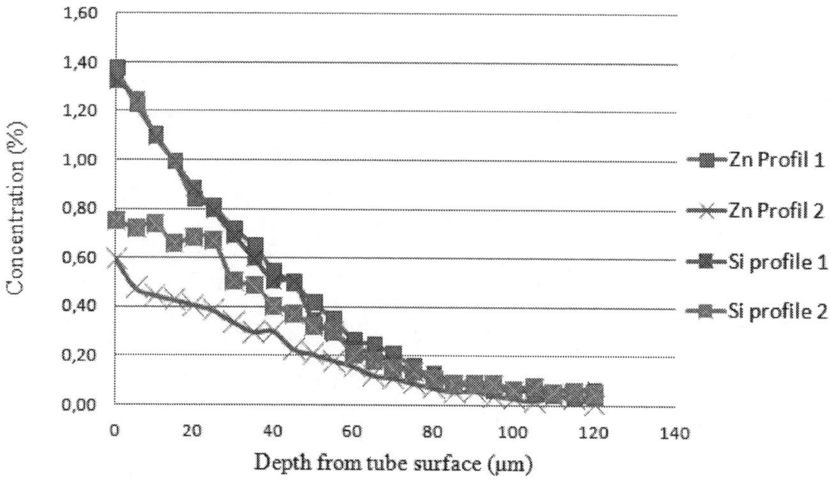

Graph 2: Comparison of Zn and Si concentrations on profil 1 and profil 2 for an exchanger brazed with type A tubes

Graph 3 corresponding to profile 2, shows the profile of Cu, Mn, Si and Zn in the area between two brazing joints. On this graph 3, peaks of Mn, Si and Cu can be seen. These peaks can be attributed to intermetallics which can decrease corrosion resistance by creating a galvanic cell between the aluminium matrix and the intermetallics or grain bondaries.

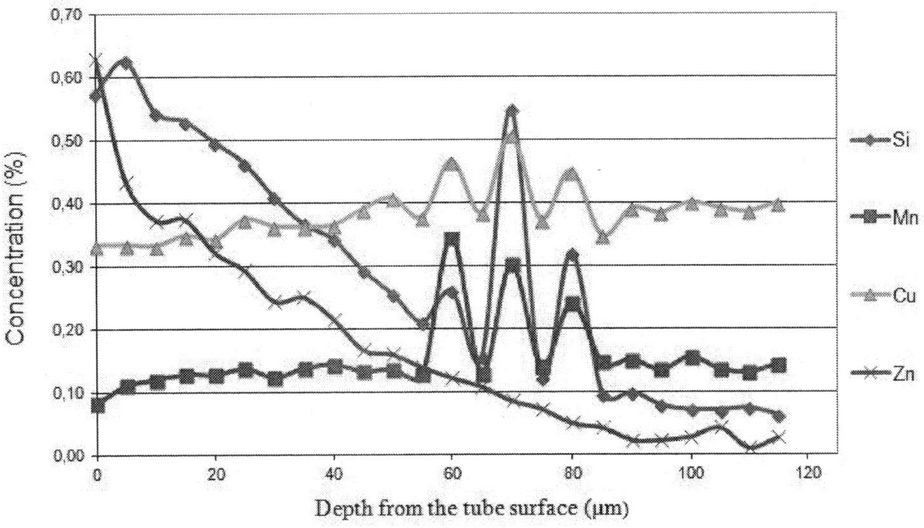

Graph 3: Elements concentration between two brazing joints (Tube Type A)

2.2.2 EPMA profiles obtained for exchangers brazed with type X alloy

Graph 4 and 5 display EPMA profiles for the surface of exchangers brazed with Type X material. At the surface of the tube, the zinc concentration between two brazing joints (0,72%) is much lower than the concentration right under the brazing joint (1,47%). And at a depth comprised between 0 and 30 µm in the tube, the Si concentration is slightly higher than the Zn concentration between two brazing joints. Peaks of Mn can be seen on both profiles and can be attributed to Al(Mn,Fe) or Al(Mn,Fe)Si intermetallics.

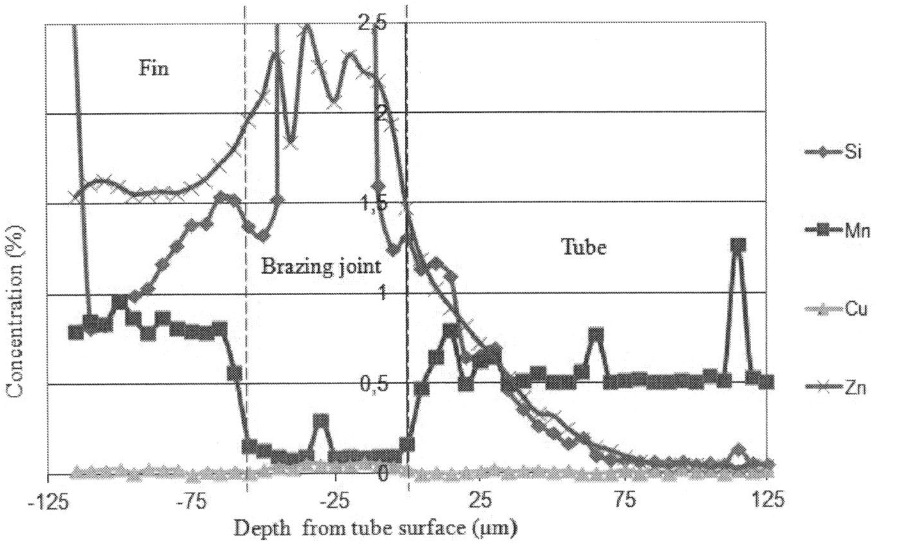

Graph 4: Elements concentration at a brazing joint with Type X aluminium

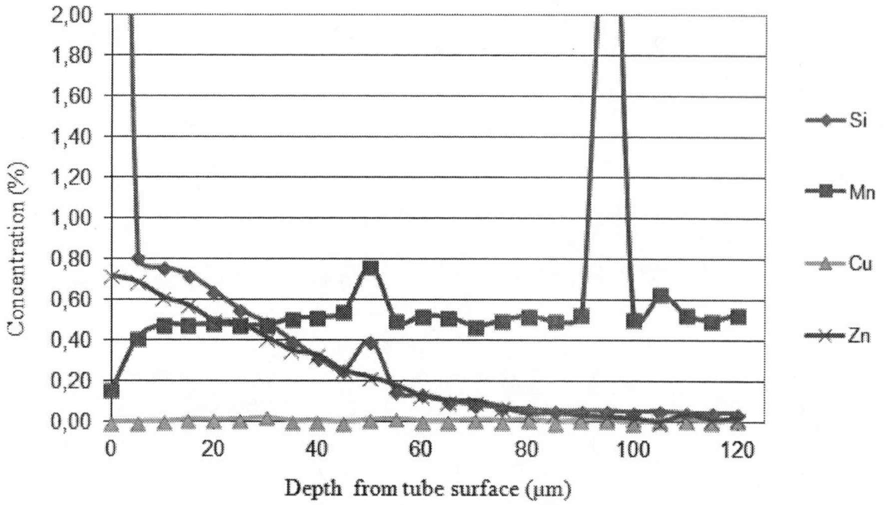

Graph 5: Elements concentration in the tube (type X) between two brazing joints

Unlike exchangers brazed with type A aluminium tubes, exchangers with type X aluminium tubes don't have copper in the brazing joints. This can explain why after SWAAT test, almost all brazing joints of exchangers brazed with type A material remain intact after 250 cycles SWAAT whereas they are all corroded when type X material is used. EPMA measurements show that the composition of the intermetallics present in the tubes is different. Similar Zn and Si diffusion profiles in the thickness of the tube are obtained with both aluminium alloys. Concentration profiles depend on the localization on the exchanger. For a same depth in the tube, the zinc concentration is higher under a brazing joint than between two brazing joints and the silicon concentration is slightly higher than the zinc concentration on the profile between two brazing joints.

To understand the consequences of these observations, electrochemical measurements were performed on brazed tubes either with or without coating in order to distinguish electrochemical properties of the core alloy and the surface which is the first protective layer against corrosion.

2.3 Electrochemical results

2.3.1 Measurements in ASTM G69 solution

Initially, Type X and Type A aluminium tube brazed without coating were characterized according to ASTM G69 standard. The test solution is highly concentrated with sodium chloride so it is very aggressive toward aluminium. Furthermore, it contains hydrogen peroxide which promotes the cathodic reaction. With this test solution, aluminium tends to passivate which leads to an increase and then stabilization of the open circuit potential. This behavior is observed for Type A tubes. However, for type X material, the stabilization is not obtained and a drop of the potential is observed before one hour of measurements. For this reason, the measurements of the open circuit potential have been extended for 20 hours in order to see the evolution of the potential in the ASTM G69 test solution (Graph 6). For type X material, the drop of the potential is significant from -730mV/SCE to -1000mV/SCE whereas the potential of Type A material remains stable.

Observations with SEM were performed on type X pellets after open circuit potential measurements in order to understand the drop of the potential. Corrosion initiates on the tube matrix around Al(Mn,Fe) intermetallics and some precipitates are undermined as shown on picture 2. Dealloying of the aluminium can possibly explain the unstability of the potential despite the important Mn concentration in solid solution. Furthermore, it might also be due to damaging of the passive film on aluminium by chlorides. In areas with the highest corrosion rate, crystallographic faceting is observed.

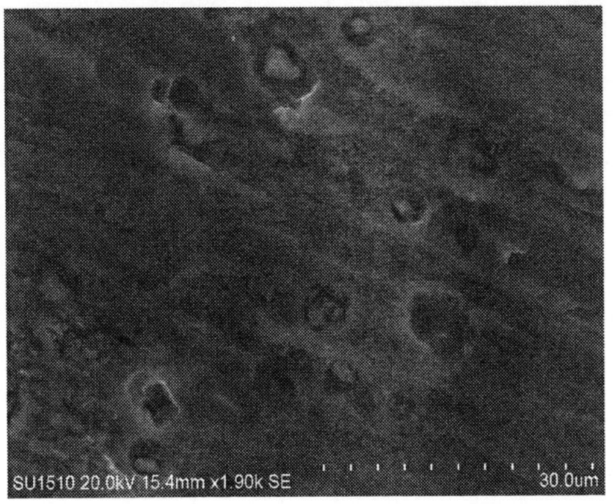

Picture 2: Type X surface after 20 hours open circuit measurements in ASTM G69 solution

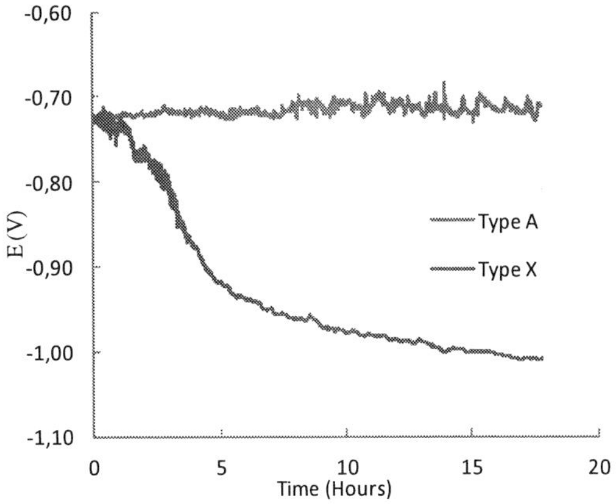

Graph 6: Evolution of the open circuit potential of Type A and Type X aluminium in ASTM G69 solution

2.3.2 Measurements in NaCl solution

Due to the reactivity mainly observed on type X material and taking into account for the field situation and the SWAAT conditions, it was decided to remove the hydrogen peroxide from the corrosive solutions, and to decrease NaCl concentration to 0,5M.

In 0,5M NaCl solution, the open circuit potential (OCP) of type A aluminium is higher than the OCP of Type X aluminium. Type A is nobler than type X aluminium despite the important Mn concentration in the latter (Graph 7). The absence of Cu in type X material and the weak oxide layer of type X material may be the main reasons for this difference of potential.

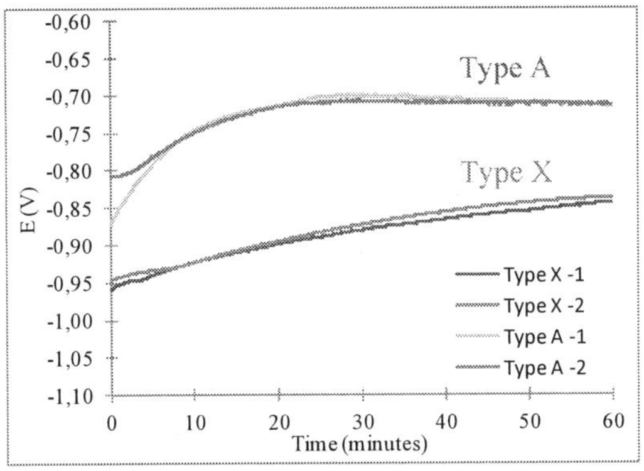

Graph 7: Graph of the open circuit potential of type X and type A material in 0,5M NaCl

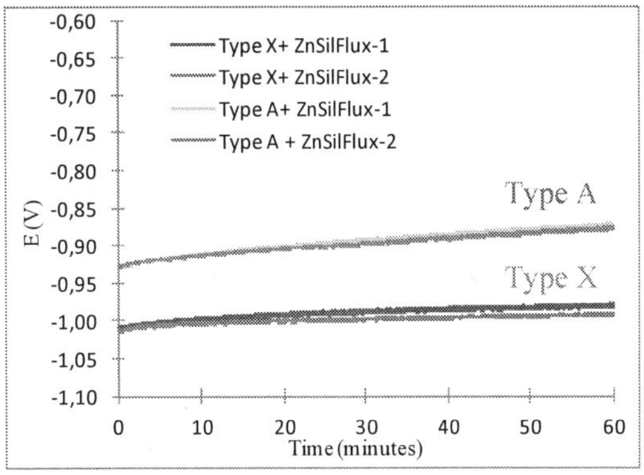

Graph 8: Graph of the open circuit potential of type X and type A material brazed with ZnSilFlux in 0,5M NaCl

OCP measurements on brazed tubes show that a potential gradient is created between the core alloy and the surface of the tubes (graph 7, 8 and table 2). This potential gradient can help to lateralize the corrosion and delay the penetration of the tube.

Table 2: Electrochemical measurements results

	OCP mV/SCE	Ecorr mV/SCE	Epit mV/SCE
Type A	-712	-711	-711*
Type A coated with ZnSilFlux	-872	-871	-709
Type X	-843	-842	-720
Type X coated with ZnSilFlux	-991	-992	-891

*As there is no passivation plateau for type A anodic curve, E_{pit} is approximated to E_{corr}.

Graph 9 displays the anodic polarization measurements for type X and type A core material. The pitting potential corresponds to the potential at which a sharp change of slope occurs. When Type A aluminium is polarized in the anodic direction, the current increases rapidly. Pitting potential of type A aluminium is approximated to its corrosion potential. Conversely, for type X aluminium, the anodic area shows a passive area where the current stabilizes at 10^{-6} A/cm^2.

Graph 10 shows the behaviour in the anodic area of the brazed ZnSilflux coated Type A and Type X tube material. When the surface of type X tubes brazed with ZnSilFlux coating are anodically polarized, currents increase very rapidly whereas there is a plateau with lower currents for polarization measurements at the surface of type A tubes brazed with ZnSilFlux. Corrosion potential of type A material brazed with ZnSilFlux is also higher than corrosion potential of tube made with type X aluminium. Thus, the tubes are brazed with the same coating but anodic polarizations performed at their surfaces give different current density responses. Type X tubes brazed with ZnsilFlux are much more sensitive to pitting than Type A tubes brazed with ZnSilFlux. This difference can be explained partially by the presence of copper in type A tube which has diffused in the coating during brazing and helps to passivate the surface and make it nobler.

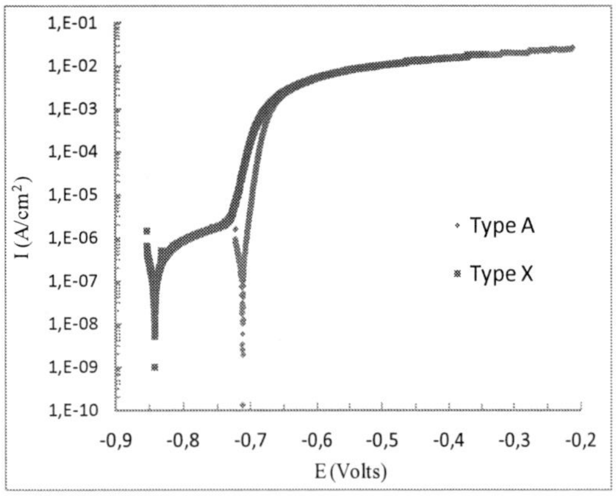

Graph 9: Anodic polarization of type X and type A material

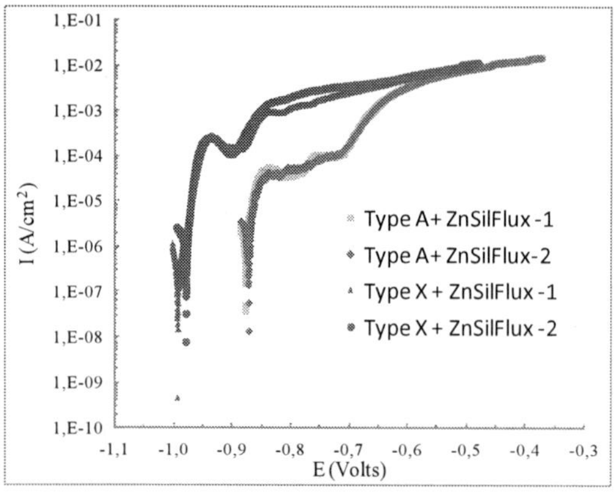

Graph 10: Anodic polarization of type X and type A aluminium tubes brazed with ZnSilFlux

Cathodic polarization curves of brazed type A and type X tubes without coating and tubes brazed with ZnSilFlux are displayed on graph 11. The core alloy is more cathodic than the external surface and low current densities are obtained for both core alloys and tube external surface. This means the overall corrosion reaction is limited by the cathodic reaction, oxygen reduction being the main reaction. It can be noticed from graph 11 that the current on Type A material with or without coating remains low in a wide potential range. Conversely, type X material shows more active surface without coating as the cathodic current increases more significantly. Furthermore, type X material with ZnSilflux coating shows first behaviour dominated by oxygen reduction and after short range of potential, more hydrogen evolution.

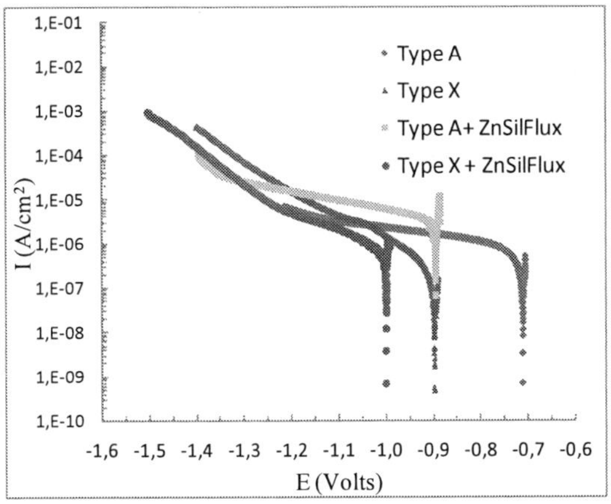

Graph 11: Cathodic polarization on core alloy of Type A and Type X tubes and on tubes brazed with ZnSilFlux

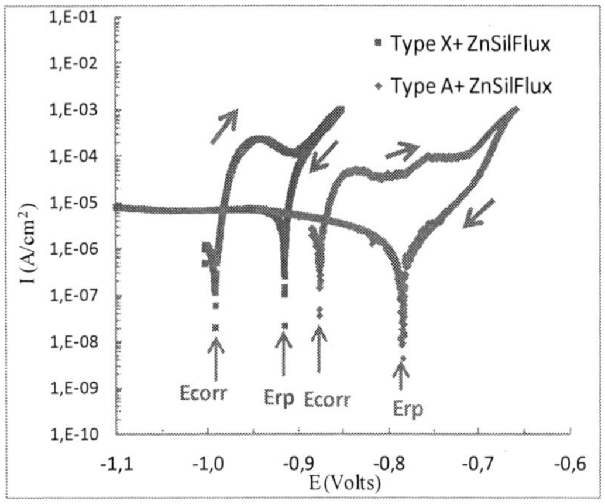

Graph 12: Cyclic polarization measurements performed on type A and type X tubes brazed with ZnSilFlux. The potential scan direction is indicated with the arrows

Further investigations by cyclic voltametry were then performed to study the difference of repassivation of each material. A positive hysteresis loop is always obtained when cyclic polarizations are performed on the aluminium tubes brazed without coating. However when cyclic polarization is performed on the tubes brazed with coating the hysteresis loop is negative for both type X and Type A configurations when the scanning direction is reversed (Graph 12). Repassivation of the surface of the tube is obtained for a potential (E_{rp}) higher than the corrosion potential (E_{corr}). This shows that the coating brings passivation properties to the surface.

Electrochemical measurements show that corrosion potential of type A aluminium is higher than the corrosion potential of type X aluminium. The presence of Zn coating permits to create a potential gradient between the surface and the core alloy. Anodic polarizations on tubes brazed with coating show that coated type A material is less susceptible to pitting and currents density measured on the surface are lower than for type X material. Cyclic polarization measurements permit to highlight that the coating provides repassivation properties to both types of aluminium alloy.

CONCLUSION

EPMA analysis, electrochemical measurements and SWAAT tests bring understanding on the impact of alloying elements on corrosion resistance of heat exchangers tubes brazed with ZnSilFlux coating. Electrochemical measurements showed that the aluminium alloy which contains copper is nobler than the aluminium alloy which contains manganese. When the extruded tubes are brazed with the coating, a potential gradient is created between the surface of the tube and the core alloy of the material. This can help to lateralize the corrosion. This corrosion gradient is due to the high Zn concentration on the surface. Although the tubes are brazed with the same coating, copper containing tubes have a higher corrosion potential and when they are anodically polarized, lower currents densities are measured at the surface than for Mn containing extruded tubes. This significant difference is due to the presence of Cu in type A material. Cyclic polarization measurements show that both copper and manganese containing aluminium have better passivation properties once they are brazed with the zinc containing coating.

EPMA analysis show that copper contained in type A aluminium diffuses in the brazing joint. This can explain why almost all the brazing joints formed with type A remain intact after 250 cycles SWAAT and why the corrosion initiates between two brazing joints. Conversely, brazing joints of exchangers brazed with manganese rich aluminium are corroded. EPMA measurements show that due to filler migration toward joints, the zinc concentration is higher in the brazing joint. Important galvanic coupling can occur between silicon eutectic needles and the primary aluminium phase which is rich with Zn and has no copper to counterbalance it. This can explain why corrosion initiates at brazing joints with type X aluminium.

REFERENCE LIST

[1] A. Gray, H-W Swidersky, L. Orman Reactive Zinc Flux – An Opportunity for Controlled Zinc Diffusion and Improved Corrosion Resistance, AFC Holcroft 11th Brazing Seminar October 2006

[2] S. Iao, S. Kuroda, M. Asano, Effects of Cu content in the core, Zn in the filler, and sheet thickness on the corrosion resistance of brazing sheet, Welding international, volume 23, No 12, December 2009, 886 – 894

[3] F.N. Afshar, J.H.W de Wit, H. Terryn, J.M.C. Mol, The effect of brazing process on microstructure evolution and corrosion performance of a modified AA4XXX/AA3XXX brazing sheet, Corrosion Science 58 (2012), 242- 250

[4] S. Tierce, N. Pébère, C. Blanc, C. Casenave, G. Mankowski, H. Robidou, Corrosion behaviour of brazing material AA4343, Electrochimica Acta, volume 52, Issue 3, 12 November 2006, 1092–1100

[5] R. Ambat, A. J. Davenport, G.M. Scamans, A. Afseth, Effect of iron-containing intermetallic particles on the corrosion behaviour of aluminium, corrosion science volume 48, 2006, 3455-3471

[6] B. Mazurkiewicz, A. Piotrowski, The electrochemical behavior of the Al_2Cu intermetallic compound, corrosion science, volume 23, No 7, 1983, 697 – 707

[7] John E. Hatch, Aluminium properties and physical metallurgy, American Society for Metals, page 26

[8] F.N. Afshar, E. Szala, A. Wittebroo, R. Mulder, J.M.C. Mol, H. Terryn, J.H.W. de Wit, Influence of material related parameters in sea water acidified accelerated test, reliability analysis and electrochemical evaluation of the test for aluminium brazing sheet, corrosion science 53 (2011) 3923-3933

A mechanistic study of aluminium controlled atmosphere brazing processes

D K HAWKSWORTH[1], D P SEKULIC[3,4], C-N YU[3], H FU[3] and R G J WESTERGÅRD[2]
[1] Diomedea Inc., Canada
[2] Gränges AB, Sweden
[3] Department of Mechanical Engineering, College of Engineering, University of Kentucky, USA
[4] School of Materials Science and Engineering, Harbin Institute of Technology, China

ABSTRACT

Aluminium brazing can only be achieved if pre-existing oxides are disrupted to allow molten filler metal wetting and flow between adjacent substrates to form a metallic bond. Mechanistic differences for various controlled atmosphere brazing processes are investigated and compared to previously described systems (vacuum, nickel, flux free). Mechanisms are predominantly surface feature controlled and therefore sensitive to surface changes. Brazing sheet clad with TRILLIUM™ composite is less susceptible to surface condition. Liberation of molten flux from within the composite acts at the metal-oxide interface disrupting the oxide, and aids filler flow by agitation thus promoting brazed joint formation.

1. INTRODUCTION / BACKGROUND

A number of brazing systems have evolved over the past 30 years for the manufacture of brazed aluminium heat exchangers for the automotive industry. Controlled atmosphere brazing using potassium aluminium fluoride flux applied to either an assembled unit or to individual components is the most common method. However some heat exchanger applications are sensitive to, or cannot tolerate post braze flux residues, or are configured such that flux application is not possible in critical areas and therefore are brazed using either vacuum brazing or Nickel brazing technology. More recently, materials that can be brazed under an inert atmosphere without the application of powdered flux have been under development. These are either complete flux-free systems or self-fluxing systems.

The basic mechanism or series of elementary steps by which a metallic bond between two adjacent aluminium substrates are formed in each brazing system vary, but essentially involve the disruption of the pre-existing oxide to allow molten filler metal to wet and flow across both substrates to form a continuous brazed joint between two faying surfaces when the filler metal solidifies as a result of isothermal solidification (Si diffusion depletes the liquid and leads to a shift in the phase diagram effectively solidifying the liquid) and cooling. The mechanism involves a number of metallurgical and chemical processes that take place on both the surface and within the material. Multiple dynamic forces play a role in brazed joint formation, namely: surface energy of the molten metal, the surface energy of the solid substrate and the interfacial energy between the solid and liquid phases. In addition, inertia and viscosity of liquid or mushy material retards the flow impacted by gravity as well. Interaction of all these forces determines the degree of capillary behaviour of the molten filler metal in the joint and hence plays a role in the mechanism of joint formation.

The Vacuum brazing mechanism [1] depends on surface oxide modifications due to the formation and growth of intrusive Mg oxides that cause oxide fracture followed by molten filler exudation, macroscopic collapse and flow of filler metal. The mechanism for fluxless nickel brazing [2] carried out under an inert atmosphere is driven by an exothermic chemical reaction between nickel and aluminium with significant localised heat generation that contributes to oxide fracture and lowering of filler metal viscosity. Filler metal behaviour and brazed joint formation for these two processes is then similar.

Disruption of the oxide film by dissolution of thin surface clad layer by an underlying molten filler metal is the key step in fluxless or exudation brazing processes [3], [4]. Exudation through the fractured oxide (resulting from differences in expansion coefficient) followed by slow wicking of molten filler metal along adjacent touching surfaces depends on the surface tension of the molten filler metal and is very sensitive to the condition of the brazing atmosphere.

The mechanism of the traditional controlled atmosphere brazing process is determined by the action of potassium aluminium fluoride brazing flux that on melting completely covers the aluminium surface dissolving the surface oxide and acts as a barrier to the brazing atmosphere to prevent re-oxidation [5]. This facilitates rapid and easy wetting and flow of molten filler over adjacent solid surfaces to form brazed joints. Recent research [6], [7] has shown that brazing sheet clad with a metal-salt composite, TRILLIUM™ Technology [8], is less susceptible to surface oxidation, the salt acting as a flux at the metal oxide interface on melting of the filler metal thus changing the mechanism for brazed joint formation.

A series of experiments conducted under inert brazing atmospheres using brazing sheet clad with different liner alloys were completed. The ability of filler metal to fill precision milled holes of increasing size was observed on a hot stage microscope, and a traditional angle on coupon test in a laboratory glass-brazing furnace used to validate these results. Energy changes associated with metallurgical and chemical changes during the brazing cycle of TRILLIUM™ composite were established using differential scanning calorimetry (DSC).

Mechanistic differences for each clad brazing sheet system under investigation are compared and principle differences in relation to previously described mechanisms discussed.

2. MATERIALS AND PROCEDURES

A series of clad AA3xxx alloy samples listed in Table 1 with similar dimensions were chosen for evaluation under an inert brazing atmosphere. These include a core alloy clad with single (sample #2) and double side (sample #3) TRILLIUM™ composite (10%Si, ±5% salt, balance Al); samples #4 and #6 which are double side clad AA4343 and AA4045 respectively; and a single side multiclad material, sample #9, (25 µm: 12% Si, 0.2% Mg; 5 µm outer clad: 1.3% Si, 0.2% Mg) for fluxless brazing.

The TRILLIUM™ metal matrix composite [8] consists of an aluminium-silicon alloy and potassium aluminium fluoride salt. The alloy grains of this spray formed composite are typically 5 – 20 µm in diameter with the majority of the spherical silicon particles <1 µm in diameter. The salt particles have an homogenous bi-modal size distribution consisting of macro sized agglomerates up to 200 µm in length and nano sized phases as small as 20 nm located at grain boundary triple points and are often found adjacent to silicon particles.

Table 1 – Brazing test materials.

Sample	Base Alloy	Clad Alloy	Clad Gauge µm	Sheet Gauge mm
#2	AA3xxx	TRILLIUM™	2 x 29	0.5
#3	AA3xxx	TRILLIUM™	1 x 25	0.3
#4	AA3xxx	AA4343	2 x 30	0.4
#6	AA3xxx	AA4045	2 x 36	0.5
#9	AA3xxx	AA4047 (25 µm) AA4007 (5 µm)	1 x 30	0.2

The experimental set-up for hot-stage microscopy is illustrated in Figure 1. Square samples (4 mm x 4 mm) were drilled with precision milled holes ranging in size from 0.2 mm up to 0.7 mm on a CNC machine, cleaned with soapy water and ethanol in an ultrasonic cleaner and dried.

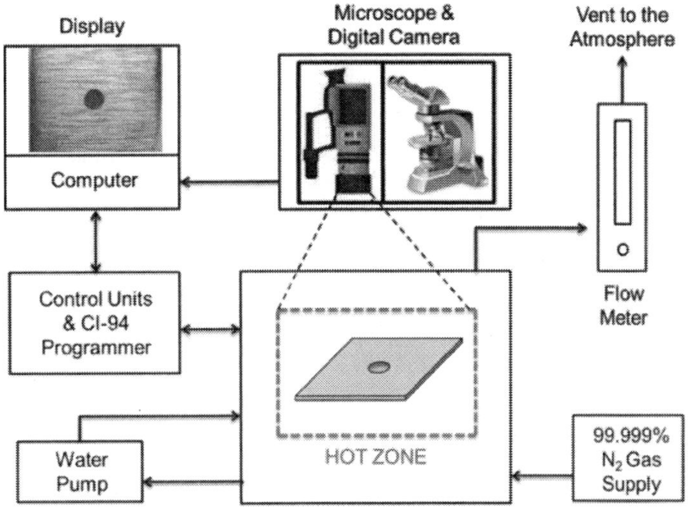

Figure 1 – Hot-stage microscopy system layout.

No flux was added to the TRILLIUM™ clad samples (#2, #3) and the flux free sample #9. A flux loading of approximately 10 g.m^{-2} was applied to the AA4343 (#4) and AA4045 (#6) clad samples. Prepared samples were placed on sapphire glass above the heating element and heated up to 600°C at a rate of 100°C.min^{-1}, held for 2 minutes at 600°C and cooled at 100°C.min^{-1} to 40°C (Figure 2). The background atmosphere in the hot stage was purged by a 99.999% ultra-high purity nitrogen gas source and monitored with an oxygen analyzer.

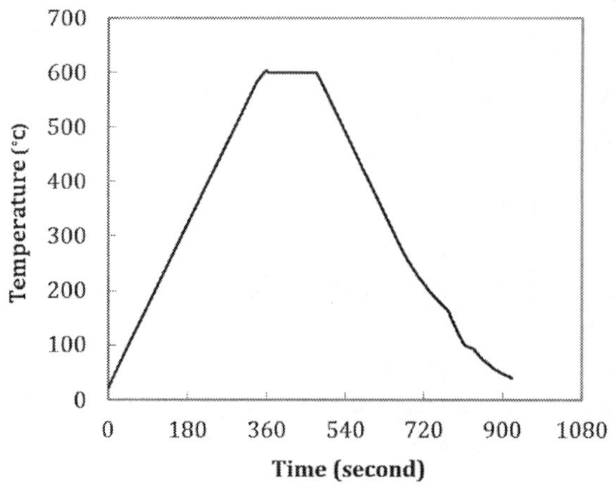

Figure 2 – Hot-stage heating profile.

The laboratory brazing furnace experimental set-up is illustrated in Figure 3. Perpendicular samples of each material 20 mm x 50 mm were brazed to a base token of AA3003, the one end elevated from the surface to create a clearance of 1.6 mm using a SS303 bar (Figure 4). No flux was added to the TRILLIUM™ and flux free samples, whereas the AA4343 and AA4045 samples were coated on both sides with ~5 g.m^{-2} flux. Prior to brazing, the furnace chamber was purged with 99.999% high purity nitrogen at a flow rate of 944 cm^3.min^{-1} for 4 hours. Dew point level was kept below -48°C and oxygen level less than 20 ppm. The same flow rate was maintained for brazing. Samples were heated to ~170°C, held for 30 minutes to remove residual moisture in the chamber, heated to 600°C with a dwell of 2 minutes, and rapidly quenched with nitrogen to 400°C (Figure 5). Data was recorded every two seconds.

Three TRILLIUM™ samples were examined by differential scanning calorimeter (Pyris 1, Perkin Elmer DSC). The instrument was calibrated with zinc (419.47°C), indium (156.60°C) and tin (231.88°C). Cleaned samples 10 – 14 mg in size were placed in an aluminium capsule for measurement. The heating rate was set to 10°C.min^{-1} from initial plate temperature of 520°C to end plate temperature of 600°C.

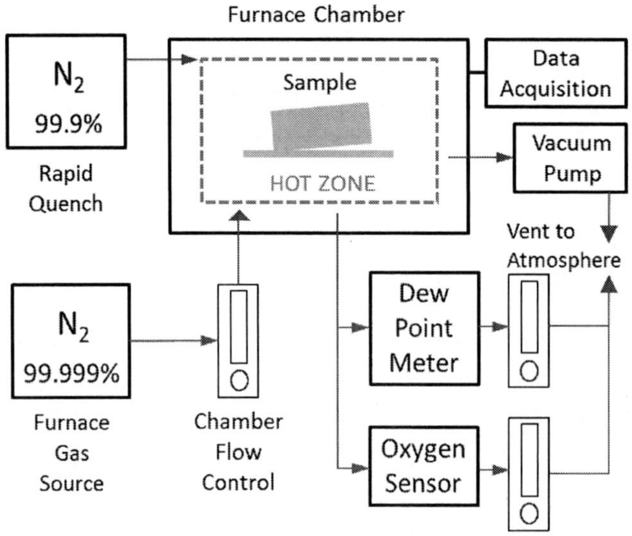

Figure 3 – Laboratory furnace brazing system layout.

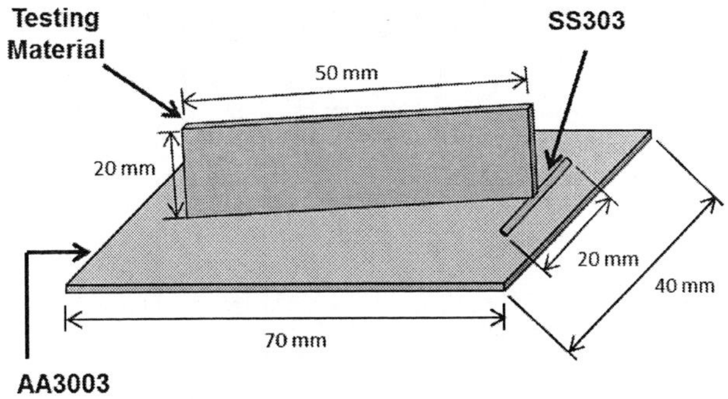

Figure 4 – Sample set up for braze clearance test.

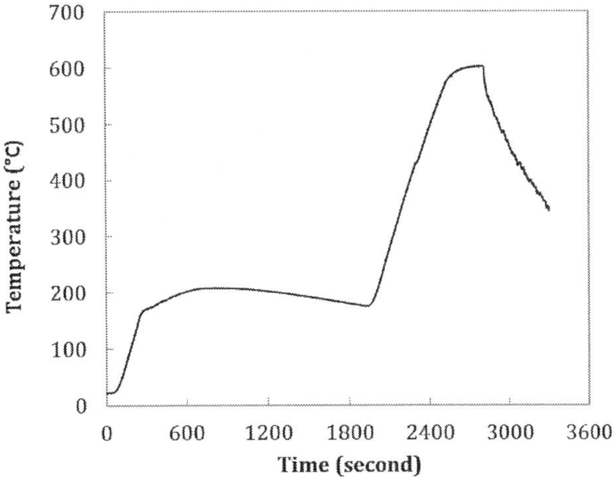

Figure 5 – Laboratory furnace brazing profile.

3. RESULTS

Distinct differences in the behaviour of each test material are observed in the hole filling experiment (Table 2). In each case a limiting hole diameter was established, with the exception of the flux free material that failed to fill even the smallest test hole of 0.2 mm. The TRILLIUM™ single side clad sample, Figure 6, (with less than half the clad thickness and hence half the salt content) and the double side clad equivalent, Figure 7, filled 0.3 mm and 0.6 mm holes respectively. The fluxed AA4045 double side clad sample, Figure 8, was similar in clad thickness and behaviour to the double side clad TRILLIUM™ sample, whereas the AA4343 sample, Figure 9, only filled holes up to 0.5 mm in diameter. The flux free sample, Figure 10, did generate an apparent oxide free melt layer, but this was clearly inactive even though the relative volume of available filler metal was considerably greater than the volume of the hole to be filled (Table 3).

Table 2 – Summary of hole filling tests.

Sample #	Clad Type	Clad µm	Hole Diameter (mm)					
			0.2	0.3	0.4	0.5	0.6	0.7
2	TRILLIUM™	2 x 29	--	●●●	●●●	●●●	●●○	○○○
3	TRILLIUM™	1 x 25	●●○	●●●	○○○	--	--	--
4	AA4343	2 x 30	--	●●●	●●○	●○○	○○○	--
6	AA4045	2 x 36	--	●●●	●●●	●●●	●●●	○○○
9	Flux Free	1 x 30	○○○	○○○	--	--	--	--

Note: ● – hole filled ○ – hole not filled

Table 3 – Ratio of available clad vs. hole volume.

	Hole size (mm)					
Sample #	0.2	0.3	0.4	0.5	0.6	0.7
2	60	27	15	10	7	5
3	41	18	10	--	--	--
4	--	34	19	12	8	--
6	--	34	19	12	8	6
9	76	34	--	--	--	--

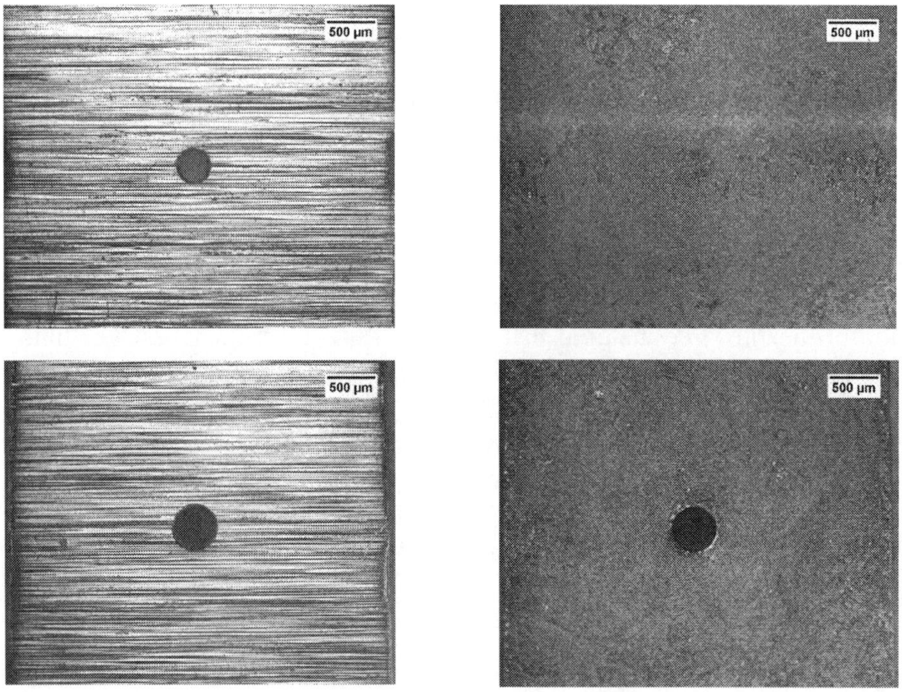

Figure 6 – Sample #3: 1 x TRILLIUM™ clad hole before and after brazing. Top shows 0.3 mm filled case, bottom shows 0.4 mm unfilled case.

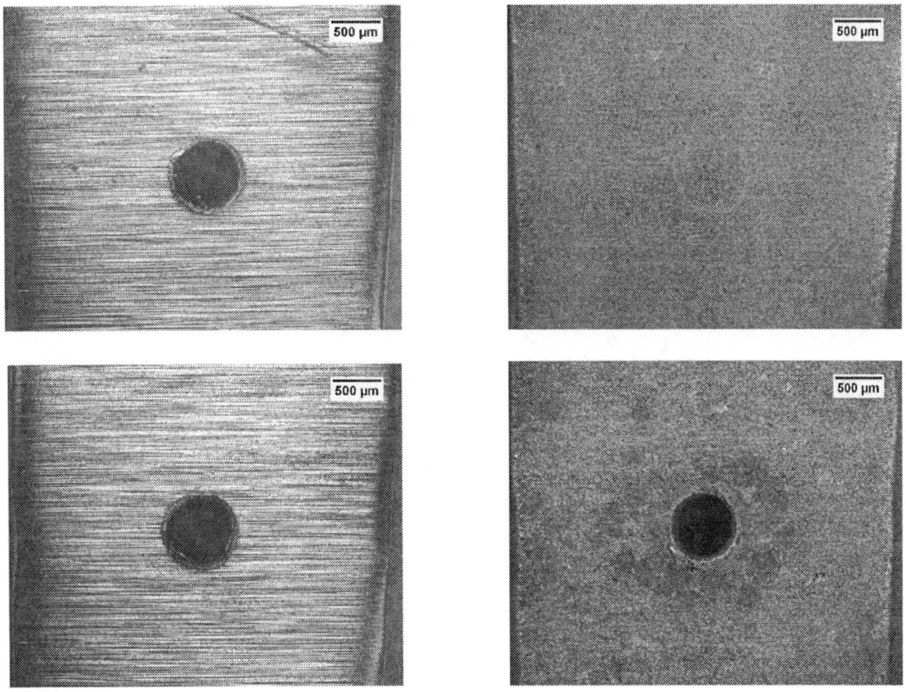

Figure 7 – Sample #2: 2 x TRILLIUM™ clad 0.6 mm hole before and after brazing. Top shows filled case, bottom shows unfilled case.

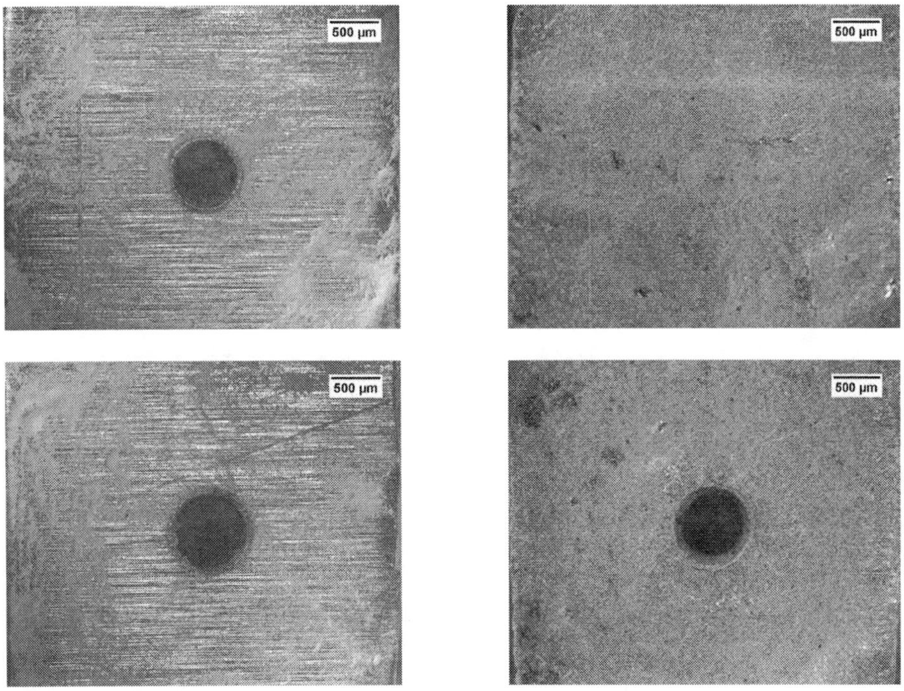

Figure 8 – Sample #6: 2 x clad AA4045 (flux loading ~10g.m^{-2}) hole before and after brazing. Top shows 0.6 mm filled case, bottom shows 0.7 mm unfilled case.

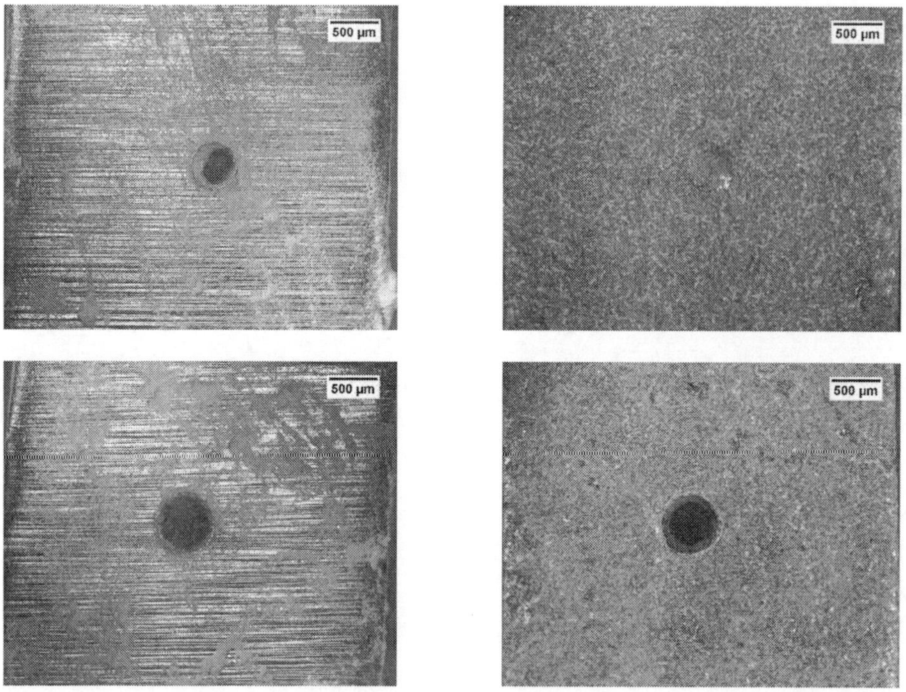

Figure 9 – Sample #4: 2 x clad AA4343 (flux loading ~10g.m^{-2}) hole before and after brazing. Top shows 0.4 mm filled case, bottom shows 0.5 mm unfilled case.

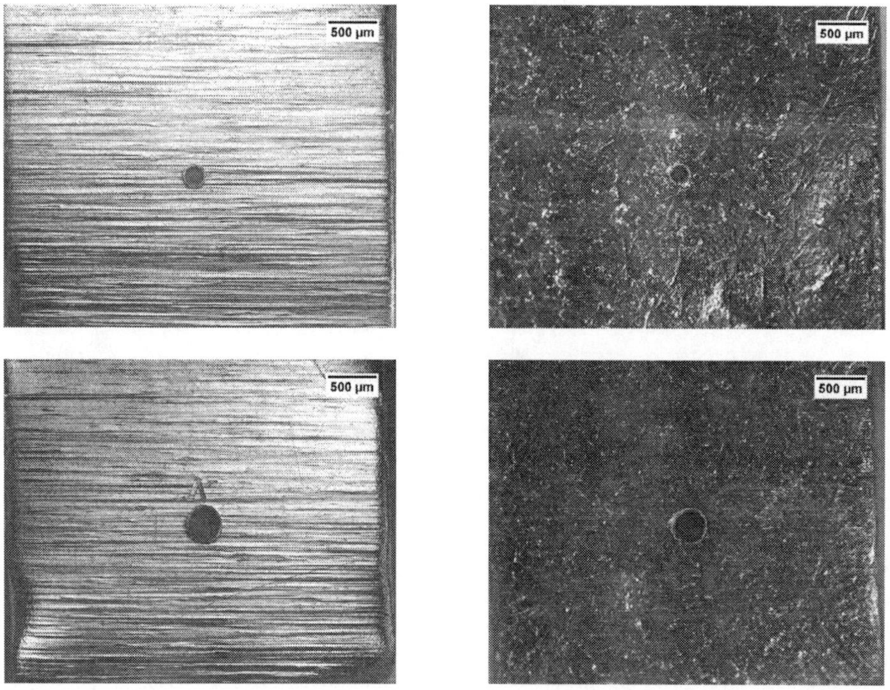

Figure 10 – Sample #9: 1 x flux free clad, hole before and after brazing. Top shows 0.2 mm and bottom shows 0.3 mm unfilled cases respectively.

Microsections through filled and unfilled holes indicate that in each case filler metal, on melting, spread onto the unclad wall of the hole, and in the case of unfilled holes the capillary force acting on the molten filler was inadequate to cause complete filling, or there was insufficient filler metal available to fill the hole completely, Figure 11 - 14. In the case of the fluxless material only slight wetting of the hole wall was observed, but forces acting on the molten filler had drawn it away from the hole creating a shiny halo devoid of filler on the top edge of the hole, Figure 15.

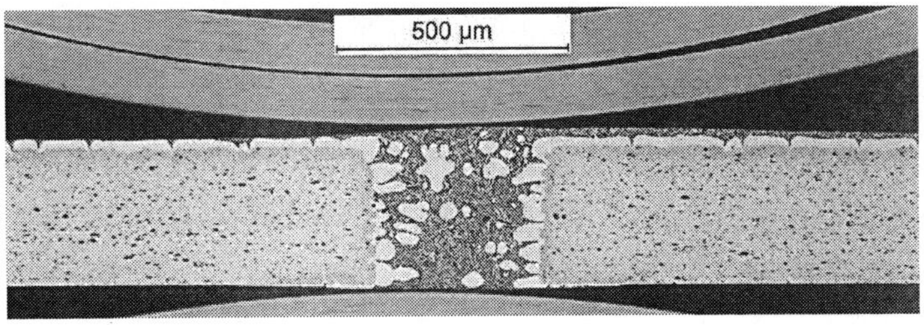

Figure 11 – Cross-section through sample #3 (1 x TRILLIUM™ clad) filled 0.3 mm hole.

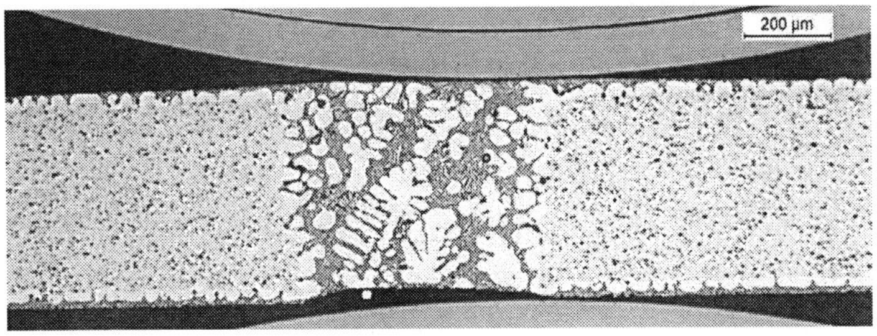

Figure 12 – Cross-section through sample #2 (2 x TRILLIUM™ clad) filled 0.6 mm hole.

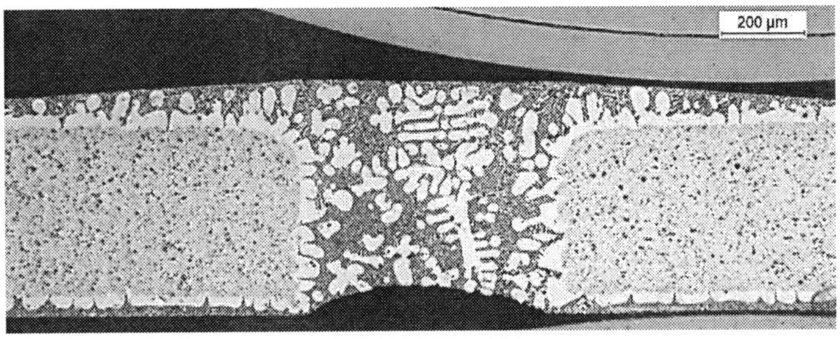

Figure 13 – Cross-section through sample #6 (2 x clad AA4045, flux loading ~10g.m^{-2}) filled 0.6 mm hole.

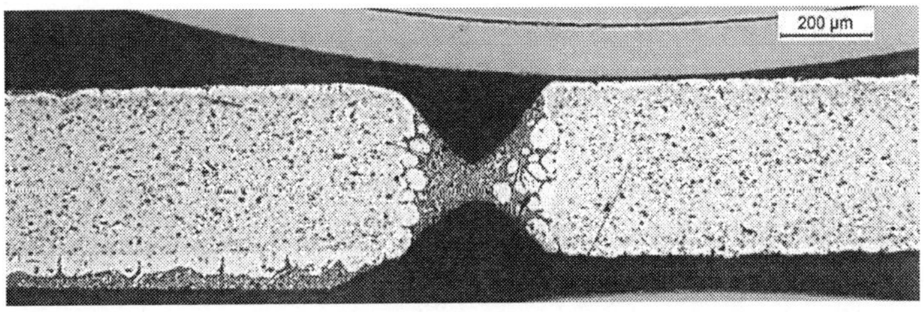

Figure 14 – Cross-section through sample #4 (2 x clad AA4343, flux loading ~10g.m^{-2}) partially filled 0.4 mm hole.

Figure 15 – Cross-section through sample #9 (1 x flux free clad) unfilled 0.2 mm hole.

Data for the gap filling braze test are shown in Table 4 and photographs showing relative fillet length produced for each test material and relative fluxing (or non-fluxing) activity are shown in Figures 16, 17, 18. A clear relationship between clad thickness and fillet length in the gap clearance braze test for AA4045 was established as shown in Figure 19. AA4343 fell below the trend line for AA4045 and flux free cladding failed to produce a brazed joint. A microsection through the flux free sample adjacent to the point of contact with the base coupon (Figure 20) illustrates how the filler metal has pulled away from the edge creating a zone of dewetting along the bottom edge and agglomeration of filler metal above.

DSC results for the TRILLIUM™ material verified previously determined data [9] showing the onset of melting of the salt at 554 °C and an additional endothermic peak, superimposed on the endotherm for metal melting at just above 577 °C was noted when plotting specific heat capacity against temperature (Figure 21). This is suspected to correspond the reaction of the salt with the surface oxide as it is released from the matrix. Reducing the heating rate, thus prolonging the time taken to reach the melting point of the filler metal gives silicon more time to diffuse from the filler metal cladding into the core alloy thereby reducing the volume of liquid filler metal that can be generated once the melting point is reached. A corresponding drop in the relative heat capacity (inversely an increase in heat flow) is observed as shown in Figure 22.

Table 4 – Brazed Fillet Length.

Sample #	Clad µm	Clad Type	Joint Length		
			Test I (mm)	Test II (mm)	Average (mm)
2	2 x 29	TRILLIUM™	25.9	24.7	25.3
3	1 x 25	TRILLIUM™	19.7	21.1	20.4
4	2 x 30	AA4343	22.0	23.0	22.5
6	2 x 36	AA4045	28.9	28.8	28.9
9	1 x 30	Flux free	0	0	0

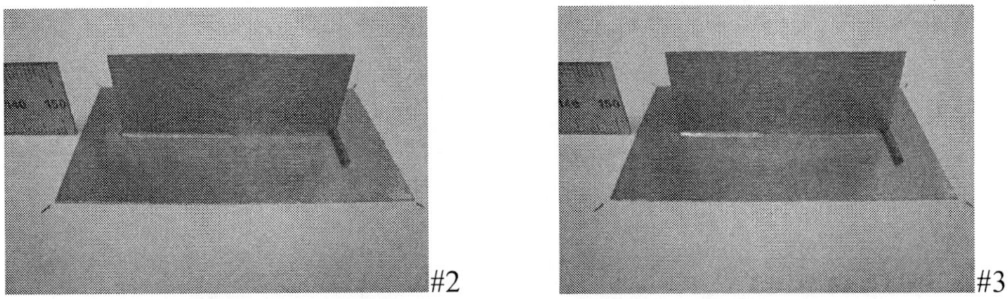

Figure 16 – Relative fillet length produced by sample #2 and #3 (double and single side TRILLIUM™ clad samples).

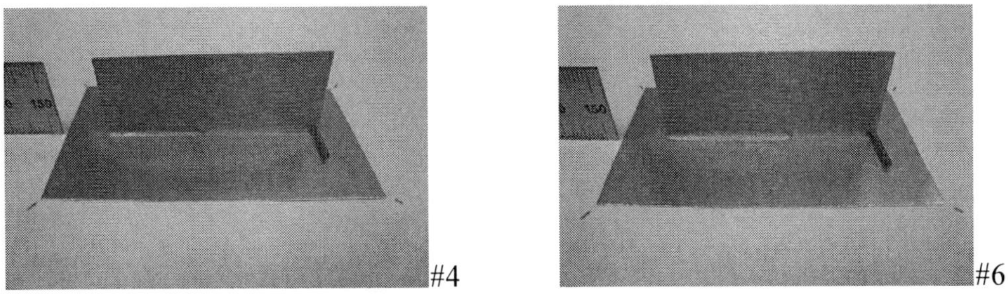

Figure 17 – Relative fillet length produced by sample #4 and #6 (AA 4343 and AA4045 clad samples).

Figure 18 – No brazing joint or fillet was produced by sample #9 (flux free clad).

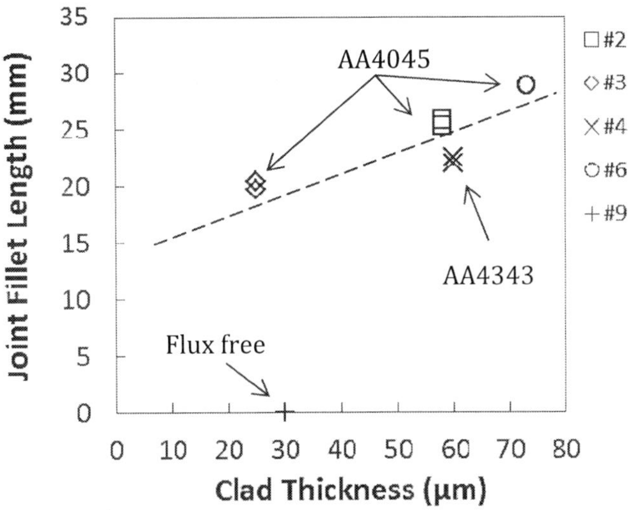

Figure 19 – Relationship between clad thickness vs. fillet length.

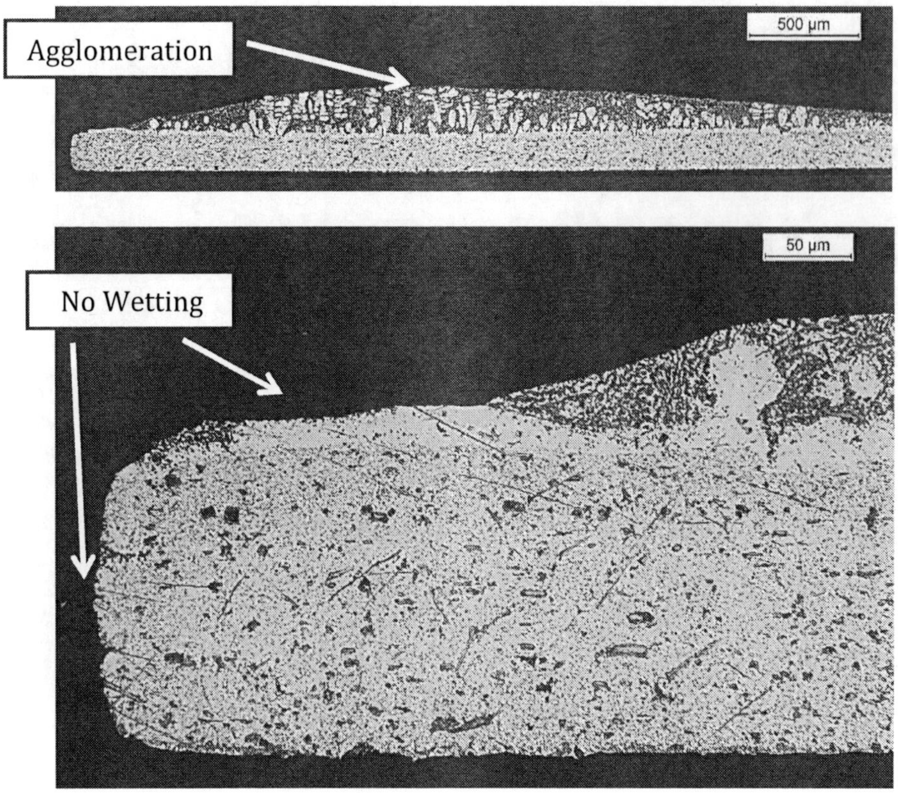

Figure 20 – Cross section through the brazed flux free sample showing agglomeration and dewetting of the filler metal.

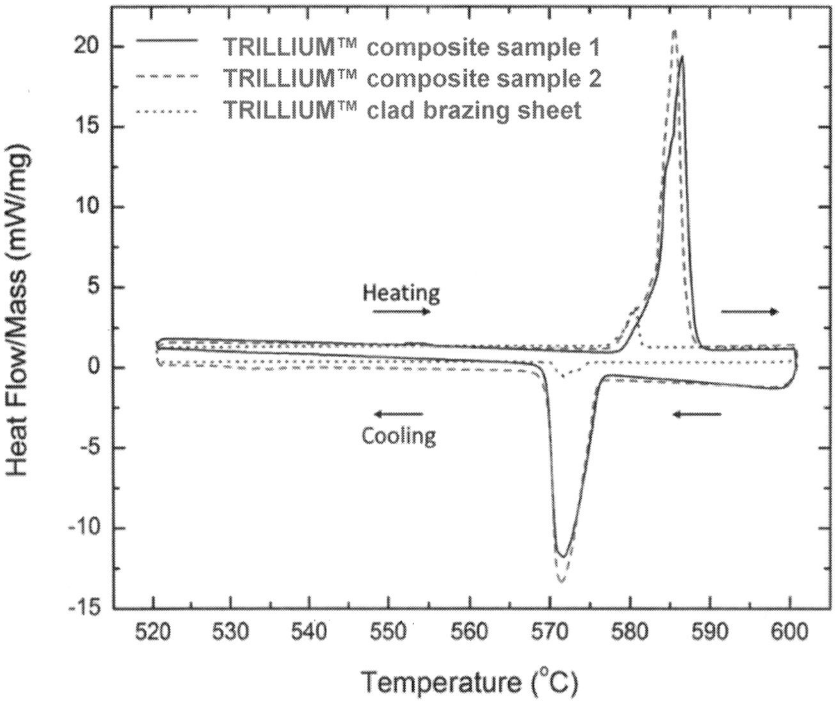

Figure 21 – Plot of heat flow/mass against temperature for TRILLIUM™ composite.

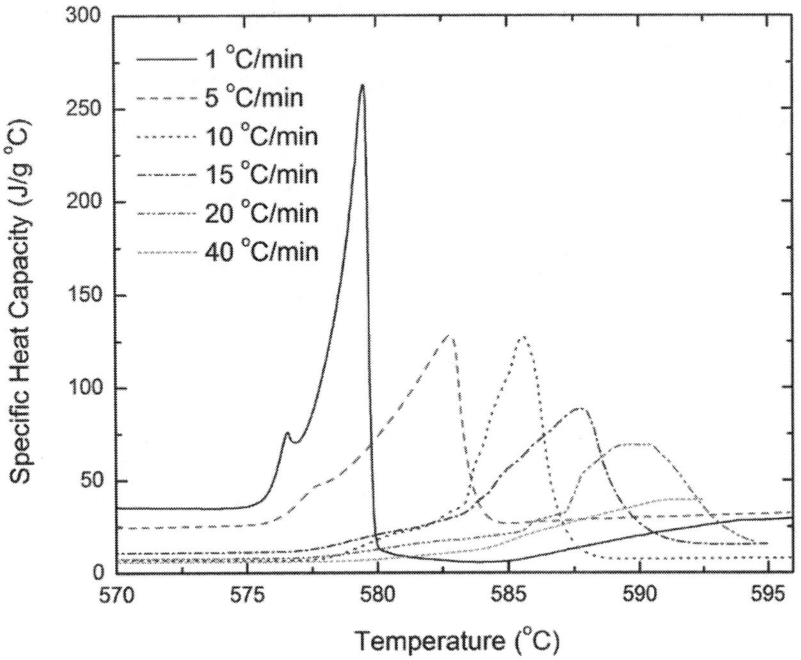

Figure 22 – Plot of specific heat capacity against temperature for TRILLIUM™ composite.

4. DISCUSSION

Each test material with the same configuration and, although subjected to the same brazing conditions, showed distinctly different brazing behaviour. This must imply that there are subtle differences in mechanism for the formation of a brazed joint.

Unlike the TRILLIUM™ and flux coated AA4045 and AA4343 clad brazing sheet, the flux free clad product did not fill holes, nor did the filler metal spread and form a brazed joint in the gap filling experiment. The oxide film on the flux free product did appear to be disrupted, but the filler metal activity was minimal, drawing away from the edges of the sample to form an agglomeration rather than flowing into the gap. Conversely the fluxing action observed with the other material aided filler metal activity and the formation of brazed joints, the TRILLIUM™ material without the addition of flux, and the traditional material with a coating of 5-10 g.m^{-2} flux.

The main difference between TRILLIUM™ composite and traditional flux coated brazing sheet is observed when material is brazed under an adverse atmosphere with high levels of oxygen and moisture as previously reported [6, 7]. Traditional brazing sheet coated with 10 g.m^{-2} of brazing flux starts to show inconsistent brazing when oxygen levels exceeded 500 ppm. TRILLIUM™ composite samples (with a salt loading of about 2 g.m^{-2}) however, produced good braze joints with oxygen levels at 2000 ppm.

DSC analysis of the TRILLIUM™ brazing sheet showed the onset of melting of salt within the matrix at about 550°C, similar to data previously reported. In addition, a flux oxidation exotherm was identified corresponding the onset of melting of the filler metal at 577°C indicating a reaction of the flux with an oxide. This is in stark contrast to previously reported data that shows a significant exothermic reaction corresponding to the onset of melting of flux at about 560°C and the reaction with surface oxide and any moisture or oxygen in the brazing atmosphere.

Two distinct factors that influence the mechanism become evident when considering this data and mechanistic postulations for each system that have been previously published. Firstly, the means by which the oxide is disrupted to expose molten filler metal is different in each system. Secondly, the dynamic properties of the filler metal in creating a brazing fillet or joint differ from system to system.

Consider the mechanisms described for the vacuum and nickel brazing processes. As the name implies under a vacuum, the brazing atmosphere is devoid of oxygen at level that can cause re-oxidation of the aluminium alloys. This mechanism is described by Creber et al. [1]. On heating, surface oxide modification of the Al-Si-Mg alloy takes place involving Mg^{2+} doping of the oxide film and the formation of crystalline magnesium oxides both directly from the oxide/metal interface and indirectly by the reduction of the air formed film. Oxides growth disrupts the film allowing Mg vapour to evaporate into the furnace atmosphere at temperatures >500°C where it reduces the concentration of reactive gases, specifically H_2O. Additions of Bi and Mg to the alloy reduced the temperature at which oxide disruption occurred. On melting of the filler alloy at 555°C for AA4104 and 575°C for AA4047, the molten filler metal exudes through the cracks in the oxide and Mg evaporation reaches a peak. As the filler metal reaches the liquidus temperature and the viscosity decreases, the growing exudates reach a point at which they collapse resulting in macroscopic flow to promote fillet formation under the conditioned furnace atmosphere. Penetration of the oxide on unclad aluminium alloys occurs when it in contact with a surface wetted with filler metal. Reactive vapour phase constituents such as oxygen and water at a sufficiently high concentration influence each step in the mechanism and can significantly reduce the efficiency of the brazing process.

The fluxless nickel brazing process carried out under an inert atmosphere described by Cheadle et al [2] differs from the vacuum brazing process in that a layer of nickel deposited on the surface of the clad brazing sheet reacts exothermically to disrupt the native oxide layer. On heating a chemical reaction between the nickel and the aluminium occurs as a result of diffusion to form a transition phase of Ni_2Al_3, which reacts further to form Al_3Ni. This strong exothermic reaction with significant heat production triggers the formation of the ternary eutectic of Al-Si-Ni at 567°C at the surface. Melting then progresses through the clad layer with further heat liberation, liquid metal mixing and fragmentation of the surface oxide. Fillet formation thereafter is similar to that of vacuum brazing with molten filler metal spreading along the oxide/metal interface of the unclad solid surface thus forming a brazed joint.

The Flux Free or Exudation brazing process, also conducted under an inert brazing atmosphere, described by Takeno et al [4] and Wittebrood et al [3] relies of the dissolution of a thin high purity alloy clad on top of an Al-Si filler metal cladding that is in turn clad to a core alloy. On reaching the eutectic point of the Al-Si alloy at 577°C the filler metal starts to melt and dissolves the thin surface cladding thus causing oxide fracture. Low concentrations of Mg in the Al-Si filler metal can aid in surface oxide fracture in same way as it does in vacuum brazing. The molten filler metal then exudes through the disrupted oxide and slowly wicks along the adjacent surfaces to form a brazed joint. This system is very susceptible to brazing atmosphere contamination and low levels of moisture and oxygen can significantly inhibit the mechanism. The system only really works successfully if two adjacent clad surfaces are in contact as exuded molten filler metal does not wet out and spread very readily on a surface oxide.

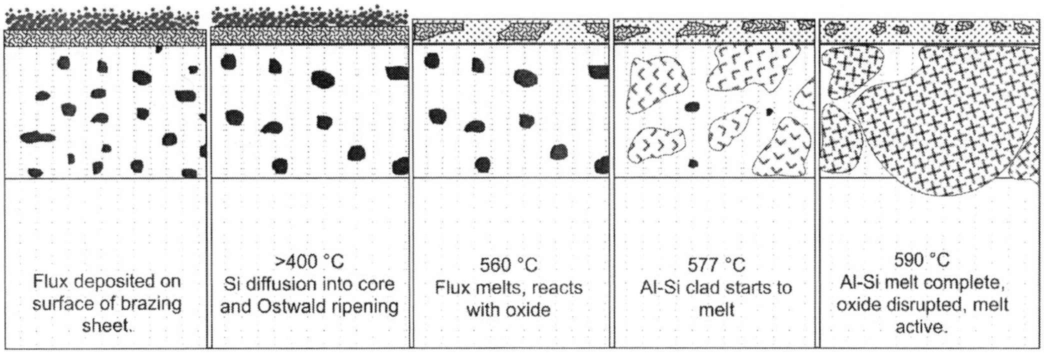

Figure 23 – Applied flux brazing mechanism.

The most effective means of dealing with oxide is by using an application of typically 5-10 $g.m^{-2}$ potassium aluminium fluoride brazing flux (a lower loading of ~2 $g.m^{-2}$ is effective if the brazing furnace has good atmospheric control). The mechanism, illustrated in Figure 23, by which the flux promotes brazed joint formation is discussed by Field et al. [5]. Molten flux is very active with a high affinity for oxygen, spreading readily across the surface and penetrating any discontinuities (cracks) in the oxide. Molten flux acts by dissolving the oxide, but also penetrates the oxide-metal interface thus lifting the oxide to aid dissolution. This permits underlying molten filler metal to wet and flow along faying surfaces, prevents oxidation of the filler metal underneath, and act as a scavenger for reactive species in the brazing atmosphere. Molten filler metal readily wets and spreads across surfaces that have been "conditioned" by the oxide dissolving flux. Flux activity can be inhibited by the presence of surface contaminants such as oils, excess oxygen and moisture in the brazing atmosphere, and reactive elements such as Mg.

The salt contained within the TRILLIUM™ matrix is only released when the filler metal melts, its presence having a positive influence on the melting mechanism, Figure 24. On melting the flux has been observed to wet grain boundaries of the still solid filler metal, and only starts to segregate when the filler metal starts to undergo a phase change at the eutectic temperature of the Al-Si alloy [9, 10]. At this point the salts separates to the surface where it reacts from beneath the surface oxide at the metal-oxide interface. Delayed release of the molten salt until the filler metal liquidus is reached provides protection from premature oxidation. The molten salt can now act as a flux by functioning from beneath the oxide and any surface contaminants, lifting and washing them away from the region in which a joint is to be formed. The release and segregation of the molten salt from the molten filler metal causes agitation of the molten filler, thus aiding in rapid flow into joints to be filled. This helps overcome surface tension that might restrain liquid filler metal flow.

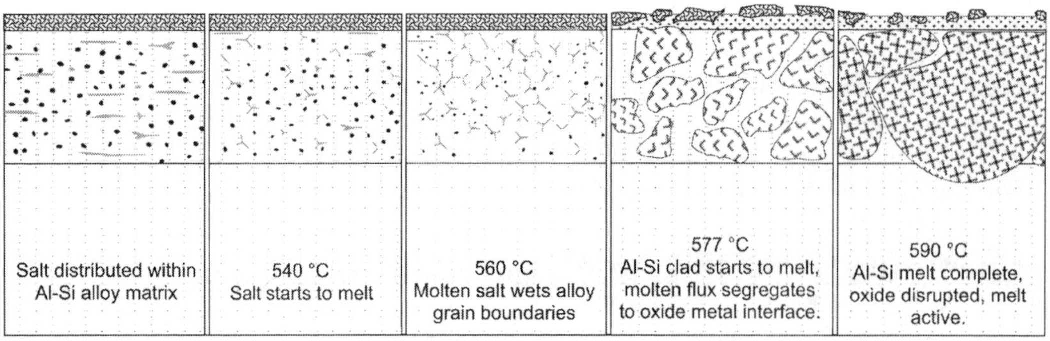

Figure 24 – TRILLIUM™ Technology brazing mechanism.

5. CONCLUSION

The mechanism for the Flux Free or Exudation brazing process is dependant on the dissolution of a thin surface clad layer to disrupt the oxide film followed by exudation and wicking of molten filler metal which has strong intramolecular forces between the metal atoms in the melt. This high surface tension and very low affinity for the solid substrate on which it is originally clad, or to which it is trying to braze, results in dewetting and poor gap filling or joint forming properties as observed in the brazing experiments. Intimate contact between components is necessary if a brazed joint is to form.

The Vacuum brazing mechanism depends on oxide modifications due to the active formation and expansion of intrusive Mg oxides. Subsequent oxide fracture allows for exudation of molten filler followed by macroscopic collapse and flow of filler metal into the brazing joint. The high affinity of the molten filler metal for the substrate and low surface tension of the filler in the absence of reactive species in the brazing atmosphere permits brazed joint formation. However, any surface contamination of either the solid substrate or the molten filler dramatically inhibits joint formation characterised by stitching and porosity in brazed joints.

Surface contamination and a poor brazing atmosphere influences the Ni brazing process just as it does vacuum brazing. When Ni coated brazing sheet reaches the brazing temperature, an exothermic chemical reaction and significant localised heat generation contributes to oxide fracture and lowering of filler metal viscosity. The alloying reaction across the solid–liquid interface and reduced surface tension enhance the wetting properties and hence brazed joint formation.

The action of molten potassium aluminium fluoride brazing flux used for the CAB and TRILLIUM™ brazing processes serve to minimise surface tension of the molten filler metal by removing the oxide, and disrupting the oxide on non-clad surfaces thus permitting intimate contact between the molten filler and the solid substrate. The good wetting properties and hence strong capillarity that draws the filler metal into closely spaced joints are the key to the brazing mechanism. This was demonstrated in the hole filling experiment where both systems showed similar brazing characteristics. The good fluxing action promoted filler metal wetting and hence flow across the solid substrate.

The main differences between the two processes are that the flux acts primarily on the surface of the brazing sheet in conventional CAB, whereas salt liberated from within the matrix of the TRILLIUM™ composite acts initially at the metal-oxide interface. Release of the molten salt (which then acts as a flux), results in further agitation of the molten filler to promote rapid brazed joint formation.

6. ACKNOWLEDGEMENTS

Permission to publish this work was granted by Gränges AB and Sandvik Osprey, joint developers and owners of TRILLIUM™ Technology.
Samples and funding in part for this research was provided by Gränges AB.
The work of D.P. Sekulic, C.-N, Yu and H. Fu has been funded in part through the US National Science Foundation, NSF-CBET-1234581 for 2014. DPS acknowledges support through the 1000 Foreign Expert Professor Talent Program facilitated at the Harbin Institute of Technology, Harbin, P.R. China.
The research team acknowledges the CNC support and technical advice by Mr. Charles Arvin (ISM, University of Kentucky).

7. REFERENCES

1. Creber D. K., Ball J., Field D. J., 1987, *A mechanistic study of aluminium vacuum brazing*, SAE Technical Paper Series, 870185, International Congress and Exposition, Detroit, Michigan, Feb. 23 - 27, 1987.
2. Cheadle B and Dockus K, 1988, *Inert Atmosphere Fluxless Brazing of Aluminum Heat Exchangers*, SAE paper 880446.
3. Wittebrood A, Jacoby B and Kirkham S, 2010, *The Evolution of Flux-free Brazing of Aluminium*, 14th International Invitational Aluminium Brazing Seminar, Novi, MI, USA, October 28 2010.
4. Takeno S, and Suzuki Y, 2005, *Development of Thermal Products Using the EB Method*, Furukawa Review No. 27.
5. Field D, and Steward N, 1987, *Mechanistic aspects of the Nocolok flux brazing process*, SAE Technical Paper Series, 870186, International Congress and Exposition, Detroit, Michigan, Feb. 23 - 27, 1987.
6. Yu, C. N., Hawksworth, D. K., Liu, W. & Sekulic, D. P.; *Al brazing under severe alterations of the background atmosphere: A new vs. traditional brazing sheet.* IBSC 2012, Apr. 22-25, Las Vegas, NV, USA.
7. Yu, C. N., Hawksworth, D. & Sekulic, D. P.; *Impact of the background atmosphere on the aluminum brazing joint formation.* ASME IMECE 2013-64776, Nov. 19, San Diego, CA, USA.
8. Ogilvy A, Hawksworth D, Abom E, 2008, *A brazing piece comprising a composite material including an inorganic flux.* UK Patent Application GB2447486.
9. Hawksworth D.K., Westergård R.G.J., Ogilvy A.J.W.; *TRILLIUM™ Technology - Aluminium brazing with a composite liner.* 7. Int. Congress "Aluminium Brazing", 22-24 May 2012.

10. Hawksworth D K (Gränges Sweden AB), Westergård R, Nilsson P (Gränges Technology, Finspång/Sweden) Brazing *properties of TRILLIUM™ composite*, 8 Int. Congress "Aluminium Brazing", June 03–05, 2014.

TRILLIUM™ Technology is protected by United States Patent No. 8871356 as well as by corresponding patents and pending patent applications in other major countries

TRILLIUM™ Technology is a registered Trademark of Gränges AB

Reaction behavior of mixtures of non-corrosive flux and non-corrosive flux containing Zn during brazing

H KUMAGAI and **N YAMASHITA**
UACJ Corporation, Research and Development, Japan

ABSTRACT

The reaction behavior of non-corrosive flux (NC-flux), non-corrosive flux containing Zn ($KZnF_3$) and the mixture of them during brazing were investigated. In regard to the mixture of NC-flux and $KZnF_3$, the different behavior from single-component flux was observed. $KZnF_3$ in the flux mixture was completely consumed above 500°C, and the pure Zn was detected at 500°C. It was considered that this behavior of the flux mixture was derived from the reactions of $KZnF_3$ with Al_2O_3 and the reduction of ZnO by $KAlF_4$.

1 INTRODUCTION

Aluminum automotive heat exchangers are generally manufactured by the controlled atmosphere brazing (CAB) using non-corrosive flux (NC-flux), such as the potassium aluminum fluoride of $KAlF_4$, K_2AlF_5 and K_3AlF_6. Recently non-corrosive flux containing Zn (Zn-flux) such as the potassium Zinc fluoride of $KZnF_3$ is also used widely. The feature of Zn-flux is that Zn diffusion layer into the aluminum multi port extrusion tube for the purpose of the protection from the corrosion perforation is formed during brazing. Therefore, the application of that flux enables to omit the conventional process of Zn arc spraying onto the aluminum multi port extrusion tube, and also enables to reduce the manufacturing cost. Furthermore, it is also allowed to apply the mixture of both NC-flux and Zn-flux.

The reactions of Zn-flux against the oxygen concentration and the moisture during brazing are more sensitive than those of NC-flux. Therefore, it is difficult to make good brazing joints without defects like discoloration by applying Zn-flux. Since the mechanism of the reactions of Zn-flux has not been reported so much, this paper showed the reaction mechanism of NC-flux, Zn-flux and the mixture of them during brazing.

2 EXPERIMENTAL PROCEDURE

2.1 Brazing of single substance and mixture of fluxes

Table 1 shows specimens, and figure 1 shows relationship between the oxygen concentration in furnace and the material temperature during brazing. $KZnF_3$ was added and mixed to NC-flux with the 0%, 50% and 100% of $KZnF_3$ fraction. These fluxes were dispersed on the organic solvent and were brushed on A3003 alloy sheets (25mm×25mm O temper). After drying, these were brazed at the maximum temperature of 400°C, 500°C and 600°C in the brazing furnace with N_2 atmosphere. The oxygen concentration in brazing furnace is over 2500ppm below 400°C, 50-2500ppm in the range of 400-500°C, and below 50ppm in the range of 500°C-600°C. The specimens after brazing were analyzed by X-ray diffraction (XRD). Regarding Z50 brazed at 500°C, the observation by the scanning electron microscope (SEM) and the elemental analysis by the energy dispersive X-ray spectrometry (EDX) were carried out.

Table 1 Specimens for brazing and XRD analysis

Sample name	Amount of flux / mg		
	NC-flux ($KAlF_4$)	$KZnF_3$	Total
Z0	4	0	4
Z100	0	4	4
Z50	2	2	4

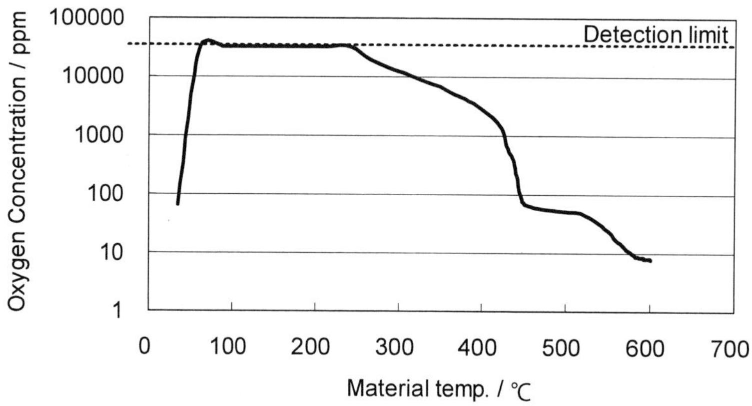

Figure 1 Relationship between oxygen concentration in furnace and material temperature during brazing

2.2 Additional examinations

2.2.1 Validation of chemical reactions

Table 2 shows specimens for the thermo gravimetric and the differential thermal analysis (TG-DTA) and XRD analysis. The after-mentioned estimated chemical reactions could be validated by TG-DTA. The specimens in Pt pan were heated to with the rising rate of the temperature at 20°C/min and the flow rate of the argon gas or the air was 300ml/min. All of the specimens were analyzed by TG-DTA. P1, P2 and P3 were analyzed by XRD after TG-DTA.

Table 2 Specimens for TG-DTA and XRD analysis

Sample name	Blend ratio / %				Flow gas species
	NC-flux ($KAlF_4$)	$KZnF_3$	Al_2O_3	ZnO	
P1	–	50	50	–	Ar
P2	–	100	–	–	Atmospheric air
P3	50	–	–	50	Ar
$KZnF_3$	–	100	–	–	Ar
NC-flux	100	–	–	–	Ar

2.2.2 Further validation for reduction of ZnO

In regard to the after-mentioned reduction of ZnO, the further validation was examined. The mixture of NC-flux and ZnO (75:25) was brazed at the maximum temperature of 600°C. The mixture with approx. 2.7mg was dispersed on the organic solvent and dropped on the center of A3003 alloy sheet (50mm×50mm, O temper). After drying, the specimen was brazed at the same condition as **2.1**. The specimen after brazing was analyzed by XRD.

3 RESULTS AND DISCUSSIONS

3.1 Brazing of single substance and mixture of flux

3.1.1 Z0 (NC-flux: 100%, $KZnF_3$: 0%)

Figure 2 shows the appearance of specimens after brazing and figure 3 shows XRD patterns of Z0 after brazing. At the maximum temperature of both 400°C and 500°C, the appearance of the specimen after brazing took on white color and powdery that was same as the specimen before brazing. In XRD patterns, $KAlF_4$ and K_3AlF_6 were detected. At the maximum temperature of 600°C, the appearance of the specimen took on transparent color and melted-solidified. It was considered that the flux was melted normally. In XRD patterns, $KAlF_4$ and K_3AlF_6 were detected. Intensity of diffraction peaks for K_3AlF_6 increased with increasing the maximum temperature of brazing. The reason why K_3AlF_6 was detected at such the low temperature as 400°C was considered that the oxygen concentration below 500°C during brazing was high, 50-2500ppm or more. It means that the atmosphere in brazing furnace contained not only O_2 but also H_2O. Therefore, $KAlF_4$ brazed in bad atmosphere reacted with H_2O to form K_3AlF_6 following the chemical reaction equations (1)-(3) [1, 2].

$$KAlF_4 \rightarrow KF + AlF_3 \qquad (1)$$

$$2AlF_3 + 3H_2O \rightarrow Al_2O_3 + 6HF\uparrow \qquad (2)$$

$$2KF + KAlF_4 \rightarrow K_3AlF_6 \qquad (3)$$

(a) 400°C (b) 500°C (c) 600°C

Figure 2 Appearance of Z0 after brazing at various temperatures

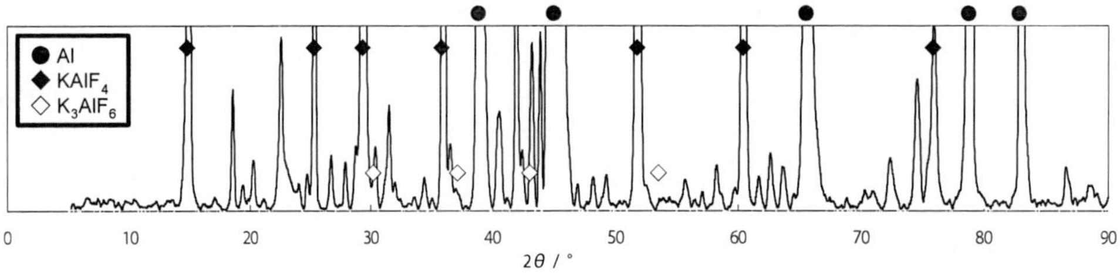

(a) 400°C

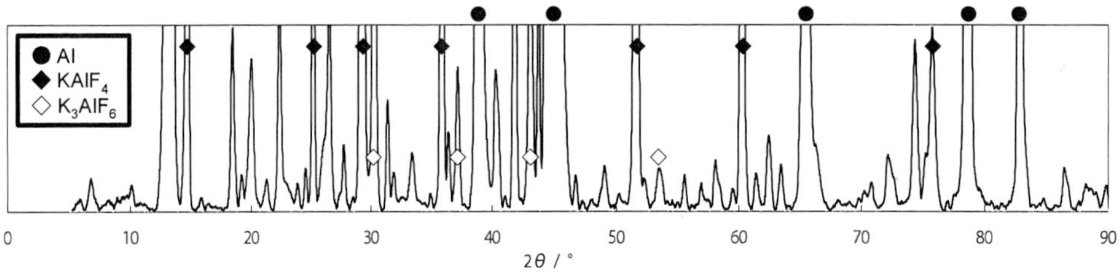

(b) 500°C

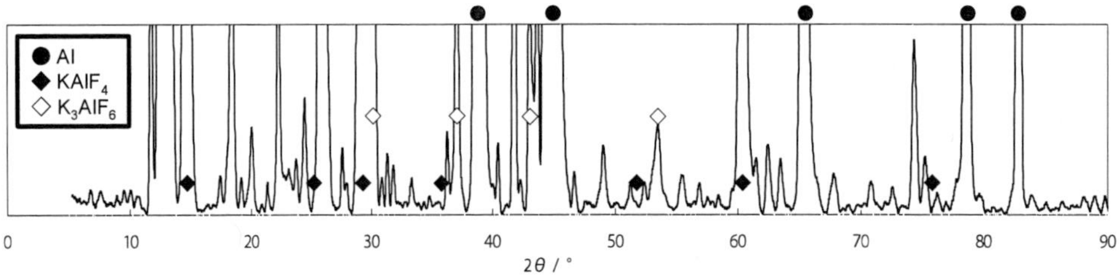

(c) 600°C

Figure 3 XRD patterns of Z0 after brazing at various temperatures

3.1.2 Z100 (NC-flux: 0%, $KZnF_3$: 100%)

Figure 4 shows the appearance of specimens after brazing and figure 5 shows the XRD patterns of Z100 after brazing. At the maximum temperature of 400°C, the appearance of the specimen after brazing took on white color and powdery that was same as the specimen before brazing. In XRD patterns, $KZnF_3$ and slight ZnO were detected. At the maximum temperature of 500°C, the appearance of the specimen took on white color and powdery. In XRD patterns, $KZnF_3$, K_3AlF_6 and ZnO were detected. At the maximum temperature of 600°C, the appearance of the specimen was transparent color and melted-solidified as well as Z0. It was considered that the flux was melted normally.

In XRD patterns, $KAlF_4$ and K_3AlF_6 were detected, but $KZnF_3$ and ZnO were not detected. $KZnF_3$ generally reacts with the aluminum at approx. 555°C following the chemical reaction equation (4) to form $KAlF_4$, K_3AlF_6 and Zn [3].

$$6KZnF_3 + 4Al \rightarrow 3KAlF_4 + K_3AlF_6 + 6Zn \quad (4)$$

Below 500°C, however, the eq. (4) could not occur. Therefore, it was estimated that chemical reaction equations (5)-(7) could occur.

$$6KZnF_3 + 2Al_2O_3 \rightarrow 3KAlF_4 + K_3AlF_6 + 6ZnO \quad (5)$$

$$KZnF_3 \rightarrow KF + ZnF_2 \quad (6)$$

$$ZnF_2 + H_2O \rightarrow ZnO + 2HF\uparrow \quad (7)$$

The eq. (5) is the reaction of $KZnF_3$ with the oxide film on the aluminum, and the eq. (7) is the reaction of $KZnF_3$ with H_2O in the brazing atmosphere. Although the amount of K_3AlF_6 was too small to detect in XRD patterns at maximum temperature of 400°C, it was considered that $KAlF_4$ formed by the eq. (5) reacted with KF formed by the eq. (1) and (6) to form K_3AlF_6 as well as Z0. It was considered that most of $KZnF_3$ out of contact with the surface of the alumimun base material could not be reacted and therefore remained, because the eq. (5) occurred on the surface of the alumimun base material. Furthermore, it was considered that the eq. (6) and (7) also occurred on overall $KZnF_3$. However, ZnO was not detected at 600°C. It was considered that the following the chemical reaction equation (8) could occur.

$$3KAlF_4 + 3ZnO \rightarrow K_3AlF_6 + Al_2O_3 + 3Zn + 6F\uparrow \quad (8)$$

Above 500°C, the eq. (4) occurred at approx. 555°C, the formed $KAlF_4$ and K_3AlF_6 were melted at approx. 562°C. Finally, Zn was diffused into the aluminum base material.

(a) 400°C (b) 500°C (c) 600°C

Figure 4 Appearance of Z100 after brazing at various temperatures

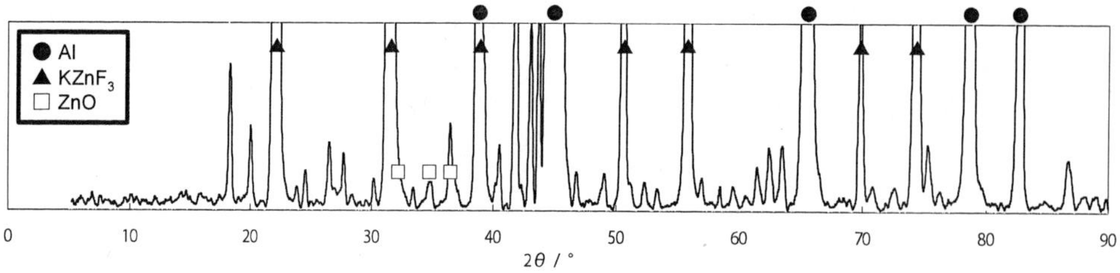

(a) 400°C

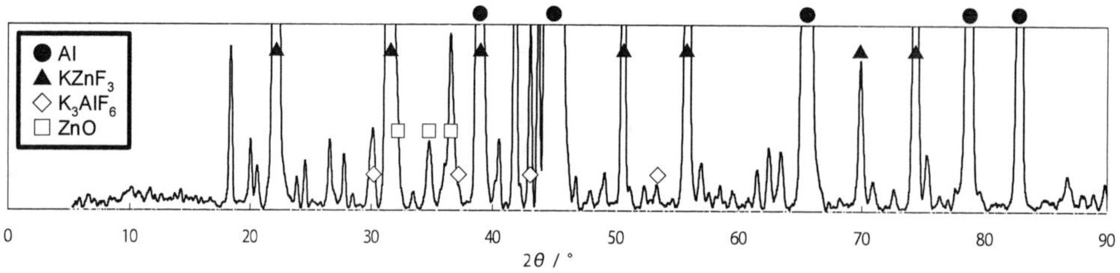

(b) 500°C

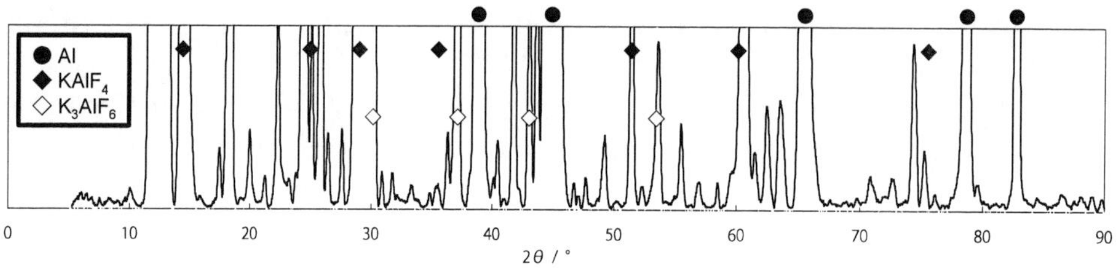

(c) 600°C

Figure 5 XRD patterns of Z100 after brazing at various temperatures

3.1.3 Z50 (NC-flux: 50%, $KZnF_3$: 50%)

Figure 6 shows the appearance of specimens after brazing, figure 7 shows the XRD patterns of Z50 after brazing and figure 8 shows the image of SEM and EDX analysis at the maximum temperature of 500°C. At the maximum temperature of 400°C, the appearance of the specimen after brazing took on white color and powdery that was same as the specimen before brazing. In XRD patterns, $KAlF_4$, $KZnF_3$ and slight ZnO were detected. At the maximum temperature of 500°C, the appearance of the specimen took on slight gray color and powdery. In XRD patterns, $KAlF_4$, K_3AlF_6 and Zn were detected, but $KZnF_3$ and ZnO were not detected. At the maximum temperature of 600°C, the appearance of the specimen was transparent color and melted-solidified as well as Z0 and Z100. It was considered that the flux was melted normally. Below 400°C, it was considered that the chemical reaction eq. (1)-(3) and (5)-(7) occurred as well as Z100. At the maximum temperature of 500°C, ZnO was not detected in the XRD patterns and the spherical Zn was observed by the observation of SEM and EDX analysis. The shape of the spherical Zn indicates that there was the melted Zn. Therefore, it was considered that the eq. (8) occurred. On the other hand, as the reason why $KZnF_3$ was completely consumed at 500°C, it was considered as follows. Firstly the eq. (6) and (7) occurred and ZnO was formed in overall Z50, after that the eq. (8) occurred in overall Z50 as the reaction between ZnO formed in the eq. (6) and (7) and $KAlF_4$, because $KAlF_4$ was included in Z50 as the one of components before brazing. Furthermore, the eq. (5) occurred as the reaction between $KZnF_3$ and Al_2O_3 of both the oxide film and the product by the eq. (8), after that both the eq. (5) and (8) continuously occurred by means of the product of the each equation as the reactant of the each equation respectively. Consequently $KZnF_3$ had completely been consumed. Next, as the reason why the spherical Zn was observed at 500°C, it was considered as follows. Below 562°C, the unmelted $KAlF_4$ and K_3AlF_6 prevented Zn from diffusing into the aluminum base material, therefore the melted Zn was agglomerated spherically. At the maximum temperature of 600°C, $KAlF_4$ and K_3AlF_6 were already melted. Therefore, Zn was diffused into the aluminum base material.

(a) 400°C **(b) 500°C** **(c) 600°C**

Figure 6 Appearance of Z50 after brazing at various temperatures

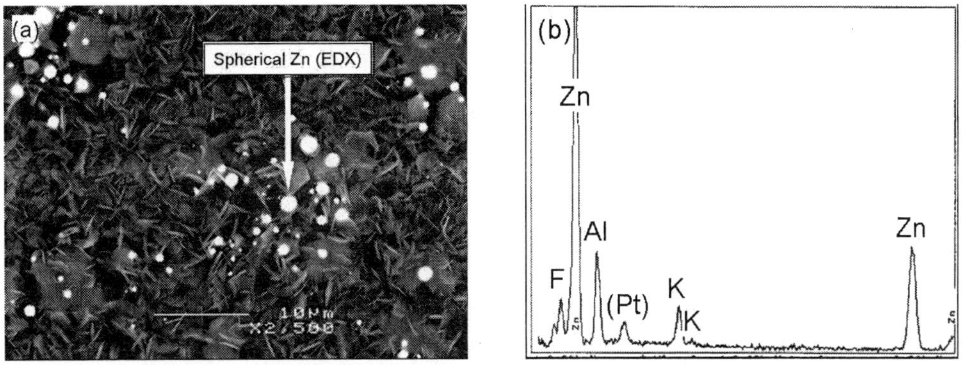

(a) Composition image of SEM micrograph **(b) EDX of spherical Zn**

Figure 7 SEM micrograph and EDX of Z50 after brazing at 500°C

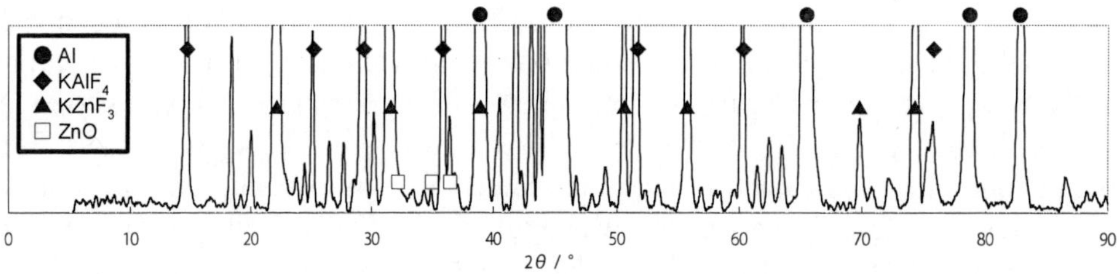

(a) 400°C

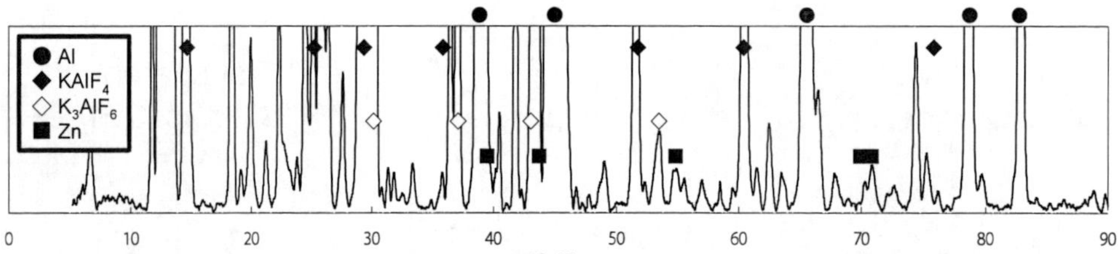

(b) 500°C

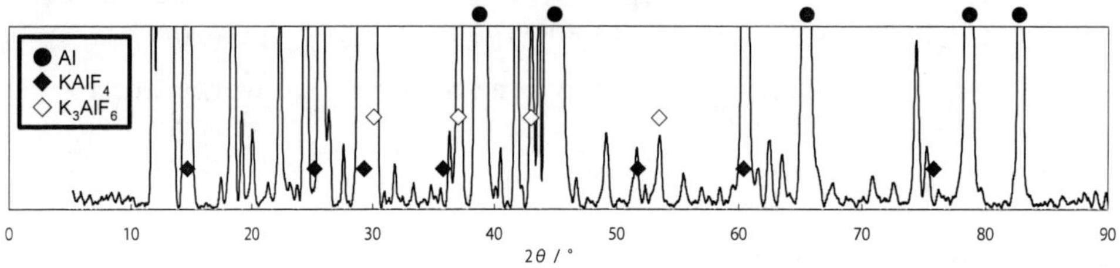

(c) 600°C

Figure 8 XRD patterns of Z100 after brazing at various temperatures

3.2 Additional examination – Validation of chemical reactions

3.2.1 P1 ($KZnF_3$: 50%, Al_2O_3: 50%, Flow gas: Ar)

Figure 9 shows the TG-DTA curves of P1 and $KZnF_3$ and Figure 10 shows the XRD patterns of P1 after TG-DTA. For the validation of the eq. (5), $KZnF_3$ and Al_2O_3 were mixed and heated in the flowing argon gas at the maximum temperature of 388°C, 492°C, 593°C and 638°C. The clear endothermic peak was observed at around 628°C in the DTA curve of P1, and the weight loss started shortly after the onset of heating in the TG curve of both P1 and $KZnF_3$. In the TG curves, the starting temperature of the weight loss of P1 was equal to that of $KZnF_3$. Therefore, it was considered that the weight loss of P1 was originated in $KZnF_3$ as a constitution of P1. But the root cause of the weight loss in such a low temperature was unclear. In XRD patterns at the maximum temperature of 388°C and 492°C, $KZnF_3$, Al_2O_3 and ZnO were detected, and at the maximum temperature or 593°C, $KZnF_3$, K_3AlF_6, Al_2O_3 and ZnO were detected. Furthermore, at the maximum temperature of 638°C, $KAlF_4$, K_3AlF_6 Al_2O_3 and ZnO were detected whereas $KZnF_3$ was not detected. According to these results, it was considered that $KZnF_3$ reacted with Al_2O_3 based on the eq. (5) below 400°C, and it was considered that the reaction of the eq. (5) completed at approx. 600°C.

As the reason why $KAlF_4$ was not detected after heating, it was considered that the eq. (8) occurred in parallel with the eq. (5) by means of $KAlF_4$ and ZnO produced by the eq. (5), consequently $KAlF_4$ had been consumed. Furthermore, as the reason why K_3AlF_6 was not detected at 388°C and 492°C, it was considered that the amount of K_3AlF_6 as the product below 500°C was too small to detect. However, the reason why Zn produced by the eq. (8) was not detected was unclear.

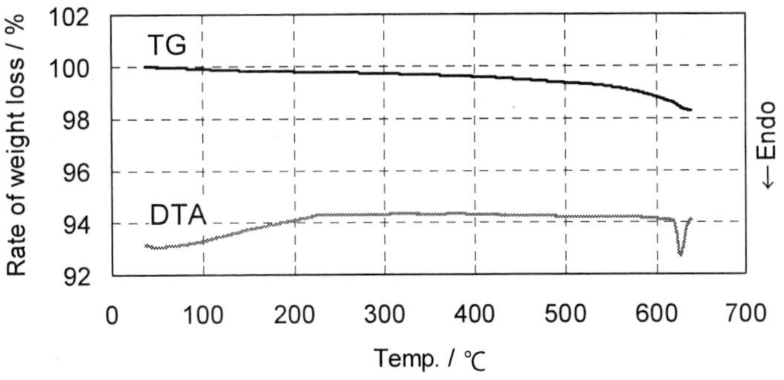

(a) P1

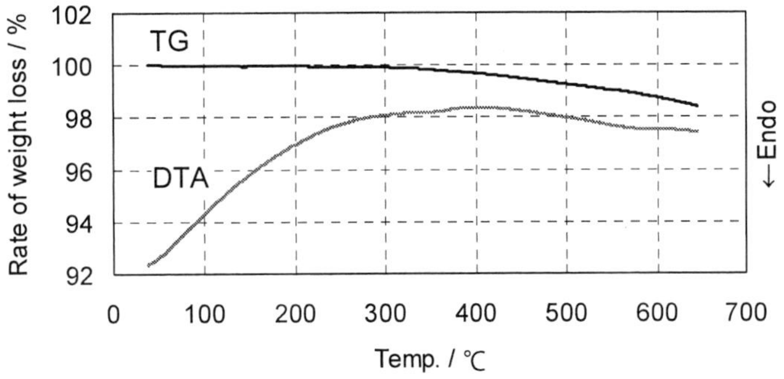

(b) $KZnF_3$

Figure 9 Results of TG-DTA for P1 and $KZnF_3$

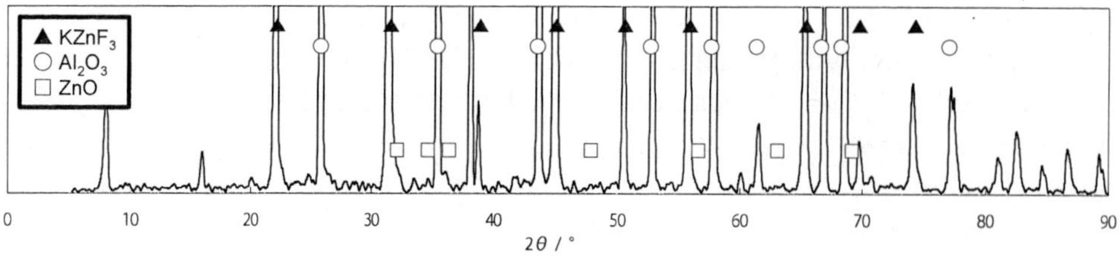

(a) 388°C

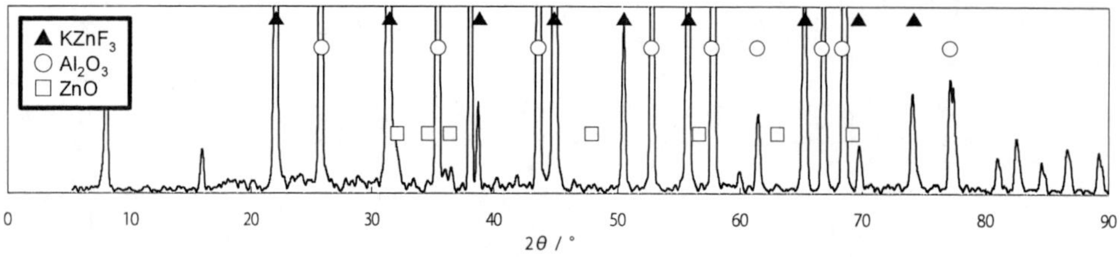

(b) 492°C

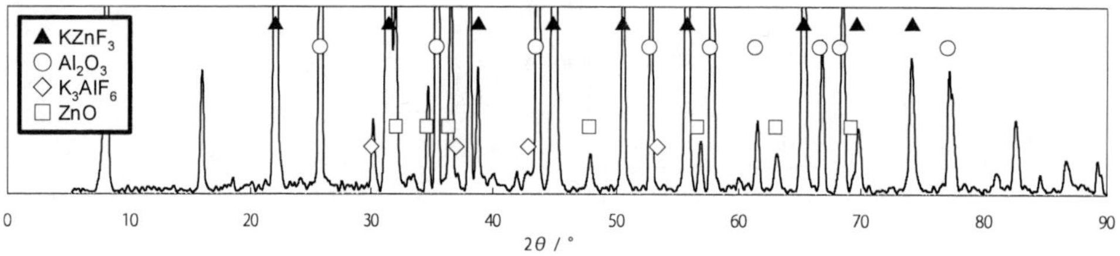

(c) 593°C

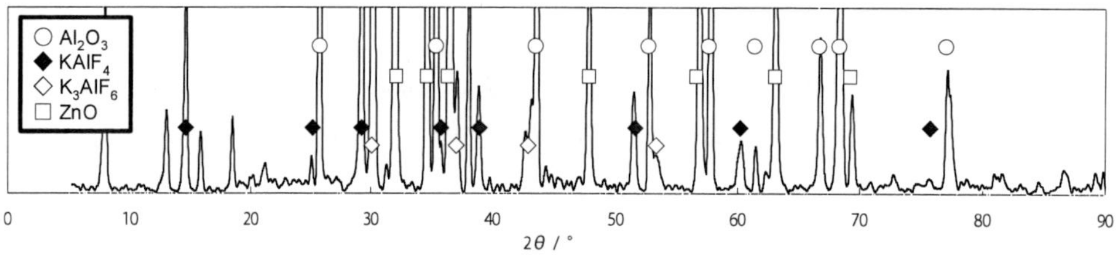

(d) 638°C

Figure 10 XRD patterns of P1 after TG-DTA

3.2.2 P2 ($KZnF_3$: 100%, Flow gas: Atmospheric air)

Figure 11 shows the results of TG-DTA and figure 12 shows the XRD patterns of P2 after TG-DTA. For the validation of the eq. (6) and (7), $KZnF_3$ was heated in the flowing air at the maximum temperature of 388°C, 492°C, 593°C and 639°C. The slight exothermic peak was observed at approx. 600°C in DTA curve, and the weight loss started at approx. 280°C in TG curve. It was similar to the result of Figure 9. In XRD patterns, $KZnF_3$ and ZnO were detected at the maximum temperature of 388°C, 492°C and 593°C. $KZnF_3$, $K_3Zn_2F_7$ and ZnO were detected at the maximum temperature of 639°C. Therefore, in the case of heating $KZnF_3$ in high concentration of O_2 and H_2O, it was considered that the reaction occurred not only the eq. (6) and (7) but also the eq. (9) of the following and especially the eq. (9) occurred at approx. 600°C.

$$KF + 2KZnF_3 \rightarrow K_3Zn_2F_7 \quad (9)$$

Therefore, the eq. (6) and (7) were validated.

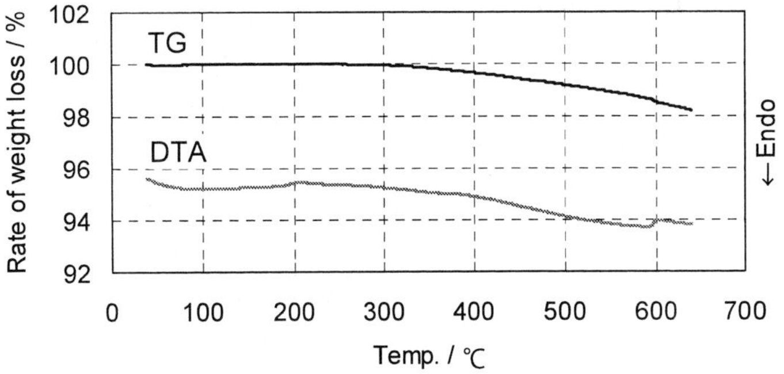

Figure 11 Results of TG-DTA for P2

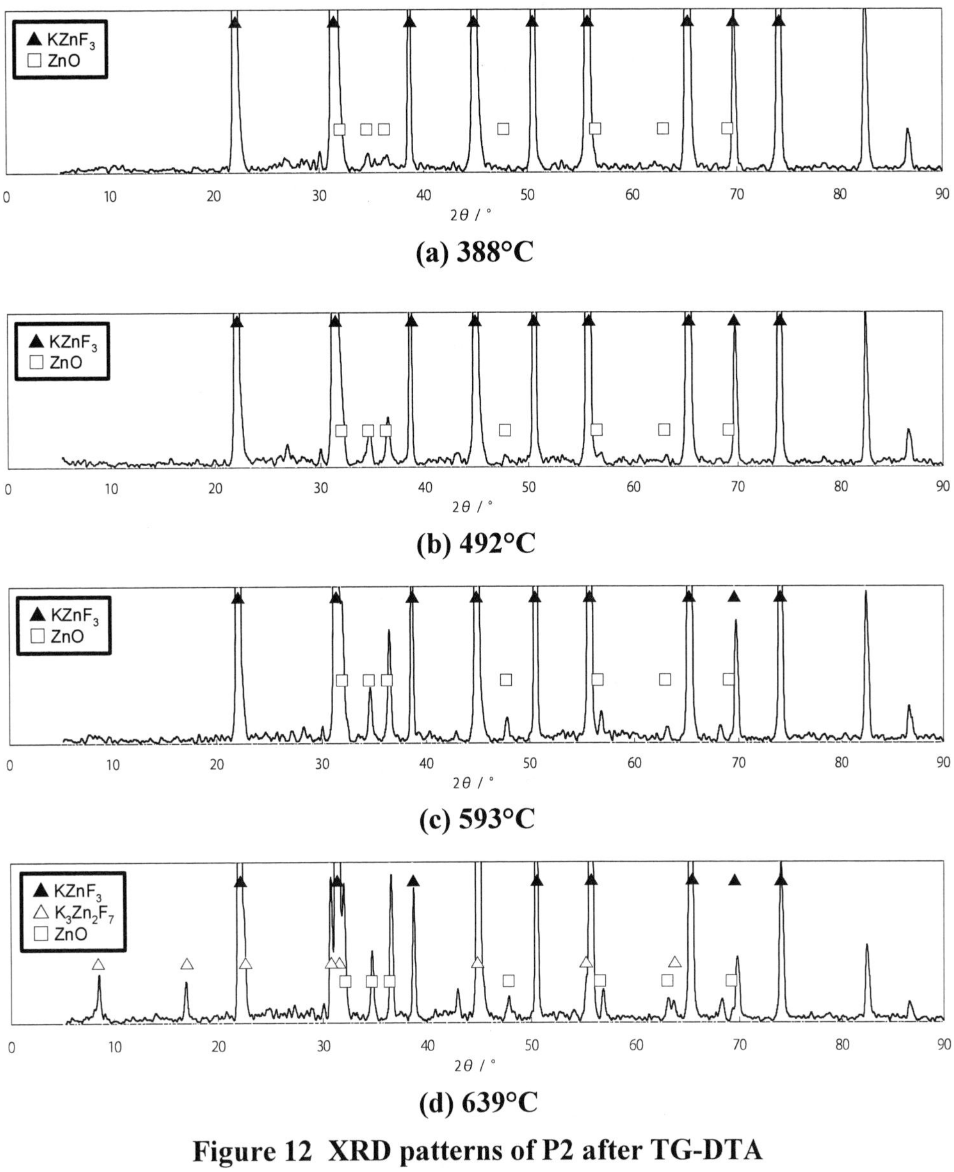

Figure 12 XRD patterns of P2 after TG-DTA

3.2.3 P3 (NC-flux: 50%, ZnO: 50%, Flow gas: Ar)

Figure 13 shows the TG-DTA curves of P3 and NC-flux and Figure 14 shows XRD patterns of P3 after TG-DTA. For the validation of the eq. (8), NC-flux and ZnO were mixed and heated in flowing argon gas at the maximum temperature of 385°C, 489°C, 589°C and 638°C. The clear endothermic peaks were observed at approx. 100°C and approx. 560°C in the DTA curve of both P3 and NC-flux. The clear weight loss was observed at approx. 100°C and above approx. 380°C and approx. 550°C in TG curve of both P3 and NC-flux. As the reason why the endothermic peak and the weight loss were observed at approx. 100°C, it was considered that the vaporization of the hydrated water of NC-Flux. As the reason why the endothermic peak and the weight loss were observed at approx. 560°C, it was considered the melting point of $KAlF_4$. In addition, the relatively slight weight loss was observed above approx. 380°C, it was considered that this weight loss indicated the emission of F gas and the vaporization of $KAlF_4$ [2]. The weight loss of the vaporization of $KAlF_4$ was especially prominent above approx. 550°C, close to the temperature near the melting point of $KAlF_4$. In XRD patterns, $KAlF_4$, K_3AlF_6, Al_2O_3 and ZnO were detected. Therefore, it was considered that the eq. (8) occurred at approx. 380°C in consideration of the weight loss at approx. 380°C. However, the reason why Zn produced by the eq. (8) was not detected was unclear. For further validation in regard to the formation of Zn, it means the reduction of ZnO, the additional brazing test was shown in the next section.

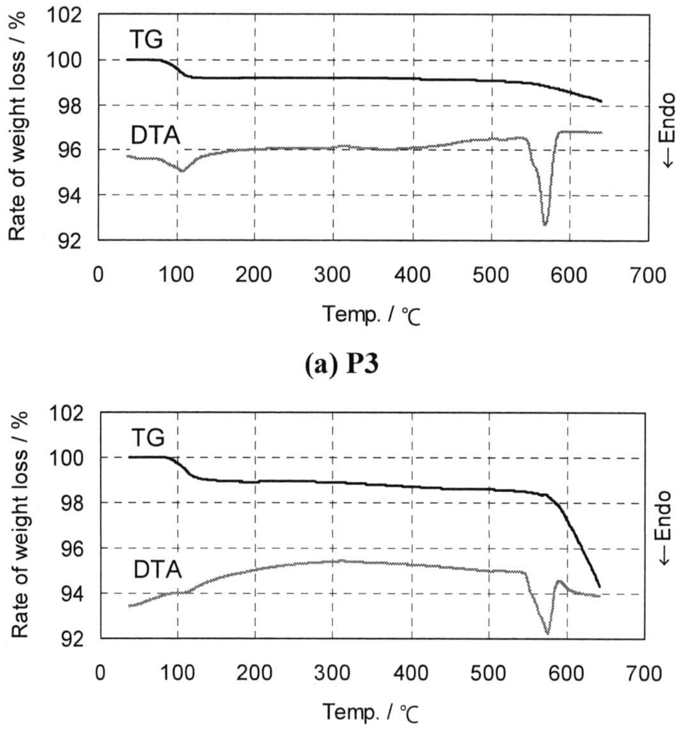

(a) P3

(b) NC-flux

Figure 13 Results of TG-DTA for P3 and NC-flux

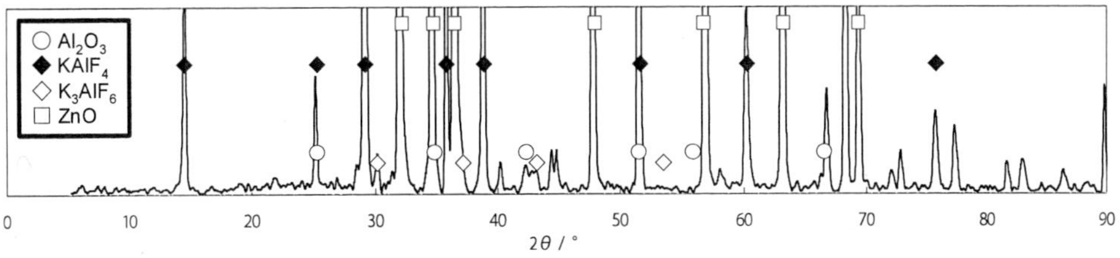

(a) 385°C

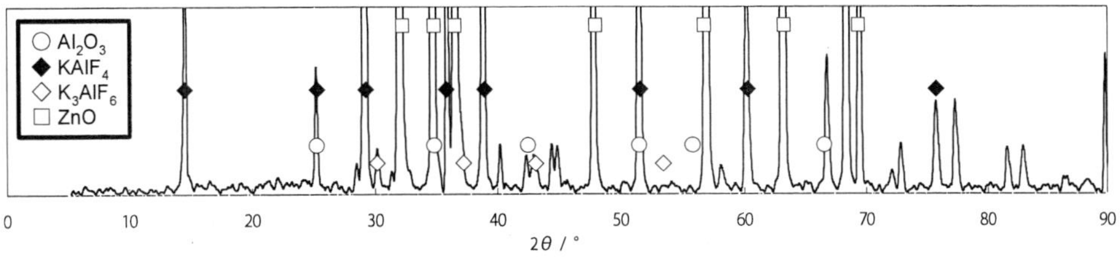

(b) 489°C

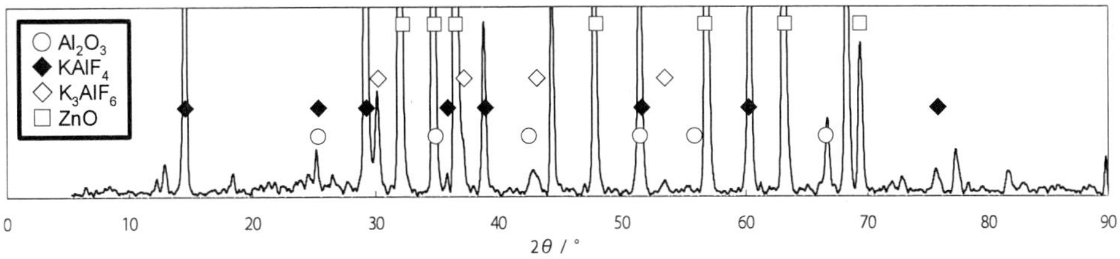

(c) 589°C

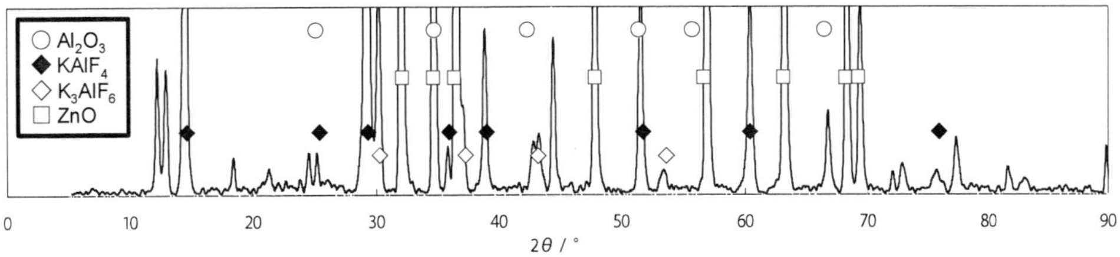

(d) 639°C

Figure 14 XRD patterns of P3 after TG-DTA

3.3 Additional examination - Further validation for reduction of ZnO

Figure 15 shows the appearance of the mixture of NC-flux and ZnO after brazing and figure 16 shows XRD patterns of the mixture after brazing. For further the validation of the eq. (8), the center of dropped flux on the A3003 alloy sheet (25mm×25mm) after brazing was analyzed by XRD. The black residue was observed on the dropped position of the specimens after brazing. In XRD patterns, $KAlF_4$, K_3AlF_6, Zn and ZnO were detected. It was considered that the black residue was the agglomerated Zn shown in Figure 7, because K_3AlF_6 prevented Zn from

Figure 15 Appearance of NC-flux+ZnO after brazing

diffusing into aluminum base material. Therefore, the eq. (8) was validated. However, the reason why Zn was not detected on P3 specimens after TG-DTA whereas Zn was detected after this brazing test was unclear.

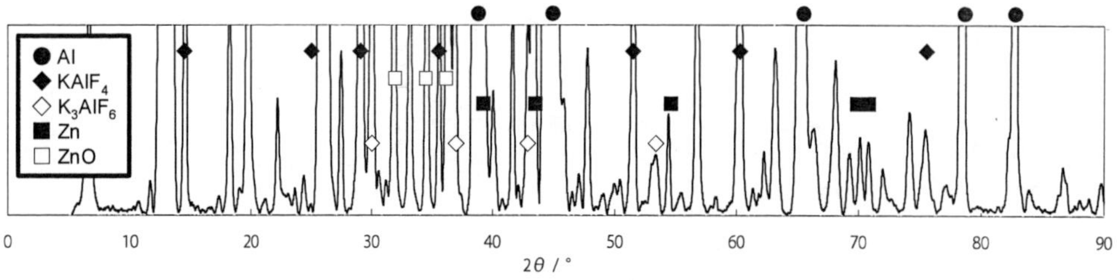

Figure 16 XRD patterns of NC-flux+ZnO after brazing

4 CONCLUSION

The reaction behavior of NC-flux, $KZnF_3$ and the mixture of them during brazing were investigated.

(1) In regard to NC-flux, K_3AlF_6 increased with the elevation of the temperature during brazing.

(2) In regard to $KZnF_3$, $KZnF_3$ reacted with oxide film on the aluminum base material and H_2O in brazing atmosphere, subsequently ZnO was formed below 500°C. Finally, Zn was formed due to the reduction of ZnO by $KAlF_4$.

(3) In regard to the mixture of NC-flux and $KZnF_3$, $KZnF_3$ reacted with oxide film on the aluminum base material and H_2O in brazing atmosphere, after that ZnO was formed below 400°C. At the temperature in the range of 400-500°C, Zn and Al_2O_3 were formed from the reductive reaction of ZnO by $KAlF_4$. These reactions occurred not only at the surface of the aluminum base material but also overall the specimen. Furthermore, Al_2O_3 formed by this reaction was used as the source of the reaction with $KZnF_3$. Consequently, $KZnF_3$ had completely been consumed in the range of 400-500°C. Finally, Zn was formed due to the reduction of ZnO by $KAlF_4$.

5 REFERENCES

[1] Thomson W. T., Goad D. W. G., Can. J. Chem., *1976, Vol. 54, p3342–3349*

[2] Takemoto T., Matsunawa A., Kitagawa A., Journal of Materials Science Letters, *1996, Vol. 15, p301–303*

[3] Juan P. G., Swidersky H. W., Schwarze T. and Eicher J., Inorganic Fluoride Materials from Solvay Fluor and their Industrial Applications, *2010, p210-211*

Recent work on low melting clad for heat exchanger production

A SCHLEGEL, B JACOBY, S KIRKHAM and **A BUERGER**
Aleris Rolled Products Germany GmbH, Germany

ABSTRACT

This paper summarises previous research activities into potential low melting clad alloys for automotive Heat Exchanger (HEx) applications, produced via Controlled Atmosphere Brazing (CAB) techniques, and also presents recent research work to support the use of such materials for mass production. Historically, several theoretical & experimental investigations on low melt cladding alloys with liquidus temperatures of below 550°C have been presented, but the practical application of these brazing approaches have been limited in usefulness due to certain material specific drawbacks (e.g. influences on corrosion performance). However the new work presented in this paper suggests there is a viable 580°C brazing methodology and material approach, suitable for modern HEx designs. Laboratory scale clad brazing material with liquidus temperatures of approx. 580°C have been produced successfully and investigated in terms of brazeabillity, corrosion resistance and strength. These results are promising and support future research and development.

1 INTRODUCTION

A literature search [1-6] was conducted to collect and collate possible low melting clad (LMC) alloy approaches from previous global research & development work for brazing around 550°C. Typical material systems for 550°C brazing require relatively high levels of Cu and/or Zn, for fluxed or flux-free brazing, and also introduce substantial material recycling issues which are difficult to support in the price sensitive Hex material market. This general lack of a viable production solution was perceived by Aleris as a technology gap for the wider industry and has driven the development of a new alloy. Customer surveys revealed that brazing at 580°C is also of interest, therefore a second assessment of the research situation focused on LMC alloys for brazing at 580°C [7-8]. R&D effort was therefore focused on investigations related to the feasibility of Aluminum brazing sheet using LMC alloys at brazing temperatures of 550°C and 580°C respectively.

2 OVERVIEW OF POSSIBLE LOW MELTING BRAZE ALLOY OPTIONS FOR BRAZING AT 550°C

Literature surveys [3-6] on Aluminum LMC alloys identified possible approaches for an alloy development with a liquidus temperature below 550°C. Therein the use of melting point depressants (for binary Al-based systems: Ag (566°C), Cu (548°C), Ge (424°C), Zn (382°C), Mg (451°C) and Si (577°C)); these investigations stated limitations due to relative properties, e.g. cost (Ag, Ge), possible formation of brittle phases (Ag, Cu, Mg), electro-chemical potential (Zn, Cu) and limited compatibility to Nocolok® type fluxes (e.g. Mg). Further promising Al-based systems were reviewed containing combinations of Si, Ni, Cu, Zn, Bi, Mg, Sr, Ca and Li. Systems investigated are as follows:

Table 1: Alloy systems for low melting clad.

Binary systems	Ternary systems
Al-Si	Al-Si-Mg
Al-Cu	Al-Cu-Si Al-Cu-Mg
Al-Ge Ge-Si	Al-Ge-Si Al-Ge-Mg
Al-Zn	Al-Si-Zn Al-Mg-Zn
Al-Ag	Al-Ag-Si Al-Ag-Mg Al-Ag-Cu
	Al-Si-Ni Al-Cu-Ni
	Al-Si-Sn

Table 2: Proposed alloy compositions and calculated melting ranges.

Alloy composition [mass-%]	Solidus temperature [°C]	Liquidus temperature [°C]
Al-30Cu-1.5Zn-1Si	521	550
Al-40Zn-5Si	483	537
Al-10Cu-10Zn-8Si	494	549
Al-20Cu-15Zn-5Si	471	520

After considerable theoretical analysis, research was focused on AlZnSi and AlCuZnSi systems, and relevant laboratory based investigations were set-up. This approach was based primarily on thermodynamic calculations. Due to reasons of costs, availability, processing, recyclability and brazeability in the CAB process with flux, it was decided that no investigations of brazing at or below 550°C were progressed at this point.

3 OVERVIEW OF POSSIBLE LOW MELTING BRAZE CLAD ALLOY OPTIONS FOR BRAZING AT 580°C

Thermodynamic calculations using JMatPro (a materials property simulation package, release version 4.0) and Factsage (version 5.5, Thermfact and GTT-Technologies) were used to propose certain chemistry ranges for an LMC alloy for brazing at or below 580°C. The following restrictions and targets were defined, based on Aleris industry experience and direct customer interviews – in order of priority:

- Similar corrosion potential compared to standard braze clad materials (e.g. 4045).
- Solidus temperature at 560°C – just below melting range of Nocolok® type flux (i.e. 565 - 575°C).
- No use of expensive or difficult to obtain / recycle alloying elements (e.g. conflict minerals).
- Limited amounts of Cu, Zn, and Mg.

Based on the binary eutectic of 12 wt% Si and Al in balance at 577°C, the optimum Si content was calculated for quasi binary systems of AlCuSi and AlSiZn. Within the binary AlCu system, the solidus reduces to 548°C. A melting range in that area can only be achieved at more than 25 wt% Cu. In Figure 1, the effect on the AlCu system of increasing Si levels is shown. To decrease solidus and melting range to reasonable levels, at reasonably low Cu levels, 12 wt% Si is needed.

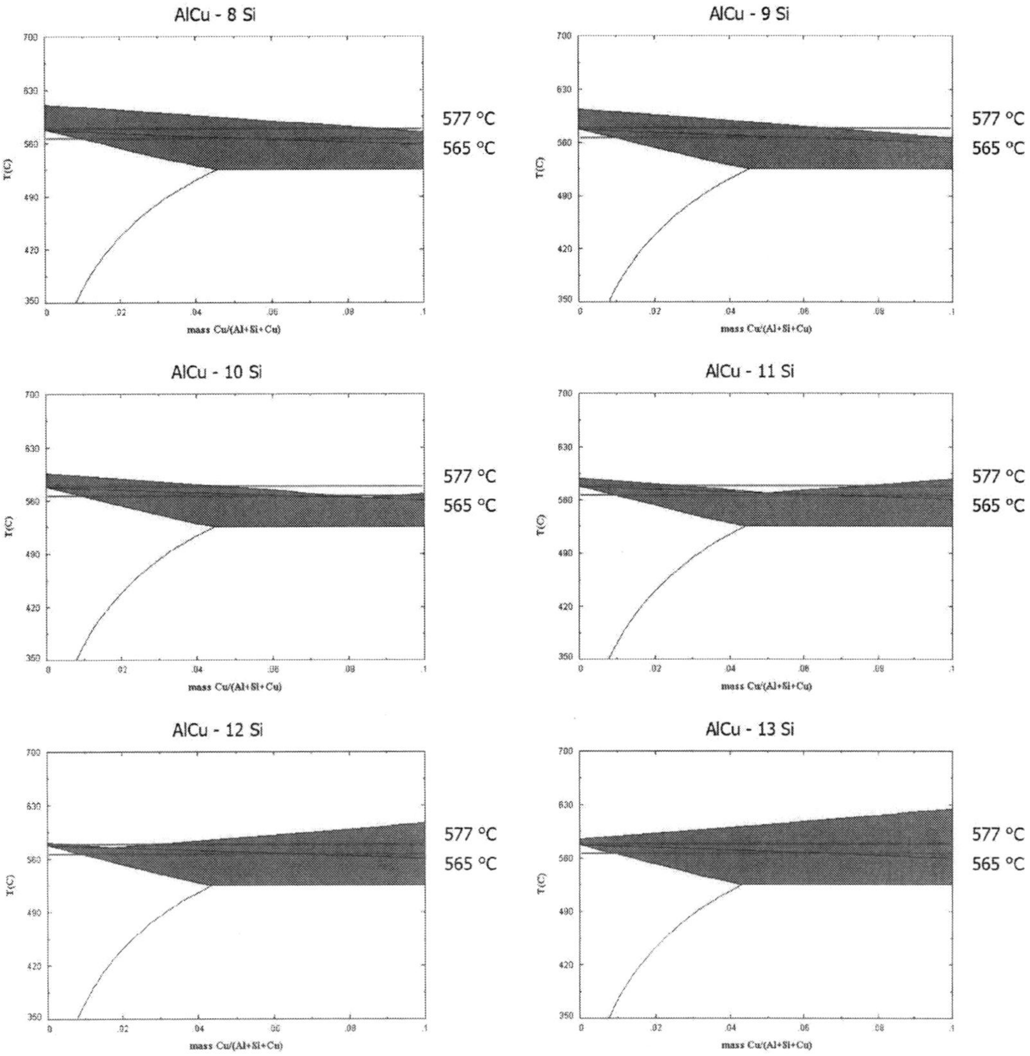

Figure 1: Calculation of quasi binary AlCuSi system from 0 to 10 wt% Cu with different Si levels. Red lines indicate solidus of flux (565°C) and the eutectic melting temperature of AlSi (577°C). The blue areas show the composition / temperature area where the clad is partially molten.

Within the AlZn system, solidus and liquidus temperatures can be significantly lowered by a Zn content beyond 40 wt%. Calculations of the quasi binary AlSiZn system showed a sufficient decrease of solidus and melting range at acceptable low levels of Zn in combination with approximately 12 wt% Si (Figure 2).

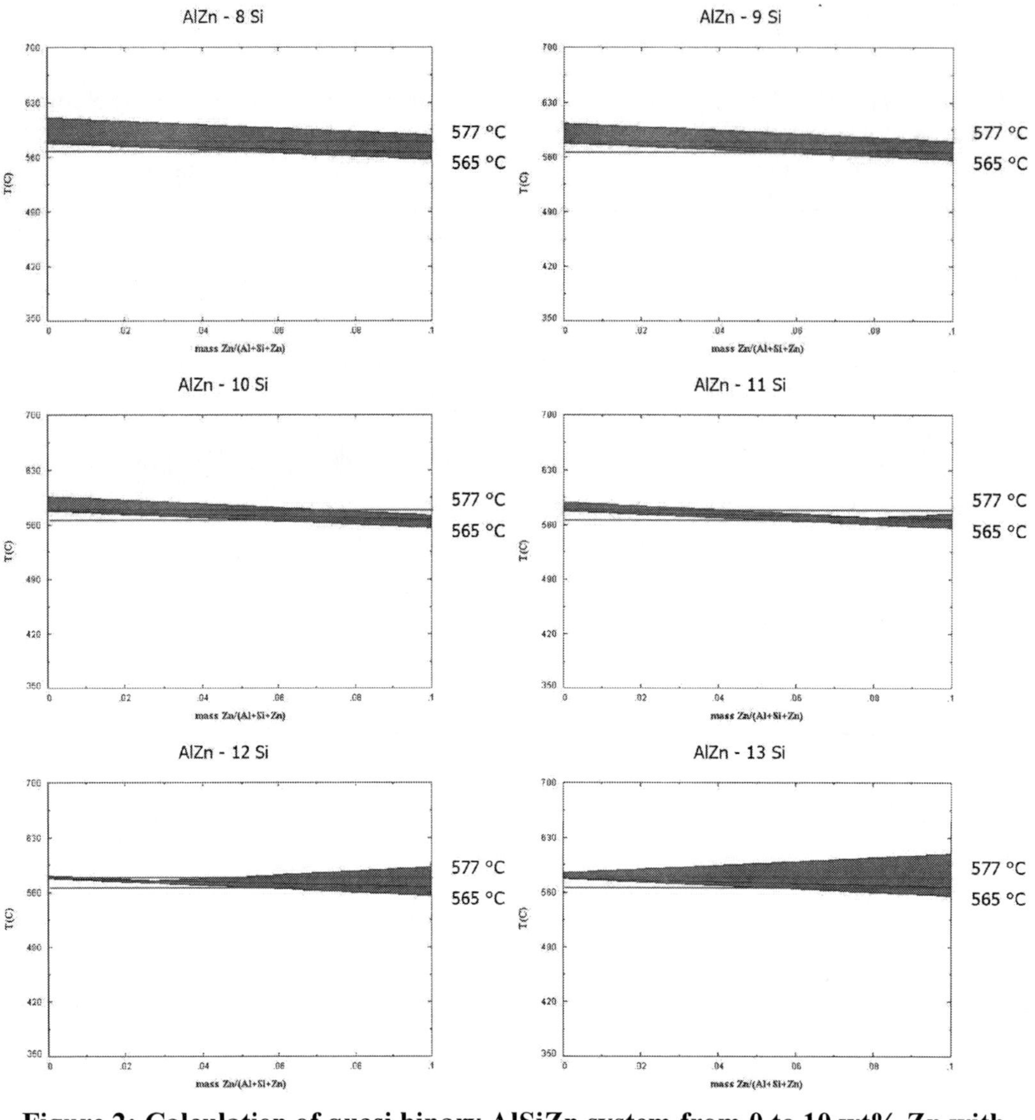

Figure 2: Calculation of quasi binary AlSiZn system from 0 to 10 wt% Zn with different Si levels. Red lines indicate solidus of flux (565°C) and the eutectic melting temperature of AlSi (577°C). The blue areas show the composition / temperature area where the clad is partially molten.

Thermodynamic calculations of the quaternary phase diagram at isothermal temperatures (Figure 3) predicted a chemistry of approximately 10 to 15% Si, 1 to 3% Cu, 0.5 to 2% Zn and Al in balance. Due to the fact, that the considered Cu and Zn levels are in solid solution and therefore affecting the corrosion potential, they will lead to galvanic corrosion. To keep the net contribution of Cu and Zn additions to a corrosion potential difference of zero, the Cu / Zn ratio was investigated and fixed. Further Fe and Mg were added to the calculations to assess their influence in lowering solidus and liquidus temperatures within the modelled chemistry range. The melting ranges of five different braze alloys, including a Flux-free variant were calculated (Table 3) and verified by DSC measurements revealing solidus temperatures of 541 to 558°C and liquidus temperatures of 579 to 583°C (Table 4).

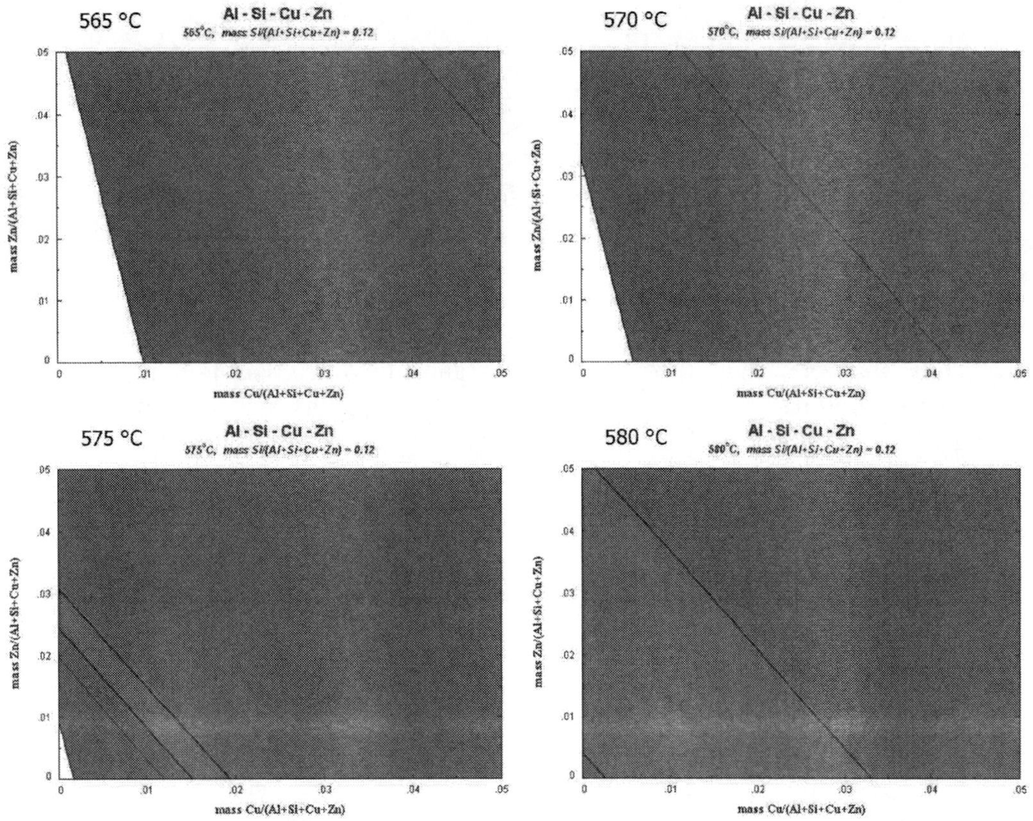

Figure 3: Phase diagrams of the AlSiCuZn system at different temperatures. Si content in all cases 12 wt%, horizontal axis wt% Cu (max 5 wt%) and vertical axis wt% Zn (max 5 wt%). Area colors: white – completely solid, blue – partially liquid, red – completely liquid.

Table 3: Proposed compositions based on the calculations for a LMC alloy. Alloy 6(*) was added as a Flux-free variant of alloy 5.

Alloy	Si	Fe	Cu	Zn	Mg	Bi	T_{sol}	T_{liq}
1	12	0.25	1.0	0.55	-	-	562	575
2	12	0.25	1.25	0.69	-	-	559	574
3	12	0.25	1.5	0.83	-	-	555	573
4	12	0.25	0.75	0.42	0.2	-	548	572
5	12	0.25	1.25	0.69	0.2	-	551	573
6*	12	0.25	1.56	0.86	0.25	0.1		

Table 4: Composition of casted alloys and DSC measurements on melting range.

Alloy	Si	Fe	Cu	Zn	Mg	Bi	T_{onset}	T_{peak}	$T_{calc.\ sol.}$
A1	11.9	0.23	1.03	0.71	-	-	558	579	561
A2	11.9	0.23	1.29	0.89	-	-	555	583	558
A3	12.0	0.19	1.55	1.05	-	-	552	579	554
A4	11.9	0.21	0.79	0.52	0.20	-	555	580	558
A5	12.2	0.14	1.29	0.89	0.20	-	548	580	550
A6	14.7	0.19	1.62	0.89	0.20	0.083	541	580	543

Based on these calculations, several braze alloys were cast, hot rolled to 4mm, including two Flux-free variants. Braze alloys were roll bonded to the Aleris core alloy 3527 (composition in wt%: 0.45-0.65 Si, 0.34-0.44 Fe, 0.6-1.0 Cu, 0.9-1.5 Mn, 0.1 Mg, 0.05-0.15 Cr and Al in balance) and further rolled down to 0.3mm.

The brazeability of these systems were then tested in a Glass tube furnace under Nitrogen atmosphere using fluxed T-samples and alloy 3003 as joining partner at different temperatures. The successfully processed laboratory material showed good brazing with good fillets at 575 to 580°C. SWAAT samples were fluxed and brazed at 580°C with a test duration of 40 days.

4 RESULTS

Alloys A1, A2, A3 and A4 were produced as two layer systems, clad alloys A5 and A6 were also processed as three layer systems. The following results are of Aleris alloy 3527 cladded with alloys A1 and A3 from Table 4 only.

4.1 Brazeability

T-samples of both variants were fluxed and brazed in a glass tube brazing furnace under Nitrogen atmosphere at different temperatures. For this work, a standard brazing cycle was used, with varying soak times. The fillets were assessed visually, results are given in Table 5. Both alloys show good brazing at 575 to 580°C. As expected clad alloy A3 showed slightly better brazeability than A1 due to its lower melting temperature.

Table 5: Visual assessment of brazing trials.

Brazing temp. [°C]	Clad alloy A1	Clad alloy A3
560	no fillet	no fillet
565	no fillet	no fillet
570	thin fillet	no fillet
573	no fillet	thin fillet
571	no fillet	thin fillet
575	thin fillet	fillet
575	fillet	fillet
579	nice fillet	nice fillet

Cross sections of the fillets are shown in Figures 4 and 5. The fillets look comparable with standard fluxed fillets, although some shrinkage is evident, which is not unusual for standard CAB materials. Regarding the brazing temperature, no differences were noted.

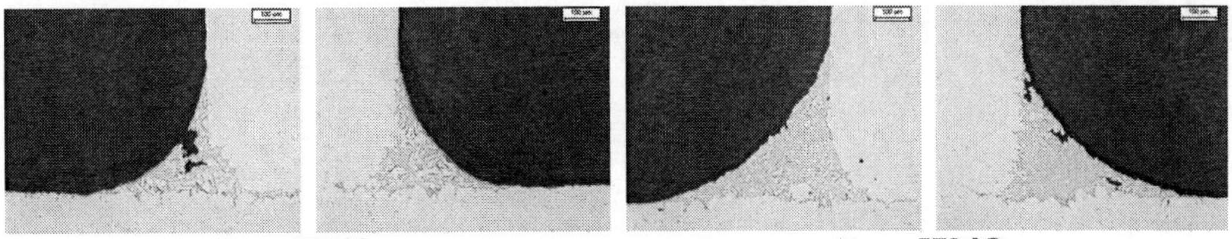

T_{braz} = 575 °C T_{braz} = 579 °C

Figure 4: Fillet formation of brazed T-sample of clad alloy A1.

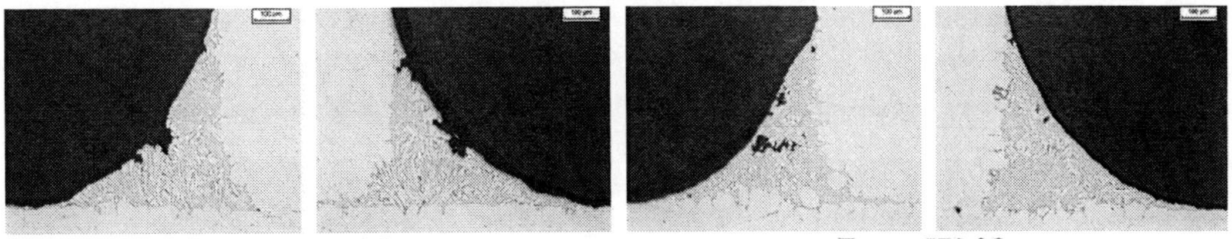

T_{braz} = 575 °C T_{braz} = 579 °C

Figure 5: Fillet formation of brazed T-sample of clad alloy A3.

4.2 Corrosion tests and mechanical properties

Coupons of both clad materials were made and brazed under Nitrogen atmosphere at 580°C. SWAAT testing was monitored and stopped after 40 days. The results are given in Table 6 below.

Table 6: Results of SWAAT in days till first perforation.

Clad alloy A1	Days	Clad alloy A3	Days
1	14.9	1	27.8
2	32.8	2	28.8
3	39.9	3	20.9
4	28.8	4	35.0
5	26.8	5	31.8
average	*28.7*	*average*	*28.9*
std. deviation	*9.2*	*std. deviation*	*5.2*

Cross-sections of corroded SWAAT samples are given in Figure 6. Corrosion starts at eutectic regions in the re-solidified clad and spreads lateral into the core-clad diffusion zone. Despite the fact that the Cu depletion profiles will be limited with this new LMC approach, some preferential attack of the diffusion area occurs. This can be attributed to a Zn enriched layer at the surface, originating from Zn diffusion from the clad into the core during brazing. There is no clear difference between the two different alloys in corrosion morphology.

Post-braze tensile properties were measured and results are given in Table 7. Compared to Aleris core alloy 3527 clad with a typical standard braze clad, the post-braze strength is considerably higher when clad with either alloy A1 or A3.

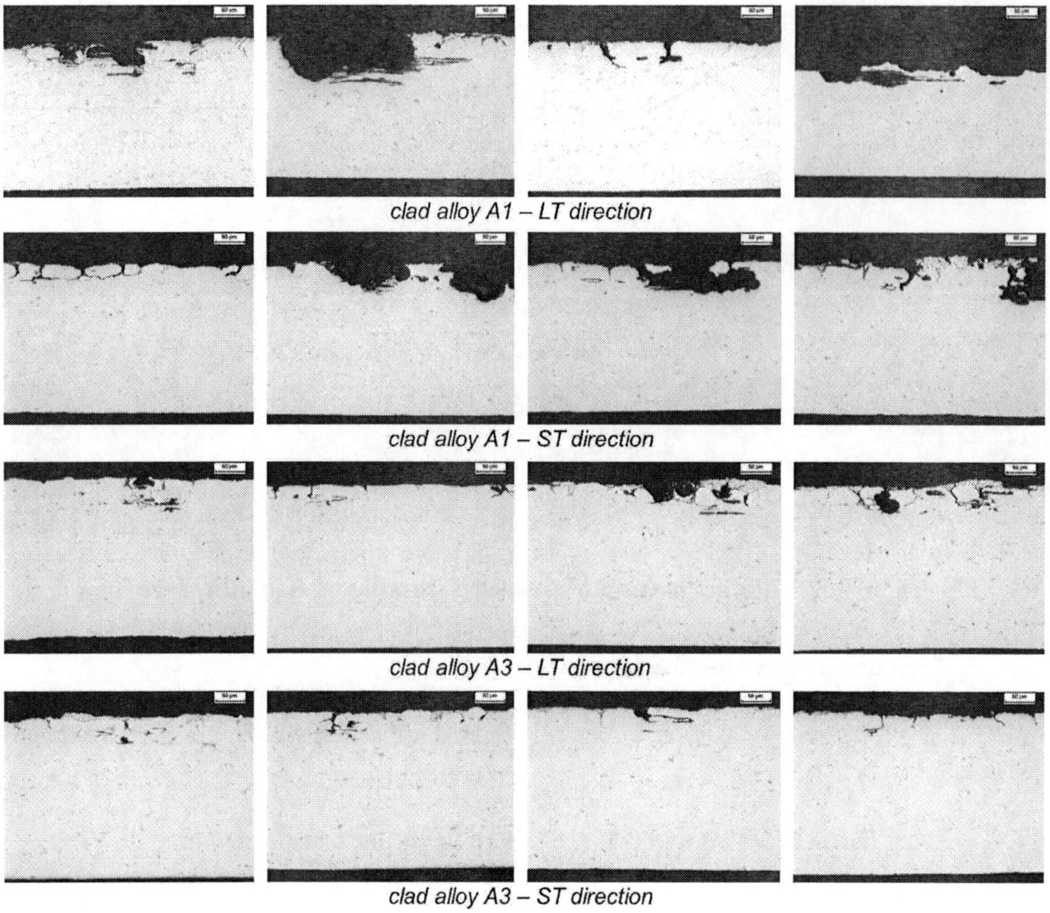

Figure 6: Cross sections of corroded samples – exposure of 40 days in SWAAT.

Table 7: Post braze mechanical properties of Aleris core alloy 3527.

Clad alloy	R_p [MPa]	R_m [MPa]	A [%]
A1	86	168	10.7
A3	83	172	11.5
EN-AW 4045	58	151	-

5 DISCUSSION

Two interesting results from this investigation are the tensile and corrosion properties. The aim of the LMC development was that these properties would not deteriorate, the actual improvements seen were greater than expected.

In addition to the main target of this development, i.e. lowering solidus temperatures, the LMC alloys were designed to be neutral in terms of corrosion potential by balancing Zn and Cu. The Zn content was elevated to compensate for diffusion losses into the core based on theoretical calculations of diffusion profiles. Calculations were maintained below the solidus temperature of the braze alloy. The ability of the molten clad to dissolve Zn and Cu is much higher compared to solid state and the driving force for diffusion is therefore different. Further investigations are needed to understand the development of Cu and Zn profiles and surface distributions during brazing and their effects on material corrosion performance.

We suggest the positive effect of LMC alloys on strength is possibly caused by the following:

1. Cu and Zn in the clad increase the strength of the re-solidified clad, contributing to the overall strength of the brazing sheet material.
2. In addition to this the core strength after brazing is enhanced due to the fact that Cu depletion out of the core is prevented.
3. The brazing sheet material has been brazed at lower temperature and consequently the overall brazing cycle was shortened. This could affect microstructure, e.g. dissolution / precipitation of dispersoids and precipitates, and thus altering the mechanical properties.

6 CONCLUSIONS

Two low melting clad braze alloy variants have been investigated based on calculated and experimentally verified chemistry ranges. Materials were processed in laboratory scale and tested successfully. A decrease of the brazing temperature was achieved of about 15 to 20°C, down to the targeted 580°C. This could enable HEx manufacturers to save time and energy costs. Possible, beneficial effects of LMC alloys for CAB Hex materials are an enhanced SWAAT performance and an increase of post-braze mechanical properties.. Further investigations on laboratory and pilot scale materials will follow to support first production testing at brazing temperatures of 580°C or lower.

7 REFERENCE LIST

[1] Gray A.: Low Temperature Brazing for Aluminium Heat Exchangers – Fact or Fiction (2005) *AFC-HOLCROFT 10th Annual Invitational Aluminium Brazing Seminar.*

[2] Tsao L. C., Chiang M. J., Lin W. H., Cheng M. D., Chuang T.H.: Effects of zinc additions on the microstructure and melting temperatures of Al-Si-Cu filler metals (2002) *Materials Characterization* **48** 341-346

[3] Quadakkers W. J.: Konstitution von Aluminiumbasissystemen – Grundlage für Verfahren und Lotentwicklung zum Hartlöten von Aluminiumwerkstoffen (1981) *Technisch wissenschaftliche Berichte, Lehr- u. Forschungsgebiet Werkstoffwissenschaften der Rheinisch Westfälischen Technischen Hochschule Aachen, Band 2* Phd thesis RWTH Aachen

[4] Kötzing B.: Entwicklung niedrigschmelzender Aluminiumbasislote zum Fügen höherlegierter Aluminiumwerkstoffe (1997) *Werkstoffwissenschaftliche Schriftenreihe, Band 15*, Phd thesis RWTH Aachen

[5] Humpston G., Sangha S. P. S., Jacobson D. M.: New filler metals and processes for fluxless brazing of aluminium engineering alloys (1995) *Materials Science and Technology, November 1995*, **Vol. 11**. 1161-1167

[6] Jacobson D. M., Humpston G., Sangha S. P. S.: A New Low-Melting-Point Aluminium Braze (1996) *Welding Research Supplement, August 1996*, 243-250
[7] EP 0637481 B1: The Furukawa Electric Co. Ltd., Nippon Denso Co. Ltd., Aluminum alloy brazing material and brazing sheet for heat-exchangers and method for fabricating aluminium alloy heat exchangers (1994)
[8] JP 06-182582: Mitsubishi Alum Co. Ltd., Aluminum alloy brazing filler metal for brazing heat exchanger and aluminum alloy brazing sheet for heat exchanger (1992)

Development of continuous cast aluminum fin stock for a smaller, lighter, and more efficient radiator

T KOKUBO and **T ANAMI**
Nippon Light Metal Co., Ltd, Japan

H TERAMOTO, S TESHIMA and **T TOYAMA**
DENSO Corporation, Japan

ABSTRACT

A new radiator that is smaller, lighter, and more efficient than previous radiators was developed. To achieve this progress, a new aluminum fin stock was developed from a Al-Mn-Si-Fe-Zn alloy using twin-belt-type strip caster. The developed aluminum alloy fin stock exhibits high strength after brazing; the post-brazed ultimate tensile strength is 145 MPa. Furthermore, the new alloy has better brazeability than previous alloys, enabling the thinnest radiator fin stock achieved to date. This study investigates the mechanism underlying the alloy's high strength and good brazeability. The improved properties strongly depend on the intermetallic compound distribution.

1 INTRODUCTION

With the aim of protecting the global environment, researchers have sought more efficient vehicle designs in recent years. One effective technology is reducing the weight of the aluminum heat exchanger. The size and weight of radiators can be heavily reduced by down-gauging the fin stock, one of the main radiator components [1]. The fins of conventional radiators are typically 50–60 μm thick [2, 3]. While manufacturing thinner fins would attain the required weight reduction, down-gauging of fin stock degrades the radiator's strength and brazeability [1]. Therefore, to realize thinner fin applications, it is necessary to increase the post-brazing strength and improve the brazeability of the fin stock alloy.

The microstructure of aluminum alloy products is radically altered by the casting process. Previous studies have reported the application of direct chill (DC) casting and strip casting to automotive fin stock development [2, 3, 4]. Strip casting is a more attractive process than conventional DC casting because of the higher cooling rate during solidification. In automotive panel sheet development, the material properties have been improved by a strip casting process known as the twin-belt casting process. Therefore, the twin-belt casting process is expected to increase the strength and brazeablity of fin stock.

The authors have developed a new Al-Mn-Si-Fe-Zn alloy fin stock that exhibits high post-brazing strength and excellent brazeability even at small thickness. The developed Al-Mn alloy fin stock is reported in this paper.

2 ADVANTAGES OF TWIN-BELT CASTING

Twin-belt casting is a strip casting technology that transforms liquid aluminum into a rollable cast ingot. Figure 1 compares the process routes of twin-belt casting and conventional DC casting. Twin-belt casting is energy-conserving because it requires no high-temperature heating operations, homogenization and hot rolling. Additionally, it achieves higher productivity than twin-roll casting [5].

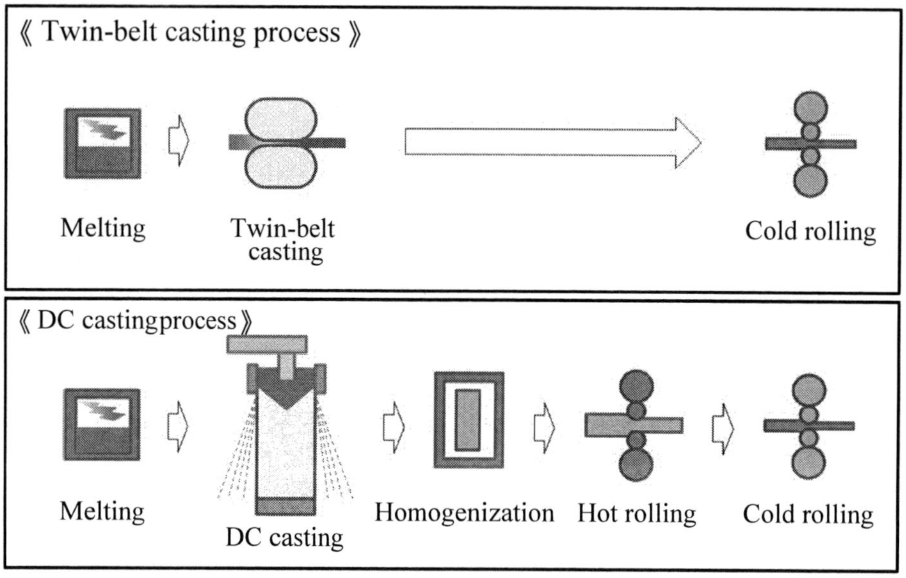

Fig. 1 The process routes for the twin belt casting and DC casting.

Figure 2 is a schematic of a twin-belt caster. The thin gauge strip of this caster was cast by continuously supplying molten metal to the narrow space between two belts. Rapid solidification forms fine intermetallic compounds in the strips, which have improved the strength and formability of Al-Mg alloy automotive panel sheets [6]. Characteristic microstructures such as supersaturated solid solutions can also improve the properties of aluminum alloy products.

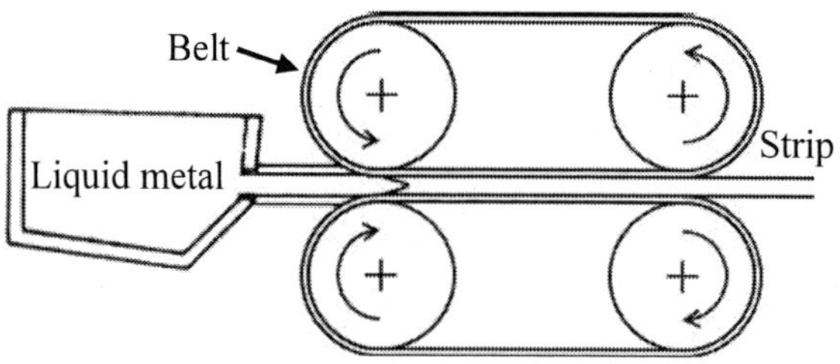

Fig. 2 Schematic of twin-belt caster.

3 EFFECT OF MN/FE RATIO ON TENSILE STRENGTH AFTER BRAZING HEAT TREATMENT

Fin stocks produced by strip casting and composed of Al-Fe alloys such as Al-Fe-Si-Mn-Zn and Al-Fe-Ni-Si-Zn have been installed in radiators [2, 3, 7]. Although thinner fin stock would reduce the weight of automotive heat exchangers, weakening and consequent failure of the heat exchanger is a major concern. Nickel reportedly increases the strength of Al-Fe alloy fin stock [3]; however, it also degrades the corrosion resistance [4]. In this study, the characteristics of Mn were exploited to improve both strength and corrosion resistance. Figure 3 shows the relationship between tensile strength after brazing heat treatment and Mn/Fe ratio in Al-Mn-Fe-Si-Zn fin stocks. The post-brazing tensile strength increases with increasing Mn/Fe ratio, saturating at higher Mn/Fe ratios. The Mn/Fe ratio is important because the Mn content strongly influences the solidified microstructure. According to one report, excess Mn addition tends to forms large intermetallic compound particles during solidification, which are detrimental [8].

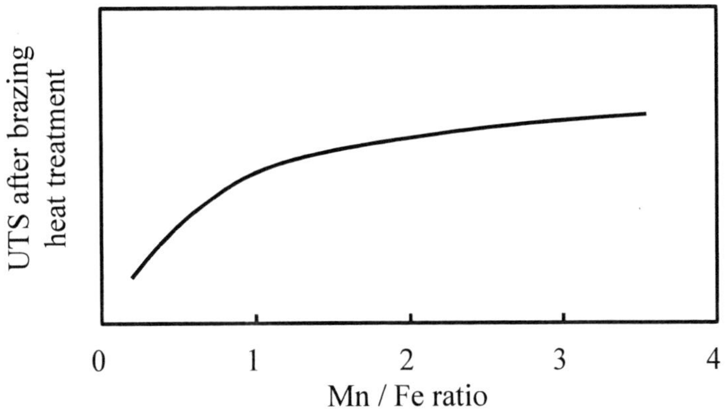

Fig. 3 Relationship between tensile strength after brazing heat treatment and Mn/Fe ratio.

4 PROPERTIES OF THE NEWLY DEVELOPED FIN STOCK

Table 1 shows the chemical composition of the specimens in this study. The newly developed Al-Mn alloy and conventional Al-Fe alloy differ mainly in terms of their Fe and Mn contents. The Mn/Fe ratio in the newly developed Al-Mn alloy is 1.3. The twin-belt cast strips were directly cold-rolled to 0.045 mm thickness as fin stocks, and intermediate annealing was applied for H14 temper.

Table 1 Chemical composition of the specimens in mass %.

	Si	Fe	Mn	Zn	Al
Newly developed Al-Mn alloy	1.0	1.0	1.3	1.0	bal.
Conventional Al-Fe alloy	0.9	1.6	0.5	0.6	bal.

4.1 Increase in tensile strength

Table 2 shows the tensile strength and elongation after brazing heat treatment of the newly developed Al-Mn alloy and the conventional Al-Fe alloy fin stocks. While both fin stocks yielded the same elongation, the newly developed Al-Mn alloy fin stock exhibited higher tensile strength after brazing (145 MPa). Therefore, the post-brazing strength is improved by changing the Mn/Fe ratio.

Table 2 Tensile properties after brazing heat treatment.

	UTS (MPa)	Elong. (%)
Newly developed Al-Mn alloy	145	9
Conventional Al-Fe alloy	135	9

High post-brazed tensile strength is associated with fine dispersed particles and solid solute atoms in the microstructure. Transmission electron microscope (TEM) images of fine particles distributed through both brazing heat treated fin stocks are shown in Figure 4.

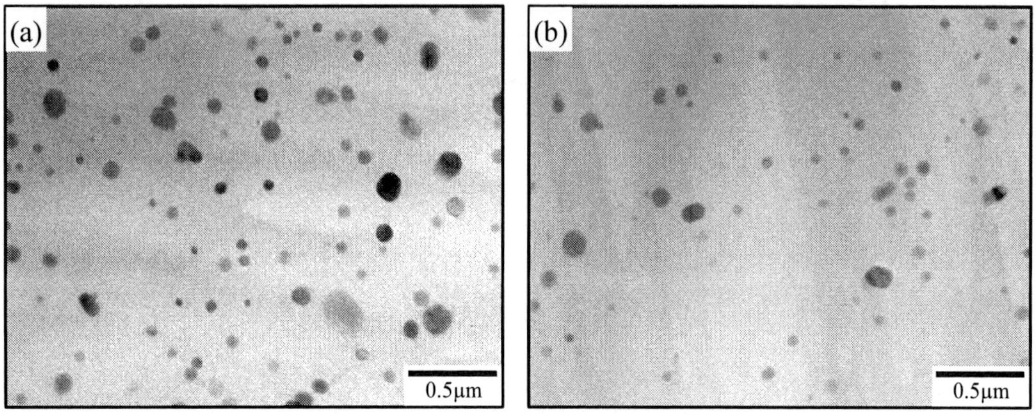

Fig. 4 Distribution of fine dispersed particles after brazing heat treatment by TEM. (a) Newly developed Al-Mn alloy, (b) Conventional Al-Fe alloy.

Fine particles such as precipitates were dispersed through both fin stocks; however, the newly developed Al-Mn alloy contained many more particles than the conventional Al-Mn alloy by virtue of the Mn addition. The increased number of fine dispersed particles increases the strength of the alloy. To evaluate the strengthening effect of the solid solution, the amount of solid solute Mn was measured in the fin stocks after brazing heat treatment. The solid solute Mn content in the aluminum matrix of both fin stocks is given in Table 3.

Table 3 Solid solute Mn content after brazing heat treatment in mass %.

	Solid solute Mn content
Newly developed Al-Mn alloy	0.12
Conventional Al-Fe alloy	0.05

The Mn solid solute was higher in the newly developed Al-Mn alloy fin stock than in the conventional Al-Fe alloy fin stock. The increased tensile strength resulting from the increased Mn/Fe ratio is associated with more dispersed particles and solid solute Mn atoms.

4.2 Improvement of brazeability in the newly developed Al-Mn alloy fin stock

In the radiator manufacturing process, fin stocks are initially corrugated, and then brazed with other components such as tube sheets and headers with the aid of filler metal. During brazing, all components are heat-treated at approximately 600°C. The behavior of fin stocks after brazing heat treatment has been previously evaluated. Good brazeability of the fin stocks is strongly required, because such failures degrade the radiator performance. In the present study, brazeability was evaluated for fin stocks formed in a corrugated structure and brazed with cladding sheets with filler metal. Fin stocks constructed from both the newly developed and conventional alloys were evaluated.

The maximum brazing temperature largely influences erosion and buckling. Stable production of heat exchangers demands highly flexible brazing temperature in fin stock manufacture. The brazeabilities of the newly developed Al-Mn alloy and the conventional Al-Fe alloy fin stocks were determined by varying the maximum temperature of brazing heat treatment between 600°C and 612°C, with a 3-min holding time at each temperature. Figure 5 shows cross-sectional microstructures of the fin stocks brazed with cladding sheets at 606°C and 609°C.

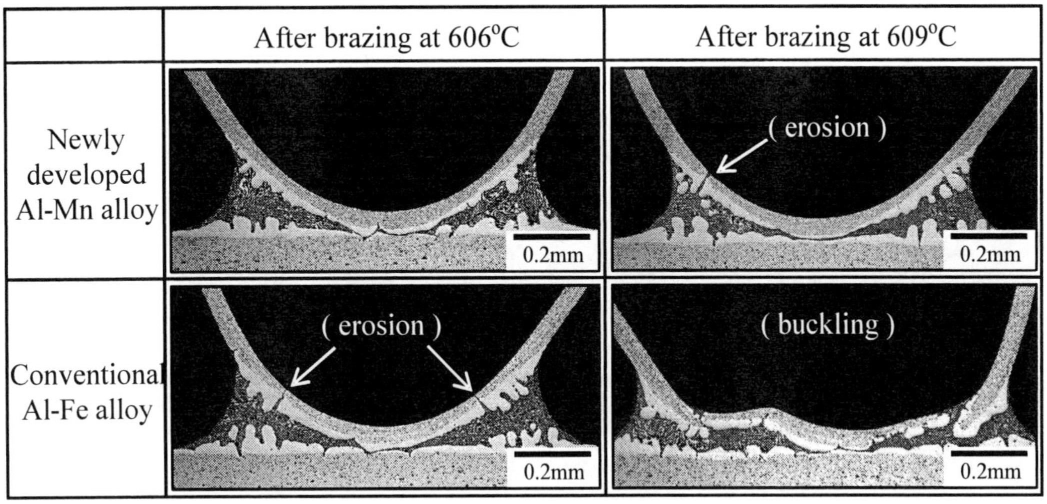

Fig. 5 Cross-sectional microstructures of the fin stocks brazed with cladding sheets at 606°C and 609°C.

The conventional Al-Fe alloy fin stock became eroded after heat treatment at 606°C, and was consequently penetrated by filler metal. Serious buckling was observed after brazing at 609°C. By contrast, the newly developed Al-Mn alloy fin stock demonstrated superior brazeability at both temperatures. No serious erosion was observed at 606°C and no buckling was observed at 609°C. From the brazeability results at all temperatures, the detected rate of erosion and buckling on the brazed fin stocks was calculated, and the number of failed corrugated fins was counted at each temperature. The detected erosion and buckling rates of brazed fin stocks are plotted as a function of temperature in Fig. 6.

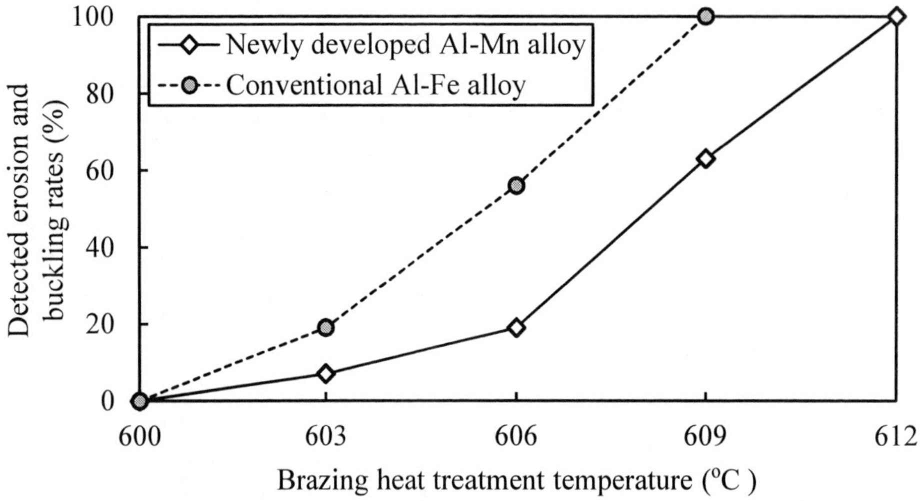

Fig. 6 Detected erosion and buckling rates of brazed fin stocks.

The detected rate was significantly lower in the newly developed Al-Mn alloy than in the conventional Al-Fe alloy. The superior brazeability of the newly developed Al-Mn alloy fin stock is clearly shown in this figure.

As is well known, brazeability is chiefly influenced by the recrystallized grain structure after brazing. Because the grain boundary is considered to be a preferential diffusion pathway of the molten filler metal [9], erosion should be suppressed by enlarging the post-brazing grain structure. Figure 7 shows the plan view grain structures of both fin stocks after brazing with heat treatment at 600°C.

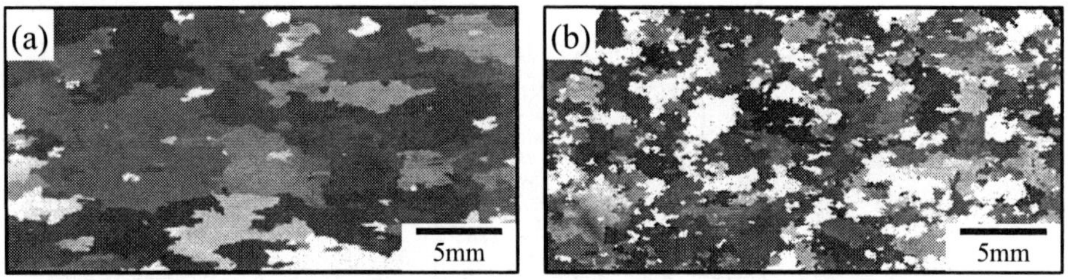

Fig. 7 Plan view grain structures after brazing heat treatment.
(a) Newly developed Al-Mn alloy, (b) Conventional Al-Fe alloy.

The recrystallized grain structure is larger in the newly developed Al-Mn alloy fin stocks than in the conventional Al-Fe alloy fin stock. The recrystallization behavior is controlled by various microstructural features; that is, relatively large secondary phase particles are required for nucleating recrystallization [10], and solid solute atoms and fine dispersed particles such as precipitates interrupt recovery and recrystallization [10, 11]. To evaluate the frequency distribution of relatively large secondary particles, the cross-sectional microstructures of both fin stocks were observed under a scanning electron microscope (SEM). SEM micrographs of the two fin stocks are shown in Figure 8.

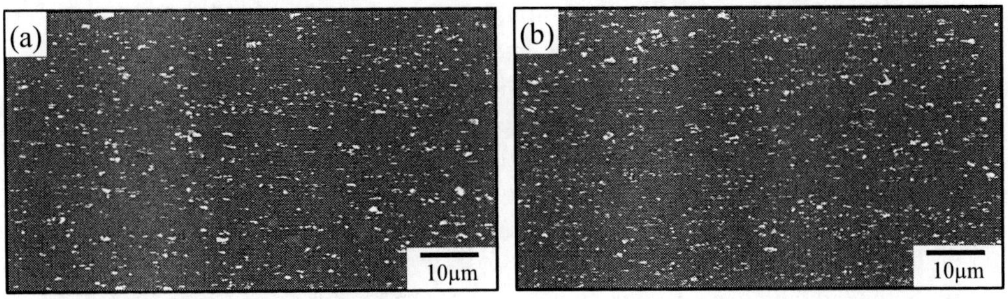

Fig. 8 SEM micrographs of cross- section in fin stocks before brazing heat treatment. (a) Newly developed Al-Mn alloy, (b) Conventional Al-Fe alloy.

The backscattered electron images were analyzed by an image analyzer (LUZEX-AP, Nireco). Figure 9 plots the frequencies of different equivalent diameters of the secondary particles. Particles exceeding 1 μm diameter are reported to become the nucleation sites of recrystallization. The number of such particles is almost identical in the developed Al-Mn alloy and conventional Al-Fe alloy fin stocks. This result suggests similar effects of the secondary particles on the recrystallized nucleation sites in both alloy fin stocks.

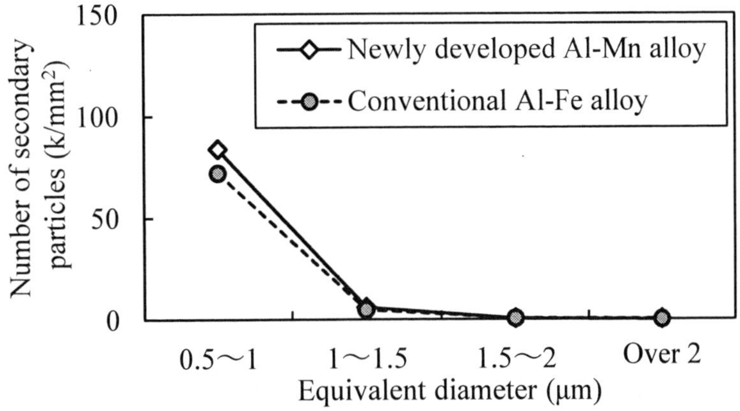

Fig. 9 Frequencies of different equivalent diameters of the secondary particles.

The recrystallized grain structure in the newly developed Al-Mn alloy is enlarged by the high content of Mn solid solute and fine dispersed particles. The number of fine dispersed particles was increased in the newly developed Al-Mn alloy fin stock after brazing heat treatment (see Figure 4), and the aluminum matrix of the new alloy contained more solid solute Mn (Table 3). The microstructural differences between the newly developed and conventional alloys may explain the larger recrystallized grain structure of the new alloy and its superior brazeability resulting from interrupted recovery and recrystallization during brazing.

5 SUMMARY

This study reports a newly developed Al-Mn alloy fin stock fabricated by a twin-belt casting process. The new fin stocks demonstrated superior strength and brazeablity, and are intended for installation in small, lightweight radiators. Examining the Mn/Fe ratio of the fin stocks, it was found that the Mn and Fe content were optimized to enhance the number of dispersed particles and solid solute Mn content. This improved microstructure not only enhanced the tensile strength of the Al-Mn alloy fin stock but also improved its brazeability by enlarging the recrystallized grain structure. Therefore, the superior properties of the newly developed Al-Mn alloy fin stock are attributable to the optimized Mn/Fe ratio along with the advantages of twin-belt casting.

The thickness of the newly developed Al-Mn alloy fin stock can be reduced to 10% thinner than conventional fin stocks which is the thinnest fin stock reported to date. Figure 10 shows the newly developed fin stock installed in a smaller, lighter, and more efficient radiator than has been achieved to date.

Fig. 10 Newly developed fin stock installed in a smaller, lighter, and more efficient radiator.

REFERENCES

[1] S. Teshima: Journal of Light Metal Welding, 50(2012), 15-21.

[2] G. Kimura, A. Kawahaara, A. Niikura, T, Doko, T. Shinoda, J. Kim and D. Mantis: Frukawa-Sky Review, 4(2008), 50-55.

[3] A. Kawahara, A. Niikura and T. Doko: Furukawa Review, 24(2003), 81-87.

[4] M. Yoshino, M. Edo, S. Kuroda, M. Asano and K. Tohma: Journal of Japan Institute of Light Metals, 59(2009), 101-107.

[5] R. E. Sanders, Jr.: JOM, 64(2012), 291-301.

[6] P. Z. Zhao, T. Anami, I. Okamoto, K. Mizusima, K. M. Gatenby, M. Gallerneault, S. Barker, K. Yasunaga, A. Goto, H. Kazama and N. Hayashi: Aluminum Alloys: Fabrication, Characterization and Application II, TMS, (2009), 11-17.

[7] Y. Oki, T. Anami, I. Okamoto and T. Shinoda: The 115th Conference of Japan Institute of Light Metals, (2008), 355-356.

[8] K. Yoshikawa and T. Sakamoto: Journal of Japan Institute of Light Metals, 33(1983), 602-609.

[9] H. Suzuki, G. Itoh and K. Koyama: Journal of Japan Institute of Light Metals, 34(1984), 708-716.

[10] K. Ito: Journal of Japan Institute of Light Metals, 30(1980), 337-348.

[11] J. C. Blade, J. W. Claire and H. J. Lamb: Journal of the Institute of Metals, 88(1959), 365-368.

Analysis of unsteady flow for vortex generator development for heat exchangers

J HARA[1)], M IWASAKI[1)] and I HONDA[2)]
[1)] Calsonic Kansei Corp., Japan
[2)] University of Hyogo, Japan

ABSTRACT

A new heat exchange improvement approach utilizing vortex mechanism with a careful numerical method was studied. One of the effective ways to improve the heat transfer performance for automotive heat exchangers is to destroy the boundary layer. In the range of various type vortices, transverse vortex is often utilized and easy to generate with a rotation axis in the direction perpendicular to the flow. However, its life time tends to be short because the velocity component of the vortex includes upstream direction of the flow. On the other hand longitudinal vortex such as the one generated at a wingtip has a rotation axis along the direction of the flow, which may lead to longer life time. With this expectation we examined the vortex generation mechanism with a triangular-shaped fin, which has conventionally been used to generate such longitudinal vortices.

As a result of CFD analysis, it was clarified that the triangular fin generates negative pressure area in the downstream side of the fin. While the negative pressure area draws a part of incoming flow over the fin, its oblique side provides a rotational component to the flow. This mechanism consequently leads to longitudinal vortex generation.

Based on this understanding, further variation of the shape including trapezoidal fin was studied to improve the heat transfer performance. By adopting the optimized trapezoidal fin to the EGR cooler, it was demonstrated that the size was reduced by as much as 36% at the same pressure drop of the EGR gas flow.

1 INTRODUCTION

There are various methods to improve cooling performance of heat exchangers. In this paper, we describe a new development approach using CFD (Computational Fluid Dynamics). Then by applying this technology to an EGR cooler, remarkably improved performance is achieved.

An EGR (Exhaust Gas Recirculation) cooler is effectively used to reduce NOx (Nitrogen Oxides) emission by decreased temperature of engine combustion. In petrol engines EGR gas has been adopted to clean the engine emission up. It has been also found that EGR improves the pumping loss in partial load condition in petrol engines.

Most petrol engines use non-cooled EGR gas which is returned from an exhaust pipe to an intake pipe directly. Then an EGR cooler has become to be widely used to reduce the temperature of EGR gas in order to restrain knocking.

Mostly stainless steel is used for an EGR cooler for its heat-resistant property since it does the heat exchanger between high temperature exhaust gas and engine coolant. Thus the thermal conductivity is lower than aluminum and then heat exchange from stainless steel fins is inferior to that of aluminum. Although to increase the heat transfer area is a way to improve, accordingly, it brings increased size, weight and cost. We then chose betterment of heat transfer of EGR on the gas side.

A straight forward way to achieve it is an offset fin with narrow cross section. It obviously increases, however, flow resistance and weight, which subsequently requires another technology to improve. In this paper we focused on various vortices which can break the boundary layer on the channel wall effectively. Through this study new fin shape that generates longitudinal vortices efficiently was identified. In the process of exploring optimum vortices, a conventional analysis method of the flow needed to be reconsidered. A new approach was identified to estimate a range of vortices and flow even in a strongly unsteady condition.

2 BASIC CONSIDERATION FOR THE IMPROVEMENT OF COOLING PERFORMANCE

Heat transfer performance of a heat exchanger is governed by the basic equation shown in Eq(1).

$$Q = K \bullet A \bullet \Delta T_m \quad \cdots\cdots\cdots \quad (1)$$

Q: *Heat Transfer Capacity* [W]

K: *Over-all Coefficient of Heat Transfer* [W/(m^2 • K)]

A : *Radiation Area* [m^2]

ΔT_m: *Logarithmic Mean Temperature Difference* [K]

$$\frac{1}{K} = \underbrace{\frac{1}{\alpha_g \bullet \eta_g}}_{\text{Exhaust}} + \underbrace{\frac{d}{\lambda_t \frac{A_w}{A}}}_{\text{Tube wall}} + \underbrace{\frac{1}{\alpha_w \frac{A_w}{A}}}_{\text{Coolant}} \quad \cdots\cdots\cdots \quad (2)$$

α_g: *Heat Transfer Coefficient (Gas side)* [W/(m^2 • K)]

α_w: *Heat Transfer Coefficient (Coolant side)* [W/(m^2 • K)]

η_g: *Total Fin Efficiency*

d: *Thickness of Tube* [m]

λ_t: *Thermal Conductivity of Tube* [W/(m • K)]

A_w: *Heat Transfer Area of Tube (Coolant side)* [m^2]

In this equation, the logarithmic mean temperature difference ΔTm varies depending on the condition of the fluid. Thus it is necessary for performance improvement to consider increasing K or A shown in Eq(2). Since A may increase the size of a heat exchanger, after all, increase of K is required.

The heat transfer coefficient K in Eq(2) is expressed by a summation of inverse numbers of heat resistances in the right side of this equation. In the case of an EGR cooler, the clause 1 in the right side contributes over 90% of the overall resistance. The exhaust gas side heat resistance is due to the thermal boundary layer on the wall of the flow channel. Then breaking this layer is essential to improvement of cooling performance.

Considering the ways to increase cooling performance in an EGR cooler as heat exchangers, there are the following approaches.

(1) Utilization of secondary flow
(2) Mixing of boundary layer
(3) Reduction of shear flow area

In this paper, a method for generating long life vortices were explored along the above approach (2).

Stainless steel is used that endures high temperature condition in 900K (Kelvin) exhaust gas of the EGR cooler. Thermal conductivity of stainless steel (SUS) is only about 1/10 of that of aluminum (Fig.1), which is used for most of the heat exchangers of automobiles. Thus the fin efficiency for heat exchangers made with stainless steel is very low and the heat dissipation via the fins is not practically expected.

Consequently a method of breaking thermal boundary layer was considered instead.

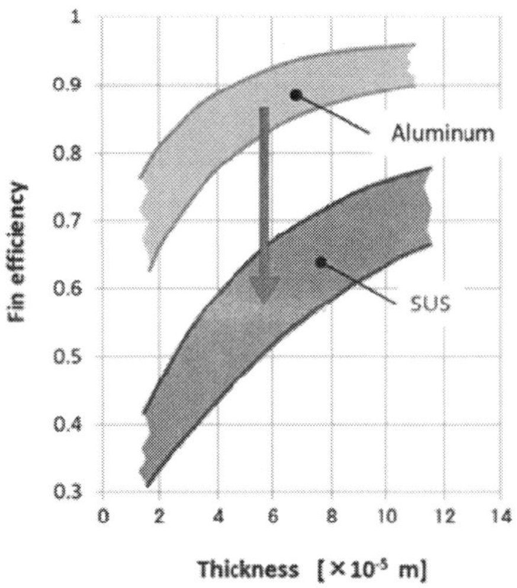

Figure 1. Comparison of Fin efficiency

3 STRUCTURE OF THE EGR COOLER

In this section the structure of the EGR cooler in our study is described. Fig.2 shows a stacked-plate-type heat exchanger. The head parts for inlet and outlet of exhaust gas are set on the both ends. The stacked plates of the core are divided into the two alternate sets of flow channels for engine coolant and exhaust gas. In the exhaust gas channel, offset fins are located to interfere with the flow. Each of the fins is formed by cutting and bending on the plates.

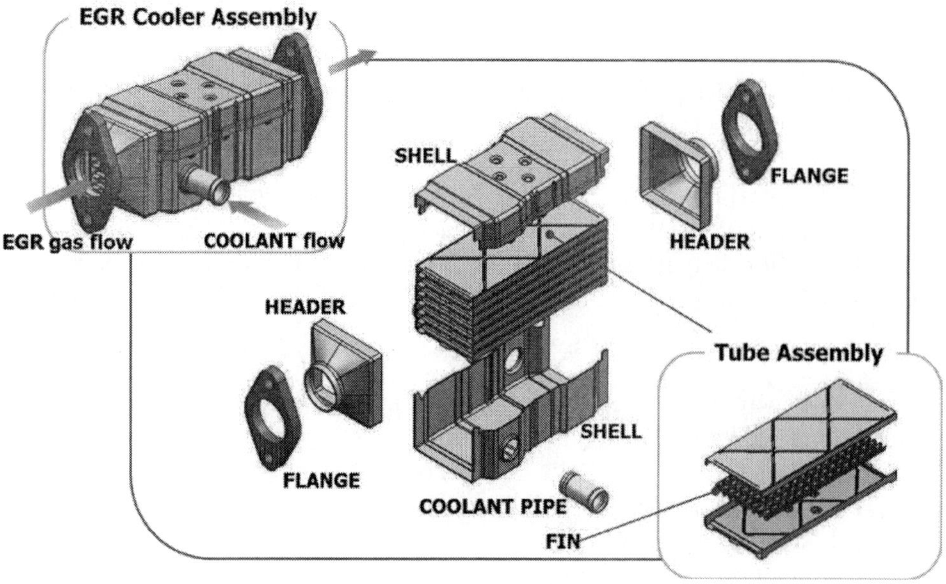

Figure 2. Structure of EGR cooler

4 CONVENTIONAL VORTEX GENERATOR

The idea to create vortices for promoting heat transfer had often been applied in conventional heat exchangers. For instance, triangular fin and hemisphere-shaped dimple were commonly seen. Fig.3 shows an example which put semi-cylindrically-formed dimples together. The vortices generated by these examples, as suggested in the attenuation pattern, are dominated by transverse vortices.

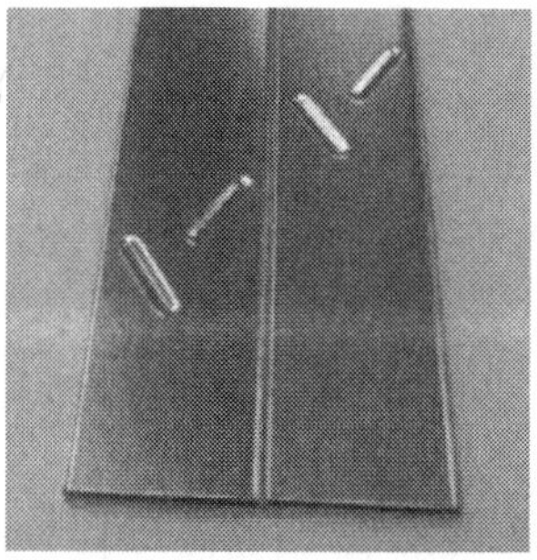

Figure 3. Conventional vortex generator

In case of this type of vortices, its rotating axis is almost perpendicular to the direction of the flow. This vortex contains both an acceleration portion and a deceleration one, which consequently forms a shear flow. In general, a shear flow in a viscous stream is one of the causes of flow resistance. Moreover it is also one of the factors to attenuate the vortices. On the contrary, a longitudinal vortex has a rotating axis in the same direction as the main stream, which is hard to attenuate due to less shearing with the surrounding flow (Fig.4).

Table 1. Comparison of vortices

Vortex	Example	Attenuation	Generation
Transverse	Kármán vortex	Short life	Easy
Longitudinal	Wingtip vortex	Long life	Difficult

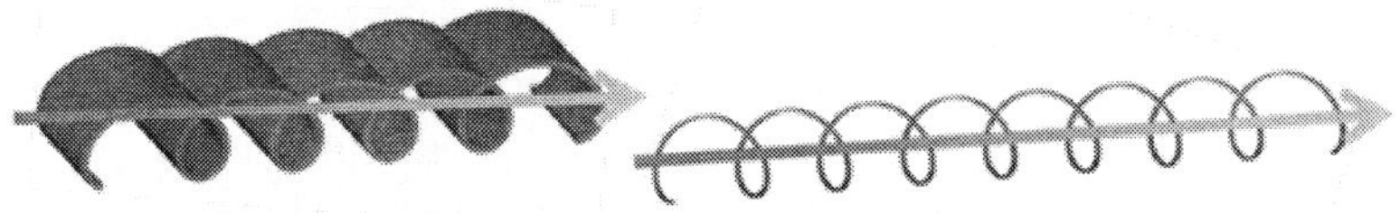

Figure 4. Transverse vortex and Longitudinal vortex

5 DEVELOPMENT OF CFD ANALYSIS METHOD

CFD is one of common method to analyze flow phenomena. However in unsteady analysis such as like vortices, the result of CFD is often different from the experimental one. Thus it was necessary to develop a method for precise enough analysis on generation and convection of a vortex. This study includes the following aspects.

(1) Number of computational grid
(2) Calculation model
(3) Unsteadiness
(4) Flow visualization

(1) Number of computational grid
There are several important aspects in CFD. In particular, the above 4 points are momentous.
The greater number of the computational grid large, the accuracy of the result is more precise. However the cost of analyzing becomes serious. Therefore, the number of the computational grids is chosen to reproduce the generation and the flow of a vortex. And its analysis is the purpose of the computation. The calculation of the velocity and the pressure of the flow is easy. On the other hand, since the analysis of the temperature field is hard, precise enough computational grid is required for this reproduction.

(2) Calculation model
There are a variety of choices in the computational model. DNS (Direct Navier-Stokes) method is the best, however the cost of computing is too expensive and this method is not practical. LES (Large eddy simulation) which models the inner of grids is better. Even, this calculation cost is large. There are various turbulence models based on Reynolds equation, and this procedure is popular. This method has some problems. In this paper, a calculation method that can be accurately reproduced in the generation of vortex is chosen.

(3) Unsteadiness
The flow containing a vortex is one of the very unsteady stream. In the selection of the calculation method described above, the choice criteria is whether it can analyze the reproduction of unsteady flow.

(4) Flow visualization
Three dimensional vortex is difficult to observe by the velocity vortex and the streak line. Thus an expression method to easily observe the vortex is selected.

5.1 Number of Computational Grid

To compare various conditions, we used a segmented part with an offset fin of a channel of an EGR cooler shown in Fig.5. In this sample, the setting angle is 60 degrees and the inclining angle is also 60 degrees. The analysis condition is shown in Table 2. This is also valid for the calculation of the other sections.

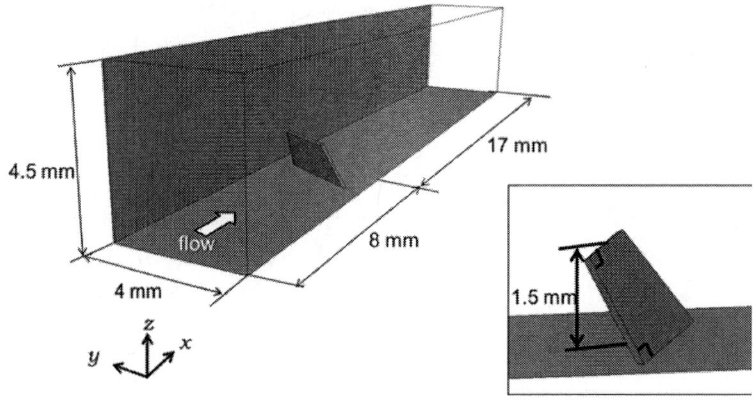

Figure 5. Computational domain

Table 2. Calculation conditions

Inlet velocity [m/s]	20	Thermal expansion [/K]	0.003354
Density [kg/m^3]	2	Inlet temperature [K]	303.15
Viscosity [Pa·s]	3.0x10^{-6}	Wall temperature [K]	403.15
Specific heat [J/kgK]	1022	Prandtl number	0.084
Thermal conductivity [W/mK]	0.03648	Time step [s]	2x10^{-5}

The representative length of Reynolds number is taken from the fin height and the reference aspect of Grashof number is based on the heat transfer plane. The time marching method of CFD is full-implicit.

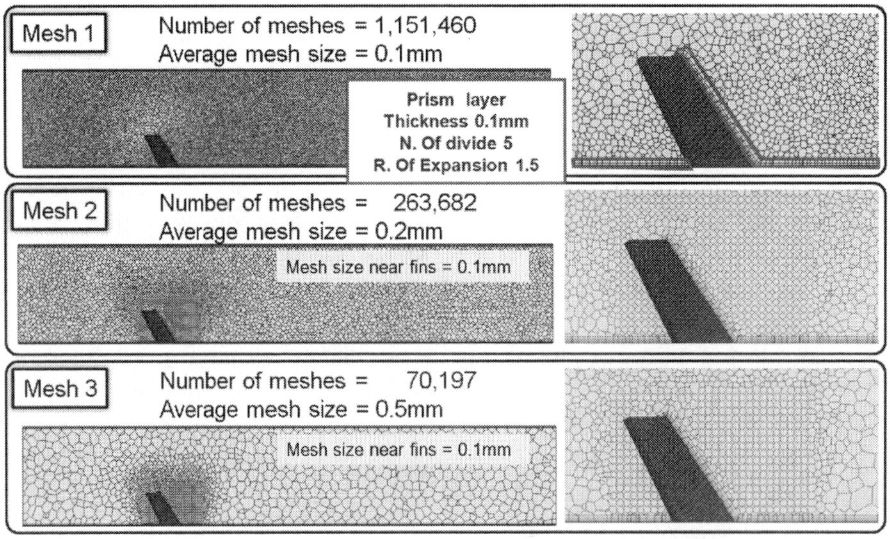

Figure 6. Comparison of computational grid

First, the difference by the number of computational grid is analyzed on 3 levels of mesh density shown in Fig.6. As shown the velocity vector in Fig.7, it is found that the coarse grid model tends to indicate restrained generation of vortices.

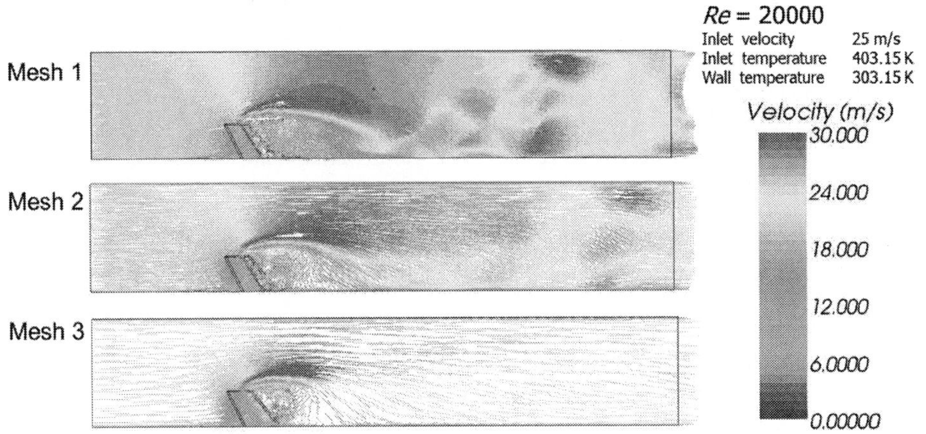

Figure 7. Comparison of velocity vector

Larger grid size brings large effect of numerical viscosity, which leads to prevent vortex in early phase from developing as it is supposed to. The convective term of Navier Stokes equation, in principle, tends to further divide the flow leading to divergence of numerical calculation. Therefore with respect to analysis of vortex generation, it is necessary that the vortex is accurately reproduced exactly while the divergence of calculation can be restrained. Fig.8 shows the temperature distribution, which also indicates the effect of grid size.

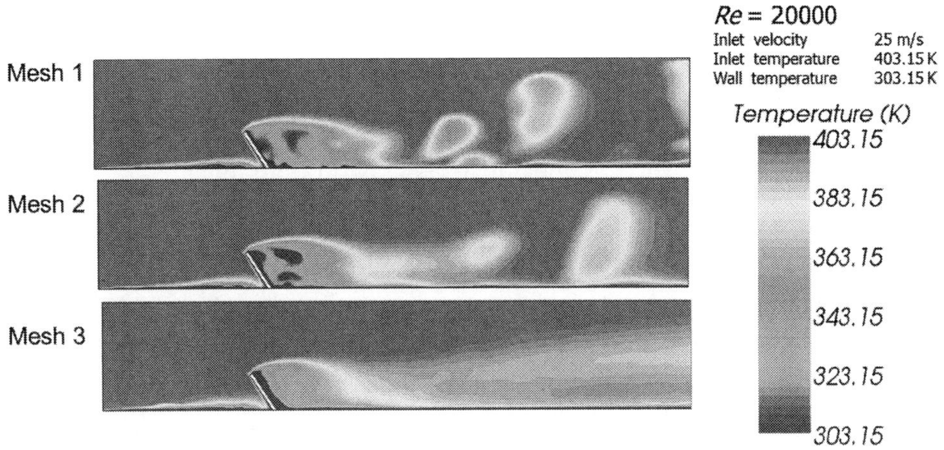

Figure 8. Comparison of temperature distribution

In addition, Fig.9 and Fig.10 show the result for pressure distribution.

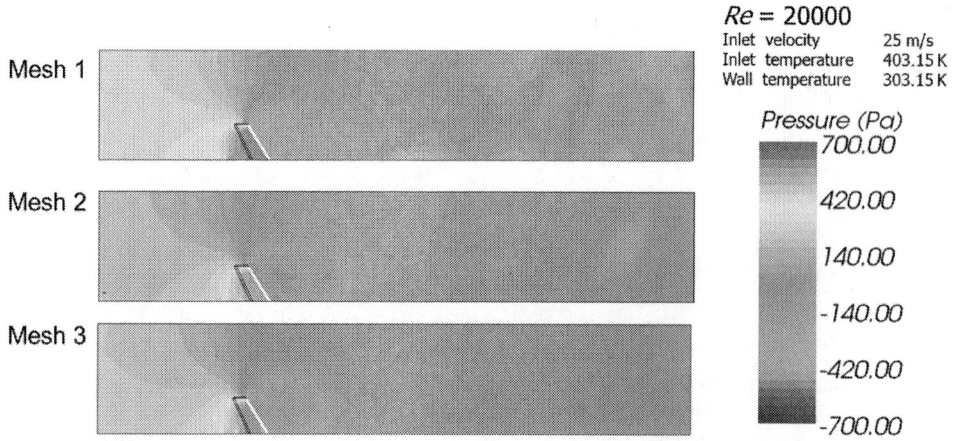

Figure 9. Comparison of pressure distribution (Side view)

Figure 10. Comparison of pressure distribution (Plan view)

Fig.10 which is observed from the upstream side of the flow indicates the formation of exceedingly complicated vortices in the downstream part of the fins.

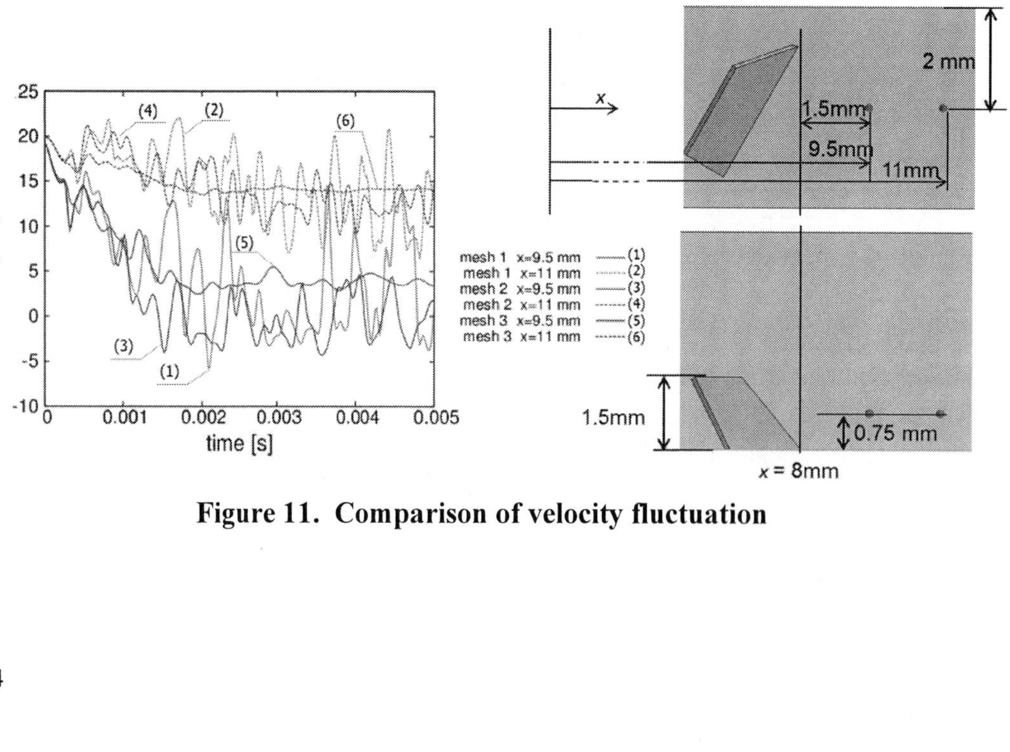

Figure 11. Comparison of velocity fluctuation

Fig.11 shows the velocity along the flow direction and the observation point behind the fin. "x=9.5mm" indicates the observation point nearby the fin and "x=11mm" corresponds to the point in further downstream side.

As shown in the figure, when the number of computational grid is small, the time fluctuation of the velocity is not observed. This is due to less generated vortices when the mesh is too coarse. Fig.12 shows the temperature distribution.

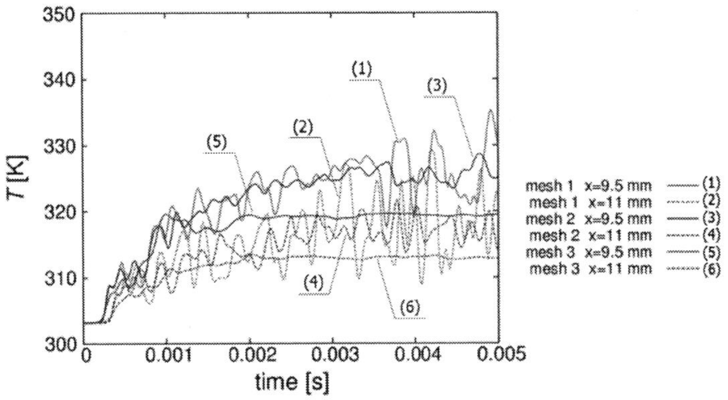

Figure 12. Comparison of temperature fluctuation

The similar tendency to the velocity distribution is observed in the temperature distribution as well. The temperature at "mesh3" with fewer grids is lower than those at the other models with more grids. This means that destruction of the boundary layer by vortices encourages heat transfer.

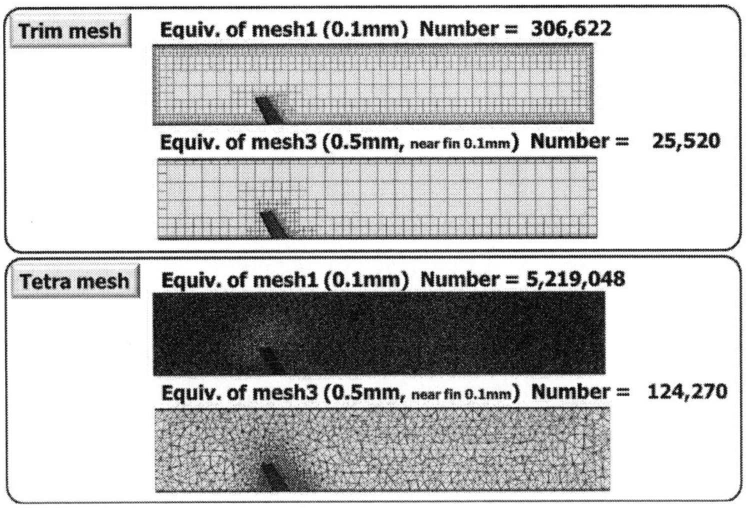

Figure 13. Comparison of Trim mesh and Tetra mesh

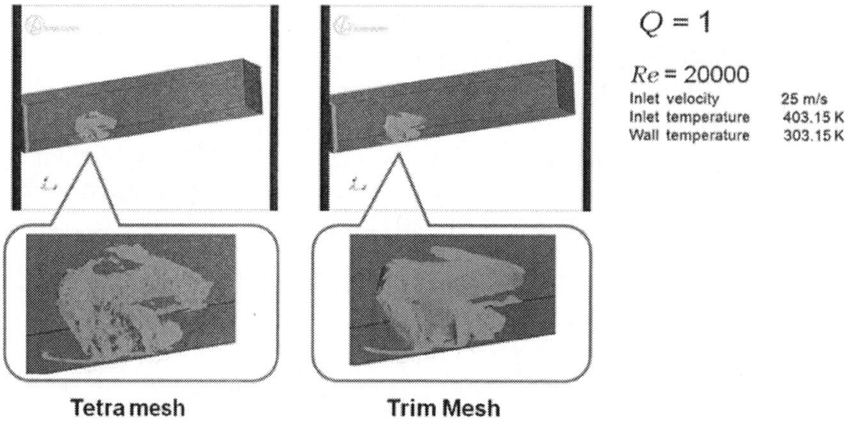

Figure 14. Comparison of meshes on shape of vortex

Then the comparison of the grid structure is studied. There are trim mesh, tetra mesh as tetrahedron and tetra mesh being applied in various cases. Comparison between trim mesh and tetra one is shown in Figs.13 and 14.

$$A_{ij} = \frac{\partial u_i}{\partial x_j} = S_{ij} + W_{ij} \quad \text{..................} \quad (3)$$

$$Q = \frac{1}{2}\left(-S_{ij}S_{ij} + W_{ij}W_{ij}\right) \quad \text{..............} \quad (4)$$

A_{ij}: *Gradient of Velocity Tensor*

u_i: *Velocity*

x_i: *Coordinate*

S_{ij}: *Symmetry Part of Gradient of Velocity Tensor*

W_{ij}: *Unsymmetry Part of Gradient of Velocity Tensor*

Q: *Second Invariant of Gradient of Velocity Tensor*

What is shown is an isosurface of second invariant of gradient of velocity tensor (Q=1). In tetra mesh, this is a complicated isosurface by concavity and convexity. On the other hand, in trim mesh, the surface is smooth. This is due to the fact that the number of grids is less for trim mesh than for tetra one at the same mesh size. Even in case of tetra mesh, if it is coarse such as "mesh3", analysis of vortex is difficult.

Lastly to acquire the vortex shape accurately, not only to make the mesh size smaller, it is required tetra mesh is adopted to calculate.

Second Invariant of Gradient of Velocity Tensor Q is shown at Eq.(3) and Eq.(4).

5.2 Calculation Model

Fig.15 shows comparison of various calculation models with the same grid. It is shown that "laminar flow model", "LES (Large Eddy Simulation)" and "AKN model" can reproduce complicated vortices. General turbulent models such as "k-ω model" and conventional "k-ε model", however, cannot simulate vortex generation with accuracy.

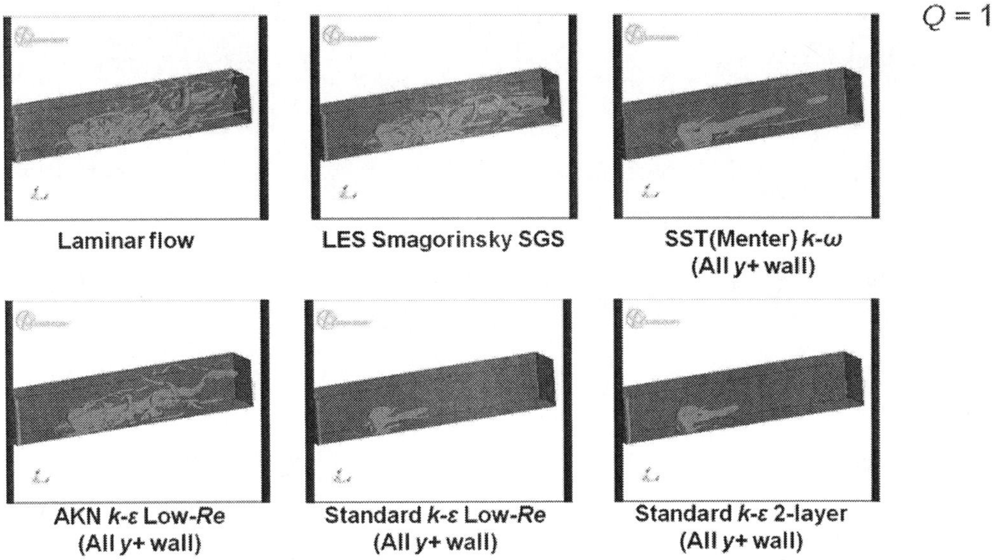

Figure 15. Comparison of calculation model

From this analysis, the calculation model such as the traditional k-ε method that obstructs the generation of the vortex are ineligible.

5.3 Unsteadiness

Unsteadiness of various calculation models is also studied. Fig.16 shows time transition of temperature. Temperature fluctuation is observed in "LES" and "laminar flow model" cases caused by vortices. In SST ("k-ω model"), the fluctuation is small. In the case of turbulent model, only AKN model can simulate certain fluctuation.

In "k-ε model" as general turbulent model, the flow is stable. Therefore unsteadiness of the flow cannot be analyzed.

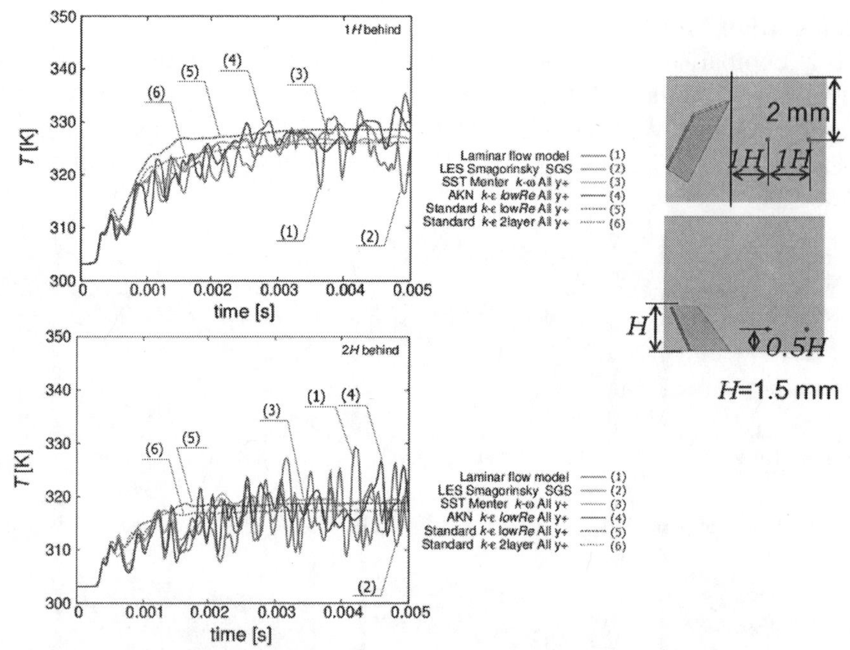

Figure 16. Comparison of unsteadiness by calculation models

5.4 Flow Visualization

To acquire a broad overview of the flow field, visualization of the flow is important.

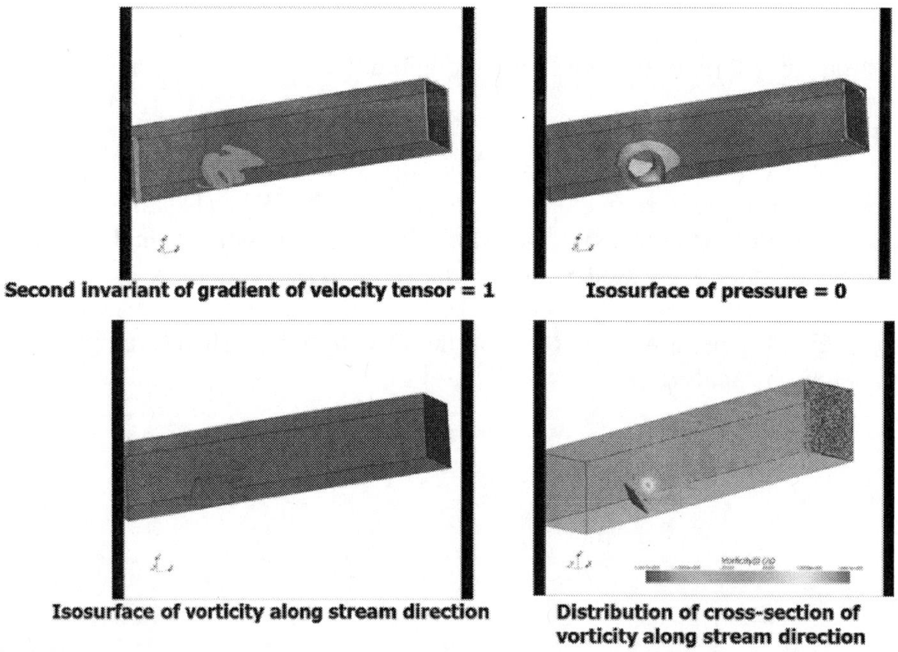

Figure 17. Comparison of flow visualization method for vortex

Fig.17 shows expression of the vortex by various physical quantities. It is found that to use "Second Invariant of Gradient of Velocity Tensor" is the most intelligible parameter to illustrate physical meaning of the flow mechanism.

From the above analysis, in studying the fins of the EGR cooler, it is clarified that the number of meshes is almost one million. As more preferred calculation model, conventional k-ε model is not to be selected. Also, to reproduce the accurate shape of the vortex, the fine tetra mesh should be chosen.

6 THE OPTIMUM VORETX GENERATOR

The optimum vortex generator for an EGR cooler for petrol engines is developed by the analytical methods described in the previous section. Conventionally triangular fin which mixes boundary layer by generated vortices behind the fin has often been used.

In this study, various calculations and experiments are conducted to optimize the shape and the arrangement of the fin in a rectangular channel. Thereby it is confirmed that the strong vortex is generated when trapezoidal fin (VG fin) is set obliquely to the stream and inclined toward upstream direction (Fig.18, 19).

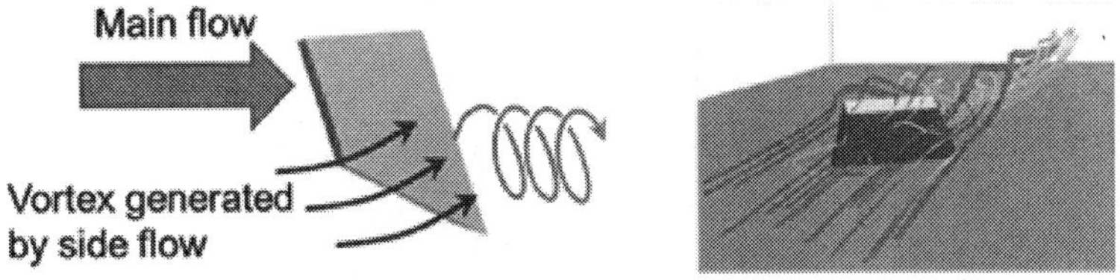

Figure 18. The optimum shape

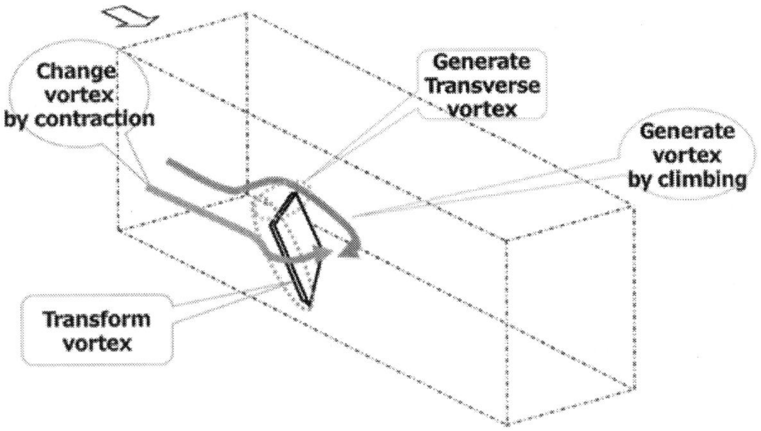

Figure 19. Mechanism of generation of longitudinal vortex

With this fin shape, transverse vortex is generated from upper side of the fin. Moreover it is found that the flow from the narrow channel between the oblique line of the fin and the vertical wall transforms the transverse vortex to the longitudinal vortex. Fig.20 shows an isosurface of Q-value for various setting angles of attack and inclined angles.

By optimizing the angle of setting and inclined to upstream side, it is observed that the strong longitudinal vortex is generated.

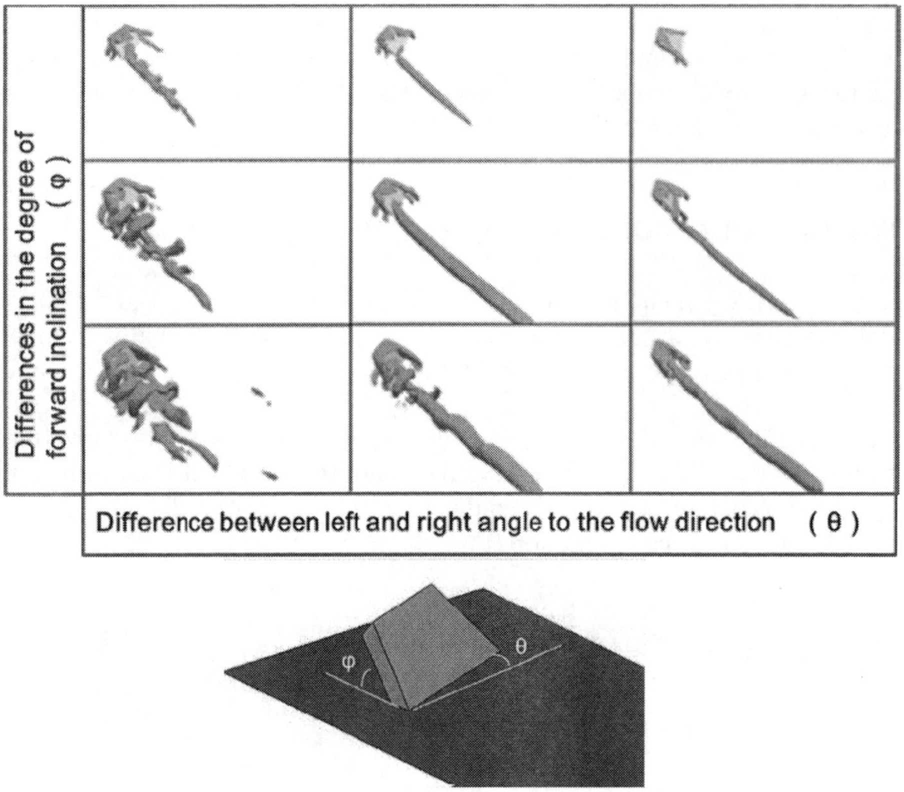

Figure 20. The optimum setting angles

7 DISTANCE FROM VERTICAL WALL

When the distance from the vertical wall of the channel to the fin is far, the space under the oblique side of the fin is not narrow enough and the velocity of the flow is not significant. Hence the effect of transformation to the longitudinal vortex becomes weak. On the other hand, the distance is too short, the flow through the clearance between the oblique side and the vertical wall is limited, then the transforming effect dilutes too. Therefore there is an optimum distance for the clearance (Fig.21).

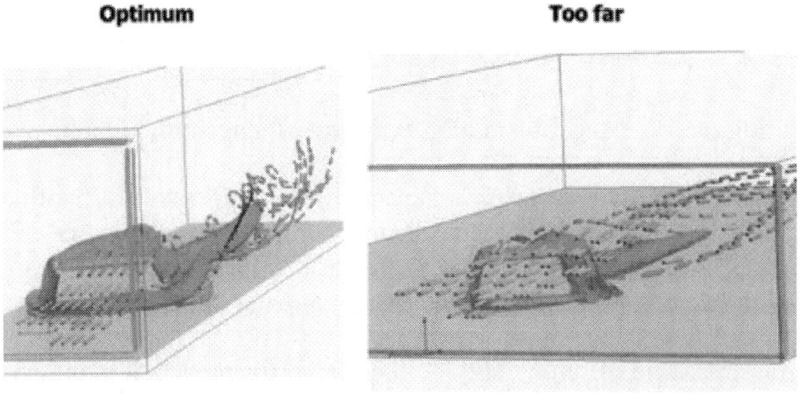

Figure 21. The Optimum distance from vertical wall

8 COMPARISON WITH TRIANGULAR FIN

Vortices by the conventional triangular fin and the trapezoidal fin (VG fin) are visualized by CFD (Fig.22). A longitudinal vortex is observed behind the triangular fin. In the trapezoidal fin(VG fin), a stronger vortex and mixing of the flow on the low wall are clearly recognized.

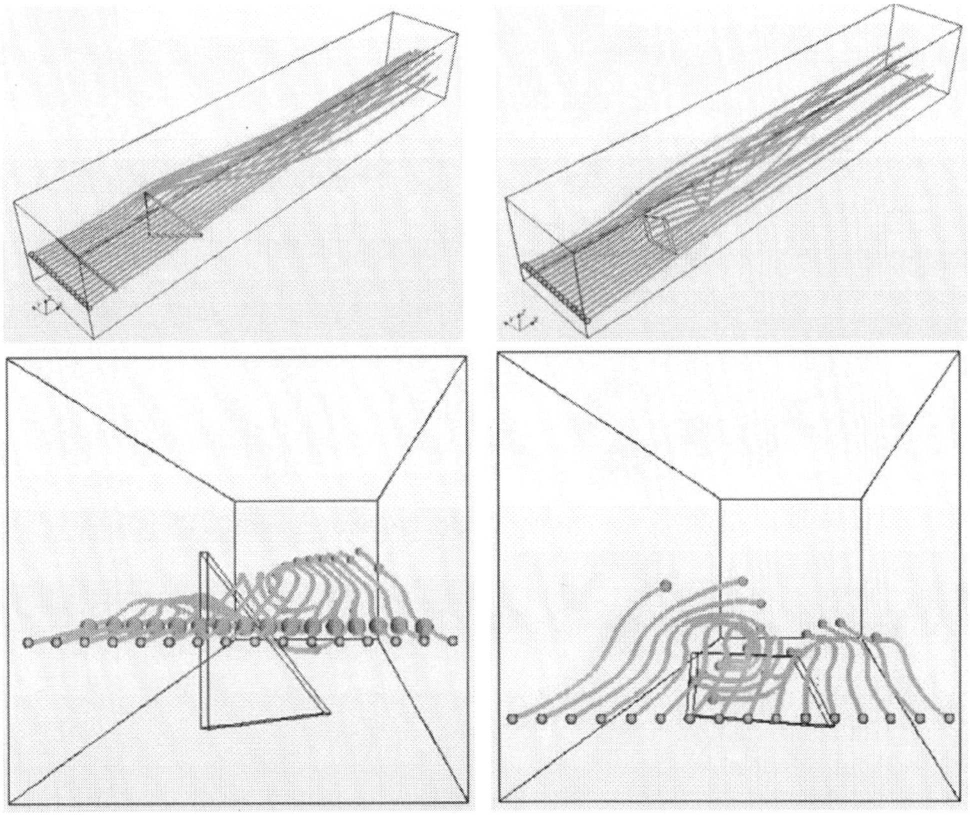

Figure 22. Comparison between triangular fin and trapezoidal fin

Fig.23 shows the distribution of the wall temperature. Since the wall is cooled, the higher temperature area of the wall means larger heat transfer. In trapezoidal fin (VG fin), a broad area shows significant heat transfer.

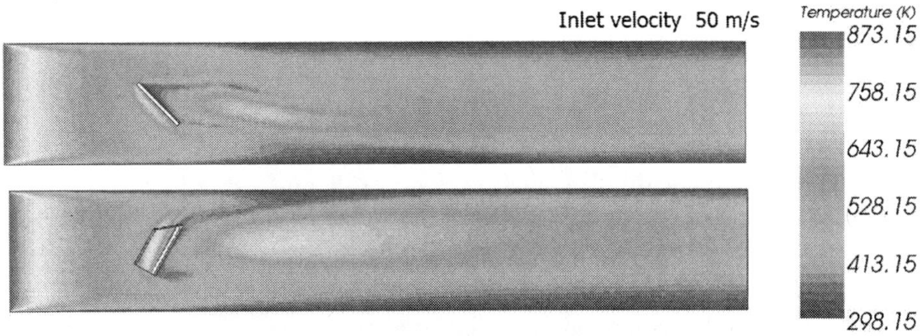

Figure 23. Comparison of wall temperature
(upper : triangular fin, lower : trapezoidal fin)

9 COMPARISON COOLING PERFORMANCE

Lastly, comparison in terms of resultant cooling performance of the EGR cooler is shown for the trapezoidal fin and the conventional fin in Fig.24. The EGR coolers with conventional fin are constructed by fine offset fin or triangular fin. At the same heat transfer level, the flow resistance of the gas side is reduced by 10-60%. When the resistance is lower, the EGR rate of engines is higher and the pumping loss and knocking are improved.

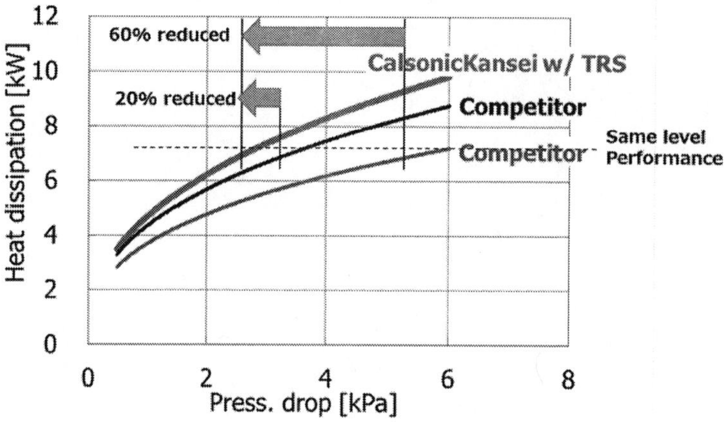

Figure 24. Comparison of cooling performance on EGR cooler

Adopting the trapezoidal fin (VG fin) to the heat exchangers, the following merits are enjoyed. These merits give the vehicle the edge.

(1) Miniaturized
(2) Light weight
(3) Lower cost

10 CONCLUSION

The new trapezoidal fin (VG fin) for an EGR cooler of petrol engines is developed. This fin can generate a longitudinal vortex with longer life, which results in the following advantages.

(1) The trapezoidal fin (VG fin) which is shaped by cutting obliquely one side edge of a rectangular fin can generate a longitudinal vortex.
(2) By setting the trapezoidal fin (VG fin) obliquely to the flow direction and inclined to the upstream of the flow, strong longitudinal vortex is generated that break the boundary layer on the wall of the heat exchanger channel effectively.
(3) By integrating the trapezoidal fin (VG fin) with offset fin and optimizing the clearance between the fin and the vertical wall, stronger vortex can be generated.
(4) By the above mentioned mechanism, the gas flow resistance can be reduced by more than 10% at the same heat transfer level compared with the conventional fin structure.

11 REFERENCES

1 Ozawa T. et al., Development of EGR cooler (in Japanese), CalsonicKansei World (2002) 37-43

2 Hiramatsu M. et al., Numerical analysis of inner-fin for intercooler (in Japanese), JSME paper(B) Vol.57 No.539 (1991) 176-182

3 A. M. Jacobi and R. K. Shah, Heat Transfer Surface Enhancement through the Use of Longitudinal Vortices, A Review of Recent Progress, Experimental Thermal and Fluid Science, Vol.11 (1995) 295-309

4 Itoh M. et al., Performance Evaluation of a Heat Exchanger with Vortex Generators for Air-Conditioners (Influence of Vortex-Generator Height) (in Japanese), JSME paper(B) Vol.68 No.665 (2002) 188-193

Development of advanced multilayer aluminium sheets solutions for Charge Air Cooler applications

L PEGUET, B CHEHAB, M PERRIER and **H NOUI**
Constellium Technology Center, France

ABSTRACT

We are developing new generations of aluminium brazing sheets to enable down gauging for light-weighting and/or to increase corrosion durability required by the car manufacturers. This process starts by laboratory scale prototyping and deep metallurgical, mechanical and corrosion characterizations (microstructure, brazeability, mechanical properties, corrosion resistance). The optimized solutions are then produced on an industrial scale and tested by our customers for qualification. For the specific Charge Air Cooler applications, a new laboratory corrosion test "CAC test" has been developed to allow a better ranking of corrosion behaviours for different material solutions. This test is performed in a climatic chamber with a synthetic condensate simulating the exhaust gas condensate formation and combining a wet and a dry cycle in order to be realistic compared to the in service situation. Solutions developed with a 3xxx interlayer show significant corrosion resistance improvement in both SWAAT (Sea Water Acetic Acid Test) and "CAC test" compared to a state-of-the-art long life solution. In this paper we will present the "CAC test" and its implementation in combination with other properties characterization (microstructure, mechanical properties, brazeability, ...) for the development of innovative multilayer solutions for Charge Air Cooler applications.

INTRODUCTION

The car heat exchangers manufacturers require advanced brazing sheet solutions in order to fulfill the more and more severe corrosion resistance and mechanical properties specifications. Charge Air Cooler (CAC) represents a particular heat exchanger application requiring increased internal corrosion performances when EGR architectures includes exhaust gases re-circulation with cooling at low pressure [1, 2]. In this case, the CAC needs to resist to acidic exhaust gas condensates [3] formation associated with wet and dry operating conditions. State-of-the-art long life solution fails to meet internal corrosion specification of such applications. In order to meet car manufacturer corrosion requirements for Charge Air Cooler, we have developed an advanced multi layered solution. Both, state-of-the-art long life and multilayer solutions were tested using a new laboratory "CAC test" and the standardized ASTM G85A3 – SWAAT test.

In this paper we will first briefly present our new multilayer solution, then we will describe the principle of the developed "CAC test" and finally we will compare the corrosion performance of both multilayer and state-of-the-art long life solutions using this new "CAC test" and the standardized SWAAT test.

EXPERIMENTAL PROCEDURE

Materials

This investigation concerns two different industrial aluminum brazing sheets: a state-of-the-art long life three layers sheet and an advanced multilayer brazing sheet. Both industrial materials were received at O temper and in two different thicknesses of 400 and 480μm.

The state-of-the-art long life solution (see Fig 1a) consists of a 3916 alloy core (see composition in table 1) cladded in both sides with 7.5% thickness ratio of 4343 alloy. The multilayer solution (see Fig 1b) consists of a 3916 alloy core cladded in both sides with 7.5% thickness ratio of 4343 and a one side interlayer with 8% thickness ratio of low Cu 3xxx alloy (see composition in table 1).

Table 1: Chemical composition in wt% of the 3916 and the low Cu 3xxx alloys.

	Si	Fe	Cu	Mn	Mg	Zn	Ti
3916	≤0.3	≤0.3	0.5-1	1-1.5	≤0.02	≤0.2	≤0.2
low Cu 3xxx	≤0.6	≤0.7	0.05-0.2	1-1.5	≤0.02	≤0.2	≤0.2

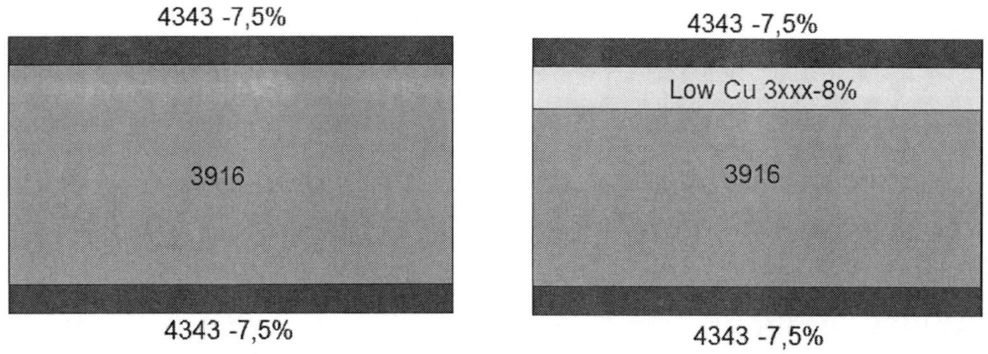

a) b)

Figure 1: Detailed architectures: a) state-of-the-art long life solution, b) multilayer solution.

The as received microstructures of the multilayer solution after Barker's and Keller's etching are given in the left and right side of Fig 2, respectively. The core presents a fully recrystallized and homogenous microstructure with small elongated grains in the rolling direction. The 4343 clad microstructure consists of an aluminum matrix with small globular crystal silicon particles. The low Cu 3xxx interlayer (on top side of the micrographs) is well revealed by a specific Keller's etching (left side micrograph of Fig 2). The laboratory brazing (600°C-2min) leads to a coarsening of the core grains which stay elongated in the rolling direction. After brazing, a brown band or "band of dense precipitates" [4,5] is also generated on the monoclad side of the brazing sheets (see left bottom micrograph of Fig 2).

Fig 3 gives the tensile properties after brazing of both solutions (state-of-the-art long life and advanced multilayer). The two solutions present similar values of yield stress, ultimate tensile strength and elongation. Thus the addition of a 8% low Cu 3xxx interlayer in the multilayer solution does not affect its tensile mechanical properties, which was expected since the major part of the material is the core (min. 77%).

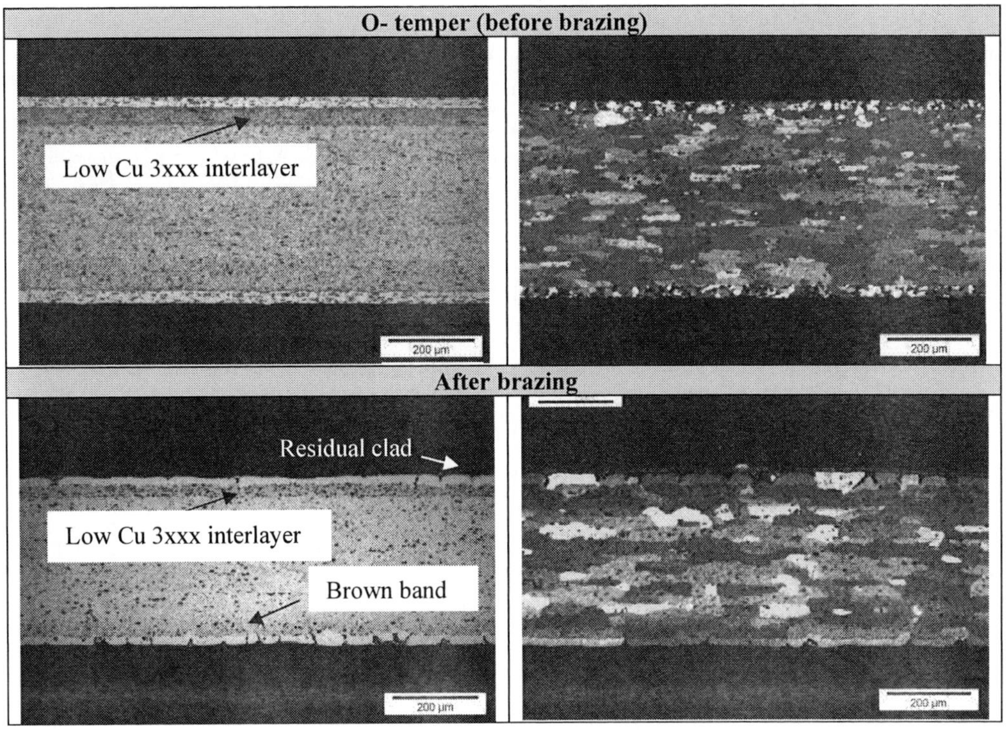

Figure 2: Microstructures after specific Keller etching and Barker's etchant before and after brazing.

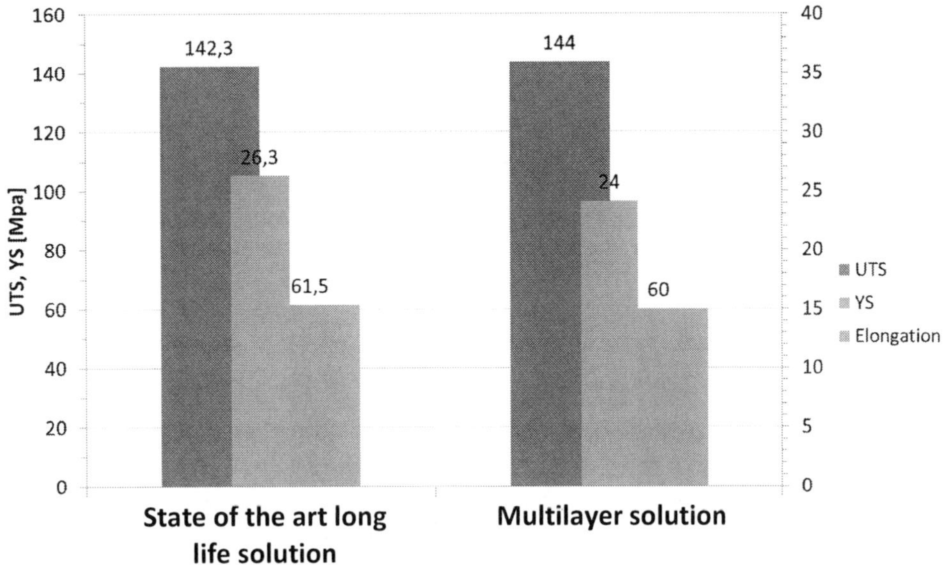

Figure 3: Tensile test properties for multilayer solution compared to a state-of-the-art long life solution.

Corrosion tests

As a first and rough evaluation of the durability of investigated materials in a corrosive environment, ASTM G85A3 – SWAAT test was carried out in a Q-FOG® chamber. The procedure is based on the basic cycle: 30min spray + 90min soak. A 5% synthetic sea salt solution at pH3 is used as the condensate. Although SWAAT test is extensively used for testing heat exchangers, this procedure is only related to atmospheric corrosion and concerns only the durability of the external side of heat exchanger such as air conditioning evaporators.

Concerning the specific case of Charge Air Cooler (CAC), SWAAT test has very limited value. Therefore, a dedicated corrosion test was developed to simulate corrosion in CAC heat exchangers. Knowing that exhaust gases circulating inside CAC consist mainly of CO_2; H_2O, NOx and SO_2 (depending on the diesel-sulfur level), if condensate formation occurs, this generates strong acid: HNO_3, H_2SO_4 [6,7] and less corrosive organic acids. Exhaust-gas condensate composition and dew point depend on fuel composition, combustion process, air ratio, load of the engine, exhaust-gas after treatment, engine start-up stage, etc… Furthermore, EGR system will frequently be exposed to successive wet and dry environments depending on the engine speed and temperature. States in which fluid acidic concentrate can dwell and dry on the components are critical. From these assessment, a new corrosion test "CAC test" based on a 3-step 4h cycle was set up (Fig 4). This internal corrosion test, aiming at assessing the corrosion resistance seen in-service includes a dry and wet cycle as well as a spraying phase using a synthetic condensate based on a sulfuric acid + nitric acid (H_2SO_4 + HNO_3) equi-molar solution. Two different levels of aggressiveness were tested: a first condensate at pH2.5 and 100ppm Cl^- and a second one at pH 2 and 1000ppm Cl^-.

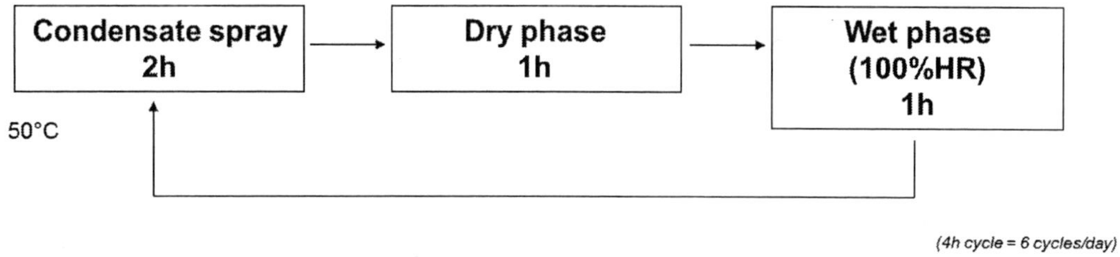

Figure 4: Schematic of the "CAC test" based on a 3-step 4h-cycle.

45mm (L) x 65mm (TL) x 0.48mm (TC) samples are cut from each reference (see Figure 5). The samples were degreased with acetone and then masked in order to expose only the front side to be tested. The edges and the back side respectively are protected with silicone and adhesive tape. Therefore, the tested surface is about 40mm (L) x 60mm (TL) leading to an exposed surface of about 2400 ± 100 mm².

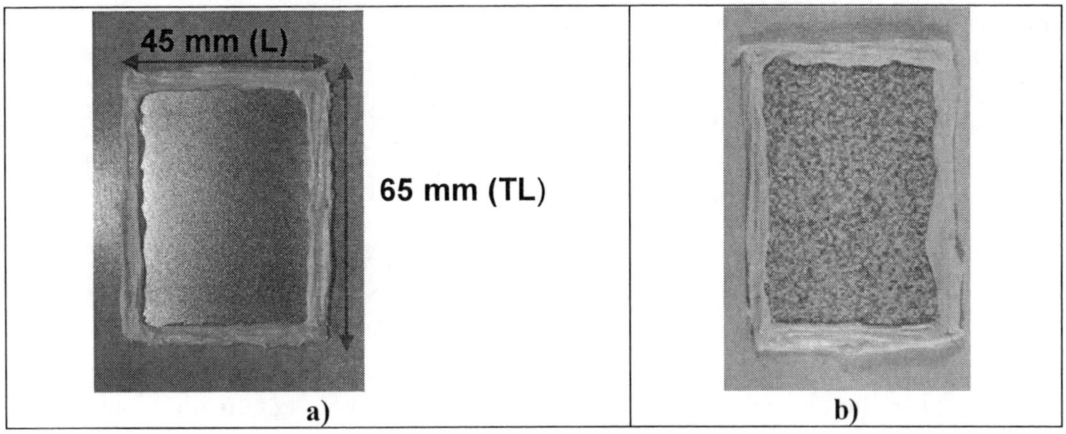

Figure 5: (a) State-of-the-art long life material before "CAC test": specimen's back & edges protected with adhesive & silicone to avoid parasite corrosion. (b) State-of-the-art long life material after 2 weeks "CAC test" & hot water rinsing.

RESULTS

The typical aspect of 400µm thick samples after a 20 days exposure in SWAAT test is presented in Fig 6. The cross-section micrographs evidence a strong improvement in the corrosion resistance of the multilayer sheet material in comparison with the state-of-the-art long life solution. On these pictures, both transgranular and intergranular (IGC) corrosion are observed for the state-of-the-art long life solution core. Regarding the multilayer product, corrosion propagation is confined in the low Cu 3xxx interlayer thanks to its sacrificial properties relative to the Cu-containing 3916 core. Hence, no further propagation of corrosion or any trace of IGC in the core alloy can be noticed.

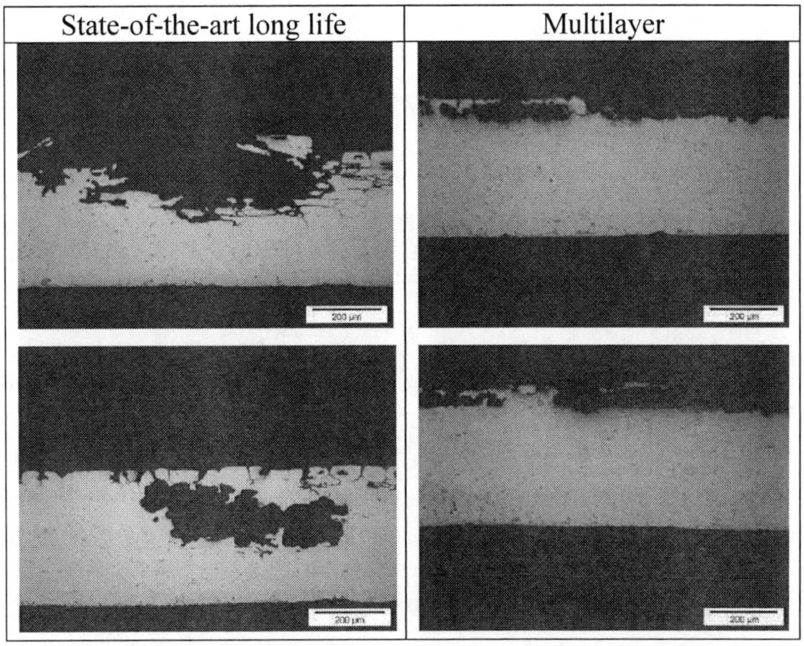

Figure 6: Cross sections of 400µm state-of-the-art long life and multilayer (low Cu 3xxx interlayer, 8%) samples after a 20 days SWAAT test exposure.

Regarding the CAC test applied on 480µm thick samples and using the condensate at pH2.5 and 100ppm Cl^-, corrosion initiates along the eutectic Si needles (Fig 7) and propagates into the core of the brazed state-of-the-art long life product within only two weeks of exposure (Fig 8).

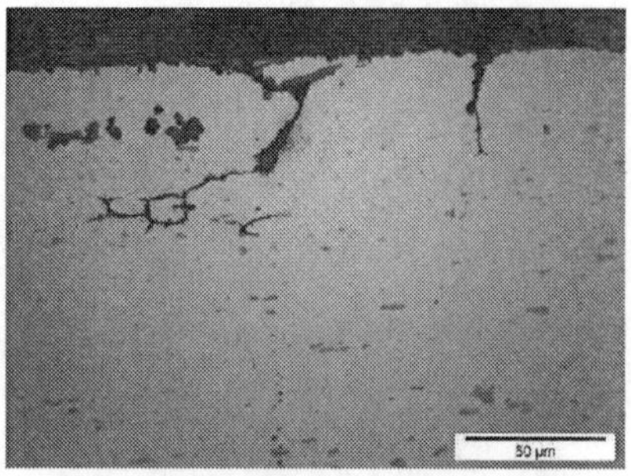

Figure 7: Detail of corrosion initiation along eutectic Si needles then IGC propagation in the core of a 480µm state-of-the-art long life product after a 2 weeks CAC test exposure.

As a comment, initiation of a corrosion path into the 4343 filler layer before brazing is much more difficult due to the homogenous distribution of spheroidized Si particles relative to needles in the "after brazing" version. (Fig 8). This hinders an easy access of the corrosive condensate to the core and explains the shallow corrosion aspect of the products before brazing.

After brazing, once the corrosive solution reaches the core alloy in the state-of-the-art long life material, extensive IGC occurs, likely to induce blistering. On the contrary, for the multilayer product, the corrosive condensate penetrating along the Si needles of the re-solidified filler alloy is stopped by the low Cu 3xxx interlayer as this Cu-free layer is IGC resistant and sacrificial vs. the 3916 core. Hence, no trace of any corrosion propagation in the core can be detected for the multilayer solution after 2 weeks test.

	Before brazing	After brazing
State-of-the-art long life	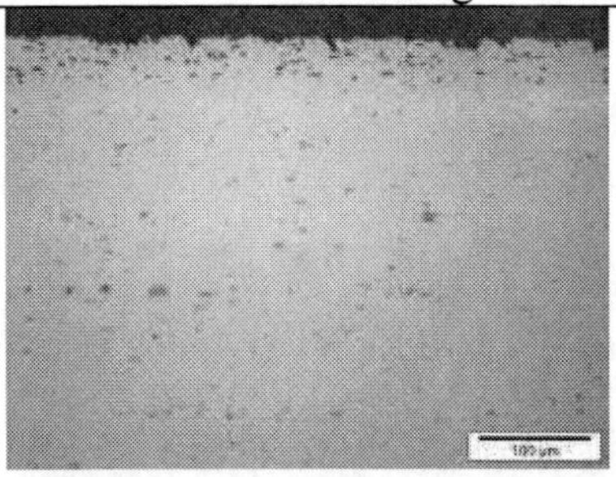	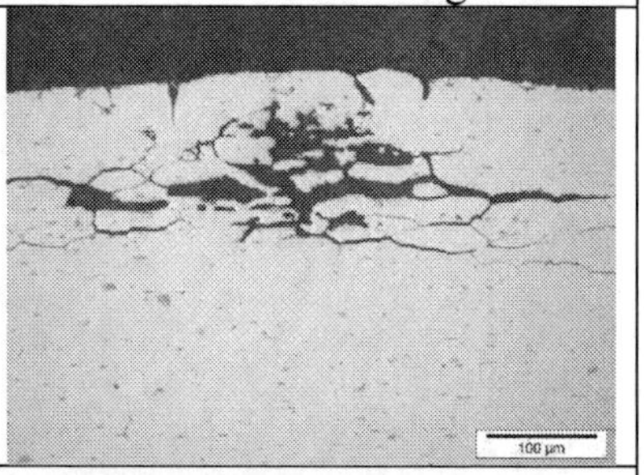
Multilayer	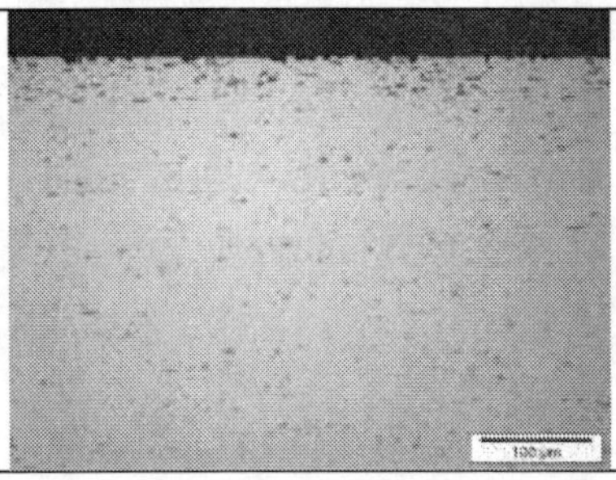	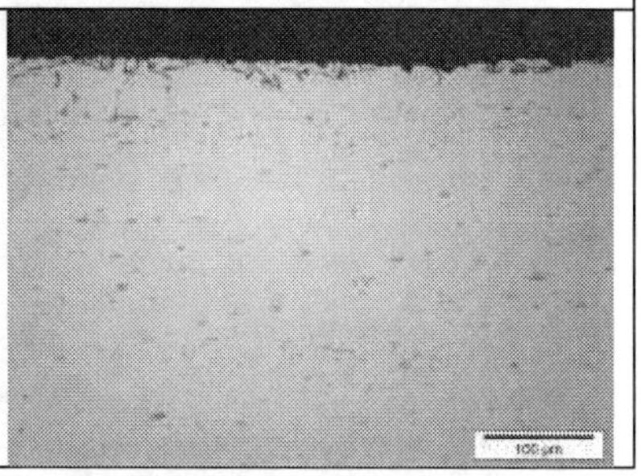

Figure 8: Cross sections of 480µm state-of-the-art long life and multilayer (low Cu 3xxx interlayer, 8%) samples after a 2 weeks CAC test exposure using a synthetic condensate pH2.5, Cl 100ppm.

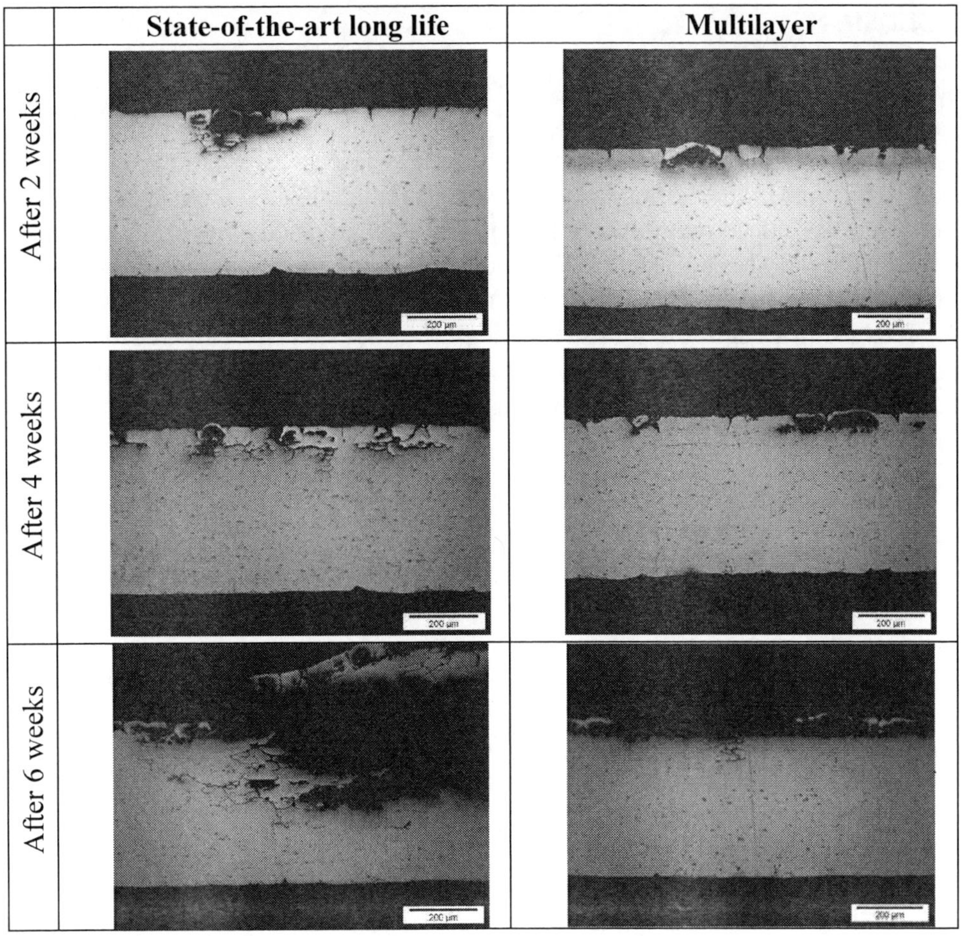

Figure 9: Cross sections of 480µm state-of-the-art long life and multilayer (low Cu 3xxx interlayer, 8%) samples after a CAC test exposure using a synthetic condensate pH2, Cl 1000ppm up to 6 weeks exposure time.

A second "CAC test" was launched using a more aggressive pH2 and 1000ppm Cl^- condensate and an exposure time up to 6 weeks (Fig 9). As it is difficult to compare accelerated corrosion tests with in service lifetime, the 6 weeks maximum duration (long exposure time) was arbitrarily chosen in order to show the durability of multilayer solution when the state-of-the-art long life solution is about to be fully penetrated and to highlight the corrosion mechanisms. After brazing, the state-of-the-art long life solution core material exhibits an IGC sensitivity by 2 weeks exposure which becomes more and more severe with the test duration leading to full destruction of the sample at 6 weeks. On the contrary, the ability of the interlayer, in the multilayer solution, to prevent any corrosion propagation in the core is evidenced at 2 and 4 weeks exposure. The low Cu 3xxx interlayer is IGC resistant and act as a sacrificial layer. Moreover, this additional layer leads to an extended Cu diffusion profile from the 3916 core to the extreme surface resulting in a very low surface Cu content which offers an optimized sacrificial behavior of the top layers. Only for further exposure up to 6 weeks, this interlayer is eventually fully corroded in some areas providing access for the corrosive solution to the core and leading to IGC initiation.

CONCLUSIONS

A new laboratory corrosion test "CAC test" has been developed to address the specific applications of Charge Air Cooler coupled with a low pressure EGR architecture. The wet and dry cycles combined with the spraying phase of a synthetic condensate is detailed in this paper and testing is carried out on a state-of-the-art long life product as well as a new developed multilayer solution. This new advanced material includes a low Cu 3xxx interlayer which affects neither the mechanical nor the brazeability properties. In addition, it shows significant corrosion resistance improvement after a standardized SWAAT as well as using the more realistic "CAC test" when compared to a state-of-the-art long life solution. The main contribution of the low Cu 3xxx interlayer is its self-resistance to intergranular corrosion, its sacrificial properties vs. the 3916 core and its ability to extend the Cu diffusion profile.

REFERENCES

[1] M. Zheng, G.T. Reader, J.G. Hawley, Energy Conversion and Management 45 (2004) 883.

[2] P. Gautier, A. Albrecht, A. Chasse, P. Moulin, A. Pagot, L. Fontvieille, Oil And Gas Science And Technology, 64 (2009) 361.

[3] S. Moroz, G. Bourguoin, J.M. Luján, B. Pla, SAE Paper 2009-01-2805 (2009).

[4] M. Nylen, U. Gustavsson, B. Hutchinson, A. Ortnas, Proceedings ICAA5, Materials Science Forum, vol. 217–222, 1703, Trans. Tech. Publication, Zurich, Switzerland (1996).

[5] G.J. Marshall, R.K. Bolingbroke, A. Gray, Met. Trans. A 24, 1935 (1993).

[6] T.L. McKinley, SAE Paper 970636 (1997).

[7] J. Garcia, SAE Technical Paper 2012-01-1947, (2012).

Alternative Powertrain

Efficient and integrated thermal management for electric vehicles

B JIANG [a,b], **Q WANG** [a] **and Y Y YAN** [a]
[a] HVACR & Heat Transfer Research Group, Faculty of Engineering, University of Nottingham, UK
[b] School of Mechanical and Automotive Engineering, Hefei University of Technology, P.R. China

ABSTRACT

In this paper, a heat pump air-conditioning system integrated with thermal management system of battery is proposed for electric vehicles to meet the requirements of safety, performance, thermal comfort and reliability. Multiple operation modes of the integrated system have been discussed. A numerical analysis has been carried out to analyze the performance of the integrated system and results show that energy used in cold climates has been reduced. A heat pipe battery thermal management system was also introduced to keep the battery temperature uniform and within the optimal range. We therefore concluded that the proposed integration could be a viable solution for EV thermal management.

Keywords: electric vehicle; heat pump; air-conditioning; thermal management

1 INTRODUCTION

The employment of electric and plug-in hybrid electric vehicles (EVs and PHEVs) has achieved great potential to significantly reduce the consumption of petroleum and other high CO_2 emitting transportation fuels. Lithium-ion battery (LIB) technology is expected to be the best choice of energy storage for EVs. However, Li-ion batteries can be used efficiently only within a specific temperature range. Higher temperatures degrade LIBs more quickly, while low temperatures reduce power and energy capabilities [1-3]. A valid thermal management system (TMS) is becoming increasingly important to keep batteries operating within optimum temperature range.

The air conditioning system plays a crucial role for that it not only provides thermal comfort (temperature and humidity condition in cars), but safe driving conditions (defrost and demist). Unlike conventional car with internal combustion engine, EV cannot harvest the waste heat from the engine to maintain a comfortable cabin environment during winter. Air conditioning system is the most energy consuming auxiliary equipment, especially in EVs. If the cooling and heating were provided by the battery packs alone, the driving mileage can be reduced up to 50% [4,5]. Therefore, it is essential to develop a high-efficiency air conditioning system to promote the practicability of EVs.

Heat pump is a well-developed high efficiency technology used for decades and many products have been designed based on that technology. It is currently being developed to work on EVs. Heat pump can be used for either heating or cooling the cabin depending on the time of the year and personal preferences of the driver and passengers. In heating mode, the system absorbs heat from the surrounding or other sources and heats up the conditioned space. Theoretically the rejected heat is significantly higher than the energy consumption and is easily to meet the requirement of cabin heating for the electric vehicles, so it will extend the mileage compared with using PTC heater by consuming battery electricity directly. A new heat pump air conditioning system integrated with thermal management system of battery for EV is proposed in this paper.

2 SYSTEM DESCRIPTION

The new integrated system is shown in Fig. 1. The heat pump air conditioning system mainly consists of a variable-speed compressor driven by a brushless DC motor, an out-car heat exchanger with a fan installed in the front of the vehicle, an accumulator, an evaporator, an in-car condenser, three expansion valves and some valves. The evaporator and the in-car condenser are installed in the different ventilation ducts and can be switched by a damper. The working fluid of the heat pump system is refrigerant (e.g. R134a, HFO1234yf). Compared with the air conditioning system of the traditional vehicle, the additional components have been installed in the system in order to obtain the manifold flows of the refrigerant. TMS utilizes the glycol solution as heat transfer medium. An auxiliary heater is installed for battery preheating during the severe cold season, and a heat exchanger is used to integrate the heat pump system with the TMS.

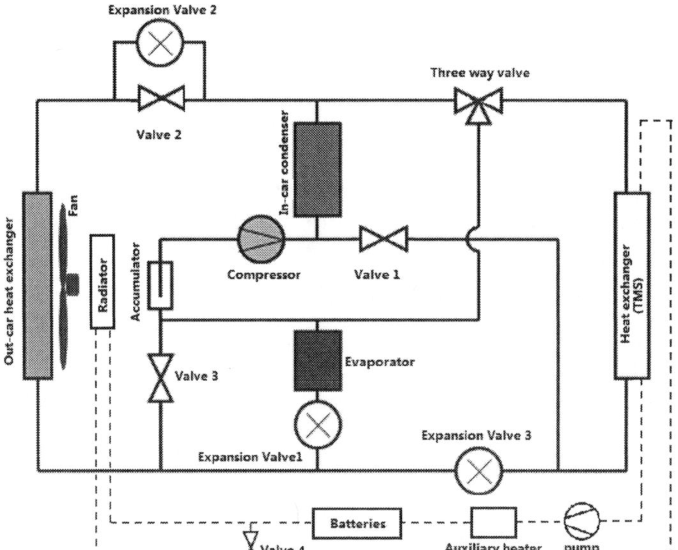

Fig.1 Schematic of integrated system

The integrated system can operate under multiple modes to meet the requirements of air conditioning and thermal management of battery.

2.1 Flow of refrigerant in air-conditioning mode

In cooling mode of air-conditioning, the compressor pressurizes the refrigerant and discharges high pressure, high temperature refrigerant vapour, the vapour flows through the in-car condenser and Valve 2 into the out-car heat exchanger, where refrigerant condenses into a high pressure liquid by forcing air over the out-car heat exchanger enabling heat to transfer from the refrigerant to the outside air. The subcooled liquid then flows from the out-car heat exchanger to the Expansion valve 1 which then changes the refrigerant into low pressure, low temperature liquid/vapour. The liquid/vapour then flows to the evaporator where the heat from either inside air or outside fresh air is transferred to the refrigerant and the refrigerant becomes slightly superheated, then flows back to compressor. The flow of the refrigerant is shown in Fig.2.

Similarly in the heating mode of air-conditioning, the compressor discharges the refrigerant into the in-car condenser, where the heat is transferred from the refrigerant to the air of cabin for heating, then flows to Expansion valve 2 and becomes low pressure, low temperature liquid/vapour. In the out-car heat exchanger the refrigerant absorbs heat from the outside air and vaporizes to be superheated, then flows through Valve 3 into compressor. The flow can also be found in Fig.2.

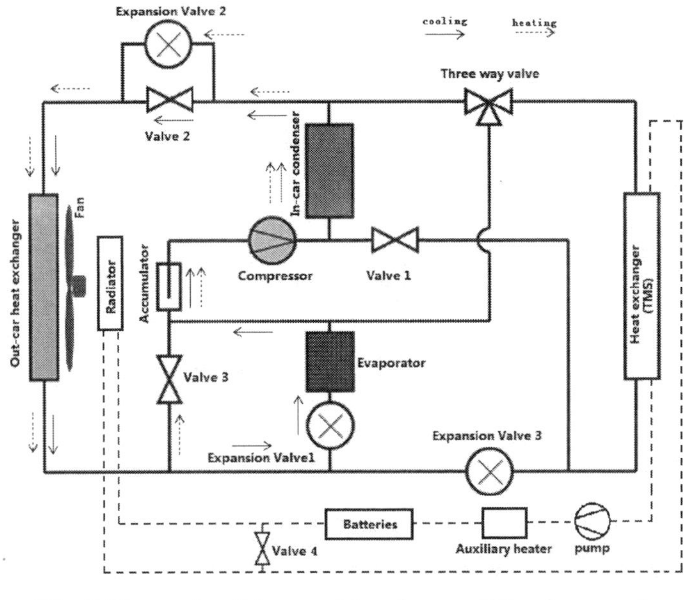

Fig.2 Flow of refrigerant in air-conditioning mode

2.2 Flow of the glycol solution

During severe cold seasons, low temperature reduces power and energy capability of the battery. The battery should be warmed up to suitable temperature before running. Turn on the auxiliary heater, heats up the glycol solution first and then transfer heat to the batteries. In order to reduce the heat loss, the Valve 4 is open to bypass the out-car heat exchanger.

In mild cooling condition of battery, the glycol solution absorbs the heat from the battery package and dissipates heat through the radiator using the air as the cooling source. After cooled, the glycol solution flows back, pumped to the batteries and the cycle repeats.

Note that there is no heat exchange between the refrigerant and the glycol solution, it only flows through the TMS heat exchanger, and Valve 4 is closed.

In hot time, the glycol solution cannot be cooled down effectively by the outside air in the radiator, and the temperature of the outside air even overtops the upper limit of optimum temperature range. In the case the system will convert to integrated mode.

2.3 Flow of refrigerant in integrated mode

Fig.3 shows the flow of refrigerant in integrated mode. The mode of cooling or heating of the air-conditioning system is the same as that of the thermal management system. For the reason of the heat loss, the Valve 4 is open during the mode.

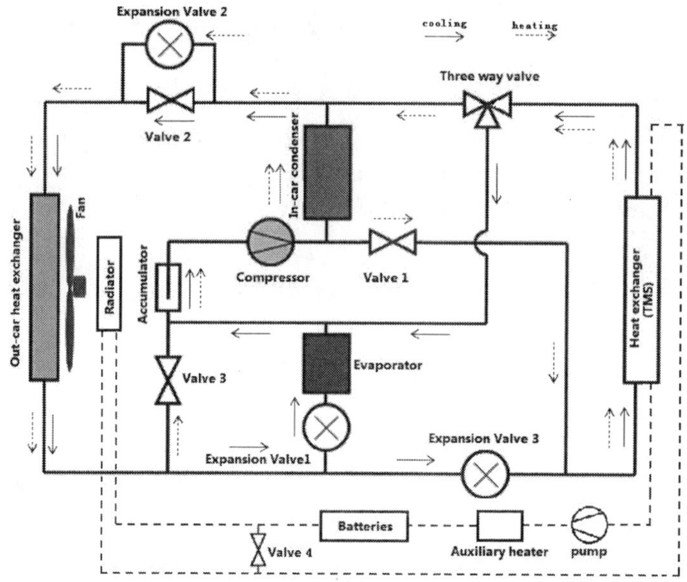

Fig.3 Flow of refrigerant on integrated mode

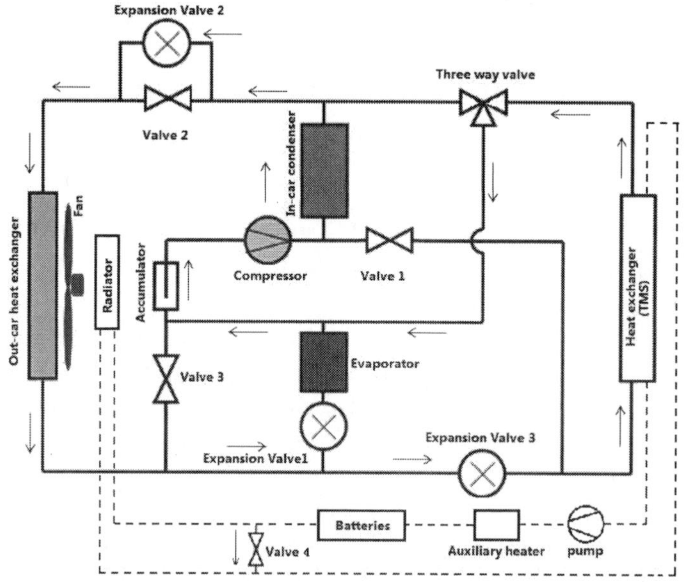

Fig.4 Flow of refrigerant on heat recovery mode

The advantage of the integrated system is that the system can recover the heat from the glycol solution in the TMS heat exchanger if the battery runs in cooling mode while the air-conditioning system operates in heating mode. It reduces the consumption of heating.The flow of the refrigerant can be found in Fig.4.

2.4 Demonstration of battery thermal management

As an important part of the integrated system, a concept of the battery thermal management (BTM) is shown in Fig. 5 [6]. Heat pipes will be inserted into each battery cell cavity, where the evaporator (cooling mode) or condenser (heating mode) of the heat pipes can be attached to an aluminium plate for temperature flattening. A liquid channel underneath the battery pack serves as a heat exchanger allowing heat to be removed from or supplied to the battery as requested as the glycol solution flows through the channel. This facilitates the bi-directional characteristics offered by the heat pipe, such that the system can provide either cooling or heating without moving parts. During cooling mode, the medium (glycol-water) in the heat exchanger performs as the catalyst to help the heat pipe effectively dissipate unwanted heat. It can also be heated supplying an adequate amount of heat to the heat pipe instantly warming up the battery.

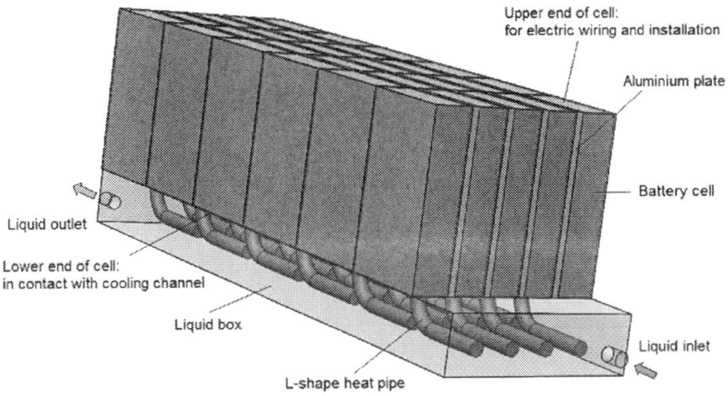

Fig. 5 Heat pipe BTM demonstration

3 RESULT AND ANALYSIS

A simulation for a five-seat EV with a 16.5 kWh lithium-ion battery package is discussed. Put aside the effect of Global Warming Potential of the refrigerant, R134a is used as the refrigerant. In cooling mode, the rated conditions are: cooling ambient temperature 35°C, in-car temperature 27°C, condensing temperature 63°C, sub-cooling temperature 0°C, evaporating temperature -1°C and superheating temperature 10°C. In heating mode, the rated conditions are: ambient temperature -10°C, in-car temperature 20°C, condensing temperature 55°C, sub-cooling temperature 3°C, evaporating temperature -15°C, and superheating temperature 2°C. The temperature of the glycol solution is designed to be 20°C. Isentropic efficiency of vapor compression is assumed to be 80%, and the heat loss of heat exchanger is neglected.

Actual battery heat generation is complex, which varies with time and temperature and depends on how current is distributed especially within large size batteries. A simplified battery thermal model [7] was used to estimate the heat generated from the battery under different charging or discharging rates. Heat generation rate during discharging is expressed with Eq.(1), Where Q_l is heat generated from positive electrode and negative electrode, I_d is battery discharge current and $R_{t,d}$ is total electrical resistance during discharging.

$$Q = 3.37 \times 10^{-2} Q_1 I_d + 3.60 R_{t,d} I_d^2 \tag{1}$$

For the Li-ion battery cell used in the simulated car, the thermal power per unit cell ranges from -0.07W to 5.5W under pre-charging and maximum charging conditions, while the heat release under 1C, 2C to 4C normal/maximum/aggressive continuous discharging varies from 3.78W to 41.27W. According to the practical driving condition, the average discharge rate is about 0.2C to 1.0C. The total generated heat of the battery package is approximately 350W, 550W and 850W under 0.3C, 0.6C and 0.8C respectively for the battery package.

According the operation conditions and using the pressure-enthalpy diagram of the refrigerant R134a, the compression works, which varies with ambient temperature of the coupling system under different battery heat dissipations, have been calculated as shown in Fig.6. The compression work decreases during heating mode but increases in cooling mode with increasing ambient temperature. The more heat the battery dissipates, the lower the compression work needed during heating mode and vice versa in cooling mode.

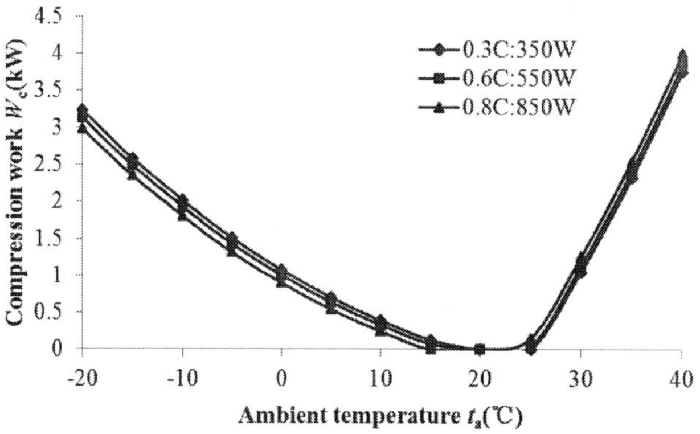

Fig. 6 Compression work varies with the ambient temperature

Fig.7 shows the energy saving rate $R_{es} = (P_{cs} - P_{cc}) / P_{cs}$ of the system with different battery heat loads under different ambient temperatures. Here P_{cc} is the energy consumption of the integrated system and P_{cs} is that of the sole air-conditioning system. The saved energy on heating mode increases with the increasing ambient temperature. The energy saving rates under three battery heat dissipation conditions are 3%, 7%, 13% respectively. It is effective to reduce the consumption for EV during cold weather. Remember that more energy will be consumed in severe cooling mode because the additional heat is rejected mainly from the air conditioning system and the compressor capacity of the integrated air conditioning system is larger than usual.

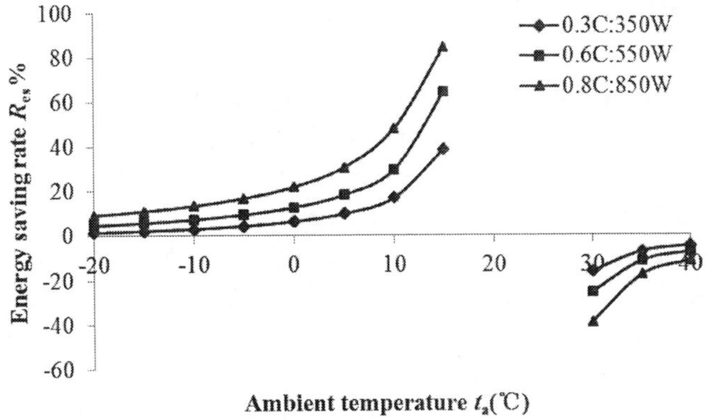

Fig. 7 Energy saving rate varies with ambient temperature

The numerical solution of the battery pack has been also obtained using pressure-based finite control volume scheme from commercial CFD package, ANSYS ICEPAK 14.5. Two conditions were simulated: 2C discharge condition for maximum continuous operation (12 W per unit cell) and 4C for aggressive discharging (41.27 W per unit cell). This covered a wide range of heat generated by the batteries revealing the degree of effectiveness of the proposed method. The temperature of the fluid supplied to the pack was set to be 20°C, and the flow rate was 0.059kg/s.

The simulation result is shown in Fig. 8. For the battery pack with 30 cells described in Fig.5, the proposed heat pipe cooling method maintains the average battery surface temperature ranging from 22.8–25.2°C and 29.8–38.2°C during 2C and 4C discharge condition. For 2C maximum continuous discharging, this method is able to maintain the cell temperature below 40°C at all times that satisfies the requirement. A maximum temperature difference of 0.5°C from cell to cell and 2.4°C across the whole battery pack under 2C discharge are resulted.

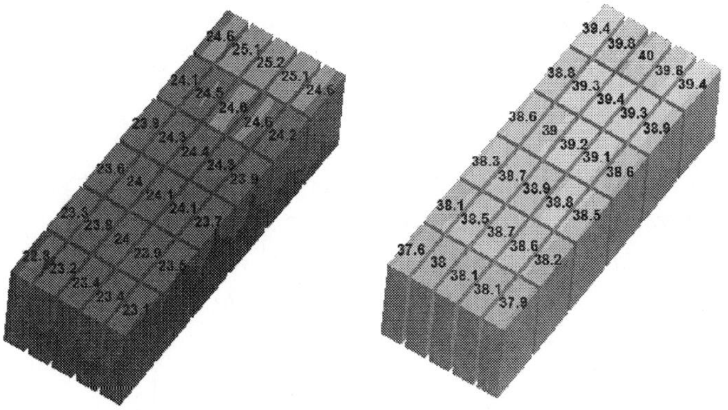

1) 2)

Fig. 8 Temperature distribution of the battery pack under:
1) 2C discharge 2) 4C discharge

4 CONCLUSION

This paper introduces a heat pump system integrated thermal management of the battery , which facilitates energy optimization. Numerical analysis indicates that the energy saving rates under three battery heat dissipation conditions are 3%, 7%, 13% respectively, but more energy will be consumed under cooling because of the extra heat from the battery. Numerical simulation suggests a good temperature uniformity achieved by the heat pipe BTMS, making the maximum temperature difference from cell to cell within 0.5°C and 2.4°C across the whole pack even under the extremely bad summer condition, i.e. 2C maximum continuous discharge condition at 35°C. These results indicate that the proposed integrated thermal management with heat pipe BTMS can be a promising solution for EVs.

REFERENCES

[1] PESARAN A.A., MARKET T, TATARIA H, Howell D. Battery Requirements for Plug-in Hybrid Electric Vehicles: Analysis and Rationale. Presented at the 23rd Int. Electric Vehicle, Symposium and Exposition (EVS-23), Anaheim, California, December 2007.

[2] PESARAN, A.A., KEYSER K. Thermal Characteristics of selected EV and HEV Batteries, Proceedings of the 16th Annual Battery Conference: Applications and Advances, Long Beach, California, January 9–12, 2001.

[3] KEYSER M., PESARAN A.A., MIHALIC A., ZOLOT M. Thermal Characterization of Advanced Battery Technologies for EV and HEV Applications, NREL Report, August 2000.

[4] HOSOZ M., DIREK M. Performance evaluation of an integrated automotive air conditioning and heat pump system. Energy Conversion & Management, 2006, 47: 545-559.

[5] CHEN J.P., SHI J.Y., ZHAO Y. International latest development of automotive air-conditioning system. Journal of Chemical Industry and Engineering (China), 2008, 59(S2): 9-13.

[6] Q.W. ANG, B. JIANG, Q.F. XUE, H.L. SUN, B. LI, H.M. ZOU, Y.Y. YAN. Experimental investigation on EV battery cooling and heating by heat pipes. Applied Thermal Engineering, DOI:10.106/j.applthermaleng.2014.09.083.

[7] N. SATO. Thermal Behaviour analysis of lithium-ion batteries for electric and hybrid vehicles, Journal of Power Sources. 99 (2001) 70-77.

Improving electric vehicle energy efficiency with co-simulation of cooling system, HVAC system and electric drivetrain

D DVORAK
Austrian Institute of Technology, Austria

C RATHBERGER and **A LICHTENBERGER**
MAGNA, Engineering Center Steyr, Austria

1 INTRODUCTION

Engineering Center Steyr (ECS) and Austrian Institute of Technology (AIT) are currently involved in a joint project (supported with funds from the "Mobility of the Future" program of the Austrian BMVIT / FFG) with the target of HVAC efficiency optimization of an electric vehicle. Both partners have their own specific background and knowledge for this topic and both partners use their individual tools for simulation tasks. In this project we have explicitly decided not to centralize simulation efforts to one of these tools, but to set up a simulation environment which combines the individual strengths of the different software products involved.

This co-simulation based approach is already quite common in vehicle development processes at different European and international OEMs, but we believe that it will gain even more importance for the development of electric vehicles:

- On the one side the individual subsystems of an electric vehicle are very closely linked to each other. For example the drivetrain of an electric vehicle is closely coupled to the HVAC system in case of active battery cooling with a chiller.
- On the other side an efficient electric vehicle can only be achieved, if all these interactions are considered and optimized in the development process. A highly efficient electric motor will not be enough to achieve decent vehicle range, if too much electric energy is used for heating the cabin.

Therefore successful development of electric vehicles requires close coordination between different domains, departments and software products. In this paper we want to identify improvement potentials for a specific electric vehicle (Mitsubishi i-MiEV) using such an approach.

1.1 Vehicle overview

The Mitsubishi i-MiEV is a compact sized electric passenger car (Japanese K-car) with a 16kWh battery pack and a 49kW electric motor. Mitsubishi specifies its NEDC range with 160km, but a strong influence of heating, air conditioning and real live customer cycles can be expected.

For characterizing the required parameters for the simulation models, the test vehicle (Mitsubishi i-MiEV) will be tested in the laboratories. Therefore different driving cycles will be used.

Figure 1: The Mitsubishi i-MiEV (picture from AIT)

1.2 Driving cycles and measurements

The "*New European Driving Cycle*" (NEDC), which is shown in Figure 2, is the current legislative cycle prescribed to determine whether a new model of light duty vehicle meets applicable EU environmental regulations for passenger cars. This test-cycle is also adopted to determine the range and the energy consumption of electric vehicles. The NEDC cycle is divided into two parts. The first part simulates the driving conditions in urban areas. The second part simulates the driving conditions in extra-urban areas.

During the tests the electrical and thermal parameters of the traction battery of the vehicle were measured by acquiring the CAN bus data via the "*On-Board Diagnostics*" (OBD) port. The data were recorded via the measurement system "*Cronos PL*" from the company "*IMC Integrated Measurement Control*" and stored on an on-board workstation. The variables recorded are:

- Time [s];
- Velocity [km/h];
- Battery state of charge [%];
- Battery voltage [V];
- Battery current [A];
- Battery pack, driver cabin and test chamber temperatures [°C].

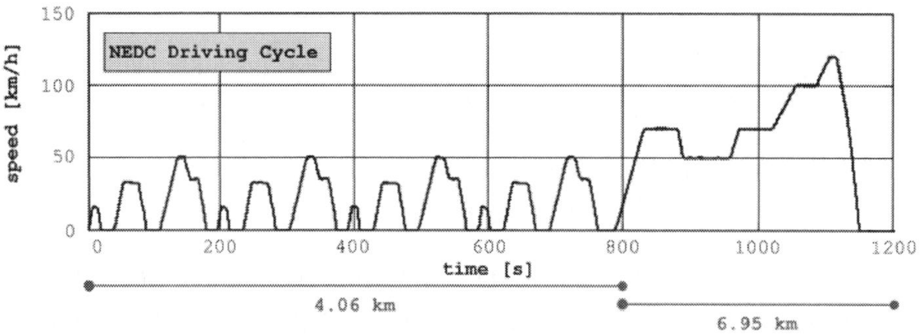

Figure 2: NEDC driving cycle and phases

2 VEHICLE SUBSYSTEMS AND SIMULATION MODELS

Thermal models for e-components, coolant- and AC-circuits and the passenger compartment have been set up in KULI, electric component models and driving simulation is done in Modelica / Dymola.

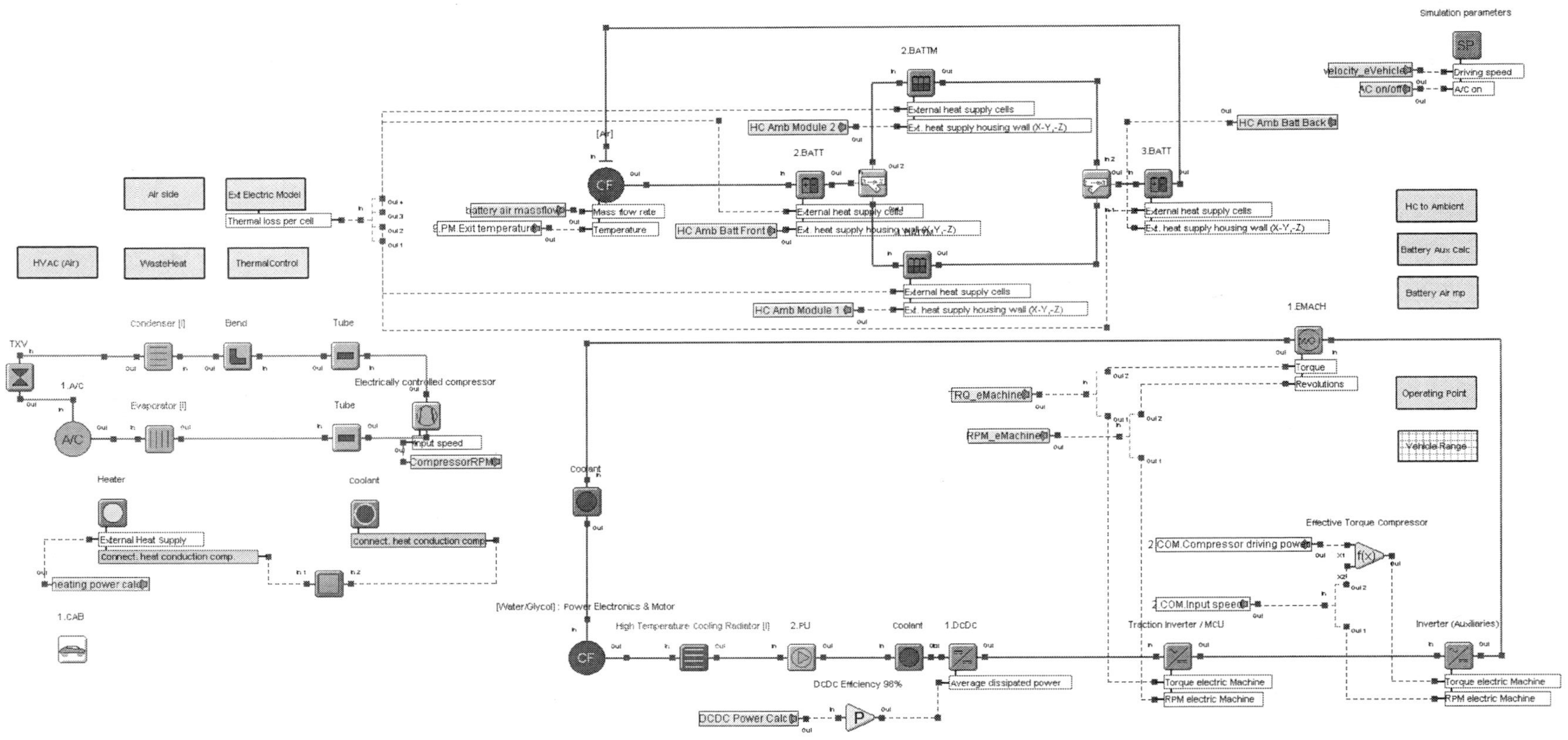

Figure 3: The thermal simulation model

In the following the various sub-systems will be explained in more detail.

2.1 Battery pack and battery pack modeling

The battery pack is located in the floor of the vehicle and consists of 88 battery cells grouped to 2 modules with 4 cells and 10 modules with 8 cells each. The cells used are Yuasa-Mitsubishi LEV 50 with the following specifications:

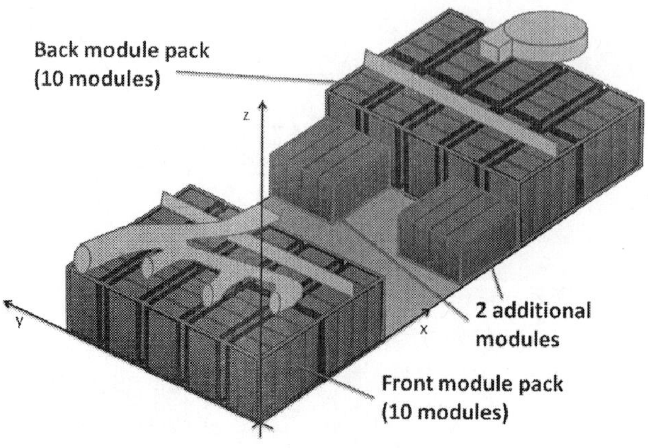

Figure 4: i-MiEV battery pack module layout

- Type: LiIon
- Capacity: 50Ah
- Nominal Voltage: 3.7V
- Weight: 1.7kg

The module is called LEV 50-4. All 88 cells are electrically connected in series which yields an overall battery pack voltage of 325.6V.

2.1.1 Battery pack thermal simulation model

The thermal model of the battery pack is set up in KULI, where the front- and back-module pack and the two additional modules are modeled separately (internally each cell is modeled as an individual component, which means that cell temperatures can be evaluated separately).

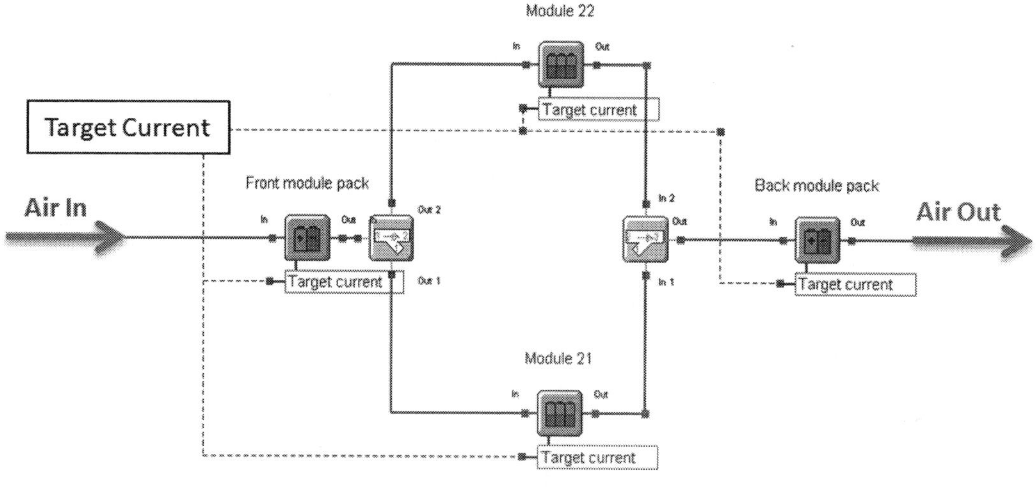

Figure 5: i-MiEV battery pack in KULI

The battery pack is air-cooled with air coming directly from the HVAC box. During fast-charging the inlet air is cooled down by the HVAC system in order to keep the maximum cell temperature below a specific limit. Otherwise air at environmental temperature level will be used for battery air conditioning.

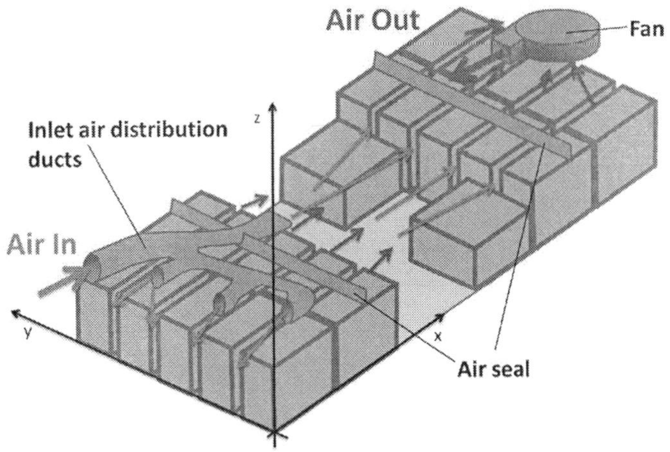

Figure 6: i-MiEV battery pack principal air flows

The cold air enters the battery pack through inlet distribution ducts, which guide the airflow both through the front module pack and directly to the back module pack. The two additional modules are cooled by the bypassing air. In the back of the battery pack there is a fan which blows the cooling air out of the vehicle.

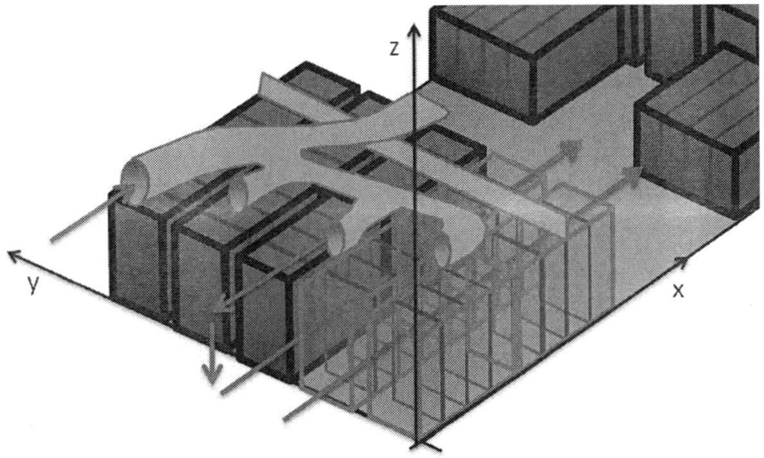

Figure 7: Cooling air flow between the front modules

This airflow causes convective cooling of the battery cells, from the layout it can be expected that the modules and cells in the back (downstream) will reach slightly higher temperature levels. Whether these temperature levels reach critical values (larger than 31°C reduces the power output, larger than 50°C damages the cells) will be one of the simulation results.

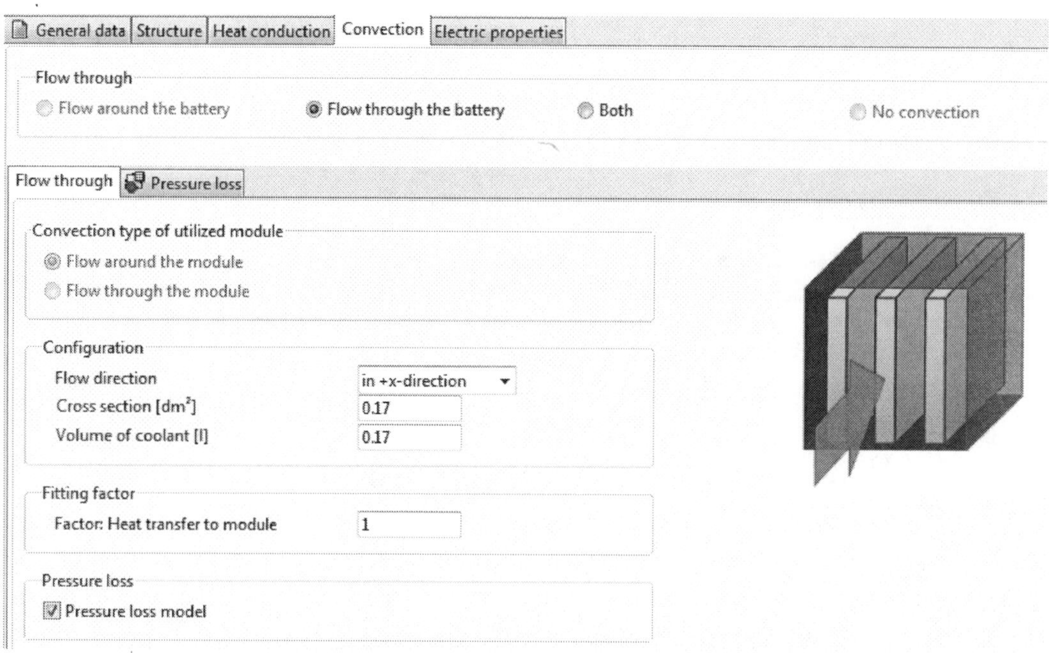

Figure 8: Air flow definition in the simulation model

As the battery cells are packed tightly inside the modules, there will be no significant air flow between the cells. From cell to cell heat conduction is modeled and additionally there is forced convection along the thin sides of the cells (i.e. along the module surface).

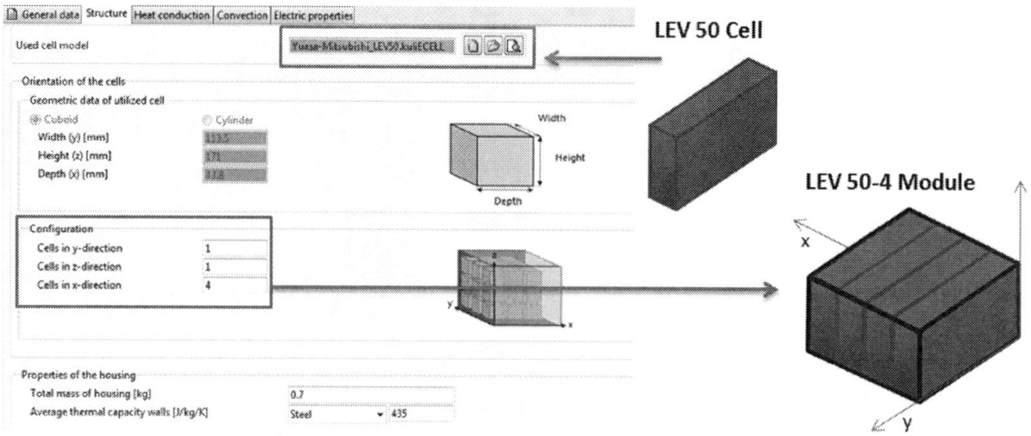

Figure 9: Module definition from cells

The battery pack will be required to provide a certain target current during the simulation. This current depends on the vehicle operating point (from driving simulation), the electric drivetrain (simulation models for electric motor, fixed transmission gearbox and power electronics) and the electric network simulation model.

This already points out the main interface between the different simulation tools in this model:

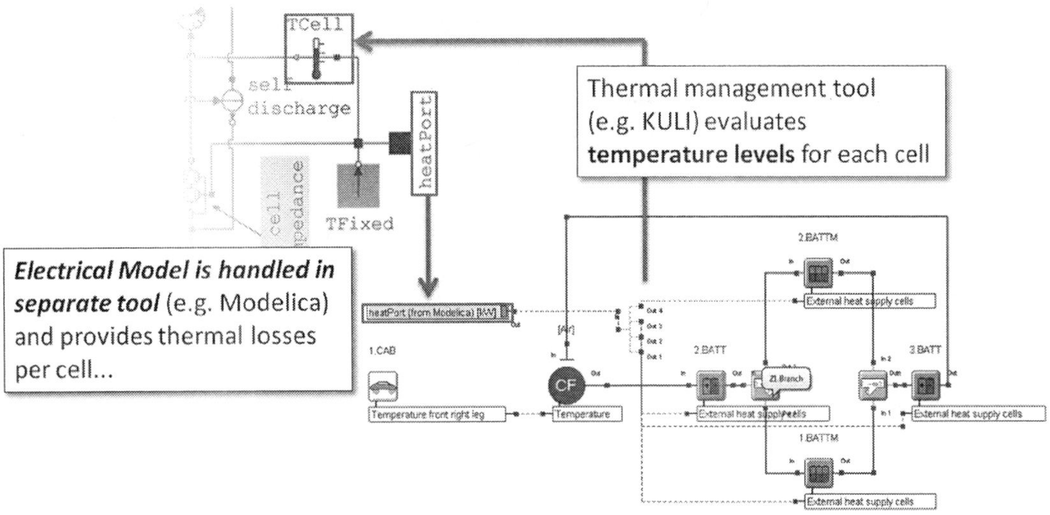

Figure 10: Interface between electric and thermal simulation models (example shows Modelica/KULI)

In our application Modelica is used to solve all electric networks and specifically determine the thermal losses per cell in the battery pack. KULI then determines cell temperatures and feeds them back to Modelica via the interface. Cell properties in the electric model will depend (among other things) on these temperature levels.

2.1.2 Battery pack electric simulation model

The battery model implemented in the Electric Energy Storages (EES) library, based on the programming language Modelica, enables simulating the electrical behavior of lithium-ion batteries on cell and stack level. Modelica allows creating complex models in different multi-physical domains, resolving either algebraic or differential equations to model a system. The graphical user interface Dymola was adopted to interpret the Modelica language. In addition to the analysis of the electrical operational behavior, the aging of the battery and its thermal behavior can be also implemented.

Two categories of electric battery models are available in the EES library:

- **Cell model**: a single battery cell;
- **Stack model**: a whole battery stack, consisting of a number of cells connected in series or in parallel.

These models can be resolved by both static and dynamic analysis, depending on the transient response of the pin voltage. The dynamics of the pin voltage depends on the number of serially connected RC-circuits used for the electric equivalent circuit. The main features of these approaches are:

- **Static model**: this model is adopted for quick analysis of a battery system, and only a simplified electric equivalent circuit is used with a minimum number of input parameters needed to configure this model. In the static models, the electric equivalent circuit of one cell consists only of a voltage source, representing the open circuit voltage (OCV), and a serially connected resistor, representing the serial resistance of the battery.
- **Dynamic model**: this model is adopted for transient operating behavior of the battery system and consists of an advanced electric equivalent circuit. In

addition to the static cell model, it includes up to two serially connected RC-circuits, to model the transient voltage drops of the double-layer effect (for short time periods) and the diffusion effect (for long term periods). Furthermore, self-discharge and aging effects can be considered.

The dynamic cell model "*DynamicImpedance*" from the EES library was used in this work, being capable to simulate the transient operating behavior of a lithium-ion battery. This model consists of an electric module and a parameters adaption module. The latter adjusts the parameters according to the different battery operating states, identified by the battery state of charge (SOC), temperature and aging status. Since the measurements in this work were performed only over a short period of time, aging effects will not be considered in this work.

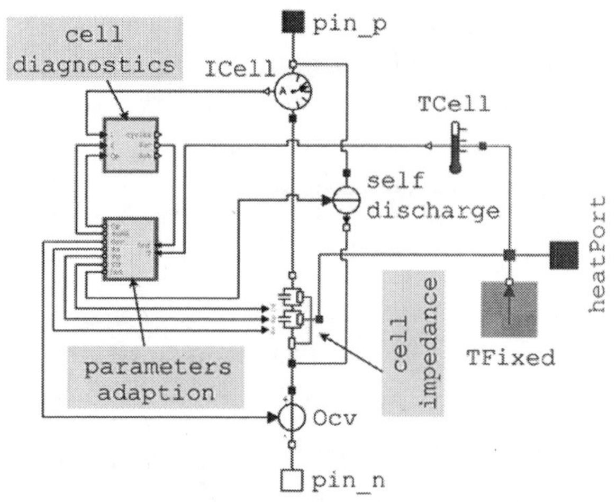

Figure 11: Diagram of the dynamic cell model

Figure 11 shows the implementation of the dynamic electrical cell model in Modelica. This model consists of the following main components:

- **Pin_p / pin_n**: positive and negative pin to integrate the battery model in a higher model layer.
- **OCV**: voltage source, representing the open circuit voltage.
- **Cell impedance**: battery impedance that consists of a serial resistance and up to two serially connected RC-circuits to model the transient voltage response.
- **Self-discharge**: current source, representing the self-discharge of the battery.
- **ICell**: cell current measurement.
- **Cell diagnostics**: model for calculating diagnostic values that cannot be measured directly (e.g. calculating the SOC).
- **Parameters adaption**: model for calculating the operational dependencies and adapting the electric parameters, related to the current state of charge and cell temperature. In this sub-model the aging mechanisms can be also considered.
- **TCell**: cell temperature measurement.
- **TFixed**: optional fixed cell temperature (it can be used when the cell temperature variation does not provide a significant value for the simulation).
- **HeatPort**: interface where the electric losses can be forwarded to an external thermal cell model to calculate the current cell temperature.

The thermal connector "heatPort" is the interface to the thermal model and hence the interface between the two simulation environments. The heat which is generated within the battery cell due to electric losses during operation is forwarded to the thermal model via this connector.

As the modeling of a battery via an electric equivalent circuit is well known in the automotive sector, we focus in the next paragraph on the SOC-OCV and parameter estimation as well as the model calibration.

2.1.2.1 Parameter estimation

For simulating the battery model, the parameters of the electric equivalent circuit over the SOC are needed. This also includes the SOC-OCV correlation from the measurement data.

Therefore the measured battery voltage and the SOC during subsequently performed NEDC load cycles (over the entire SOC) is analyzed for each point in time where the load current equals 0 A, and then it is correlated to the corresponding SOC value. A schematic overview of this strategy for SOC-OCV determination can be seen in Figure 12.

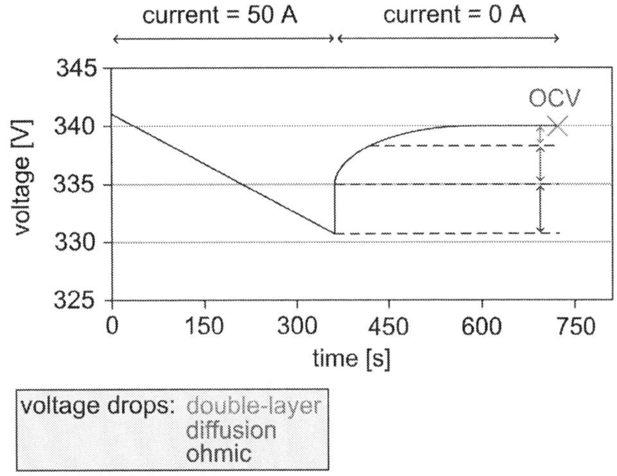

Figure 12: Voltage drops during relaxation interval and OCV determination

Single values that differ too much from the average trend are removed. The SOC-OCV points are then best-fitted with a curve, as depicted in Figure 13.

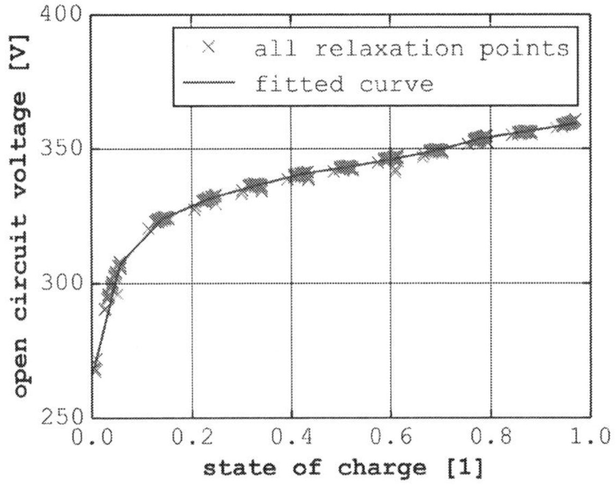

Figure 13: Fitted SOC-OCV graph from discrete states during real-life measurements

Based on the different time constants of the serial resistance and both RC-parallel circuits the remaining parameters could be extracted by analyzing the dynamic operating behavior of the battery voltage during operation. The model parameters (except from the OCV) have been extracted as constant values, i.e. they are not dependent on the SOC. The validity of this assumption will be analyzed in chapter 3.2.

2.2 Electric Components liquid cooling circuit

Apart from the battery, additional components which require cooling are:

- Electric motor:
  The i-MiEV has a 49kW permanent magnet synchronous motor which (even though the efficiency is between 80 and 90% for most operating points) requires active cooling (up to 10kW thermal losses)
- Motor Control Unit:
  This is an integrated unit which includes the traction inverter (converting 330V DC to 3-phase AC) and the motor control.
- DC/DC:
  An additional converter providing low voltage DC for charging a secondary (auxiliary) 12V conventional lead battery. This battery supplies power e.g. for pumps and on-board electronics.
- Further cooling requirements arise from the on-board charger (which will not be relevant during driving operation, though).

Each of these components will be subject to thermal losses (the highest thermal loads will occur in the electric motor, the lowest in the DC/DC) and thus have to be cooled. Compared to the main battery (operating preferably below 31°C and getting damaged above 50°C), power electronics components tolerate a much higher temperature level. Therefore a secondary cooling circuit is used.

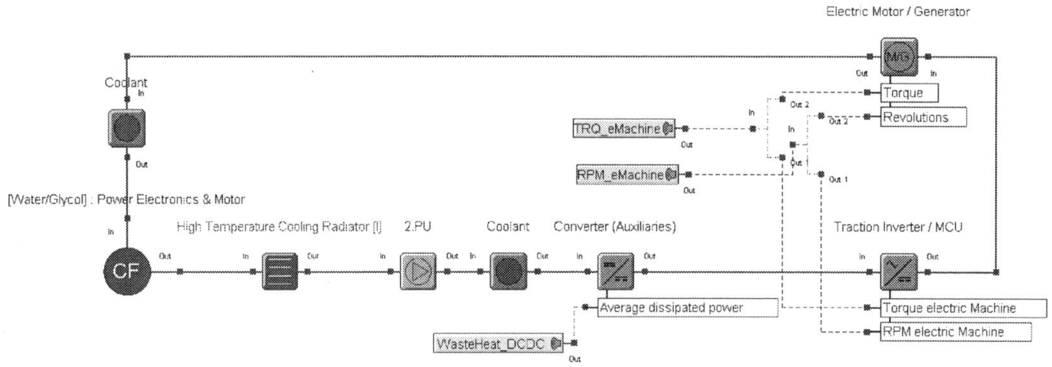

Figure 14: Power Electronics Cooling and High Temperature Radiator

The thermal models for the power electronic components are kept relatively simple, as this work focuses on system integration topics.

The electric motor is modeled in KULI as two point masses (total motor mass including gearbox 65kg).

- The first mass models stator and motor casing and also contains the channels of the cooling jacket. As we have a permanent magnet motor, the main currents and thus the main thermal losses occur in the stator.
- The second mass models the rotor, it will by coupled to the stator thermally by heat conduction and will see only little direct warm-up.
- As the motor is located in the rear underbody of the Mitsubishi i-MiEV, some convective air cooling is modeled as well (in addition to the active coolant cooling).

The traction inverter and the motor control unit (MCU) are one integrated, water cooled component in the Mitsubishi i-MiEV and are modeled as a DCAC component with respective mass and performance in our simulation model. This component and also the DCDC converter are modeled as simple lumped masses each.

2.3 HVAC systems

The Mitsubishi i-MiEV is heated by an indirect PTC element. Water is heated up in a dedicated circuit and the hot water is then used to transfer heat to the cabin air-flow via a radiator. This setup has the advantage that ICE variants of the i-MiEV can use the same heating system with the combustion engine as the main heat source.

In our simulation model we have so far decided not to model the coolant circuit explicitly (currently no coolant mass flow information is available), but to consider both the (electric) heat source and the thermal capacities involved by alternatively implementing a setup of two thermally connected point masses.

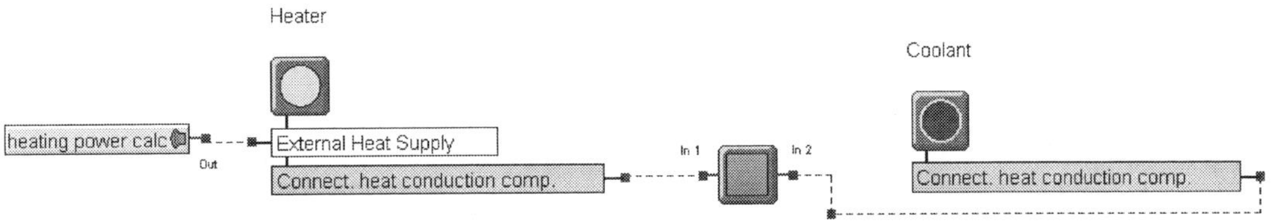

Picture 1: Indirect PTC heater system

The air conditioning system is a standard AC circuit driven by an electric compressor.

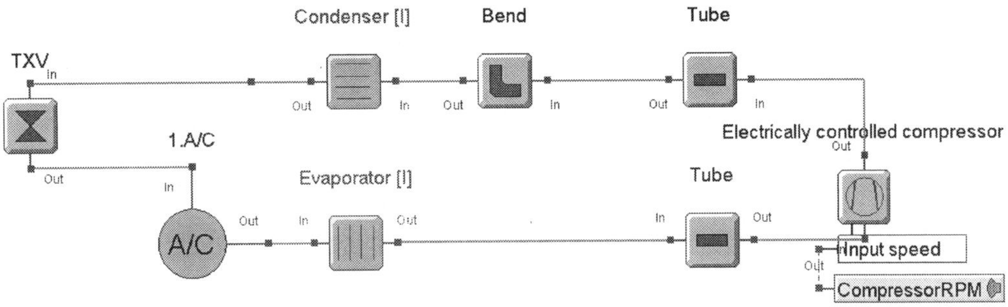

Figure 15: The air conditioning system

The control strategies for the HVAC systems (modeled in separate sub systems) have to fulfill the following requirements:

- The passenger compartment has to be heated up to a specified interior temperature (for a certain cabin zone, e.g. “driver head”) and kept at this level in winter conditions.
- The passenger compartment has to be cooled down to a certain temperature level and kept there in summer conditions.
- During fast charging the HVAC system has to provide sufficient cooling for the air cooled battery pack.

Finally a multi-zone cabin model is used to simulate the effects of these systems on passenger comfort.

3 SIMULATION MODEL VALIDATION AND ADJUSTMENT

In order to prove the validity of the simulation models and the extracted parameters, a validation must be performed. Therefore, simulation results will be compared to real measurement data.

3.1 Validation of the thermal simulation model

For the validation and fine-tuning of the simulation model, data was available from a dynamometer measurement under warm-up conditions. The complete vehicle including cabin was preconditioned to -5°C, the battery was preconditioned to 1°C and the ambient temperature was kept around -5°C as well.

The velocity profile shows alternating phases of movement (vehicle driving at 50km/h, 90km/h and 100km/h) and halt, the heating is started after 400s at different power settings during the cycle.

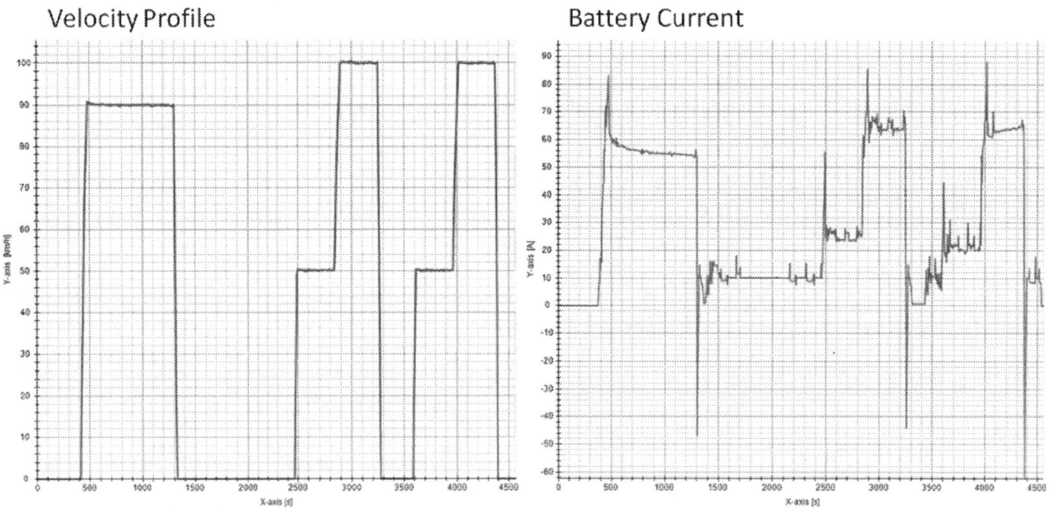

Figure 16: The adjustment cycle

Especially during the high velocity / high load phases of the cycle the battery pack warms up (for the idle phases the temperature levels are almost steady), which can also be reproduced in the simulation model.

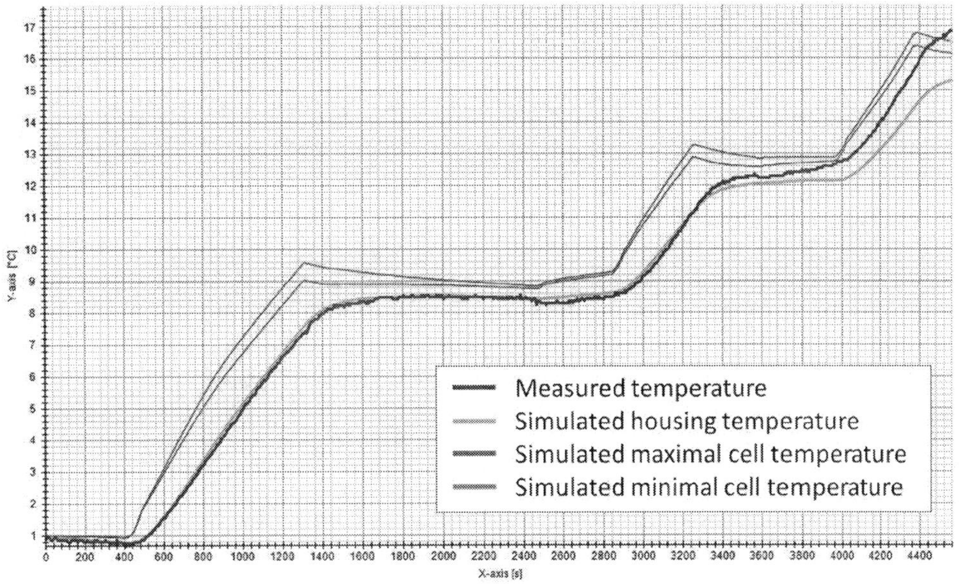

Figure 17: Measured and simulated battery temperature levels

In the measurement the temperature sensor was placed in the service port of the battery pack, therefore the measured temperature correlates mainly with the local housing temperature. The cell temperatures are slightly higher, and in the end temperatures in the battery pack reach around 17°C. Small deviations between measurement and simulation are most likely due to uncertainties of the exact sensor position and model simplification.

Another focus was put on the temperature simulation inside the cabin, the basis for this being of course the temperature in the inlet ducts of the ventilation system (between heater and cabin air vents).

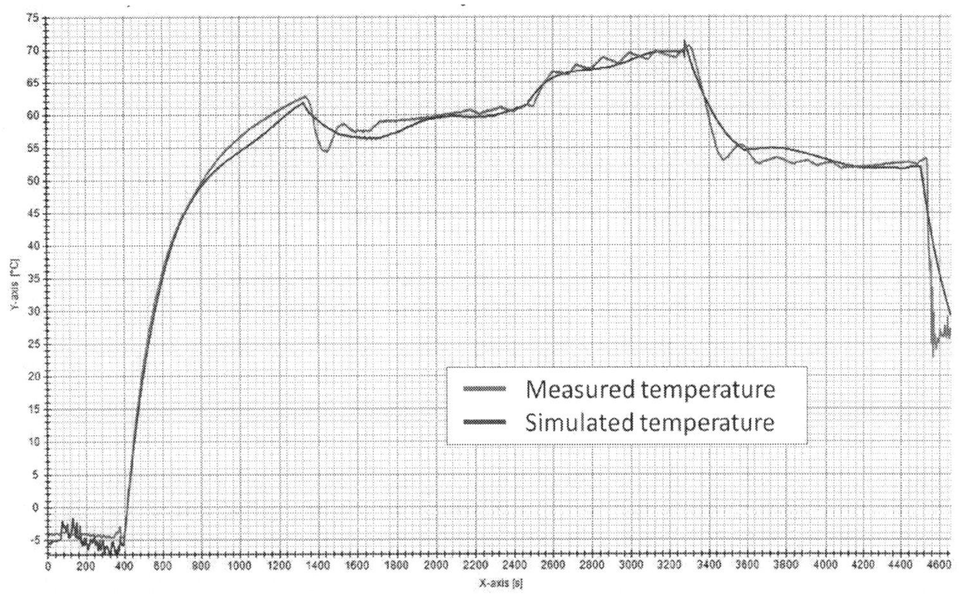

Figure 18: Measured and simulated inlet duct temperatures

These values show good overall agreement with small deviations again mainly due to model simplifications.

Finally temperatures inside the cabin were evaluated. For these topics there is always some deviation, because the measurement sensors generally log the air temperatures at specific positions, whereas the simulation model can only provide average temperature levels for certain zones of the cabin model.

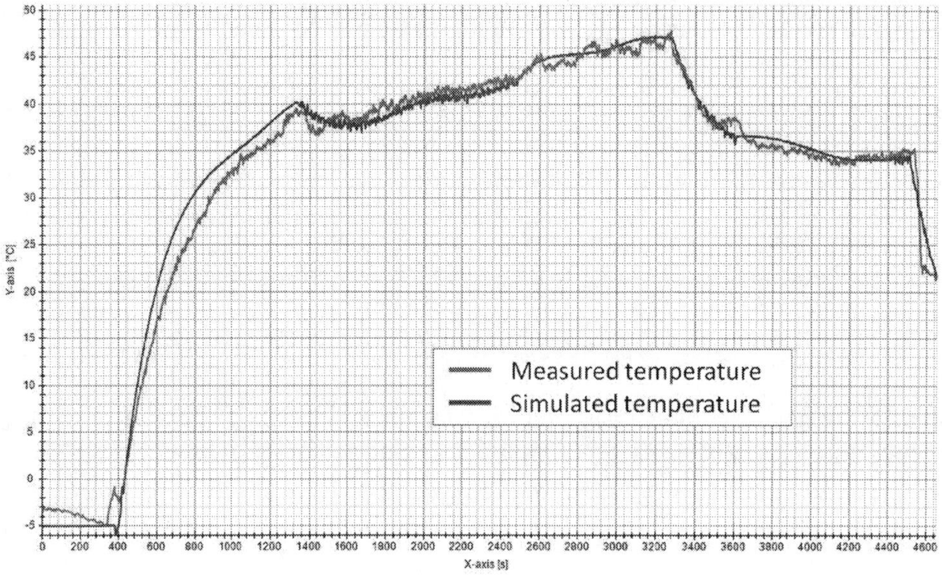

Figure 19: Measured and simulated air temperatures front head middle

As the focus in this investigation is on the driver and co-driver head and chest positions, this is where most effort was put in the model adjustment. Apart from the initial warm-up, where some deviations are present between simulation and measurement, the agreement is quite good.

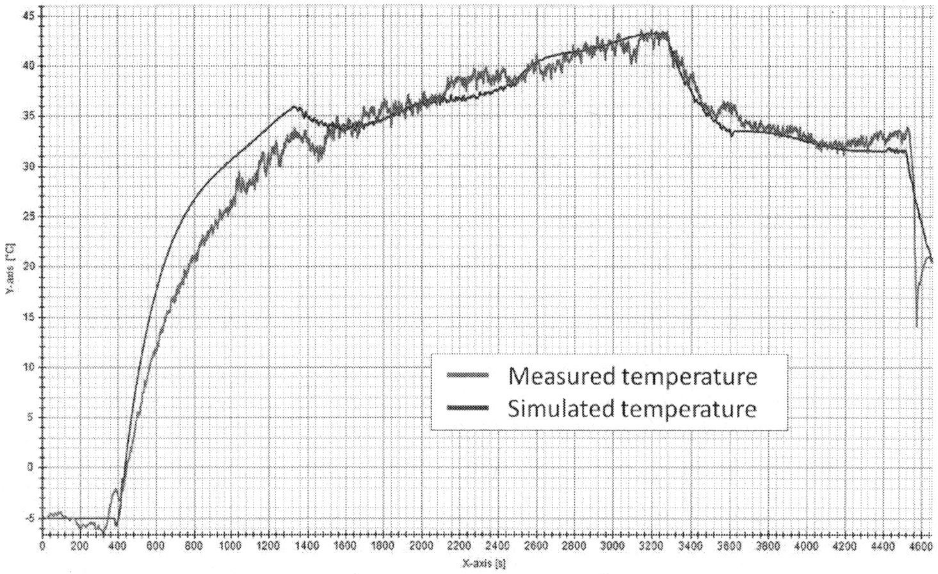

Figure 20: Measured and simulated air temperatures front chest right

Based on this adjusted model the next task was to investigate the influence of heating and identify improvement potentials for a different standard driving cycle.

3.2 Validation of the electric battery model

For validating the electric battery model, a specific current profile has been applied to the battery. The graph of the load cycle over time is depicted in Figure 21.

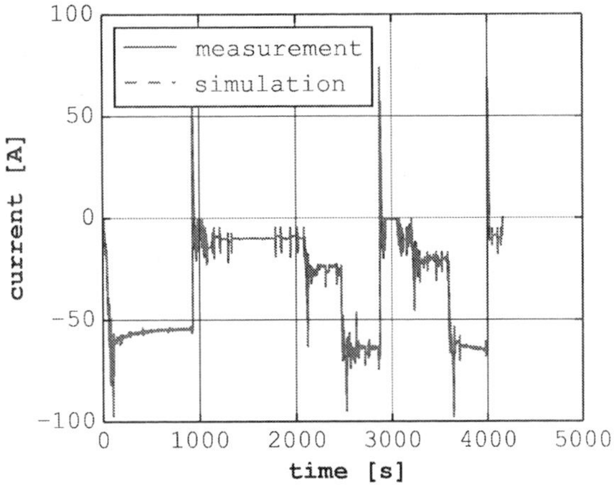

Figure 21: Current profile used for validating the electric battery model

During validation the battery voltage and SOC will be analyzed. Figure 22 shows the measured and simulated SOC during the current load cycle. The figure shows a high coherence between the simulation and the measurement.

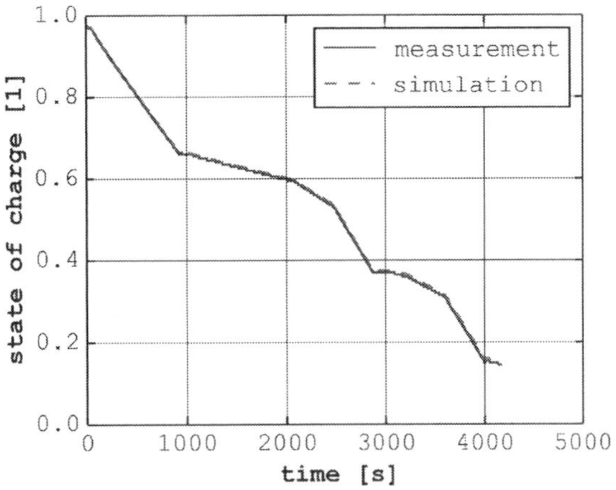

Figure 22: Measured and simulated SOC trend during the validation load cycle

As the battery voltage is a more dynamic measure than the SOC, it is much more difficult to reproduce its real behavior than for the SOC. Figure 23 illustrates both the measured and simulated battery voltage during the load cycle. It can be seen that the transient behavior could be reproduced at high precision in the simulation. The voltage offset which can be observed in the figure can be explained by an inaccurate extraction of the SOC-OCV. Anyway, the maximum deviation between the measured and simulated voltage is below 4 V. As this means an approximate relative error of 1.2% (relatively to 330 V average voltage) the parameters are considered as sufficiently accurate for an entire vehicle simulation. Furthermore, it must be taken into account that the simulation results represent a battery stack of 88 cells connected in series.

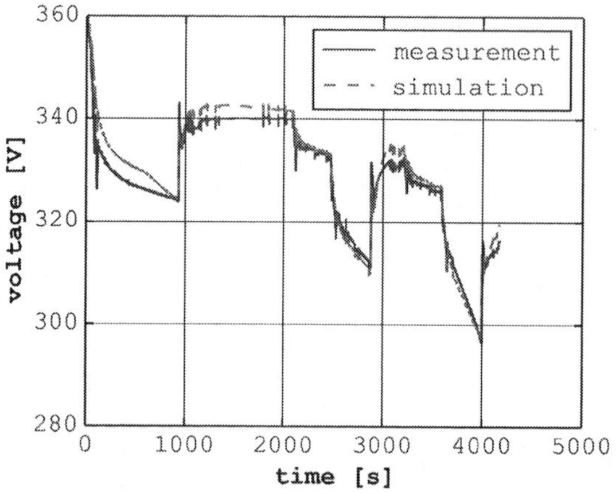

Figure 23: Measured and simulated battery voltage during the validation load cycle

4 RANGE INVESTIGATIONS FOR DIFFERENT HEATING CONCEPTS

For our next investigations the NEDC cycle was used. The reason for this choice was the fundamental importance this cycle still has for legislative issues and the good availability of related data.

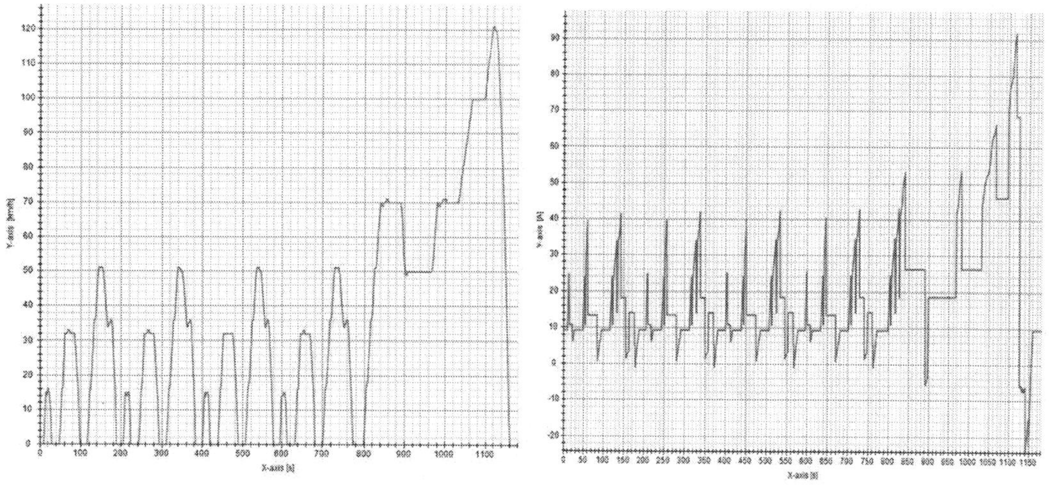

Figure 24: Velocity profile and simulated electric load for the NEDC

We have chosen boundary conditions similar to the original adjustment case (ambient temperature and vehicle initial temperature at -5°C, battery pack preconditioned to 1°C). As the NEDC generally is a very low-load driving cycle (apart from a short slightly higher velocity phase close to the end), only very little thermal losses are available for heating (even for vehicles with internal combustion engines). The first simulation results show transient temperature levels in the battery pack.

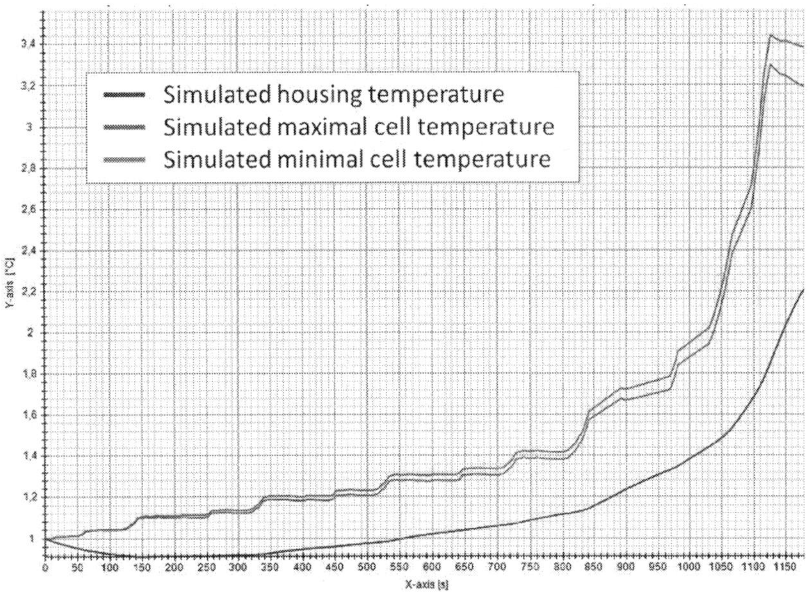

Figure 25: Simulated housing- and maximum and minimum cell temperatures

As expected we see only very slow warm-up of the battery pack, obviously it will not be possible to harvest any useful thermal losses for additional cabin heating (at least not in a straight forward way).

Turning our attention to the cabin temperatures, we have compared two different simple heating concepts: A heater providing constant 3kW of heating power and a second, controlled heater set to achieve temperature levels of around 21°C in the cabin interior (with a maximal available heating power of 3kW). In both scenarios the recirculation rate is set to 0% (fresh air mode). This means that more heating power is required, but this is also necessary to prevent condensation of accumulated air humidity on the cabin windows (due to the low ambient temperatures).

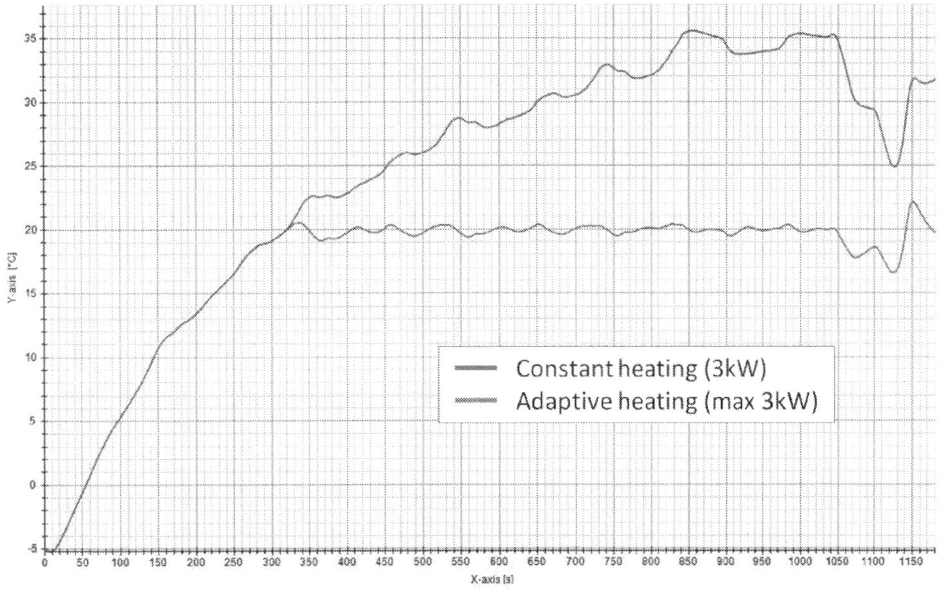

Figure 26: Warm-up for front chest right zone in NEDC

We can see that the target temperature is reached after around 300s so that obviously the full heating power is not required for the complete cycle. This becomes especially important, if we look at the implications this has on the vehicle range.

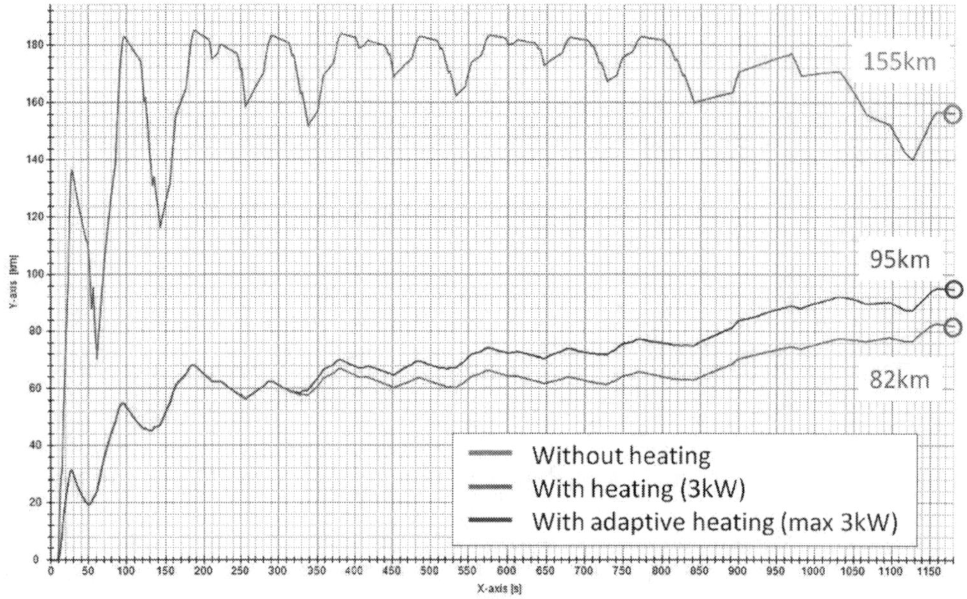

Figure 27: Transient average based range prediction for NEDC

While for a simulation run without heating we get a simulated NEDC range of 155km (which correlates well with available data for the Mitsubishi i-MiEV), having the heater set to full 3kW for the complete cycle reduces the range to 82km (approximately 53%). A more realistic controlled heater usage leads to 95km NEDC range (61%).

If we want to increase the vehicle range (while still providing the same passenger comfort), one possibility is to make use of additional heat sources:

- As the battery generates only minor thermal losses and due to the very high total thermal capacity of the battery pack only to low temperature levels are reached. Using them in a heater (e.g. to pre-heat the ventilation air for the cabin) is not practicable due to too low temperature gradients (although more complex applications like e.g. a heat pump might still be possible).
- There might be some salvageable thermal losses in power electronics and electric motor. We will investigate this.
- Finally the high thermal capacity of the battery pack can also be used in a positive way. If the battery pack is pre-heated to a high (but still safe) temperature level before the driving cycle starts (e.g. by fast-charging the battery and reducing the cooling efforts in a controlled way), it turns into a quite significant thermal storage. We will precondition the battery pack to 30°C, switch on the battery air-cooling system and use the warm air leaving the battery pack to pre-heat the air entering the cabin (before the PTC heater itself). Safety and legal considerations (battery gassing must not contaminate the cabin interior, of course) have to be taken into account for practical application, but are not considered for this first potential analysis.

In the following we compare warm-up and range simulations for these different heating variants.

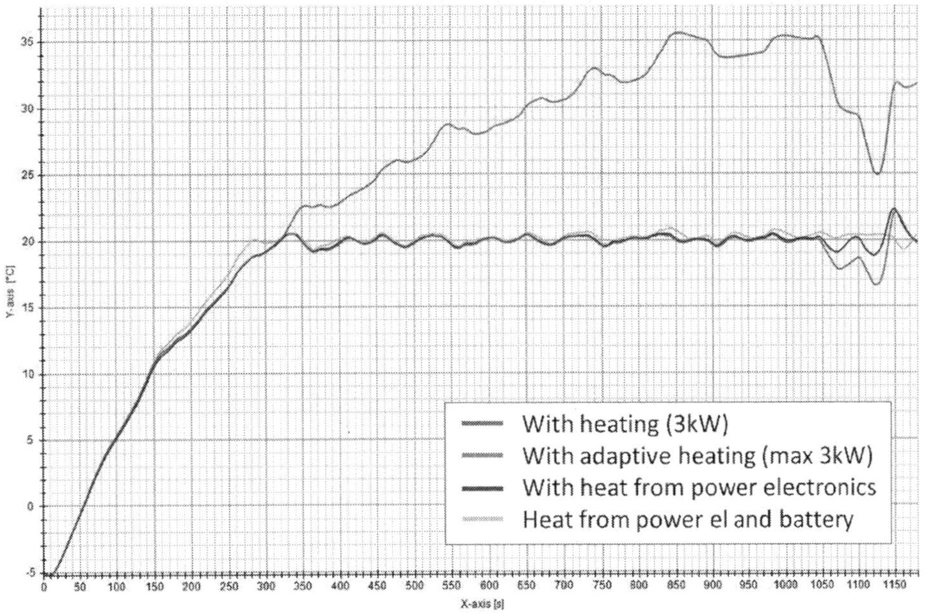

Figure 28: Simulated temperature levels front chest right

This first figure shows that the different variants show hardly any influence on cabin temperatures. Using the additional heat sources only slightly speeds up the warm-up and makes temperature levels slightly more independent from air flow fluctuations. Most importantly we can rule out any negative impacts on passenger comfort.

Vehicle range simulation on the other hand shows some clear benefits.

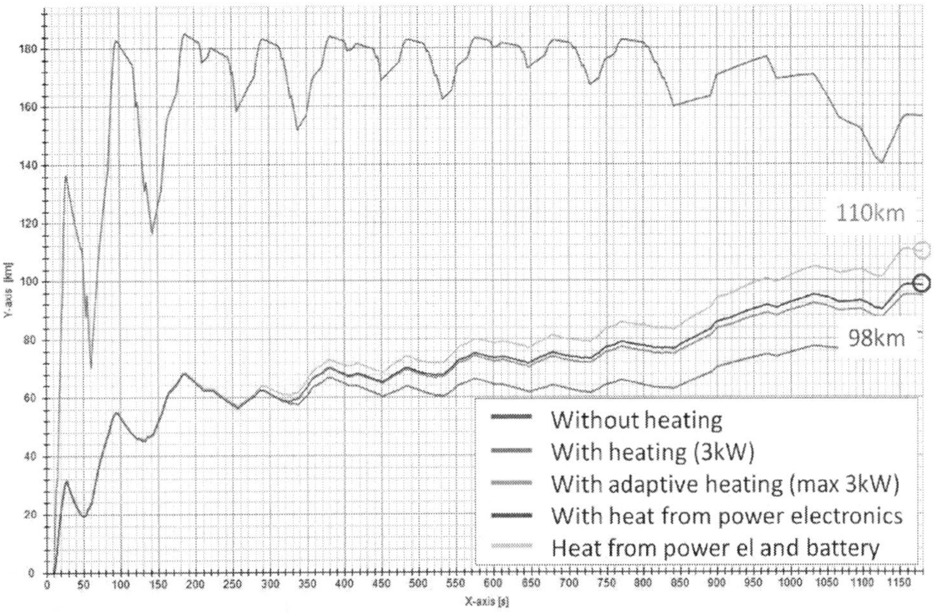

Figure 29: Transient average based range prediction for NEDC (using thermal losses for heating)

Using the thermal losses from power electronics and electric motor yields only a small range improvement to 98km (63%). This is no big surprise due to the low load operating points of the NEDC, we might still gain better benefits for higher system loads.

Using the battery pack as a thermal storage on the other hand leads to a significantly increased vehicle range of 110km (71%). If safety concerns can be solved in practical applications, this might be an interesting basis for further improvements.

5 SUMMARY AND OUTLOOK

We have set up a comprehensive simulation model describing electric components, drivetrain, cooling circuits and HVAC systems including a cabin model of the Mitsubishi i-MiEV. The initial models were calibrated using available testbench measurements for a warm-up run. Finally the calibrated models were used to identify improvement potentials for energy efficient vehicle warm-up during an NEDC.

While several approaches show relevant benefits, both intelligent control strategies and using thermal storage (e.g. in the battery pack) for initial vehicle warm-up promise some potential.

This paper merely represents the first steps in an ongoing research project, the next steps will include:

- Further validation runs for the simulation models
- Investigation of thermal storage in PCM elements
- Investigation of possible benefits from heat pump applications.
- Extension of investigations to different driving cycles (e.g. WLTC)

LITERATURE

[Abi14] Kälte ist der Feind des Elektro-Autos, 04.01.2014, http://www.bild.de/auto/auto-news/elektroauto/wintertetst-reichweite-34053450.bild.html (online: 17.01.2014)

[Ack10] M. Ackerl, G. Breuer, G. Pichler, A. Türtscher: Battery Thermal Simulation and Energy Management of Hybrid Commercial Vehicles, Battery Technology Expo 2010, San Jose, US

[Azz13] Dr. K. Azzouz, Dr. G. De Pelsemaeker: Thermal Comfort, Engine Cooling and Warm-up, SAE TMSS World Congress 2013, Detroit, US

[Boh13] F. Bohlender, H. Reiss: Elektrische Innenraumheizung von E-Fahrzeugen per PTC-System, ATZ 02/2013

[Bou10] C. Bouvy: Kälte aus Wärme; Adsorptionstechnik für die Klimatisierung im Automobil, ATZ 04/2010

[Cap13] C. Cap, C. Hainzlmaier: Schichtheizer für Elektrofahrzeuge, ATZ 06/13

[Fat11] M. Konz, N. Lemke, S. Försterling, M. Eghtessad: Spezifische Anforderungen an das Heiz-Klimasystem elektromotorisch angetriebener Fahrzeuge, FAT-Schriftenreihe 233, Forschungsvereinigung Automobiltechnik e.V. (FAT)

[Gam12] C. Gamsjäger, J. Hager, V. Kordesch, Dr. W. Öller: Intelligent Heat Management – A Chance for further Reduction of Emissions and Fuel Consumption, 21st Aachen Colloquium Automobile and Engine Technology 2012, Aachen, D

[Gop11] Goppelt G.: Dekra ermittelt Reichweite von Elektroautos bei Minustemperaturen, 14.11.2011, http://www.springerprofessional.de/dekra-ermittelt-reichweite-von-elektroautos-bei-minustemperaturen-14865/3951516.html (zuletzt als verfügbar geprüft am 17.01.2014)

[Hag12] J.Hager, Dr. C.Stroh, A.Lichtenberger: Investigation of Alternative Heating and Cooling Strategies for EVs, SAE TMSS World Congress 2012, Scottsdale, US

[Pay11] J. Paya, J. M. Corberan, B. Torregrosa-Jaime, C. Vasile-Müller: Innovative Air-Conditioning Systems for conventional and electric vehicles, 13th European Automotive Congress, June 2011, Valencia, Spain

[Pro11] G. Prokop, P. Lewerenz: Thermomanagement – Lösungen für neue und alte Herausforderungen, ATZ 11/2011

[Rat13] C. Rathberger, A. Lichtenberger: Scalable Complexity Simulation in the Electric Vehicle Thermal Management Development Process, Fünftes Symposium Entwicklungsmethodik, Wiesbaden, D

[Str12] Dr. C. Stroh, S. Schnörch, C. Rathberger: Thermalsimulation eines Hybrid-LKWs, Tagung Wärmemanagement des Kraftfahrzeuges, Haus der Technik, Berlin, D

[Waw13] Dr. M. Wawzyniak, M. Kohl, W. Seewald, Dr. R. Trapp, Dr. D. Nagel: Induction Based Coolant Heater for Hybrid and Electric Vehicles, SAE TMSS World Congress 2013, Detroit, US

Engines

The split cycle engine and its impact on the vehicle cooling system

R E MORGAN, G DONG and **M R HEIKAL**
Centre for Automotive Engineering, University of Brighton, UK

ABSTRACT

The split cycle engine represents a radical departure from conventional internal combustion engines. In the split cycle engine, the compression and combustion processes occur in different cylinders, enabling extreme charge air cooling, intra cylinder waste heat recovery and independent optimisation of the compression and combustion chambers. Combining these benefits, thermal efficiencies of over 50% and approaching 60% in some applications are theoretically possible. The split cycle engine is of particular interest in commercial vehicle applications where load factors are high and savings in fuel consumption of major economic benefit to end users.

The energy balance of the split cycle engine is markedly different to conventional engines with significantly more energy being rejected through charge air cooling but correspondingly less through the engine cooling system. How the charge air is cooled will affect how thermal energy is rejected to the environment and the consequential impact on the vehicle cooling system.

In this paper, the split cycle engine will be described and the energy balance compared with a conventional heavy duty diesel engines. Methods of charge air cooling; through multi stage compression with intercooling, direct cooling with a water spray and by the injection of a cryogenic liquid spray will be discussed. The overall energy balance, impact on powertrain efficiency and heat exchanger sizing for the water spray cooling case is compared with a comparable diesel engine. Finally, the consequences of the split cycle on component thermal loading will be discussed and strategies for controlling critical component temperatures described.

1 INTRODUCTION

The long haul commercial vehicle presents a number of challenges in improving the fuel economy of the vehicle. The mean load factor on a typical journey is higher than other on road applications, such as passenger cars and light commercial vehicles, limiting the benefit of downsizing and hybridisation of the power train [1]. As such, the Internal Combustion Engine (ICE) is likely to remain the primary power source in this application for the foreseeable future [2]. Improvements to the IC engine have been proposed, including waste heat recovery and advanced thermodynamic cycles. [3, 4]. Inevitably, recovery of waste heat for conversion to power and modification to the ICE cycle will impact the energy balance of the system and the vehicle cooling system.

In this paper, we describe a new thermodynamic cycle with integrated waste heat recovery and the impact of the new cycle on the system energy balance and vehicle cooling system. The new cycle, referred to as the 'split cycle', separates the compression and combustion strokes into separate chambers. This has a number of benefits including:

1) Reduction of the compression work through quasi-isothermal compression.
2) Decoupling of the compression and expansion strokes enabling a Miller cycle and independent optimisation of the chambers, in particular the surface temperatures to manage thermal losses.
3) High pressure waste heat recovery between the compression and combustion cylinders by recuperation of exhaust energy.

The split cycle engine is described, focusing on the impact of the cycle on the vehicle cooling system. Three methods of achieving isothermal compression are described and assessed in the context of the commercial vehicle transport application. The energy balance of the split cycle engine is then compared with a conventional heavy duty diesel engine and the relative sizes of key components are presented. Finally, a comparison is made with other methods of waste heat recovery.

2 THE SPLIT CYCLE ENGINE

2.1 Description of the Cycle

The split cycle engine was first described by Coney [5], and later by Jackson [6] and Dong [7]. The variant of the split cycle engine considered in the present work is shown in figure 1. Air is compressed in a supercharger (1-2) and then cooled (2-3) in an air-air intercooler. The charge air is further compressed in a compression cylinder (3-4), into which water is injected to cool the charge air during the compression stroke. The water is separated from the charge air (4-5) and the charge air heated in a recuperator (5-6) using heat from the engine exhaust.

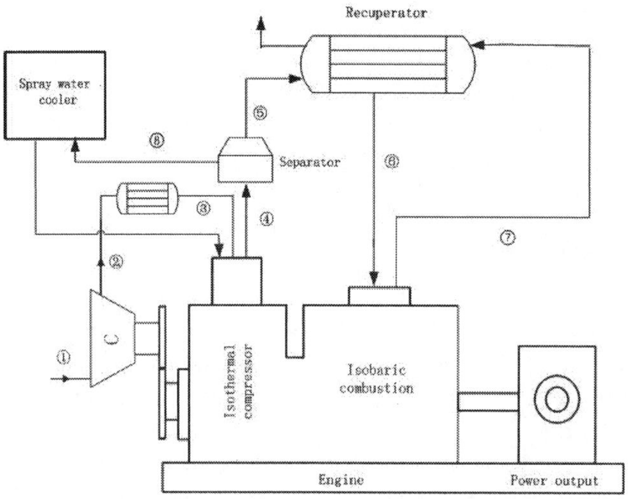

Figure 1: Schematic of a split cycle engine

The hot high pressure air is then transferred to a combustion cylinder, mixed with fuel which is burnt and work is extracted through the piston (6-7). After expansion, the combustion products are removed from the combustion cylinder and used to pre-heat the charge air in the recuperator before being exhaust to atmosphere. The recovered water from the compressor cylinder is cooled and re-injected into the compressor cylinder completing the split cycle. Heat is therefore rejected at low temperature through the spray water cooling, but recovered at high temperature from the exhaust between the compressor and combustor cylinders. There are other methods of achieving isothermal compression, which will be described in section 4.

2.2 Cycle Optimisation

A comprehensive analysis of the split cycle engine was recently reported by Dong [7], using the AMESim software [8]. The key assumptions of the modelling results used in the current work are listed in table 1 and optimised engine parameters for a 350kW shaft output engine in table 2.

Table 1: Modelling assumptions

Supercharger Efficiency	85%
Recuperator Effectiveness	0.73-0.86 (upon to engine load conditions)
Combustor chamber wall temperature	610 K
Phase of heat release CA 50	14 ° ATDC
Air Fuel Ratio	21:1

Table 2: Optimised engine parameters (derived from [7])

Stroke Type	2 stroke	Intake valve opening (IVO)	17° BTDC (compression chamber) 5° BTDC (expansion chamber)
Number of cylinders	6 (3 combustor +3 compressor)	Intake valve closing (IVC)	174° BTDC (compression chamber) 10° ATDC (expansion chamber)
Bore [mm]	136	Exhaust valve opening (EVO)	5° BTDC (compression chamber) 152° ATDC (expansion chamber)
Stroke [mm]	143	Exhaust valve closing (EVC)	10° ATDC (compression chamber) 2° ATDC (expansion chamber)
Compression ratio	23	Design speed [rpm]	1200
Expansion Ratio	27.5	Target power	350 kW
Compressor discharge pressure	110 bar		

The temperature of the spray water is a key factor which affects the isothermal compression process. The impact of spray water temperature on the cycle efficiency is presented in figure 2, using the theoretical analysis method described by Dong [7]. Maintaining the spray water temperature below 72°C has a minimal impact on the overall efficiency, whereas the efficiency significantly degrades above this value. The reason for this is the effectiveness of heat recovery is diminished as the compressor outlet temperature approaches the exhaust temperature of the combustor cylinder. For the stationary application described by Coney [5] a spray water temperature of 35°C was assumed, which is practical for such an application. However, for the vehicle application where packaging and aerodynamic constraints limit the area available for heat rejection to the environment, it is beneficial to raise the spray water temperature and a value of 72°C was selected for further analysis.

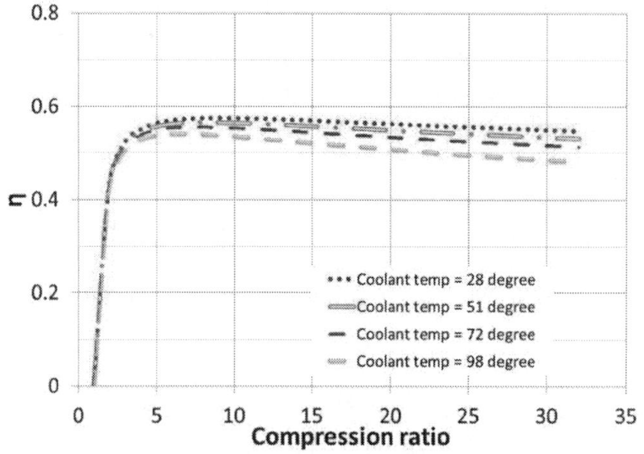

Figure 2: Effect of spray water temperature on cycle efficiency as a function of compression ratio

3 ANALYSIS OF COMPRESSION COOLING OPTIONS

The achievement of effective isothermal compression is essential to realising the potential of the split cycle engine. Isothermal compression could be achieved in a number of ways including:

1 Multi stage compression with intercooling
2 Direct cooling during the compression stroke through water injection
3 Direct cooling during the compression stroke through liquid nitrogen injection

3.1 Multi Stage Compression With Intercooling

To practically achieve the required combustor inlet pressure of 110bar using conventional turbo-machinery, at least three stages with a compression ratio of 3.3 would be required. More stages are desirable to reduce the overall compression work but the number of stages must be traded off against complexity of the intake system and therefore cost. A simple analysis was undertaken to investigate the impact of the number of compressor stages with (a) compression work and (b) heat rejection to the intercoolers. An ambient air temperature of 15°C was assumed and intercooler outlet temperature of 44°C. The compressors were assumed to be adiabatic, with an efficiency of 85%. The results of this analysis, compared to the ideal isothermal case are presented in figure 2. A 20% work saving was achievable using four compressor stages. The final gas temperature will also be of the order of 44°C and so significant heat will be recoverable from the vehicle exhaust. Although theoretically viable, this solution would require a complex chain of compressors and large intercoolers. The high pressure stages and intercoolers would also be heavy components to contain the high pressures, increasing cost and weight. This option was discounted from further analysis for these reasons.

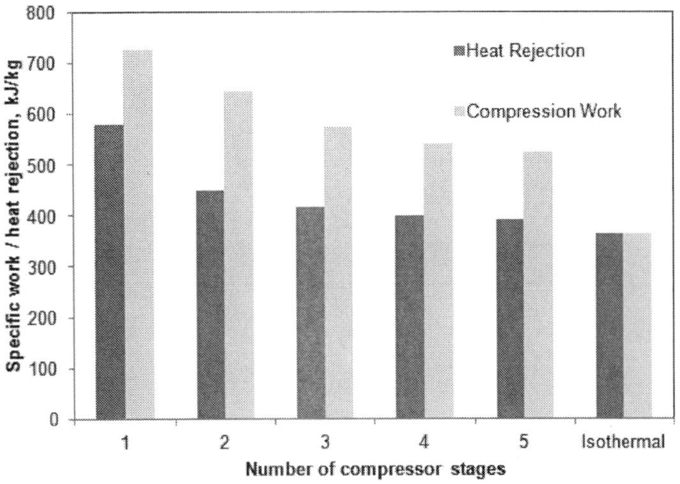

Figure 3: Comparison of work and heat rejection from multi stage compression to achieve 110 bar pressure

3.2 Direct Cooling with a Water Spray

Cooling of the charge air by the injection of a water spray directly into the cylinder during the compression stroke was proposed by Coney [5] and more recently by Iglesias [9] and Qin [10] . The former work was on an early split cycle engine concept, known as the isoengine whereas the later work was for a high efficiency compressed energy storage concept. To be effective the heat of compression must be rapidly absorbed by the spray water. The process can be modelled using the following relationship, relating the initial charge air temperature T_1 and the temperature, T_2 after a compression process of ratio CR:

$$\frac{T_2}{T_1} = CR^{(\gamma-1)exp\left[-\frac{ua}{C_p}\left(1-\frac{T_w}{T_1}\right)\Delta t\right]} \quad (1)$$

Where u is the heat transfer coefficient between the water spray of temperature T_w and the charge air, a is the total surface area of the water droplets and Δt is the incremental time over which the compression process occurs. C_p is the specific heat capacity and γ is the ratio of the specific heats of the charge air. Inspection of equation (1) shows the process tends to the adiabatic case when the value Ua is small, and the isothermal case when Ua is high. In the isothermal case, the end of compression temperature, T_2 will converge to the injected water temperature T_w. To achieve effective isothermal compression it is therefore necessary to:

1. Ensure good mixing of the air and water through charge air motion and/or high water droplet velocity to maximise the heat transfer (u).
2. Maximise atomisation of the water spray to increase the effective area for heat transfer (a).
3. Maximise time for heat transfer (Δt) by limiting the maximum speed of the compressor.

The spray water must not change phase during the compression process as this would lead to a significant volume change as the water changes phase and rise in pressure and compression work. Assuming a 10°C rise in temperature during the compression process this would mean the maximum water temperature is 82°C under normal operating conditions. Referring to figure 4, as the compressor inlet pressure is 3bar, the maximum spray water temperature is still 50°C below the boiling point at the minimum cycle pressure. This provides sufficient margin to avoid boiling in a local hot spot due to uneven distribution of water in the cylinder or during transience where the water injection rate may drop below the target value.

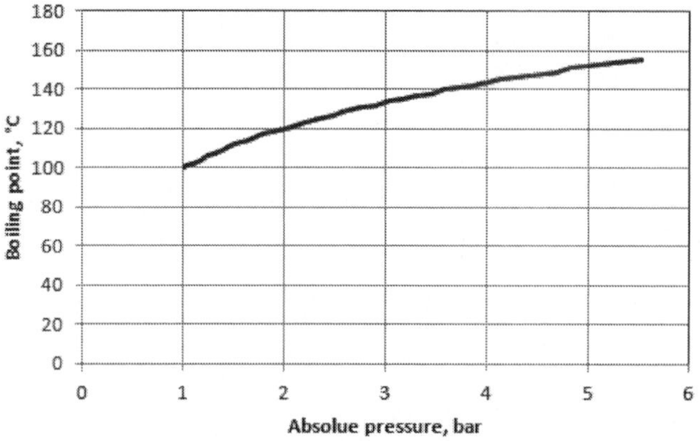

Figure 4: Relationship between boiling point and pressure for water

3.3 Direct Cooling through Liquid Nitrogen Injection

A novel concept was proposed by Jackson [6] using a cryogenic fluid, such as liquid nitrogen, as the cooling media. The cryogen is injected directly into the compressor cylinder, not only absorbing the heat of compression but also contributing to the compression through the change of phase of the cryogen from liquid to gas. The thermal energy required to achieve isothermal compression is therefore stored in 'a tank of cold'. The cryogen can be treated as a fuel, and manufactured in a stationary refrigeration unit or possibly on the vehicle, using energy regenerated during breaking. This promising concept is at an early stage and the subject of ongoing research. Based on the effectiveness and technical maturity the spray water cooling option was selected for further analysis.

4 HEAT BALANCE AND IMPACT ON VEHICLE COOLING

4.1 Comparison of Heat Balance

Based on the research in reference [7], the heat balance of a 350kW heavy split cycle engine with spray water compression cooling was calculated and summarised in table 3. For comparison, the heat balance for a heavy duty Euro 6 diesel engine using a mixed strategy of Exhaust Gas Recirculation (EGR) and Selective Catalytic Reduction (SCR) after-treatment for NOx control is also presented.

From table 3, the heat rejection requirements for the vehicle cooling pack can be calculated and are summarised in table 4. It was assumed the EGR heat was transferred to the vehicle cooling system and then rejected to the environment through the radiator in the diesel engine case. Also from table 3, it can be seen that 20.7% less thermal energy is rejected from the split cycle vehicle compared to the diesel baseline. A new cooler is required to reject the heat transferred to the spray water from the isothermal compressor.

Table 3: Comparison of heat balance in terms of kW and percentage of fuel energy for a Euro 6 heavy duty diesel engine and equivalent split cycle engine

	Diesel Engine		Split Cycle Engine	
	kW	%	kW	%
Fuel Energy	875	100	701.4	100.0
Break Power	350	40	350	49.9
After-cooler	61	7	32.4	4.6
Spray Water Cooler	0	0	118	16.7
Water jacket (radiator)	149	17	65	9.3
EGR (addition to radiator)	61.3	7	0	0
Exhaust	315	36	137	19.5

The reduction in vehicle thermal load is unsurprising given the 19.8% reduction in fuel consumption and hence primary heat input into the combustion chamber. Waste heat is recovered internally to the cycle, unlike in external waste heat recovery concepts such as the Organic Rankine Cycle (ORC), where significantly more thermal energy needs to be rejected to the environment to recover work from the vehicle exhaust. This will be discussed in more detail later in the paper. The after-cooler heat rejection is significantly lower for the split cycle engine as less air is required at lower boost than for the equivalent diesel engine. As the after-cooler is often as big as the radiator in trucks, significant frontal area would be freed up for the spray coolers.

Table 4: Comparison of cooling requirements

	Diesel Engine kW	Split Cycle Engine kW	Change in Heat Rejection
Radiator	210	65	-68.9%
Spray water cooler	0	118	-
Aftercooler	61	32.4	-46.9%
TOTAL	271	215	-20.6%

To understand the impact of the changes to the vehicle heat balance the temperature at which heat is rejected must also be considered and will be discussed in the following sections.

4.2 Spray Water Cooler

The spray water cooler must cool the water recovered in the phase separator, this could be achieved by passing the water through a fin and tube type radiator. However, the spray water is recovered at high pressure which would lead to a significant increase in the weight of the radiator and loss of performance due to increase in tube thickness to contain the elevated pressure of the spray water. It is advantageous to maintain the water at the recovered pressure to facilitate the injection of water back into the isothermal compressor and reduce the risk of boiling. Reducing the pressure was therefore considered impractical. A supplementary cooling loop (figure 5) is proposed between the spray water cooler and radiator. The spray water cooler can therefore be a compact plate type heat exchanger. The secondary cooling loop then circulates coolant from between the spray water cooler and radiator at low pressure, comparable with existing engines.

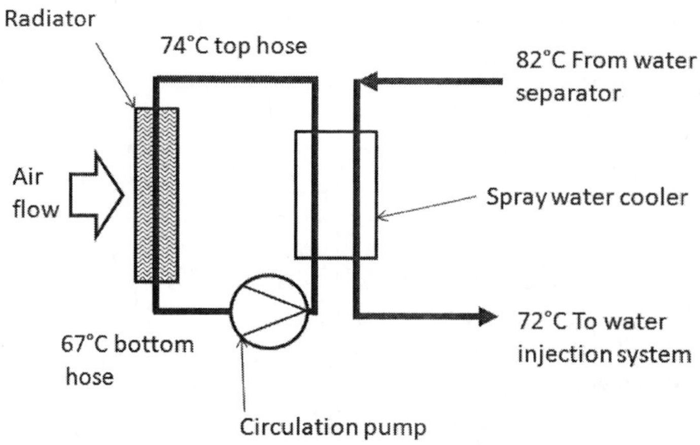

Figure 5: Spray water cooling circuit

The coolant in the spray water loop is water. A temperature rise of 10°C and an approach temperature of 5°C was assumed in the spray water cooler. The spray water radiator bottom hose temperature of 67°C is lower than the 90°C more typical of heavy duty engine radiator cooling systems. To estimate the impact on the vehicle front end cooling pack the Log Mean Temperature Difference (LMTD) method was used:

$$LMTD = \frac{\Delta T_A - \Delta T_B}{ln\left(\frac{\Delta T_A}{\Delta T_B}\right)}$$

Where ΔT_A is the temperature difference across the hot side and ΔT_B across the cold side. Assuming the thickness and design of the heat exchanger is unchanged, the change in frontal area can be calculated using:

$$A_2 = A_1 \frac{Q_2\, LMTD_1}{Q_1 LMTD_2}$$

Where A is the frontal area of the radiator, Q is the heat rejection and subscripts *1* and *2* refer to the two radiators in question. The resulting LMTD values for the main heat exchangers, compared with values calculated for the baseline diesel engine, are presented in table 5.

4.3 Cooling System Design

The LMTD and heat rejection data presented above was used to calculate the change in cooler sizes compared to a diesel engine for the split cycle engine. For this analysis, the air conditioning condenser and transmission oil coolers were not considered as these would be unchanged. The change in heat exchanger area are summarised in table 5.

Table 5: Comparison of relative cooler sizes

	LMTD	Change in required area relative to diesel case
Aftercooler	45.5	-47%
Radiator	47	-69%
Spray water cooler	22.9	+115% (relative to the diesel radiator)

From table 5, it can be seen that the spray water cooler is a substantial cooler, requiring 15% more frontal area than the Diesel vehicle radiator. However, the substantial reductions in the engine coolant radiator (-69%) and after-coolers (-47%) would provide sufficient space to package the spray water cooler. In most heavy duty vehicles, the after-cooler is positioned in front of the radiator. The reduction in after-cooler frontal area would provide a more direct air path to the spray water cooler of lower temperature air, further reducing the size of this component.

Conventional engine coolants would be used for the engine cooling circuit and the secondary spray water cooling loop. However, water would be used for the injected spray which could freeze in cold climates when the vehicle was not in use. Other fluids with a lower freezing point could be used but as the fluid sprayed into the compressor is mixed with the combustion air any unrecovered fluid would be exposed to the hot recuperator heat transfer surfaces. For these reasons, water is probably the best choice but some form of freezing protection will be required to protect the water injection circuit in cold climates.

4.4 Engine Thermal Design

Although not the main focused of the current work, some comments can be made on the thermal design of the engine. The thermal loading on the combustor cylinder is significantly higher than conventional engines;

1. High intake air temperature, increasing loading on the intake valves and loss of charge air cooling of the combustion chamber
2. Two stroke cycle, increasing power density and specific thermal loading
3. Increased combustion chamber wall temperatures to reduce chamber heat losses, increasing mean metal temperatures

Aspects of the thermal design of a large (megawatt scale) split cycle engine was reviewed by Coney [11]. In particular, the cooling of the top areas of the liner is of concern to preserve the integrity of the oil film. Coney concluded these issues could be addressed on a large engine format, which is encouraging considering the downsized engine for a heavy duty transportation application is likely to be easier to cool.

In the previous work by Coney the piston and cylinder head were manufactured from high temperatures alloys. This was necessary to contain the high mean combustion temperatures rather than mechanical loading, which will be lower for the split cycle engine due to the quasi-isobaric combustion process. As with the liner cooling, the smaller format engine will reduce the heat path between the combustion chamber wall and cooling channels and so more conventional materials may be feasible.

To investigate component thermal loading further a more accurate measure of the combustion heat release characteristics is required. This and further research on the impact of the slit cycle on engine cooling is the subject of ongoing research between Ricardo and the University of Brighton.

5 COMPARISON WITH OTHER HIGH EFFICIENCY CONCEPTS

The split cycle engine is a very novel concept with excellent potential, but other routes to achieving high efficiency in the heavy duty vehicle application have been proposed. Incremental improvements in friction, combustion, vehicle aerodynamics have been proposed, with waste heat recovery being the most significant single improvement. The most promising waste heat recovery solutions, and an assessment of the impact of the system on vehicle cooling is summarised in Table 6:

Table 6: Comparison of waste heat recovery technologies

	Efficiency improvement	Impact on vehicle cooling system
Turbo compounding (mechanical)	5%	Reduced overall heat rejection due to efficiency improvement
Turbo compounding (electrical)	10%	Reduced overall heat rejection due to efficiency improvement
Organic Rankine Cycle (ORC)	5%	Significant increase in heat rejection from ORC condenser
Thermoelectric Generation (TEG)	5%	Significant increase in heat rejection
Split cycle engines	25-36%	Change in heat balance but not overall package size

From table 6, it is apparent that the two thermal energy recovery methods (TEC and ORC) will impact the vehicle cooling system. The analysis presented in this paper suggests the impact of the split cycle engine on the overall cooling requirements is minimal, which is a significant benefit of the split cycle over other waste heat recovery methods.

6 CONCLUSIONS

A novel split cycle engine concept is described with separate compression and combustion chambers and intra cylinder heat recovery. Energy efficiencies of over 50% are possible for the concept, in excess of the combined improvements of conventional diesel cycles with exhaust heat recovery. Three concepts for compression cooling were presented and the spray water concept selected for further analysis. Multi stage turbocharging was discounted due to the high pressures and impact on cost and weight. Direct cooling using a cryogenic fluid is interesting and merits further research.

The heat balance of the engine was compared with a conventional diesel engine and a 20.6% reduction in overall thermal loading to the vehicle cooling system was predicted. A detailed review of the impact of the new cycle on heat exchanger components showed the vehicle cooling pack would be largely equivalent in overall dimensions to a conventional powered diesel vehicle of equivalent rating. The reduction in heat rejection to the radiator and after-cooler heat exchangers is sufficient to accommodate the new spray water cooler.

Further analysis of the component thermal loading is required and is the subject of ongoing research.

ACKNOWLEDGEMENTS

The authors would like to thank the UK's Innovate UK for their co-funding of the Cryopower Project under the IDP8 programme.

REFERENCES

1. Stanton, D.W., Systematic Development of Highly Efficient and Clean Engines to Meet Future Commercial Vehicle Greenhouse Gas Regulations. SAE Int. J. Engines, 2013. 6(3): p. 1395-1480. doi:10.4271/2013-01-2421
2. Tai, C. High Fuel Economy Heavy Duty Truck Engine, in 2011 Anual Merit Review 2011. http://energy.gov/sites/prod/files/2014/03/f11/ace060_tai_2011_o.pdf accessed 20/3/2015.
3. Stanton, D.W. The Role of Waste Heat Recovery in Meeting Phase 2 US EPA Greenhouse Gas Regulations. Diesel Emissions Conference & Ad Blue Forum 2013, Dusseldorf, Germany.
4. Rossi, R., et al., Simultaneous Reduction of Soot and NOX Emissions by Means of the HCPC Concept: Complying with the Heavy Duty EURO 6 Limits without Aftertreatment System. 2013 SAE2013-24-0093. doi:10.4271/2013-24-0093
5. Coney, M.W., C. Linnemann, and H.S. Abdallah. A thermodynamic analysis of a novel high efficiency reciprocating internal combustion engine—the isoengine. Energy, 2004. 29(12–15): p. 2585-2600.
6. Jackson N.S, Atkins, A , Split Cycle Reciprocating Piston Engine. WO2010067080 A1
7. Dong, G., Morgan, R., Heikal, M. A novel slit cycle internal combustion engine with integral waste heat recovery. In Press 2015 doi:10.1016/j.apenergy.2015.02.024
8. AMESim, Mentor Graphics.
9. Iglesias, A. and D. Favrat, Innovative isothermal oil-free co-rotating scroll compressor–expander for energy storage with first expander tests. Energy Conversion and Management. 2014. 85(0): p. 565-572. doi:10.1016/j.enconman.2014.05.106
10. Qin, C. and Loth, E. Liquid piston compression efficiency with droplet heat transfer. Applied Energy, 2014. 114(0): p. 539-550. doi: 10.1016/j.apenergy.2013.10.005
11. Coney, M.W., Linnemann, C., Morgan, R., Bancroft, T., Sammut, R., A novel internal combustion engine with simultaneous injection of fuel and pre-compressed pre-heated air, in ASME ICED Fall Technical Conference 2002, ASME: New Orleans.

Development of a high efficiency liquid-air engine for cooling, heat recovery and power

N OWEN[1], H CLARKE[1], D CHARTERS[2], J TREMBLEY[3], C GARNER[4], S MOHR[4], A WILLIAMS[4] and H ZHAO[4]
[1] Dearman Engine Company, UK
[2] MIRA Limited, UK
[3] Air Products, UK
[4] Loughborough University, UK

ABSTRACT

The Dearman Engine is an efficient Rankine-cycle expander which uses liquid air or nitrogen as a working fluid. There is a cost, air quality and carbon case for the technology in transport refrigeration, as a waste-heat hybrid for buses and trucks, and as a prime-mover for small urban vehicles.

The paper describes development of a practical unit based on adapted ICE practice. A single-cylinder research engine has been modelled, constructed and integrated into a truck refrigeration unit. Further work is ongoing to apply the unit alongside a diesel engine in a bus, providing a hybrid-like function for waste heat recovery.

GLOSSARY AND ABBREVIATIONS

CA: Crank Angle, in degrees after Top Dead Centre (TDC)
DE: Abbreviation for a Dearman Enginem a Rankine-cycle expander which uses an in-cylinder Heat Exchange Fluid to maintain a nearer to isothermal expansion process
HEF: Heat Exchange Fluid, a fluid such as water or propylene glycol that is introduced to the cylinder of the Dearman Engine in order to maintain nearer to isothermal expansion
LiAir: Liquid Air, a potential working fluid for a Dearman Engine
LiIon: Lithium-Ion battery, considered as a benchmark for zero emission vehicle energy storage
LiN: Liquid Nitrogen, used as the working fluid for the Dearman Engine in this paper
TDC: Top Dead Centre
TRU: Truck Refrigeration Unit, a device used to refrigerate the goods compartment
WHR: Waste Heat Recovery, in this context from the coolant or exhaust streams of a vehicle prime mover such as an ICE
Wh: Watt-hours
Wh/kg: Specific energy delivered by a Dearman Engine, in Watt-hours of mechanical work per kilogram of LiN or LiAir; used as a proxy for fuel efficiency
ZEV: Zero Emission Vehicle, with no emissions at point of use other than water, air or nitrogen

1 INTRODUCTION

Growing pressures on energy use, carbon and air quality emissions are driving great change across the transport sector. Regulation of tail-pipe emissions on the basis of test cycles is being reinforced by growing awareness of well-to-wheel and lifecycle emissions, and of city air quality issues in both the developed and developing world.

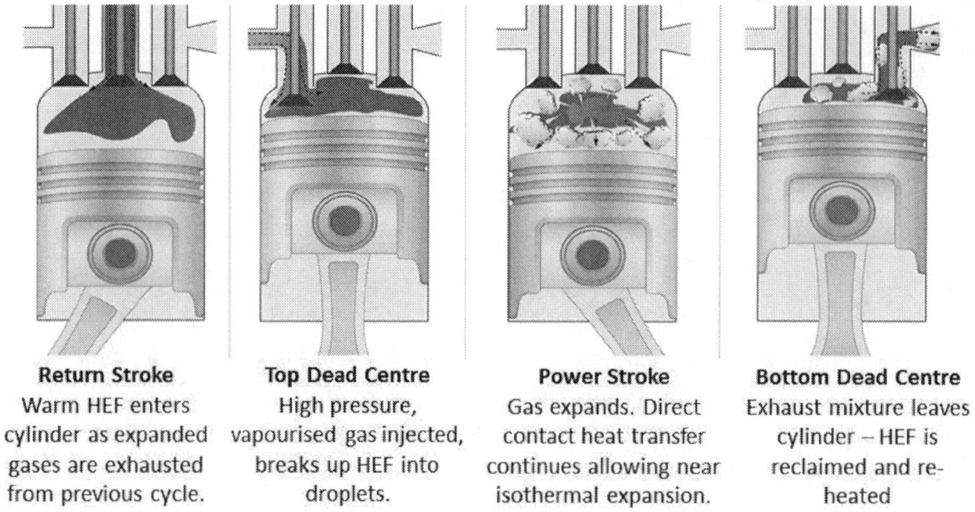

Figure 1: Dearman engine operating cycle

The Dearman Engine is a high efficiency Rankine-cycle expander engine, developed specifically for use with liquid air or nitrogen. It uses the same pressurise-vapourise-superheat-expand-exhaust cycle as other open-cycle Rankine devices, but is differentiated by an in-cylinder heat transfer process [1], in which a Heat Exchange Fluid (HEF) is used to store heat and maintain closer to isothermal conditions during expansion (Figure 1).

Early analysis was based upon conceptual modelling and lab experimentation [2, 3]. The maximum work that can be extracted from a cryogenic fluid can be expressed using Exergy [3], which is calculated from the working fluid's change in enthalpy, entropy and temperature:

$$\Delta\Psi = \Psi in - \Psi out = (Hi - Hf) - To(Si\text{-}Sf) \quad (1)$$

Where Ψ is the exergy, H the enthalpy, S the entropy and To the temperature of the cryogen; subscripts i and f denote initial and final states of the fluid.

Using this analysis, liquid nitrogen offers a specific exergy or useable energy of 771 kJ/kg, or 214 Wh/kg. In a practical application, considerations such as working pressure limits, the effectiveness of in-cylinder heat transfer and parasitic losses suggested that 100Wh/kg may be achievable, comparable to a contemporary Lithium-Ion battery. Further analysis showed that the engine could be cost-effective in several transport applications, offering the user a pay-back of 3 years or less on capital spend, and the prospect of immediate carbon savings that improved as the manufacture of electricity is de-carbonised, and enhanced by good life-cycle footprint [4,5,6]. Analysis has also shown that the liquid nitrogen or air can be supplied cost-effectively, and that its production offers synergies with renewables through demand-side management of the liquefaction plant, and with imported LNG, by using waste cold [6]. Truck refrigeration is an early market, as the coldness of the working fluid can be fully used, and there is a synergy between re-warming the heat-exchange fluid and cooling the condenser of a vapour cycle driven by the engine (Figure 2); there are wider opportunities for reinforcing the security of the cold-chain and its energy system, especially in developing economies [7,8].

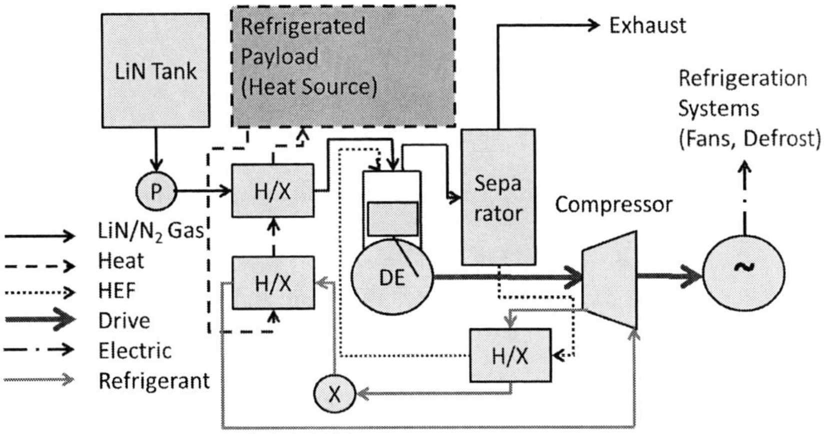

Figure 2: Dearman engine refrigeration circuit

However, the engine can also make a case for itself where cooling is not required, especially as a waste heat recovery (WHR) device working alongside an internal combustion engine – or indeed a fuel cell or battery-electric system in future (Figure 3).

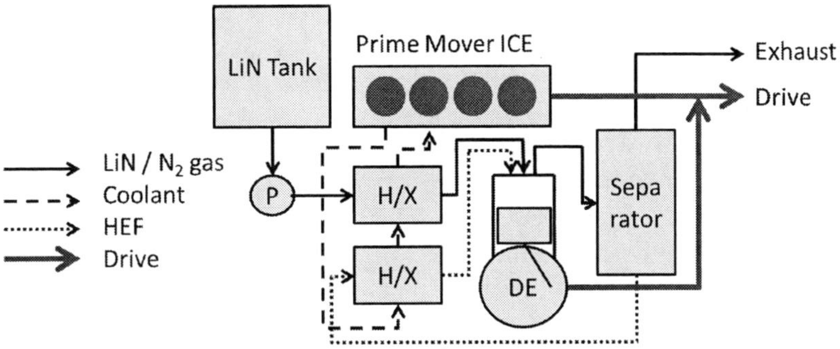

Figure 3: Dearman Engine as waste heat recovery system for an ICE

Initial analysis has shown that this configuration is especially effective in urban-use heavy duty vehicles such as buses, where conventional WHR technologies are less effective; also by using coolant heat, the complication of exhaust heat recovery is avoided [4,5,6].

2 ENGINE AND SYSTEM HARDWARE

2.1 Dearman Engine

As with any Rankine cycle expander, the efficiency of a Dearman engine is strongly impacted by the expansion ratio that can be achieved in practice. A piston expander was selected, as it offered a basis for good expansion ratio in a single stage compared to rotary types (plus a high degree of synergy with ICE manufacture). In such a device, expansion ratio is dictated by clearance volume at "top dead centre" (TDC) and by intake valve timing. As the intake event happens just after TDC, the length of the intake event is in direct conflict with expansion ratio. However, as the HEF is added during the exhaust stroke, it can have a beneficial impact on reducing TDC clearance, if hydraulic lock is avoided (Figure 4). In practical terms it is difficult to create a valve event of less than 20-30 degrees duration, meaning that expansion ratio is limited to around 20 per stage. Given the need for some exhaust backpressure to drive flow through a HEF separator, and the likelihood of losses across the intake valve, this equates to a working pressure of 30-50 bar for a single stage device.

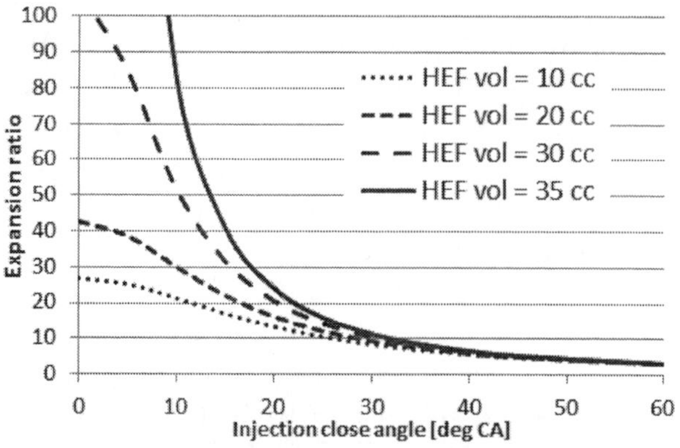

Figure 4: Impact of intake timing & HEF trapped volume on expansion ratio

Gas is admitted via a pair of poppet valves, which open outward into the cylinder like an ICE intake valve; however, because these are forced open by gas pressure, a counter-piston on the valve stem is used to maintain sealing at a reasonable spring load (Figure 5). In the prototype, these valves are variably driven using an adapted Fiat MultiAir hydraulic lost-motion system [9]; a subsequent study has shown that an open period of around 25 crank degrees is achievable using a simple mechanical cam. Exhaust valves are conventionally operated.

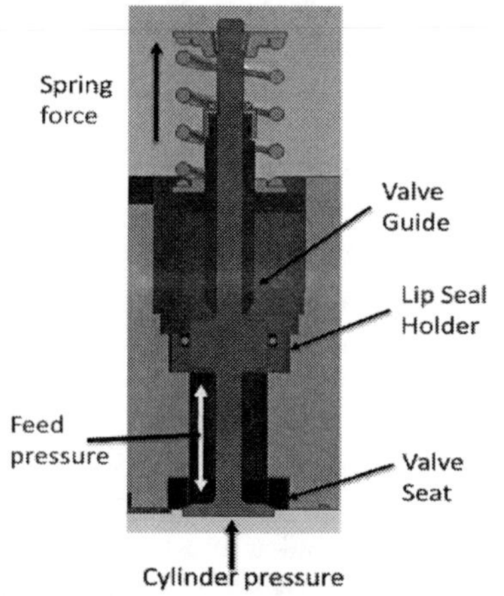

Figure 5: Nitrogen intake valve

The HEF is admitted by a third, self-acting poppet valve that opens only during the exhaust stroke where cylinder pressure is low. For initial experimentation, pure water and water / propylene glycol blends were used as the HEF – water offers outstanding heat capacity, and its fundamental non-toxicity means that the reclaim system in the exhaust can be simplified if regular top-up is practicable. Figure 6 shows the prototype engine.

Figure 6: Engine on laboratory dynamometer

2.2 Cryogenic systems

The engine is supported by three key cryogenic handling systems:

- The storage tank for the working fluid, which must be sufficiently insulated to prevent excessive losses to boil-off
- A pressure feed pump, which must efficiently pressurize the liquid to the system's working pressure at reasonable efficiency
- A vapouriser, effectively the "boiler" which converts liquid to gas

All work to date has been performed using liquid nitrogen (LiN) as a working fluid, as it is readily available from industrial gas suppliers in quantities that are adequate for early volumes of Dearman engines [6]. Higher uptake, use indoors or underground, or use in conjunction with a small local liquefaction scheme, could precipitate a switch to liquid air, which is thermodynamically similar in the engine itself, though harder to handle in the cryogenic stages due to the different boiling points of its components. Although the liquid nitrogen or air is sometimes referred to as the engine's "fuel", this is only correct in the economic sense – meaning that it is consumed, and has to be paid for. Strictly, the "fuel" is ambient or waste heat, although the cryogen can be considered to carry an absence of energy.

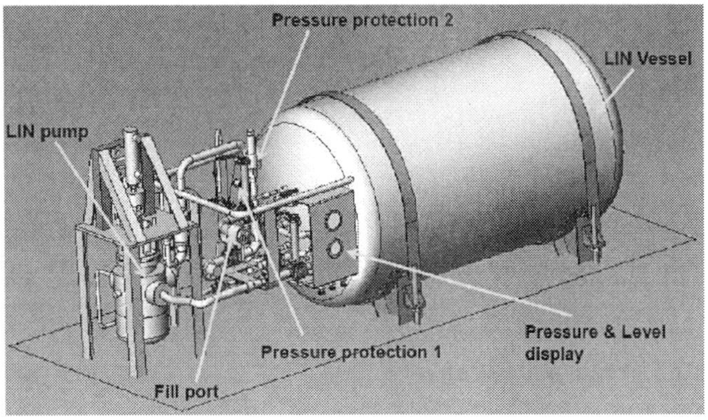

Figure 7: LiN tank and pump

LiN is stored in a conventional, vacuum insulated tank at just below its boiling point (Figure 7), such that the tank self-pressurises to 3-4 bar due to the small amount of heat penetrating the insulation. Unlike the case for LNG (which has a high global warming potential), small quantities of boil-off are only an issue if they impact the economics of operation. As most applications have potential for a daily fill, such losses are expected to amount to less than 3% of LiN consumed.

Pumping the LiN has proved to be the greatest challenge of the cryogenic systems. The first issue is that the fluid is so close to its boiling point that there is a great risk of cavitation in the intake to any pump; secondly, unless the pump is kept permanently cold, both LiN and time will be lost in cooling it down upon re-start; and finally, because of these first two factors this pump must be located in or near the tank, not on the engine, so an indirect drive introduces energy conversion inefficiencies. For the prototype, an industrial piston pump was employed, mounted adjacent to the LiN tank and encased in its own self-venting, vacuum insulated reservoir of LiN (Figure 7). The pump operates at relatively low speed (up to 10 strokes per second). In the laboratory this pump is driven by independent mains electric motor; on the vehicle it is hydraulically actuated via a slave piston which takes hydraulic power from an engine-driven hydraulic pump.

As parasitic losses from this system were a risk to overall efficiency, early laboratory tests were undertaken to establish the power consumption of the pump at a range of pressures and flows. The results, shown in figure 8, are expressed in Wh of mechanical pumping energy per kg of LiN displaced. The specific work should be considered in the context of overall engine efficiency, as discussed in the next section, of around 30Wh/kg before this loss is subtracted; it should also be borne in mind that a hydraulic or low voltage electric drive is unlikely to have an efficiency much greater than 50%. An un-optimised LiN pump, with an overall efficiency (including drive from the engine) of around 10%, can consume up to a quarter of the engine's gross output. However, from this data two things are apparent: firstly, there is clearly a "sweet spot" of pump efficiency at mid flows; secondly, higher working pressures are delivered more efficiently by this type of pump. Further work is ongoing in this area and future prototypes will have significantly lower pumping losses.

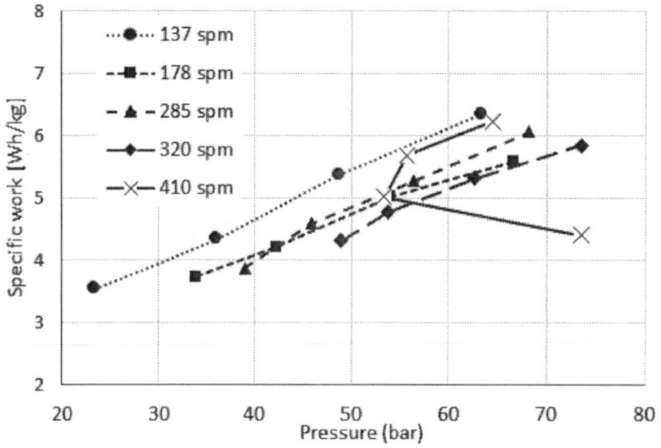

Figure 8: Effect of pressure and flow on measured LiN feed-pump work

From the pump, LiN passes through a tube-and-fin heat exchanger where it is vapourised and superheated. This component is relatively conventional and has proved effective, despite operating around the critical-pressure of nitrogen (34 bar) in the first prototypes.

3 LABORATORY TEST RESULTS

3.1 Performance and Thermodynamics

A series of parametric tests were conducted on the prototype, embracing variation of speed, feed pressure, valve opening period and flow of HEF. As well as recording steady-state data such as nitrogen consumption and key temperatures, in-cylinder and port pressure recordings were made on a crank-angle resolved basis.

A peak power of 11.5kW was obtained at 1200 rev/min, equating to 16.8kW/l, a similar level of performance to a moderately rated naturally aspirated industrial diesel engine. Valve-train limits prevented running at higher speeds, although the response of efficiency to speed suggests that higher speeds are not beneficial. Figure 9 illustrates the impact of speed and feed-pressure on power and efficiency, the latter being cited as Wh of gross flywheel power (before feed-pump power is subtracted) per kg of nitrogen consumed. This indicates that a speed of 1000 rev/min is optimum for efficiency, with choking of the nitrogen intake valves, and possibly HEF-to-gas heat transfer rates being suspected as limiting factors at higher speeds. A feed pressure of 40 bar offers highest efficiency, here the expansion ratio is the limiting factor and the loss of efficiency at 50 bar is entirely expected.

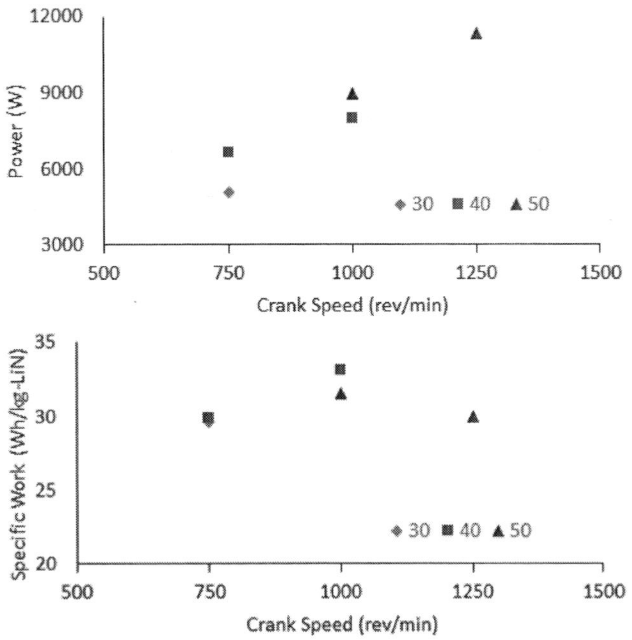

Figure 9: Performance curves, initial build 30, 40 & 50 bar feed pressure

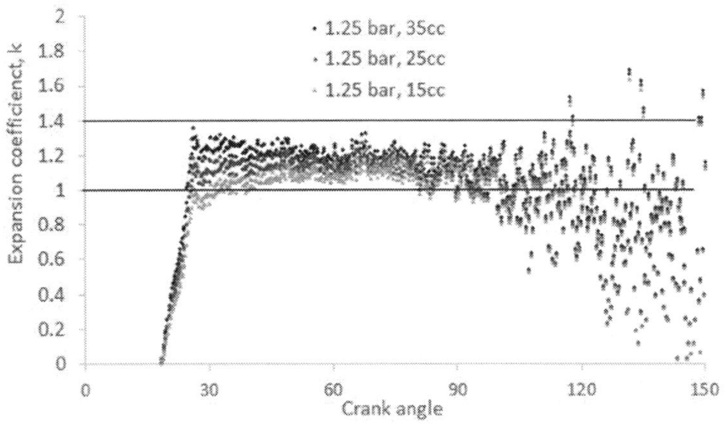

Figure 10: Polytropic index for a range of assumed trapped HEF volumes – 1.25 bar exhaust backpressure (abs)

Analysis of cylinder pressure data was performed to establish a polytropic index for the expansion process. This analysis is complicated by inaccuracy in establishing both the amount of HEF trapped immediately before the intake valve opens (hence the clearance volume and true expansion ratio), and the pressure around BDC where transducer errors are highest. So a range of assumptions were input to the analysis, as illustrated in Figure 10. Data from a number of test conditions suggests a typical index if 1.15, compared to 1.4 for perfect adiabatic expansion, and 1.0 for perfect isothermal. The expansion phase of an ICE achieves an index of 1.3 to 1.35. This suggests that the HEF is having a significant beneficial effect, but that there is some scope for improving the heat transfer between HEF and gas, which is most likely to be achieved by development the intake port shape, to develop gas motion and mixing with trapped HEF. This, plus optimisation of trapped HEF to maximise expansion ratio, is the topic of current research.

3.2 Friction and parasitic losses

Because specific power is moderate, friction and parasitic losses of this type of engine are important to its efficiency. Measurements of parasitic power (principally the liquid nitrogen feed-pump, discussed above, and the HEF pump) were combined with a motored friction tear-down, to understand where improvement can be made. Figure 11 shows a breakdown of major losses. The LiN feed-pump is by far the biggest contributor, as discussed above, and therefore offers the greatest potential for improvement – the actual work done on pressurising the liquid is only around 3% of the engine's gross power. Ongoing research is looking at developing pump efficiency, including different pump geometries and drive arrangements, and it is expected that power consumed in this system can be reduced by 50%.

The HEF pump is also a significant source of loss. In this prototype, the self-acting valve meant that the pump (again an electric industrial unit not optimised for efficiency) is effectively pushing the valve open. A new concept is in development using a cam to actuate the valve. This is much more efficient, because the cam recovers some of the energy put into the valve spring, the HEF pressure can be lower, and the HEF inlet event can be timed to reduce flow straight out of the exhaust. These changes again reduce losses by 50%, but require optimisation of valve timing.

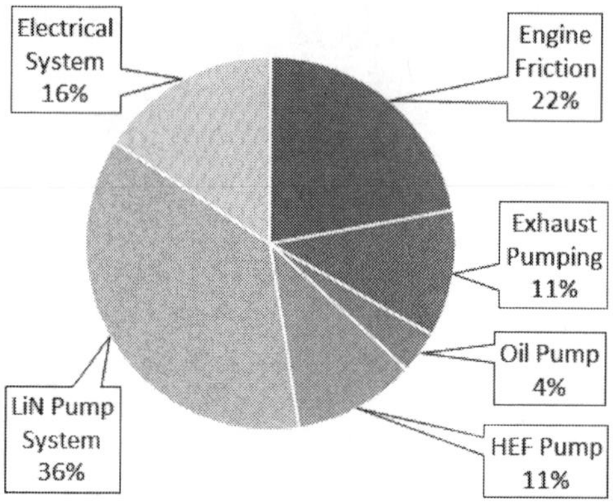

Figure 11: Breakdown of measured friction and parasitic losses, 750rev/min

The breakdown of internal friction is similar to a gasoline ICE, with around 60% arising from crank and piston, the remainder from valvetrain, balancers and oil pump. The prototype was based on a Yamaha quad-bike engine with roller main and big-end bearings, so there is little room for improvement here; however study of other friction sources suggests small improvements. In all, a reduction of 800-1000W of losses is expected in the next generation of engine, resulting in a very significant 36% increase in net power and efficiency of this 5kW unit.

4 ENGINE MODELLING

4.1 Modelling approach

High-level modelling of the Dearman Engine has been undertaken in order to achieve a more detailed characterisation and allow a deeper understanding of the sensitivities of the engine to engineering factors. The modelling strategy consists of a modular structure in which sub-models written in the engineering software AMESIM are being integrated to give a comprehensive heat transfer, fluid flow and work transfer model.

The schematic of the model developed in AMESIM is shown in Figure 12, indicating the input and output parameters. The model used the Dearman engine geometries and architecture, valve lift profiles and injection conditions (taking into account the mechanical, pneumatic and thermal effects). In addition to the dedicated IFP engine library within AMESIM, the thermo-physical properties of N_2 including phase changes were calculated according to NIST recommendations [10]. The engine inlet and outlet ports were modelled with the in-built 1D-CFD Gas Dynamics library with the implementation of a 1D solver into the AMESIM environment. Non-isentropic effects were calculated by this solver by taking into account the effect of compressibility, friction and heat exchange.

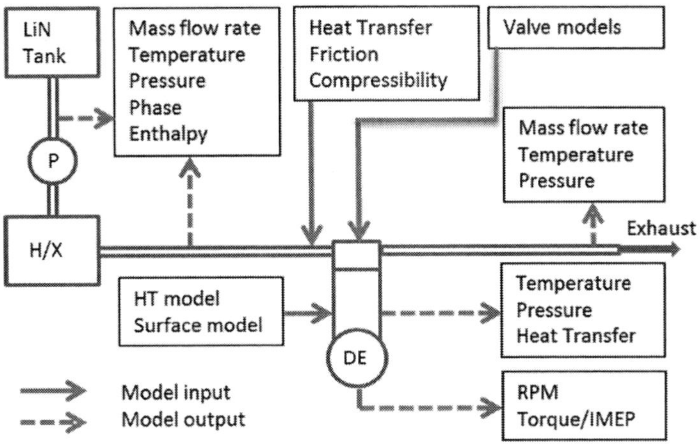

Figure 12: Schematic diagram of the model developed in AMESIM

Accurately predicting the in-cylinder thermo-physical mechanisms and heat transfer characteristics during the mixing of N_2 and HEF is one of the most significant challenges. There is currently a significant amount of uncertainty associated with this, meaning that particular heat transfer and surface progression models have to be developed to better quantify the effect of this process.

The presence of HEF was simulated with models of variable N_2-HEF surface area progressions and in the first instance; the in-cylinder heat transfer was described by a model similar to the one of Eichelberg [11]:

$$\dot{Q} = \alpha V_P S_{HEF} (T - T_{HEF}) \sqrt{PT} \quad (2)$$

with P : the pressure in cylinder [Pa]
T : temperature in cylinder [K]
V_p: mean piston velocity [m/s]
S_{HEF}: surface area of heat exchange [m^2]
T_{HEF}: temperature of HEF [K]
α : user defined coefficient

Its physical meaning (such as inertial effects of moving piston and the effect of in-cylinder pressure and temperature on the heat transfer coefficient) can be applied to this non-combustion process presented here and the heat transfer coefficients were adjusted through the scaling factor α to represent that of convective heat transfer from N_2 to moving water [12].

The N_2/HEF contact surface area during injection and the expansion stroke is needed to adequately describe the in-cylinder heat transfer. Figure 13 shows surface progression models hypothesised as a function of crank angle. They show the extremes of near constant contact area as well as progressions that account for increasing surface area as the piston moves and the N_2 expands.

4.2 Modelling results

Since the expansion process during the power stroke is most efficient if it can be carried out isothermally, the polytropic indices plotted for the expansion process of the surface models are shown in Figure 14. The results show an almost adiabatic expansion for the smallest surface progression (dash-dot line) and near perfect isothermal expansion for the near constant surface area of around $6m^2$ (dashed line). The model for the progressing surface area (dotted line) is adapted to fit the experimental results presented in Figure 10 and shows a similar decline of the expansion coefficient k with increasing crank angle. Establishing a semi-empirical basis for this heat transfer and surface model is subject of further current research and validation.

A distinctive result for all models is the initial increase in temperature due to the filling of the cylinder and is due to heat generation by the compression of the N_2 gas. In fact, heat is initially flowing from the N_2 to the HEF in all cases. This is more pronounced for the near isothermal case as shown in Figure 15 for the total heat exchanged between N_2 and the HEF as a function of crank angle.

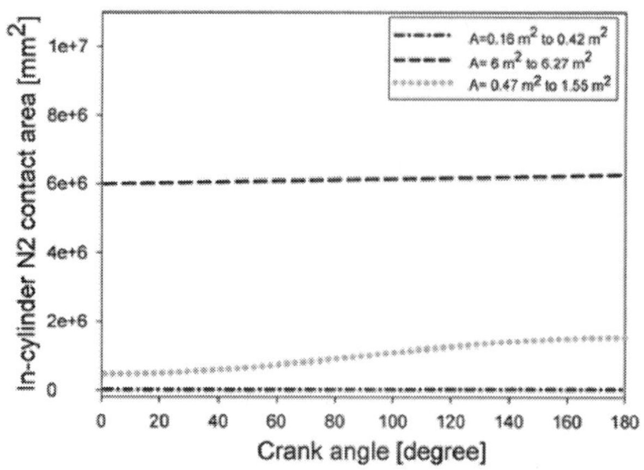

Figure 13: Model contact areas of injected N2 with HEF as a function of crank angle

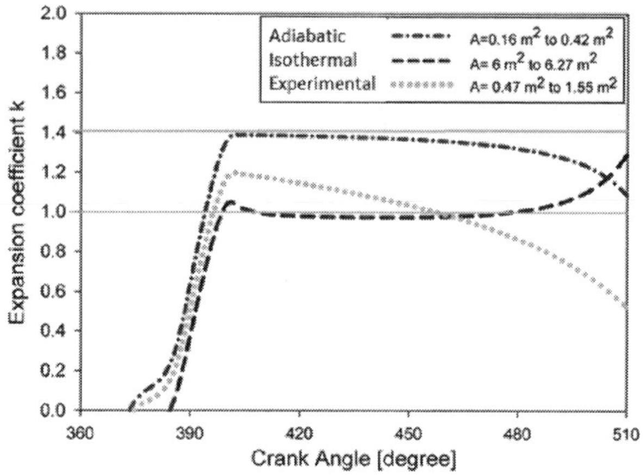

Figure 14: Polytropic index of the N_2 expansion for the model N_2-HEF surface area progression shown in Fig.13

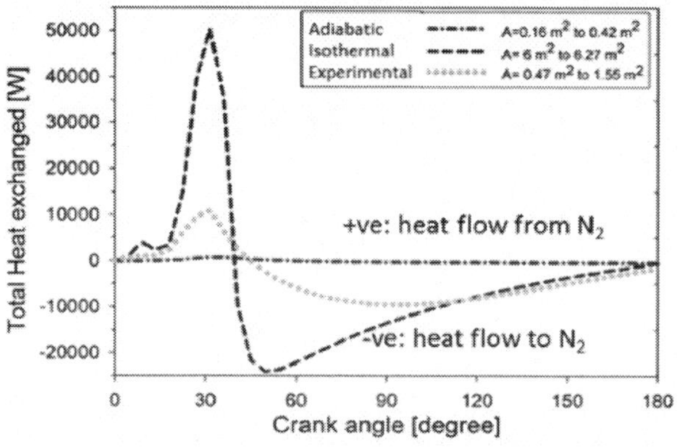

Figure 15: Total heat exchanged in the cylinder as a function of crank angle

It shows that the in-cylinder heat transfer during the N_2 valve open phase can significantly affect the maximum nitrogen flow rate during the cylinder filling by its effect on the density of the nitrogen within the cylinder. The presence of the HEF therefore acts to increase the power capability of such LiN engines as well as increasing the efficiency through reduced clearance volume and continued heat transfer to the working fluid during its expansion.

Parametric sensitivity studies were carried out for various engine conditions with input parameters accessible to the Dearman engine for future validations. As an example, Figure 16 shows the effects of N_2 and HEF temperature variations on the engine output torque, power, specific work and the engine N_2 mass flow rate. These are of prime interest for the study of thermal efficiency and performance of the engine. The highest power and torque output is delivered with the maximum temperature difference of N_2 and HEF. A high HEF temperature can be achieved by tapping into an above-ambient heat source such as the cooling system of an IC engine and can (combined with a colder N_2 injection temperature) be useful for applications such LiN-IC hybrid vehicles.

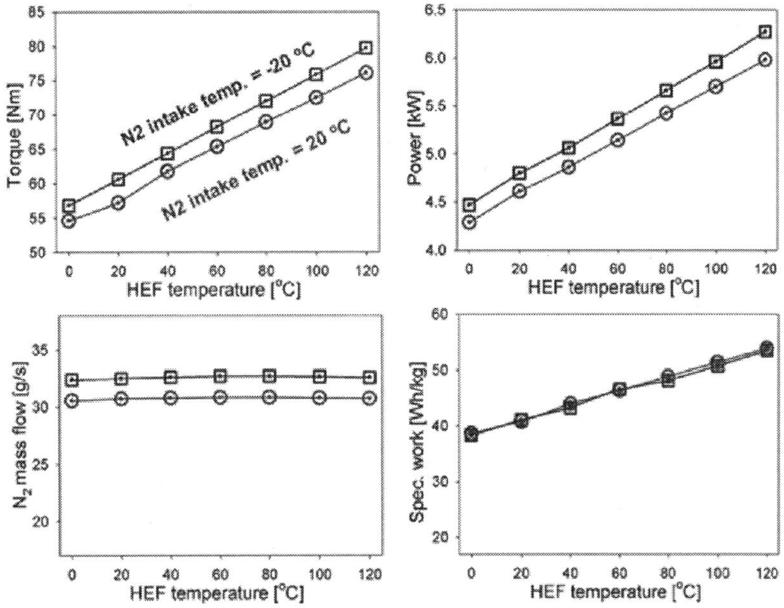

Figure 16: Effect of N_2 and HEF temperature on torque, power, N_2 mass flow and specific work. The engine conditions are: P_{N2}=30bar, Speed=750rpm

Future model improvements will include further study of in-cylinder heat transfer, pneumatic effects on inlet valve flow and two-phase exhaust flow. Validation experiments are subject of current on-going research.

5 SYSTEMS INTEGRATION

5.1 Truck refrigeration unit

As described in the introductory section and Figure 2, refrigeration is delivered by two mechanisms: the use of the LiN vapouriser to extract heat, plus the use of an engine-driven, conventional vapour-cycle heat pump. This system takes advantage of a particular synergy between the engine and heat-pump. Strong cooling of the vapour cycle condenser is achieved using HEF, which significantly improves the coefficient of performance of the vapour cycle compared to conventional truck systems – a CoP of 2 was indicated for the first, non-optimised prototype system on a calibrated test rig, compared to a CoP of around 1 for a conventional system. The heat removed from the condenser makes a significant contribution to warming the HEF; as shown in the previous section, this improves the efficiency of the Dearman engine.

System modelling suggests that the first, non-optimised prototype will deliver a performance of 14kW heat removal, with a specific work of over 110Wh heat removal per kg of LiN, and that a second generation system embracing some of the improvements described in this paper will deliver 145Wh/kg, with 165Wh/kg being ultimately possible. These levels of performance are sufficient to create a significant payback against diesel-driven or basic LiN-evaporation systems.

Figure 17 shows the system installed in a truck – the LiN tank is on the opposite side of the vehicle. Early testing of this system has indicated that the condenser-HEF heat transfer synergy works well, and that the specific LiN consumption is as expected. Further development will take place through 2015, with a first unit entering field trials at the end of the year.

Figure 17: Truck refrigeration installation

5.2 Waste Heat Recovery

Conventional WHR systems almost all use exhaust energy, and are best suited to applications where the IC engine is operating at a steady, high load condition. The Dearman Engine offers the ability to use coolant heat, as its quality is easily sufficient (and indeed exhaust heat can be too warm for the HEF). This introduces a new, helpful factor: the ability to use the thermal inertia of the ICE and its cooling circuit as an energy store. This means that the Dearman WHR system can be applied in the same manner as an electric hybrid system, being used as a transient assistant to, or substitute for the ICE, enabling the ICE to be right-sized and operated at a more steady state condition. This enables new fields of application such as buses and urban trucks to be considered [6], and because the savings in primary fuel are greater with this mode of operation, the Dearman system is most attractive in precisely the type of application where conventional WHR is not. Figure 18 illustrates the essential principle of this hybrid-like operation, using the Millbrook London Bus cycle, and showing how the Dearman system can assist with power peaks, and / or operate as a sole provider of lowest powers. Right-sizing the ICE for the intermediate power levels is practically limited by driveability (especially off-cycle), but up to 40% engine downsizing may be possible. Other control strategies also offer advantages, for example zero-emission launch to improve air quality at bus stops, or a more extended zero emission mode using a larger reserve of gasified nitrogen to avoid short term heat depletion.

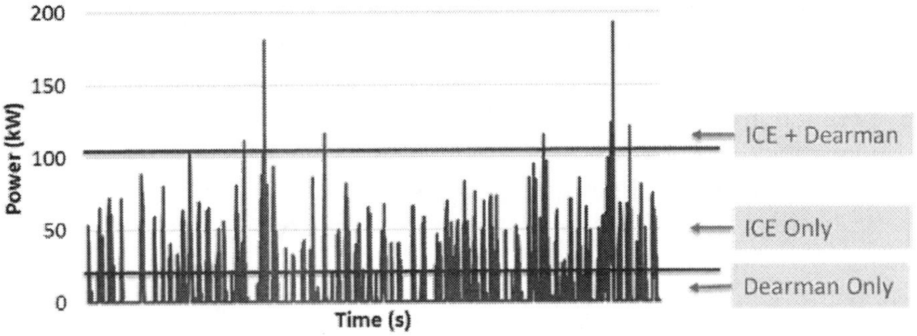

Figure 18: Principle of operation of the "Heat-Hybrid" WHR system

A system is being developed for evaluation in an urban bus. The principal of the energy and fluid circuits is shown in Figure 3 at the start of the paper, and the physical layout in figure 19. The Dearman Engine is coupled to the ICE at the gearbox input via a fixed gear ratio and clutch, enabling it to disengage when not in use. Hardware evaluation results will not be available until 2016, but initial cycle simulations have indicated potential to save up to 25% in diesel consumption. If the bus has an air-conditioning requirement (which a Dearman system can meet for no extra fuel cost), this saving can rise to over 40%. However, to be successful such a system requires a higher efficiency Dearman Engine than the simple refrigeration unit described in the previous sections.

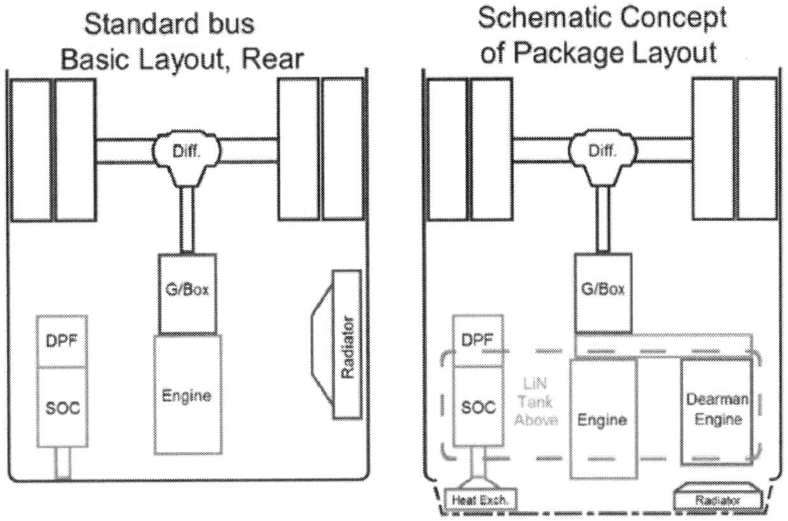

Figure 19: Packaging in a low-floor, rear-engine bus

6 ENGINE AND SYSTEM IMPROVEMENTS

6.1 Truck refrigeration engine

Truck refrigeration requires only a moderate level of efficiency to be commercially attractive, because the extraction of heat performs the primary function of the engine. Efficiency is beneficial because the secondary, vapour cycle system driven by the Dearman Engine achieves a CoP of well over 1, and because shaft power can meet a variety of electrical needs including both the refrigeration system fans and defrosters, and in future other vehicle electrical systems. Test data and validated modelling from the first prototype, described above, has been used to conceptualise a second generation engine optimised for this application, incorporating improvements described such as fixed valve timing with mechanical cams (figure 20), cam-operated HEF valves, and a more efficient, tank-immersed LiN pump. Table 1 compares key performance metrics for the two generations, illustrating the efficiency improvements that are possible with just a year of development – and postulating targets for a further refined, third generation that could take advantage of improved heat transfer and faster acting rotary intake valves.

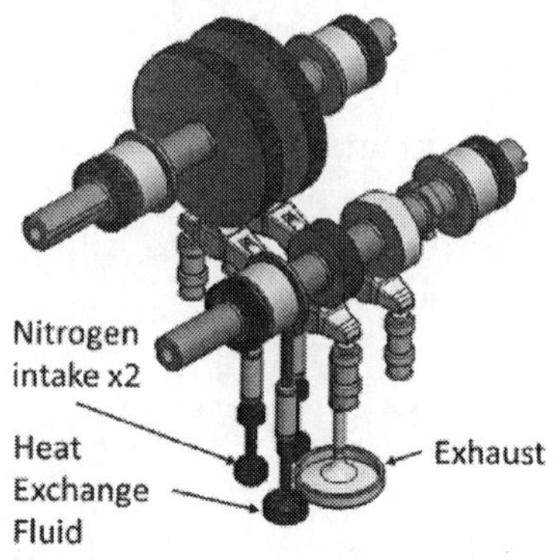

Figure 20: Mechanical valvetrain

Table 1: Refrigeration engine characteristics

	First prototype	Second generation	Ultimate potential
Expander capacity	680cc,1 stage	470cc, 1 stage	~500cc, 1 stage
Intake valve period	35-40 deg crank	25 deg crank	<20 deg crank
Power (Mech/Cool)	4kW / 14kW	4.5kW / 16kW	~4kW/16kW
Specific Energy, Flywheel (Gross)	30-35Wh/kg (Measured)	45Wh/kg (Predicted)	>60Wh/kg (Target)
Specific energy, Net of engine systems	16-19Wh/kg (Measured)	30Wh/kg (Predicted)	>50Wh/kg (Target)
Specific cooling energy (Net)	110Wh/kg (Predicted)	145Wh/kg (Predicted)	>165Wh/kg (Target)

6.2 Waste heat recovery engine

In contrast to the refrigeration application, any use requiring only mechanical power as an output requires the highest possible efficiency and specific work per unit of LiN used. Techno-economic analysis [6] suggests that good pay-back, of less than 2 years, is achieved by a bus system with a net specific work of 100Wh/kg or better, in contrast to the refrigeration application which is viable from around 20Wh/kg Net. Significant improvements in efficiency require a package of measures:

- Reduction of parasitic losses, for example the LiN feed pump: Ultimately, it is possible to eliminate this loss by using waste heat to self-pressurise a volume of gas, although such a system presents new challenges in pressure regulation, so in the interim a more efficient pump is likely to be the most promising route
- Maximisation of expansion ratio, meaning minimization of intake valve opening period and clearance volume: Analysis has shown that achieving an open period

of less than 25 degrees for a poppet valve is infeasible, primarily due to the cam diameter required to achieve reasonable profile concavity for manufacture (figure 20); this limits expansion ratio to around 20:1 with effective use of HEF to absorb clearance (figure 4)

- Improved heat transfer from HEF to gas, to achieve a more isothermal expansion: modelling (figure 14) indicates that a finer break-up of the HEF by incoming gas is possible, with a ten-fold increase in heat transfer area being required

A thoroughly optimised but practical single stage, using rotary valves, improved heat transfer, reduced parasitic losses and harvesting heat at ICE coolant temperature, has been shown by modelling to have potential for around 70Wh/kg net specific work. To progress further, multiple expansion stages are by far the most practical option.

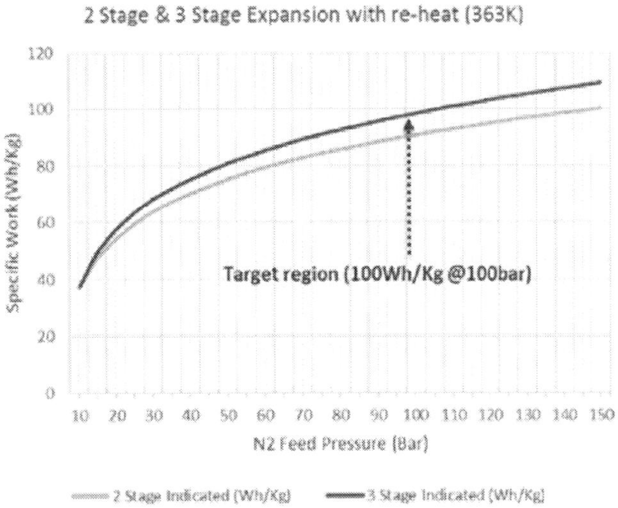

Figure 21: Specific work of 2 and 3 stage expansions versus feed pressure

Figure 21 shows a prediction of specific work energy against feed pressure for two and three expansion stages, based on assumed stage expansion ratios for two poppet-valve piston stages plus a third, optional rotary expander. At a feed pressure of 100bar, the efficiency target of 100Wh/kg can be met on a gross specific work basis. For early applications, it is likely that hardware will adopt just two stages, and seek applications where a supplementary air-conditioning need or other factor makes the economics attractive.

Direct injection of liquid nitrogen is the final step in the efficiency routemap. The cryogen is injected straight into the cylinder at the lowest feasible pressure, where it meets an enlarged quantity of HEF which provides all the required heat for vapourisation, then resumes its function in maintaining isothermal expansion. This approach enables the shortest possible intake event, of 10 degrees or less, hence higher expansion ratio; the introduction of an isochoric-isothermal expansion, and elimination of feed-pump work, both further aid efficiency. Laboratory experiments have shown that direct injection of LiN into a HEF can achieve a suitably fast vapourisation and pressure rise rate (similar to the pressure rise of a gasoline engine), but development of a controllable, efficient direct injector presents many challenges:

- Insulation of the cryogenic liquid is difficult in the space constraints of a cylinder head
- Fast actuation of the injector requires sophisticated approaches such as electro-hydraulics, which carry parasitic losses
- Efficiency requires fast pressure rise, but this can force early injector closure

Direct injection will be a topic of future research.

7 CONCLUSIONS

This paper has described many aspects of the Dearman Engine, from the physics of its operation to practicalities of implementation and application. As with any innovation, all these aspects are important to its chances of success. Key conclusions that emerge from the work so far are:

- The Dearman Engine has been proven to offer greater efficiency than a conventional Rankine-cycle expander due to the effects of its unique Heat Exchange Fluid
- An analytical exploration of in-cylinder thermodynamics, validated by experimental work, has shown that the Heat Exchange Fluid acts in several ways, including two-way heat transfer and improved expansion ratio by reduction of dead volume
- Application requires a holistic, system approach to understanding the flow of heat and other forms of energy, as illustrated by the HEF-condenser cooling synergy in refrigeration, and by the use of thermal inertia as an energy store in the waste heat recovery "hybrid"
- Sufficiently high levels of system efficiency are achievable with optimization, to make the Dearman Engine attractive in a number of applications. As a simple rule, efficiency must be high enough to realize over 100Wh of useful effect (cooling, shaft work) per kg of LiN to be attractive at today's market prices; routes have been identified to achieve this in a transport refrigeration application, and achieve it in a propulsive waste heart recovery application

8 ACKNOWLEDGEMENTS

The authors would like to thank InnovateUK for their support to the Cool-E and CE-POWER projects funded under the Low Carbon Vehicles Innovation Platform, which have made this paper possible.

9 REFERENCE LIST

1. Patent no EP1257733; JP4647872; US6983598
2. "Cryogenic heat engines for powering zero emission vehicles", Ordonez, Plummer & Reidy, University of Texas, from Proceedings of the 2001 ASME International Mechanical Engineering Congress, IMECE2001/PID-25670, 2001

3. “A novel cryogenic energy system for zero emission vehicles”, PhD, Dr H Clarke, School of Engineering and Materials Science, Queen Mary University of London, 2011
4. “Liquid Air in the energy and transport systems - Opportunities for industry and innovation in the UK”, Centre for Low Carbon Futures (CLCF), 2013
5. “Liquid Air Technologies – a guide to the potential”, CLCF, 2013
6. “Liquid Air on the Highway - The environmental and business case for liquid air commercial vehicles in the UK”, CLCF and Liquid Air Energy Network, 2014
7. “A Tank of Cold: Cleantech leapfrog to a more food secure world”, Institution of Mechanical Engineers, June 2014
8. “Exploring the Potential for Cold Chain Development in Emerging and Rapidly Industrialising Economies through Liquid Air Refrigeration Technologies”, Dr. Lisa Kitinoja, Postharvest Education Foundation, 2014
9. “Inside Fiat’s innovative MultiAir system”, Society of Automobile Engineers, 2010, http://articles.sae.org/8703/
10. NIST Standard Reference Database 23 (2010), http://www.nist.gov/srd/fluids.cfm
11. “Some new investigations on old combustion engines problems”, G. Eichelberg, Engineering, Oct. 1939, pp 463-547
12. “Direct Contact Gas-Liquid Heat Exchanger for Energy Recovery”, J. R. Fair, Trans. of ASME, Journal of Solar Energy Engineering, vol. 112, (1990), pp. 216–222

Simulation / Energy Management

Thermal gasoline concept vehicle, simulation and control

X LIU and **R RASTELLI**
BorgWarner Inc., USA

ABSTRACT

Simulation and control are fundamental aspects in BorgWarner's efforts to develop the Thermal Gasoline Concept Vehicle (TGCV). An integrated system model was developed in GT-SUITE including a real time mean value engine model, vehicle model and thermal system model with coolant and oil circuits. In addition, a thermal management controller in Simulink was developed and integrated in co-simulation with the plant model to assess flow/heat distribution among different coolant/oil circuits for fuel economy improvement. Co-simulation allowed the structuring and virtual test of complex control strategies before implementation. It also facilitated the optimization of this multi-variable system. The thermal controller was implemented in the actual TGCV vehicle and initial vehicle test results are presented.

1 INTRODUCTION

The advantageous cost-benefit relationship offered by some well-known technologies has been a blessing in terms of its effectiveness and at the same time a curse for the commercialization of new approaches that seek improvements in fuel economy for internal combustion engines (ICEs). One of these is the ubiquitous centerpiece of pressurized cooling systems control, namely: the wax thermostat. The same argument can apply to mechanically driven coolant pumps. This combination of control and hardware has reigned and remained fairly unchanged since the 1950's. Still found in the majority of the cooling systems of ICEs in production, it allows a reasonably short warming period and protects the engine from overheating. The system operates reliably and at a relatively low cost.

Lately, the increased pressure for improved efficiency has promoted technologies like electric cooling pumps to compete as a plausible component in more advanced thermal management systems.

Many excellent works exploring the potential of minimizing internal combustion engines warm up time through advanced hardware and control have been published during the years [1-6]. Recently, Wenzel, Shutty and Becker [1] have presented a comprehensive study on the adoption of advanced technologies in diesel engines, specifically: dual mode coolant pump (DMCP) and electric coolant control valve (ECCV). DMCP can be either driven by an integrated electric motor or through a mechanical link to the crank. ECCV allows the adoption of thermal management strategies independent of temperature. Additionally, the exhaust heat recovery system (EHRS) represents an additional mean to speed up engine warm up by utilizing more of the exhaust energy that would be otherwise wasted.

A simple inspection of Sankey diagrams for ICEs shows how, approximately and on coarse average, more heat is transferred to the coolant (30% vs 25%) and more heat is released with the exhaust (35% vs 25%) by comparing gasoline vs. diesel [7]. This leaves more room to look for efficiency improvements through advanced thermal management strategies when we apply them to the Otto cycle. The aim of BorgWarner's Thermal Gasoline Concept Vehicle (TGCV) is to explore the potential of these technologies and implement them in a popular mid-size (D-segment) vehicle.

The current paper focuses on describing the effort that went through the development of simulation and control models as an integral part in the development of the gasoline powered thermal management vehicle demonstrator. Results from simulation are reviewed as well as their influence on control design, the system hardware implementation and the test program itself. It will be also shown how this is an iterative process that enabled the evaluation of effective thermal management approaches, before and in parallel to test.

2 OBJECTIVE AND APPROACH

2.1 Objective

The overall objective was to develop an in-vehicle engine thermal system model and controller to assess the impact of components and control strategies and explore their impact on fuel economy. Simulation and control were integral aspects in the development of a gasoline powered thermal management vehicle demonstrator in which the hardware was put to test.

2.2 Approach

From a broad perspective, not only the models were validated with test results, but also the simulation was used to design the controls that then were used in the vehicle. Sort of like an iterative loop that is depicted in Figure 1.

A co-simulation approach that integrated a Simulink controller model with a GT-SUITE plant model was adopted throughout. The high fidelity plant model was created based on first principles along with physical characteristics of the engine and vehicle (e.g. displacement, gear ratios, and heat exchanger characteristics). The target vehicle was instrumented and tested and the data was used to calibrate the simulation model and the model based thermal controller. A validation process that compared simulated results to measurements was conducted. Tests were performed both virtually on simulation model and on vehicle in order to gain a deeper understanding of the thermal system effects along with the benefits of the advanced thermal management approaches. As previously mentioned, this process was iterative, i.e. vehicle results led to further refinements in the simulation model which then led to improvements in the thermal management system (layout, components, control) that needed further validation through test and so on.

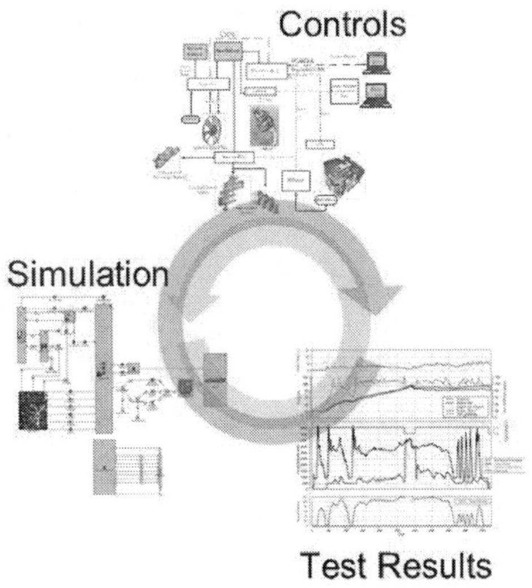

Figure 1. Development approach

3 ENGINE COOLING LAYOUT AND KEY THERMAL COMPONENTS

3.1 Target vehicle

A Ford Fusion 2013 was selected as the base test vehicle. It has a 2 liter, inline 4-cylinder EcoBoost engine with one single-scroll turbocharger. The compression ratio is 9.3. Key vehicle characteristics such as peak power and peak torque are listed in Table 1. The material of the engine head and block is cast aluminum.

Table 1. Test vehicle specifications

Manufacturer	Ford Motor Company
Model	2013 Ford Fusion
Engine	2.0 liters, inline 4, VVT, CR: 9.3:1
Turbo	Single stage
Power	172 kW @5500 RPM
Torque	366 Nm @3000 RPM
Drive Type	Front wheel drive,
Transmission	6-speed automatic

3.2 Baseline cooling layout

Figure 2 shows the cooling system layout of the baseline engine. A mechanical pump driven by the engine circulates coolant through engine block and head. A wax thermostat controls coolant flow through bypass or radiator based on coolant temperature. Also coolant flows through turbo charger, cabin heater, and transmission oil cooler. Transmission oil cooler allows oil flow only when transmission oil is hot. The two radiator cooling fans are electric motor driven. They can be operated in series or in parallel.

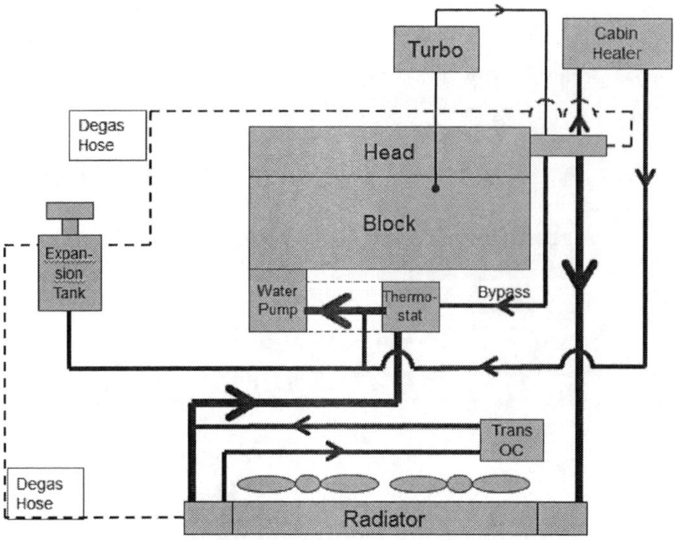

Figure 2. Baseline engine cooling layout

3.3 Advanced thermal components

The baseline vehicle was modified to incorporate advanced thermal components with the goal of shortening the engine and transmission warm up period, leading to better fuel economy. Figure 3 shows advanced and standard thermal components adopted in the update build. The stock mechanical pump was replaced by a dual mode coolant pump (DMCP). The wax thermostat was replaced by an Electric Coolant Control Valve (ECCV) in which opening can be controlled independent of coolant temperature. At engine cold start, the ECCV is fully closed enabling zero-engine flow strategy. An exhaust heat recovery system (EHRS) was added to reuse waste heat from exhaust gas.

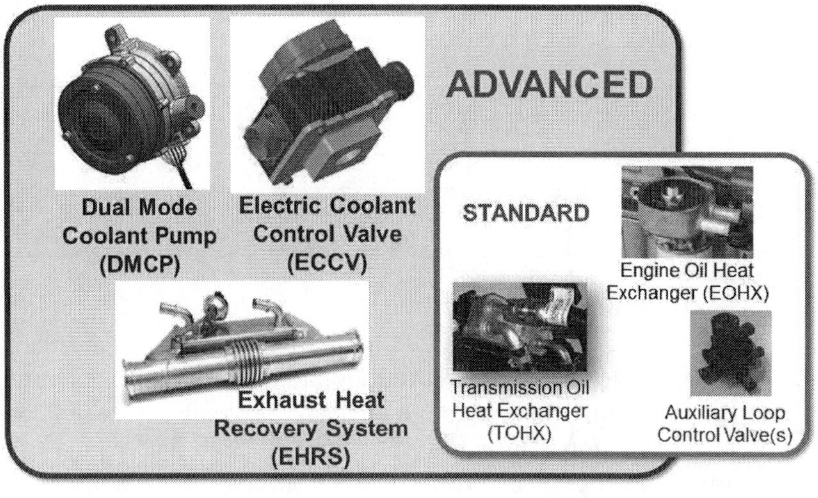

Figure 3. Advanced and standard thermal components

The majority of the cooling system was composed of standard components. However several advanced thermal management components were added and/or used. The DMCP, ECCV, and EHRS are BorgWarner products. A description of these advanced components is given below.

3.3.1 Dual mode coolant pump (DMCP)

The dual mode coolant pump is a prototype part developed by BorgWarner. It replaced the stock mechanical coolant pump as a drop in replacement. The pump was located in the same location as the stock water pump and connected to the accessory drive belt in the same manner. The DMCP has a brushless DC motor and friction clutch built in, which enables the DMCP to operate in two modes: in mechanical mode, the pump is driven by the engine, like a conventional water pump. In electric mode, it is driven by the integrated DC motor. The pump speed can be controlled independent of engine speed. The DMCP allows overspeed, underspeed, zero-engine flow strategy and turbo aftercool. The failsafe operation mode is mechanical mode.

The interface to the pump was via CAN to control both electrical pump speed and clutch state. More details of the working principle and the operation map were provided in [3].

3.3.2 Electric coolant control valve (ECCV)

A prototype level electric coolant control valve developed by BorgWarner was used in the update build in place of the stock wax thermostat. The valve is actuated by a brushless DC motor. The additional functionality this device offered over the stock wax thermostat is that it can be controlled to any opening regardless of coolant temperature. The ECCV is located at engine outlet, in contrast, the stock wax thermostat is at engine inlet.

3.3.3 Exhaust heat recovery system (EHRS)

The exhaust heat recovery system was placed after the catalyst and flex coupling as shown in Figure 4. The heat exchangers were designed with high heat transfer efficiency from exhaust gas to coolant and minimal gas pressure drop across the unit. There is a bypass valve on the gas side to control EHRS in either heat exchanger mode or bypass mode. The bypass valve is actuated through a vacuum solenoid. To avoid disruption to the vehicle vacuum supply, a reservoir was added between the EHRS vacuum solenoid and the vacuum source.

EHRS has benefit to reuse exhaust heat to warm up engine coolant and oil, but it also introduces extra thermal inertia (coolant volume and metal mass). Therefore efficiency and compactness are key attributes in EHRS design. It is important to minimize added thermal inertia in coolant layout design.

3.4 Standard thermal components

In order to implement a functional thermal system, the advanced components were combined with some standard components. The Engine Oil Heat Exchanger (EOHX) is standard in some EcoBoost versions (e.g., South America). The Transmission Oil Heat Exchanger (TOHX) is standard in all versions. Three auxiliary control valves are production parts from other vehicles and are strategically placed to allow some unconventional modes of operation, discussed below.

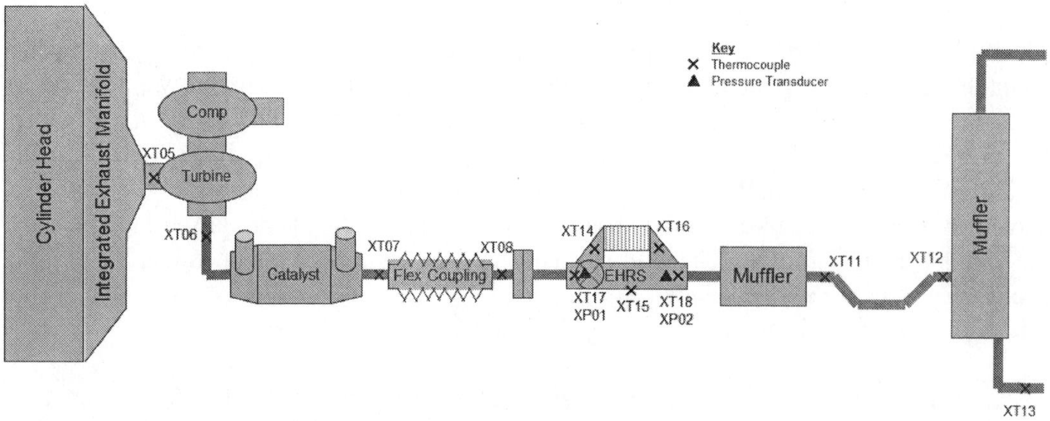

Figure 4. EHRS installation location

3.5 Update build cooling layout

Figure 5 shows the cooling layout of the update build. The DMCP, ECCV, and EHRS are highlighted in darker color. The ECCV is at engine outlet, which is different from baseline engine, where the wax thermostat is at engine inlet. There is a bypass valve on the EHRS gas side to control EHRS in either heat exchanger mode or bypass mode. There are four auxiliary coolant valves to control coolant flow and direct heat to desired components. The key philosophy behind the layout design is to enable the engine block zero flow strategy at cold engine start. At the same time, utilize the EHRS to warm up coolant and oil.

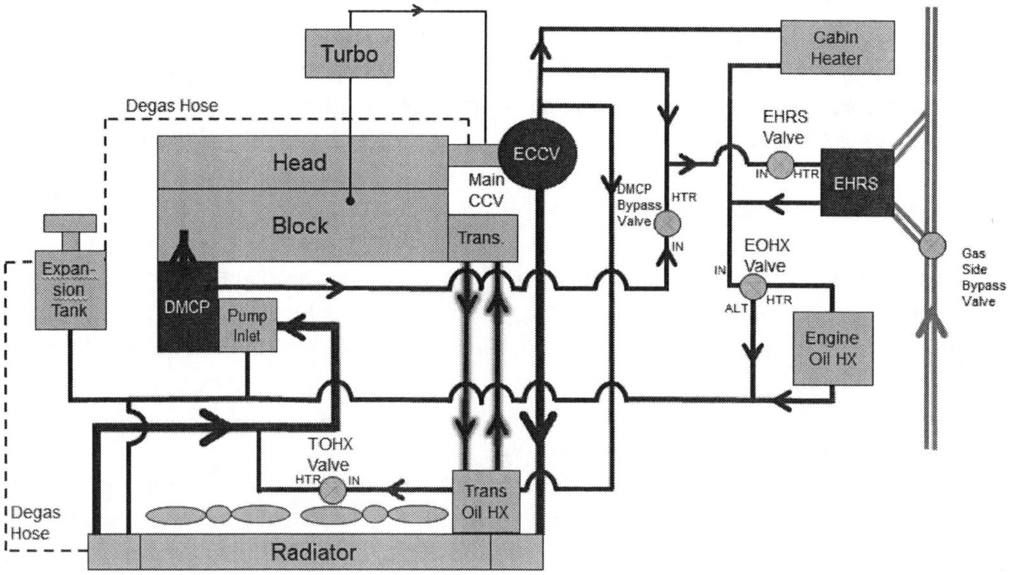

Figure 5. Update build cooling layout

3.6 Operating modes of update build

There are six operating modes in the update build, these are:

Mode 1 (Figure 6): cold start – zero engine flow. ECCV is fully closed. There is no flow through engine block, allowing rapid engine warm up. DMCP pumps coolant through DMCP bypass valve, then cabin heater, EHRS and EOHx. There is no coolant flow through TOHx. EHRS is in heat exchanger mode, actively transferring heat from exhaust to coolant outside engine block.

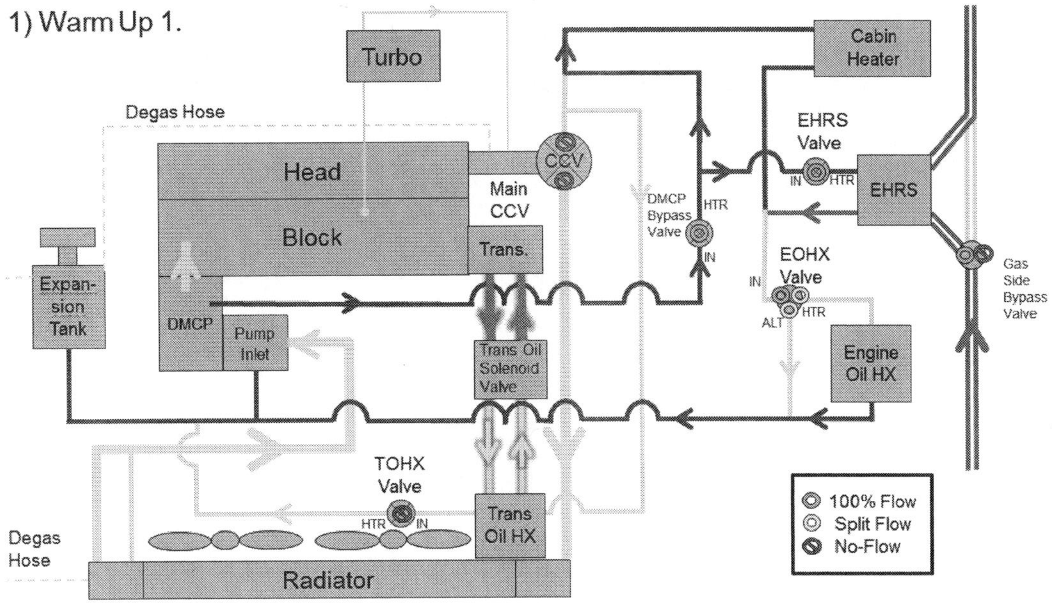

Figure 6. Operating mode 1 of update build

Mode 2 (Figure 7): ECCV bypass port is open, and DMCP bypass valve is closed. It allows coolant flow through engine block. EHRS is still in heat exchanger mode. There is no flow through TOHx.

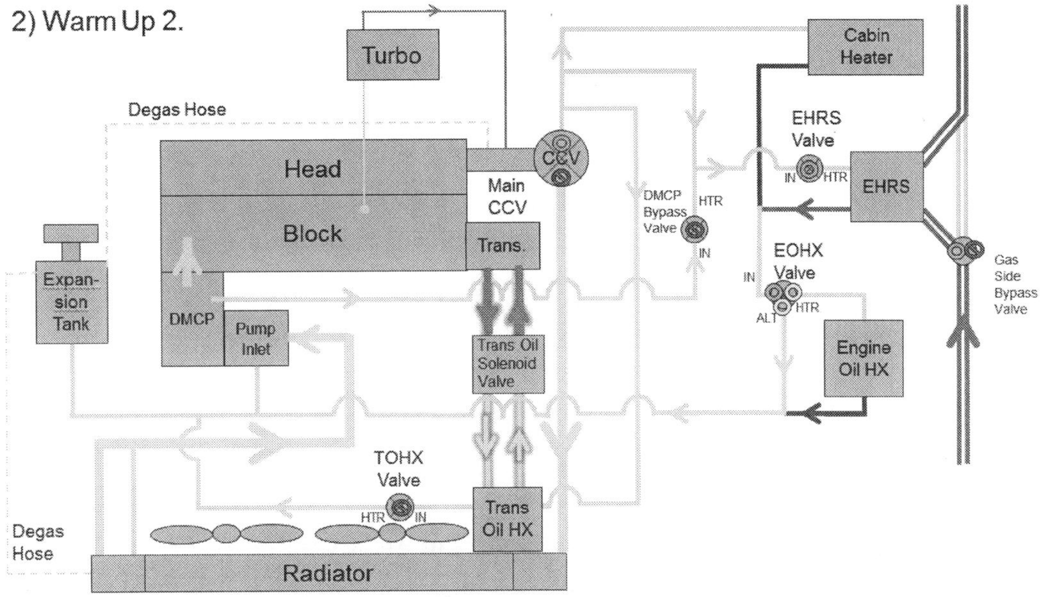

Figure 7. Operating mode 2 of update build

Mode 3 (Figure 8): TOHx valve is open allowing coolant flow through transmission oil heat exchanger. Other valve positions are same as in Mode 2.

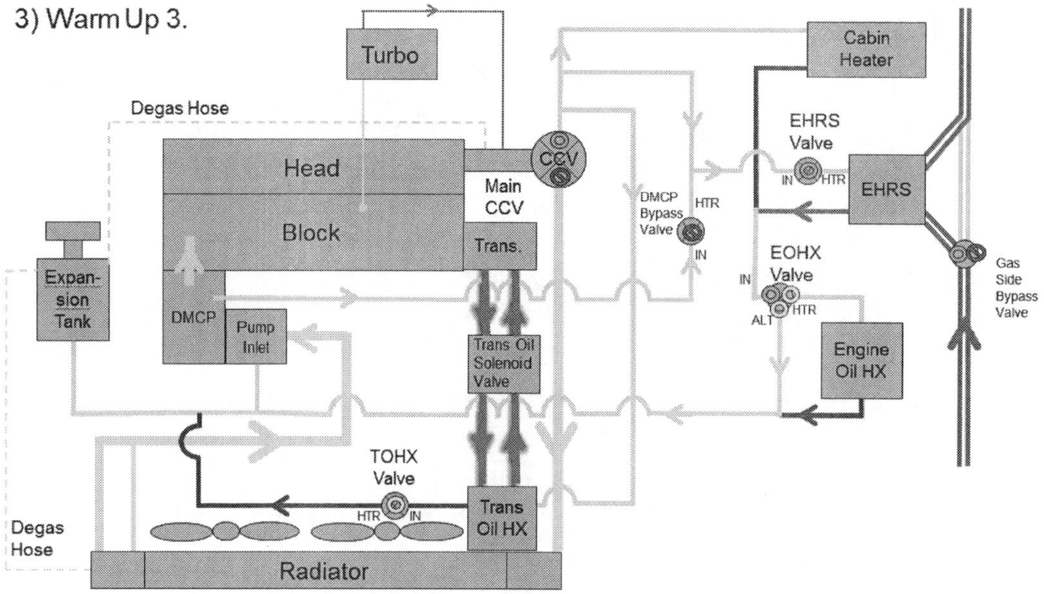

Figure 8. Operating mode 3 of update build

Mode 4 (Figure 9): hot operation with thermostat closed. ECCV radiator port is still closed. EHRS can be in bypass or heat exchanger mode depending on engine oil temperature. If in bypass mode, the coolant flow through EHRS is reduced as well.

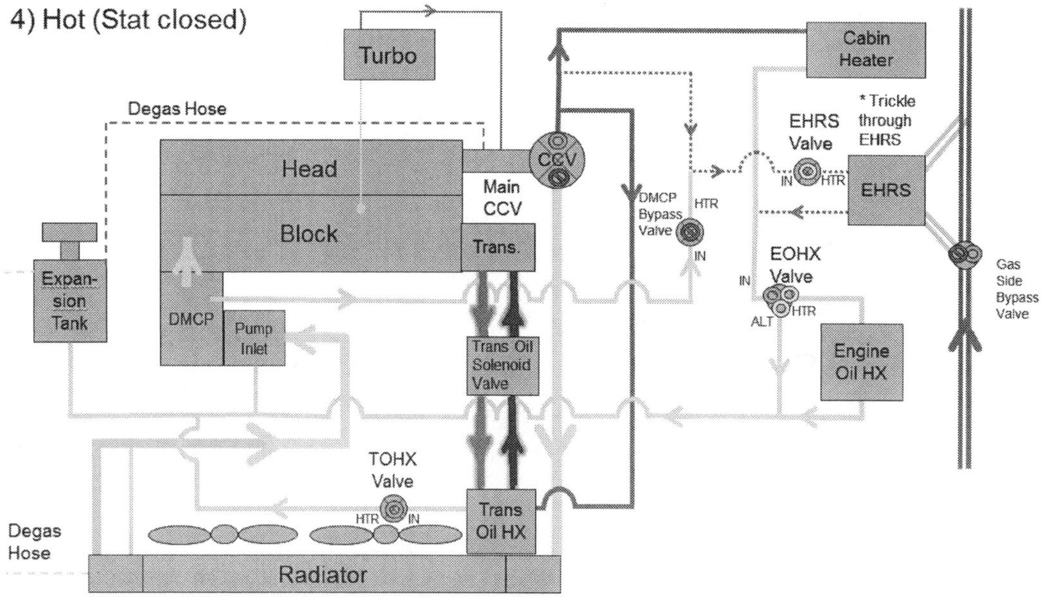

Figure 9. Operating mode 4 of update build

Mode 5 (Figure 10): hot operation with thermostat open, flow through radiator. Typically EHRS is bypassed in this mode. However, if engine oil temperature is still low, the bypass operation mode can be delayed to allow more heat from exhaust gas be transferred to coolant and oil. One could question whether it makes sense to keep EHRS active while radiator loop is open and removing heat from engine coolant. With EHRS in heat exchanger mode, the coolant temperature at EOHx is higher and more heat is transferred to engine oil. But if the coolant temperature at EHRS outlet is close to boiling point and DMCP is at its limit to further increasing coolant flow rate, EHRS will be switched to bypass mode. So EHRS can be in heat exchanger operation in mode 4 or even in mode 5, depending on engine oil temperature and coolant temperature.

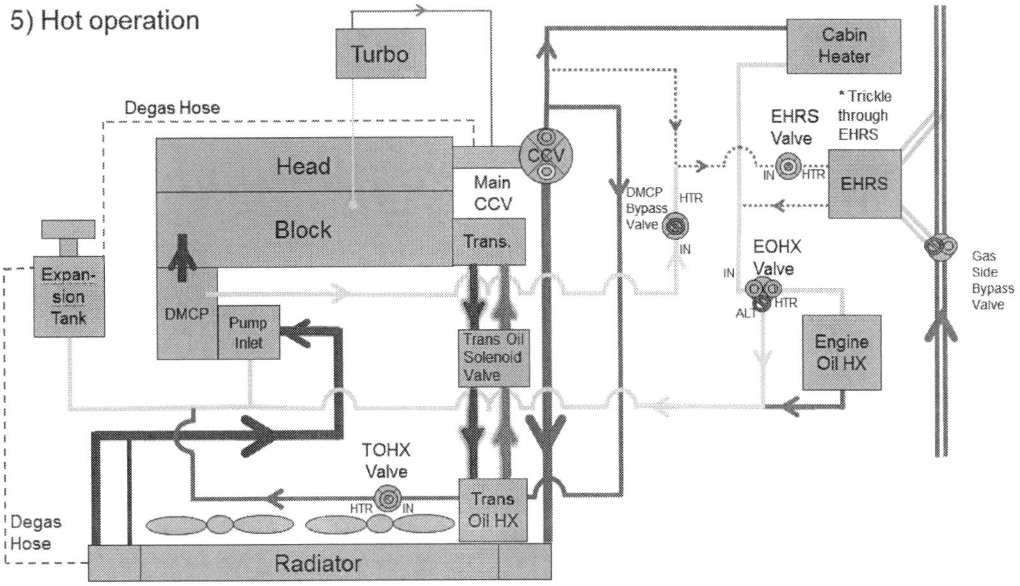

Figure 10. Operating mode 5 of update build

Mode 6 (Figure 11a-b): engine off, pump after-run. It includes two sub modes, hot soak or cold soak. The engine is off but the pump is still running. Coolant flows through turbo to allow bearings cooling. Radiator port of ECCV is open or closed depending on coolant temperature.

4 SIMULATION MODEL AND FEATURES

A full vehicle simulation model was developed in GT-SUITE. The model includes a mean value engine model, vehicle model and thermal system model with coolant and oil circuits. A thermal controller was implemented in Simulink. The interface between the GT plant model and Simulink controller is a Simulink-Harness block. Co-simulation of Simulink controller and GT plant model has been adopted throughout the project development cycle.

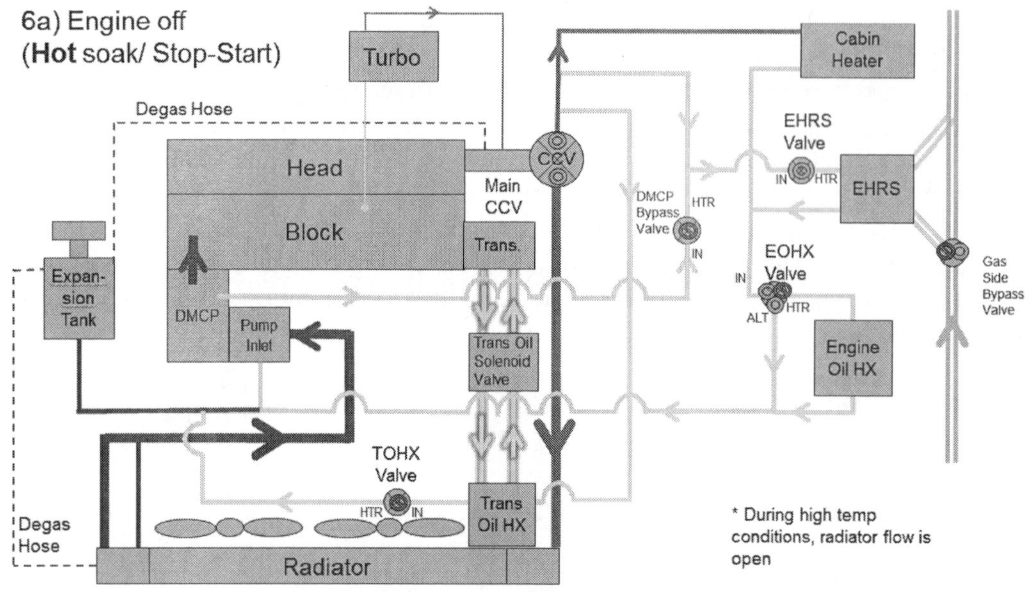

(a)

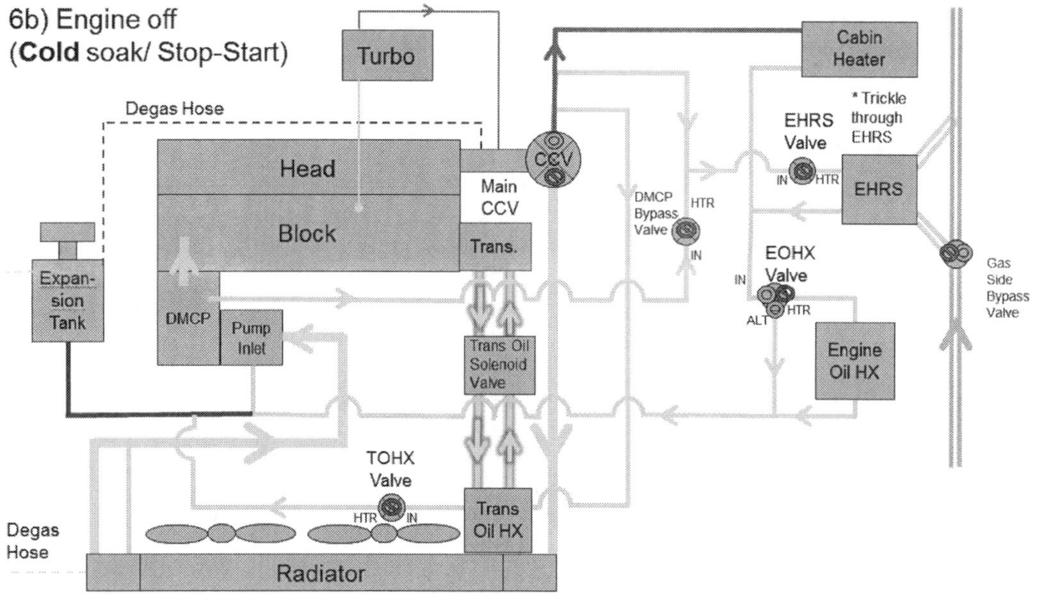

(b)

Figure 11. Operating mode 6a-b of update build

4.1 Integrated model

Figure 12 shows a schematic of the simulation model. The integrated models are based on the setups presented in sections 3.2 and 3.5. For controls development, the host environment is Simulink with a GT plant model running as an S-function. For general and optimization work, a Dynamic Link Library (DLL) of the Simulink model is created and called by GT-SUITE through the Simulink-Harness. Descriptions of the major elements of the integrated model are given in the following sections.

4.1.1 Mean value engine model

The mean value engine model was converted from a detailed engine model, which was developed based on engine parameters and calibrated with measurement data. The conversion used a neural network and significantly reduced simulation run time. The neural network looks up volumetric efficiency, IMEP, exhaust temperature and heat rejection at different engine operating conditions. Input variables are intake manifold (boost) pressure, intake temperature, engine speed, and exhaust pressure.

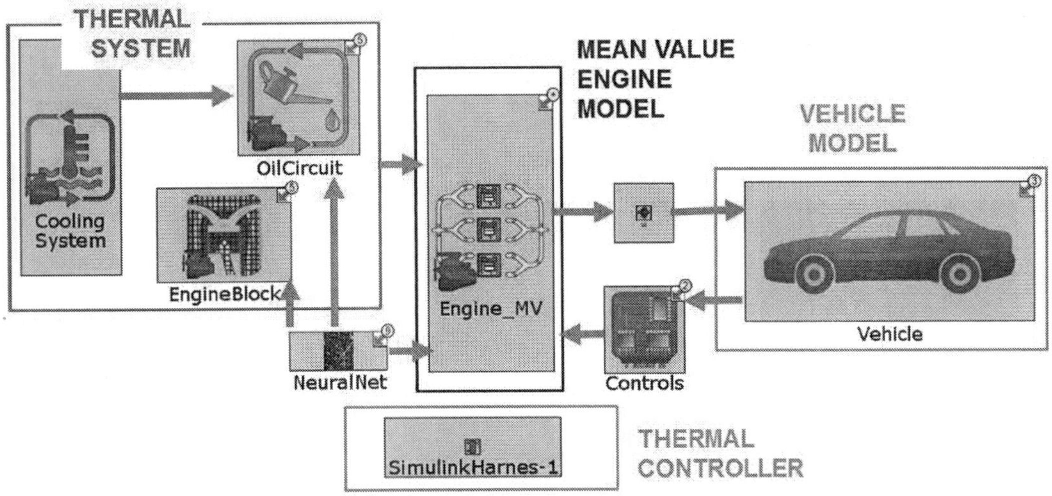

Figure 12. Integrated simulation model diagram

The thermal impact on engine friction is mainly dependent on inner block metal temperature and engine oil temperature. The base FMEP which was essentially a speed-load map, was considered to be representative of a warm condition. Then a correction was implemented as function of oil temperature and inner engine block temperature, with input signals received from separate models. The signal split, 40% oil Temperature 60% inner block (or temperature at the spark base) was chosen as it allowed the best correlation. The details are highlighted in figure 13.

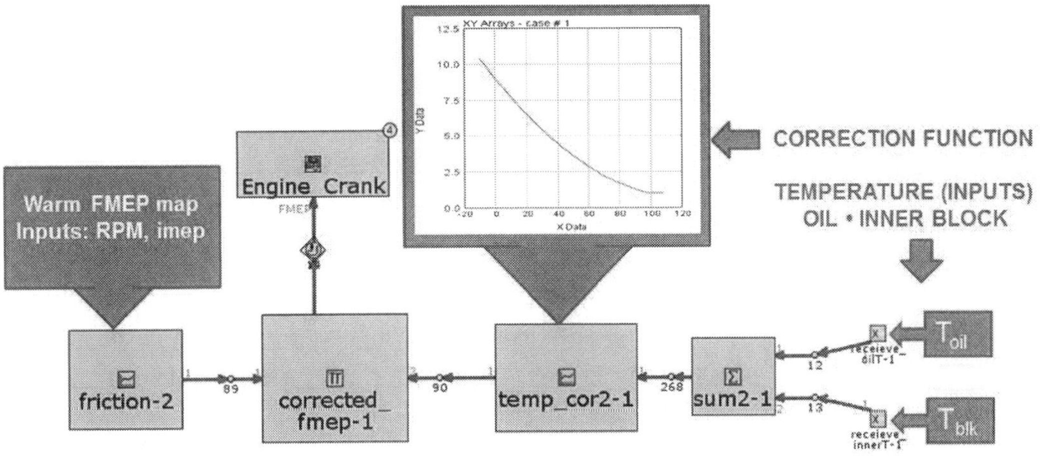

Figure 13. Engine friction model detail

Vehicle dyno testing was used to determine the engine friction temperature. Ideally, data collected on an engine dyno would have been a desirable alternative for FMEP correlation, but acquiring data in this way was out of the project scope. Instead, the in-vehicle engine was instrumented with a piezoelectric pressure transducer integrated into a spark plug to measure IMEP. Data was collected at several speeds during warm-up in neutral gear, see Figure 14(a). The IMEP in addition to the warm (base) IMEP is then attributed to increased friction due to cold temperature. Since vehicle dyno data was used, loses other than what comes from the engine had to be subtracted. These losses, in neutral, account to the transmission spin losses that were measured as well (see transmission section). Finally, the correction factor was normalized by the warm IMEP. The factor approaches 1 as the engine warms up, as shown in Figure 14(b).

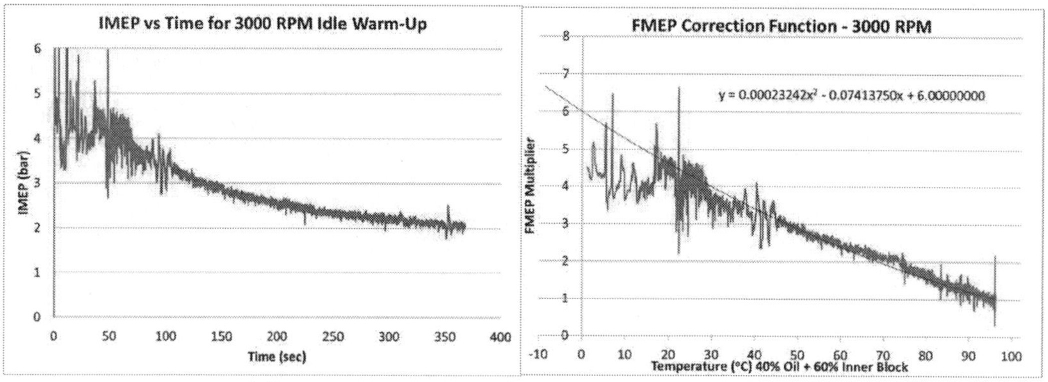

(a) (b)

Figure 14. Neutral warmup IMEP (a) and FMEP correction (b)

4.1.2 Thermal system model

The engine block in the thermal model was built with 3 masses: head, inner-block and outer-block. Thermal loading of head and inner block is a function of engine operating conditions (engine speed, intake manifold pressure, intake manifold temperature, and exhaust pressure). Exhaust valve bridge temperature is also modelled and correlated with testing data. Ideally, this could have been implemented with a more detailed model, based in CAD data. The math data was not available, so the effectiveness shown by this simplified approach rested on the accuracy of the bulk masses.

Heat transfer from engine block to coolant is flow dependent, it ties into the cooling model by means of the coolant flow going through the block and the head. The signal to the friction model is taken from the Inner Block that represent a zone close to the cylinder liners.

The oil system model is thermally loaded by three sources:

1. Conduction heat from the outer block mass.
2. Heat from the piston, 50% of heat rejection coming directly from neural network.
3. Heat from the engine crank friction.

The oil system model ties to the cooling system through the engine oil heat exchanger. This model provides the second signal to the friction correction, i.e. the engine oil temperature at the sump.

A full cooling system was modeled with pump, thermostat, EHRS, radiator, EOHX, TOHx and coolant control valves. It also includes a simple under flow model with radiator fan. Heat exchanger performance maps are based on steady state testing data. All other thermal models tie into this uniting system model where the thermal components and warmup strategies are actually implemented.

4.1.3 Transmission

A 6-speed automatic transmission was modeled in GT-SUITE. The transmission efficiency is based on test data that depends on input torque, input speed and ATF temperature. Transmission oil circuit with an oil/coolant heat exchanger was also modeled. Gear shift schedule was developed and correlated with vehicle test data.

4.1.4 Vehicle model

The vehicle model included measured rolling resistance, aerodynamic drag, and road grade. A throttle controller was set up to follow pre-defined drive cycles. The engine speed and load were determined to follow the vehicle speed profile in the cycle.

5 MODEL VALIDATION

The baseline model output was first compared to vehicle data at steady state conditions for model validation and parameter tuning. It was then compared to vehicle data during various drive cycles to ensure a good transient performance match, with further model correlation and parameter tuning performed. During these comparisons, the vehicle was operated in stock configuration, i.e. wax thermostat, mechanical pump. Figure 15 shows key thermal parameters during an FTP 75 cycle. The red solid lines are from the simulation while the blue dash lines are from the vehicle. The correlation of engine inlet coolant temperature, engine outlet coolant temperature, engine oil sump temperature, transmission oil temperature, and radiator inlet coolant temperature was quite good.

6 THERMAL CONTROLLER

A model based thermal controller was developed in Simulink to handle the various thermal actuators and allow transient cycle simulation or vehicle test. The Simulink controller was used in both PC based co-simulation environment and rapid prototype controller setup on vehicle.

6.1 Objectives

The project objective was focused on how to manage thermal system to minimize fuel consumption. It aims to achieve fast engine/transmission warm up by controlling coolant/oil flow and heat transfer, to maintain optimal temperatures by coordinating thermostat/coolant pump/radiator fan actions, and to minimize thermal system parasitic losses.

6.2 Controller structure

Figure 16 is a top level block diagram of the controller structure. Key input and output signals are shown.

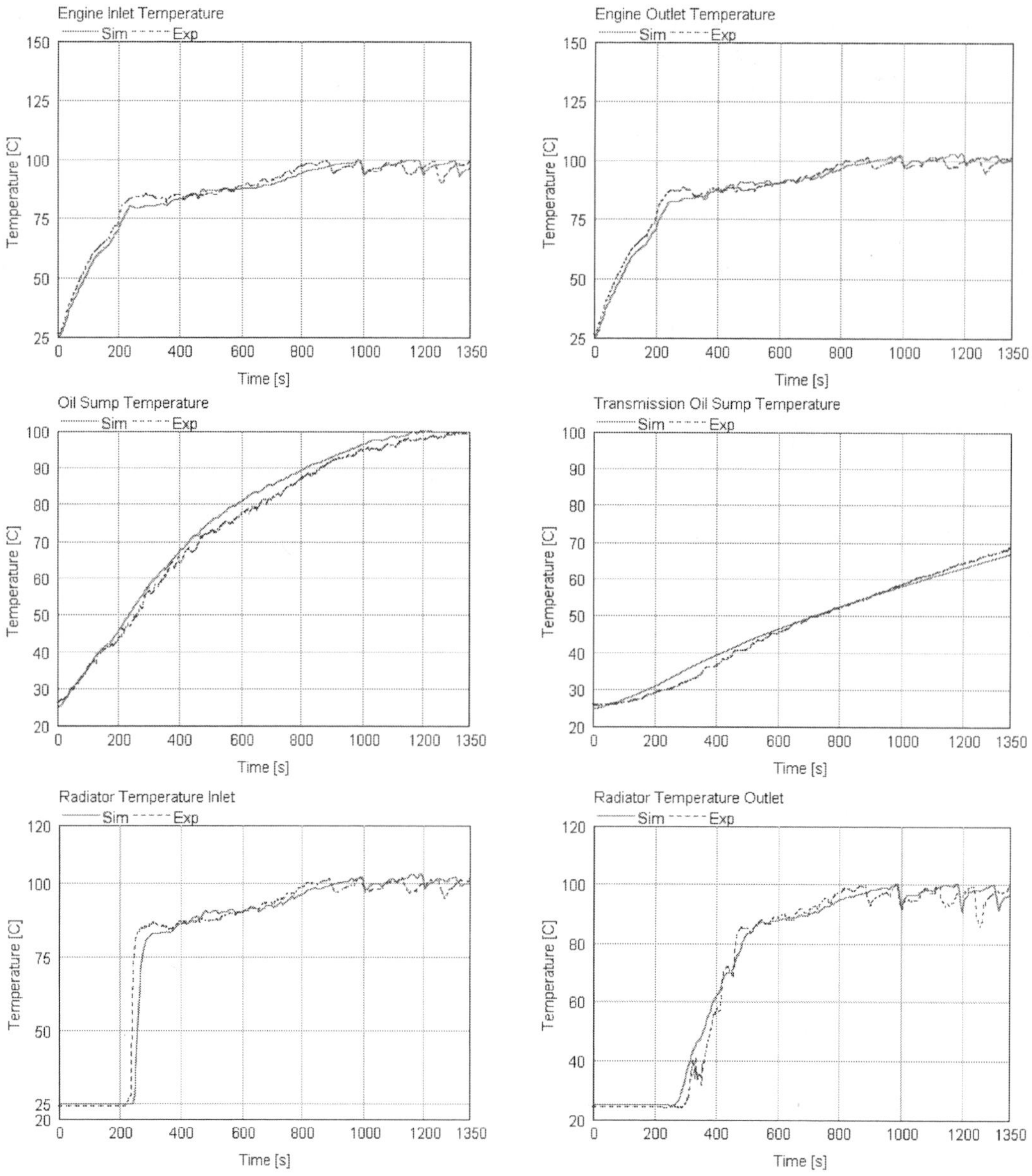

Figure 15. Baseline simulation vs. vehicle measurement – FTP75

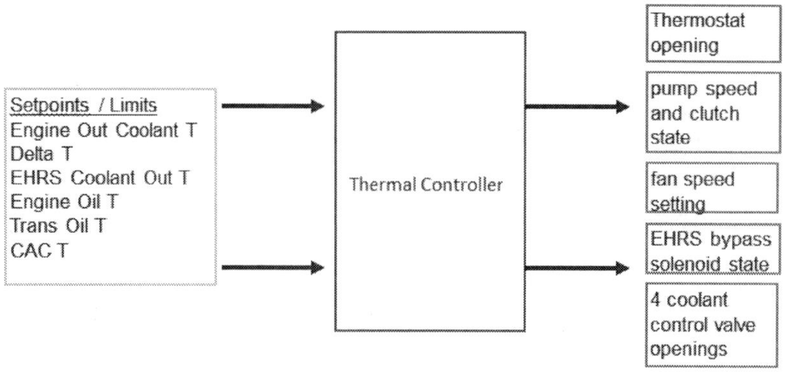

Figure 16. Top level thermal controller

The actuators in the coolant layout are shown in Figure 17.

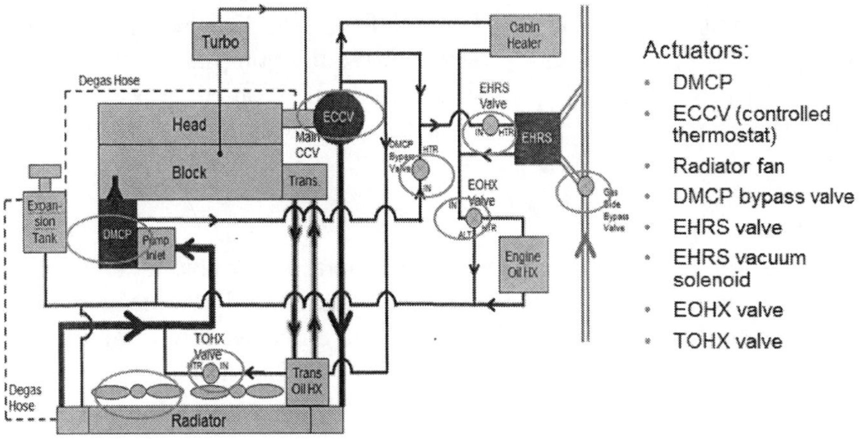

Figure 17. Actuators in the update build cooling layout

A high level view of the inner structure of the thermal controller is shown in Figure 18. In hot operating conditions, the controller coordinates thermostat/pump/fan actions to maintain engine out coolant temperature. It utilizes thermostat first, then pump or fan based on power consumption. Also, it controls pump speed to regulate maximum coolant temperature (typically EHRS downstream coolant temperature or turbo outlet coolant temperature) and ΔT across engine. Fan is also used to limit charged air cooler (CAC) downstream temperature.

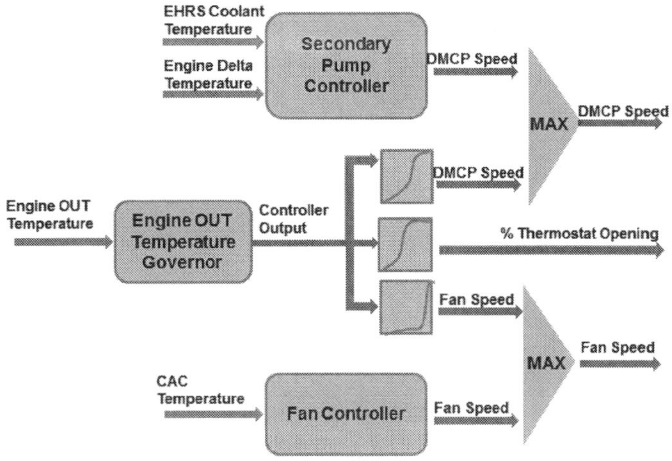

Figure 18. Thermal controller block diagram

Different operating modes of the update build layout were discussed in early sections. A Stateflow chart was developed to represent modes & transitions.

7 VEHICLE TEST RESULTS

The high level thermal controller was implemented in a dSPACE MicroAutoBox. A Mototron ECU was used for actuator control, e.g., ECCV and auxiliary valve position control, EHRS vacuum solenoid control, and radiator fan relay control. The DMCP contained its own controller for clutch and brushless DC motor. The command and sensor feedback communication between all three controllers was via CAN bus. Data acquisition was implemented with SPARTAN from imc DataWorks, and some ECU signals was accessed via CAN bus. Power to drive the pump and fan was drawn from the vehicle electrical system while power for data acquisition equipment was taken from an auxiliary battery to allow a fair comparison. Vehicle tests were performed on a chassis dyno.

7.1 FTP test results

Figure 19 shows key temperatures, pump speed, thermostat opening, and vehicle speed traces over bag 1 and 2 of the FTP cycle of the update build vehicle. The set point for the engine outlet coolant temperature is 100°C. The pump was set in mechanical mode, so the pump speed was proportional to engine speed. The radiator fan did not turn on over the period. For the first 180 seconds of the test, the operating mode was in warm up mode 1 with no coolant flow through the engine block.

The middle graph shows the thermostat (ECCV) opening, pump speed and operating mode. A rapid engine warm up can be observed in the engine outlet coolant temperature trace during mode 1 operation. When the thermostat began to open at around 400 seconds, the engine inlet coolant temperature dropped quickly due to cold coolant from radiator rushing in. The thermostat cycled open/closed several times to regulate the outlet coolant temperature. Once the coolant temperature in the radiator warmed up, the engine outlet coolant T was well regulated around the set point.

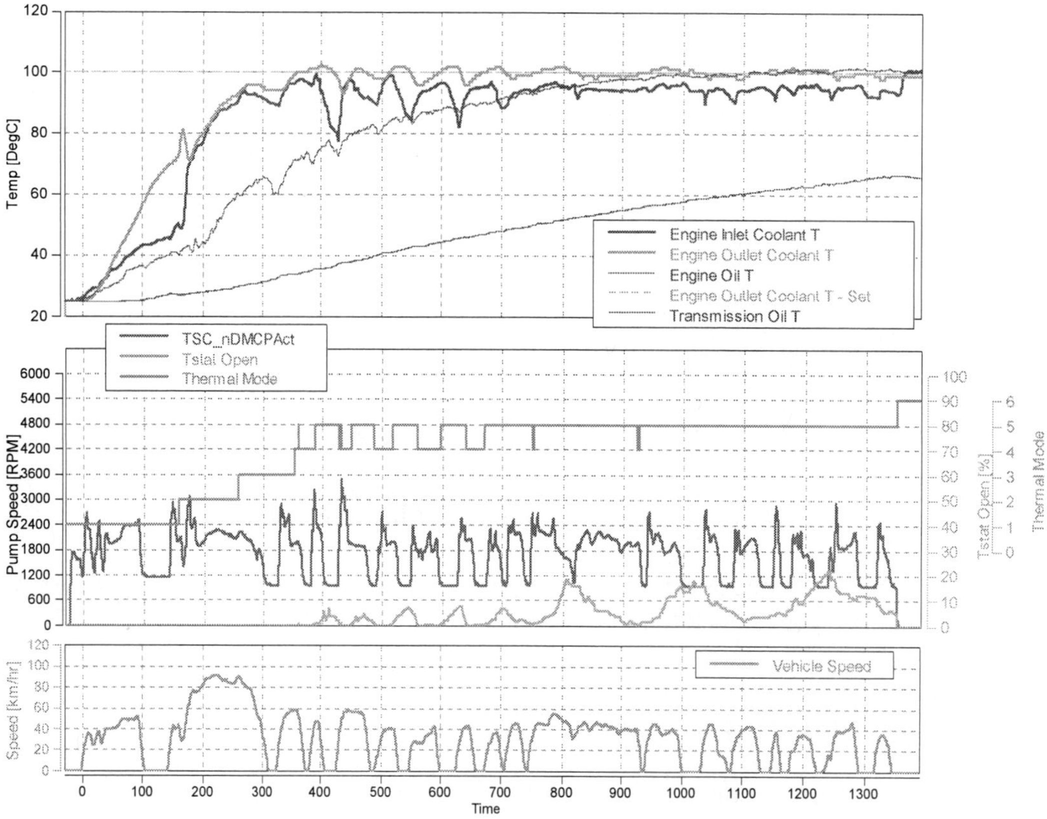

Figure 19. An FTP trace of update build

Figure 20 shows the comparison between a partially optimized update build (solid lines) and the system configured to mimic the baseline operation (dashed lines). The baseline had a wax thermostat, mechanical coolant pump and no EHRS. The TOHx was not active.

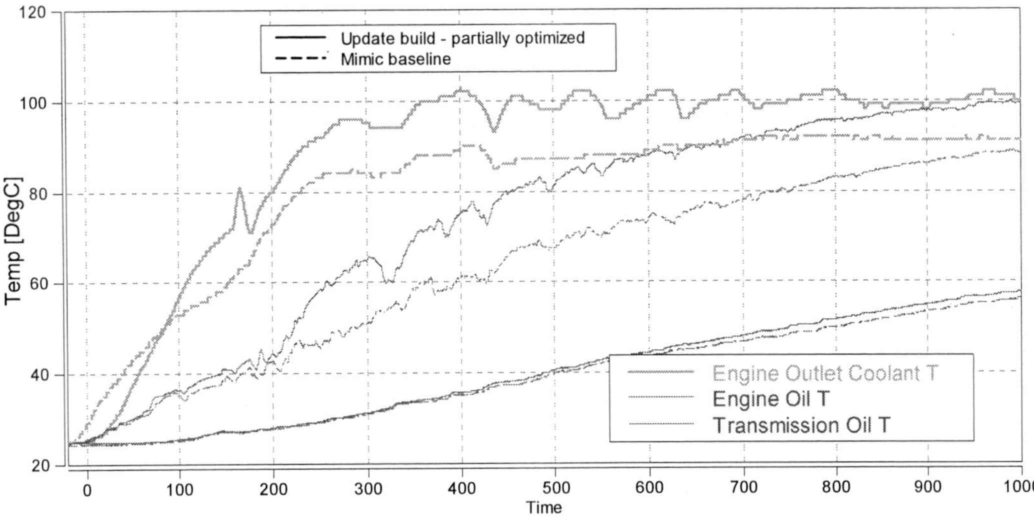

Figure 20. Comparison of engine/transmission warm up

8 SUMMARY/CONCLUSIONS

An integrated system model for a thermal concept gasoline vehicle with several advanced thermal components was developed and correlated with test data. A thermal management controller was also implemented and tested on vehicle. Vehicle test results showed faster engine warm up, achieved through the following approaches:

- Initial zero engine block flow to allow fast engine block warm up. Although a long duration in this mode has the side effect of lengthening engine sump oil warm up because engine friction heat is not available to the coolant flowing through EOHx.
- The use of ECCV which can raise the engine outlet coolant temperature by allowing delayed opening and overall better control of engine coolant temperature. This is in contrast to the stock wax thermostat with non-adjustable opening vs. temperature characteristic curve.
- Faster engine oil warm up by the joint effects of the EHRS and EOHx.

The simulation and control platform we have developed enables the evaluation of effective thermal management approaches before and in parallel to test. The thermal controller is an effective regulation tool and enables system optimization.

Future work includes additional fuel economy improvement through cooling layout refinement, controller fine tuning and thermal component design optimization. The cooling operating modes will be further optimized in several aspects, including zero flow duration and DMCP combined mechanical and electric operation.

ACKNOWLEDGMENTS

The authors would like to thank Andy Sutherland, Emily Sharpe for the work on the concept vehicle, Brad Tillock for the modeling work, and John Shutty for the guidance.

REFERENCES

1. Shutty, J., Wenzel, W., Becker, M., Bohan, S. et al., "Advanced Thermal Management of a Light Duty Diesel Vehicle," SAE Technical Paper 2013-01-0546, 2013, doi:10.4271/2013-01-0546.
2. Wenzel, W., Buchholz, T., Shutty, J., Tsai, J., "Thermal management for light-duty-vehicle with diesel engine: Evaluation of an optimized cooling system with variable cooling components," 14th Stuttgart International Symposium, Automotive and Engine Technology, Volume 1, p 61, 2014.
3. Keller, P., Wenzel, W., Becker, M., Roby, J., "Hybrid coolant pump with electric and mechanical drive," ATZ, Volume 10, p52, 2010.
4. Cortona, E., and Onder, C., "Engine Thermal Management with Electric Coolant Pump", SAE Technical Paper 2000-01-0965, 2000.
5. Allen, D., and Lasecki, M., "Thermal Management Evolution and Controlled Coolant Flow", SAE Technical Paper 2001-01-1732, 2001.
6. Keller, P., Wenzel, W., Becker, M., and Roby, J., "Coupled Thermal-Engine Simulation for a Light Duty Application", SAE Technical paper 2010-01-0806, doi:10.4271/2010-01-0806.
7. Stanton, D., "Systematic Development of Highly Efficient and Clean Engines to Meet Future Commercial Vehicle Greenhouse Gas Regulation," SAE Technical paper 2013-01-2421, 2013.

Charge air sub-cooling for improved transient response

N FRASER, I REYNOLDS and **J MILLER**
MAHLE Powertrain Ltd, UK

P WIESKE and **R DINGELSTADT**
MAHLE International GmbH, Germany

A EILEMANN
MAHLE Behr & Co. KG, Germany

ABSTRACT

Downsizing in combination with turbocharging is an effective method for improving vehicle fuel economy. This is especially true for light-load operation. However, the transient response of the engine generally decreases with reduced engine displacement. Here, the potential of charge air sub-cooling for transient response improvement is examined.

Typical A/C systems are designed for a rapid cabin cool down and are oversized for the continuous operation, once cool-down has been achieved. Thus, there is spare capacity within the A/C system, which could be employed to cool the engine charge air below the ambient temperature, where necessary. By this mechanism the A/C system is acting as a 'thermal compressor'.

This paper focusses on the transient response and fuel economy benefits achievable with such a 'thermal compressor' when taking a bespoke A/C compressor strategy into account. The required components are packaged within the engine bay of a C-segment vehicle and results from engine and vehicle testing are presented.

NOTATION

A/C	Air Conditioning
BMEP	Brake Mean Effective Pressure
BLD-1	Borderline Detonation minus 1 °CA
BSFC	Brake Specific Fuel Consumption
°CA	Crank Angle
CADC	Common Artemis Drive Cycles
COP	Coefficient of Performance
(i)CAS	(indirect) Charge Air Sub-Cooling
dCAC	direct Charge Air Cooler
ECU	Electronic Control Unit
iCAC	indirect Charge Air Cooler
HVAC	Heating Ventilating Air Conditioning
MFB50	Mass Fraction Burned 50%
MFE	MAHLE Flexible ECU
NA	Naturally Aspirated
T/C	Turbocharger
r.H.	relative Humidity
RDE	Real Driving Emissions
RUL	Run Up Line
WOT	Wide Open Throttle

1 INTRODUCTION

Downsizing in combination with turbocharging is an effective measure to improve fuel economy [1-4]. However, the transient response of such aggressively downsized engines generally decreases with the reduced engine displacement, particularly at low engine speeds. Numerous technologies are in series production (or development) to improve the driveability [5-9]. An additional, or alternative, measure to improve low-speed transient response could be the use of a 'thermal compressor'. Here, the charge air cooling system is coupled to the air-conditioning (A/C) system and the charge air is cooled to below the ambient temperature where necessary [10-12]. This paper focusses on the transient response, and fuel economy benefits, achievable with such a 'thermal compressor'. The system has been designed and packaged within the engine bay of a C-segment vehicle and results from engine and vehicle testing are presented.

2 CHARGE AIR SUB-COOLING CONCEPT

2.1 A/C System Sizing

Modern automotive A/C systems are designed to be able to quickly drop the cabin temperature to a comfortable temperature within a short amount of time (pull-down period). Consequently, the cooling power capability of the system must be sized to achieve this temperature pull-down within the desired period. For the remainder of the time, after pull-down has been achieved, the system has excess cooling capability.

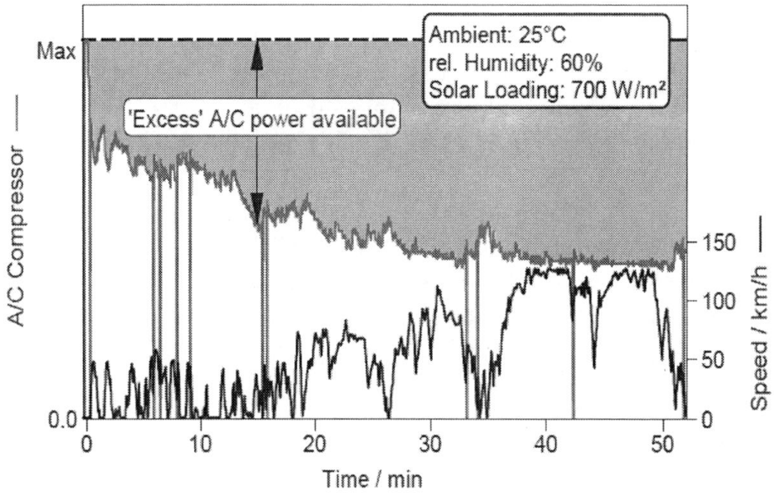

Figure 1: Climatic wind tunnel measurements of a vehicle with a variable A/C compressor in moderate ambient conditions on the common Artemis drive-cycle (CADC)

Figure 1 shows measurements, taken in a climatic wind tunnel, for a vehicle with a variable A/C compressor in moderate ambient conditions. The drive-cycle used for these measurements is the Common Artemis Drive Cycle (CADC) in order to represent a real driving scenario [13]. The A/C control is set to AUTO with a target cabin temperature of 22 °C. An A/C compressor load below maximum indicates the availability of excess cooling power of the A/C system. It can be seen that the target cabin temperature is reached shortly after the start (~1 minute) and after this time there is considerable excess A/C power available. The vehicle results shown in Figure 1 are taken from a vehicle with a peak HVAC evaporator cooling power of 6.4 kW. Depending on the ambient conditions as well as cabin temperature, generally more than 3 kW of 'excess' cooling power is available. It is also worth noting that the A/C compressor is decoupled during harsh accelerations to enhance the transient response of the vehicle.

2.2 System Layout and Vehicle Integration

Figure 2 shows a schematic layout of an indirect charge air sub-cooling system (iCAS) and Figure 3 displays the packaging of the required components around the test engine. Downstream of the conventional direct charge air cooler (dCAC), the air path is split into either a sub-cooled path or bypassed using a pair of shut-off valves. An indirect charge air heat exchanger is placed in the sub-cooled path and connected with a water/glycol circuit and auxiliary fluid pump to a chiller in the A/C circuit. Thus, the A/C circuit can be used for sub-cooling the charge air.

The system also incorporates an iCAS bypass air path and two valves that allow switching between the two inlet tracts. This allows bypassing of the iCAS when sub-cooling provides no benefit (part-load) as colder charge temperatures would result in higher fuel consumption, due to the increase in pumping work that is incurred from a concomitant increased amount of throttling required to maintain charge density. The bypass path also allows the cooler to be ‘pre-loaded’ with sub-cooled air, further improving engine transient performance and reducing the cool-down time of the charge air during an acceleration event. Additionally, because the A/C compressor is disengaged during accelerations, when maximum cooling is required by the iCAS system, a certain amount of thermal inertia is required to enable the charge air to be cooled when the A/C compressor is off. Thus, the coolant surrounding the by-pass system provides an effective way to provide the thermal inertia required to be achieved.

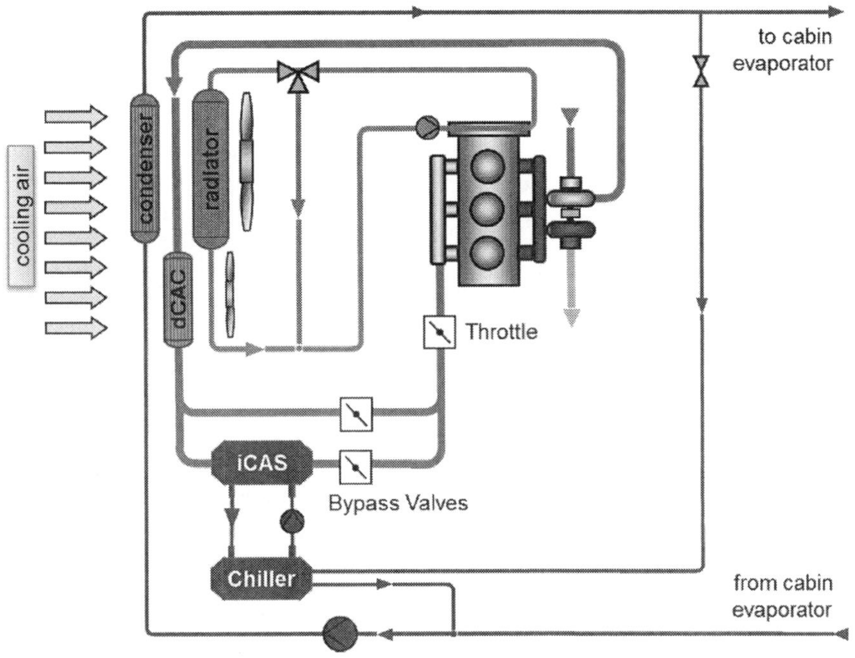

Figure 2: Schematic layout of iCAS System

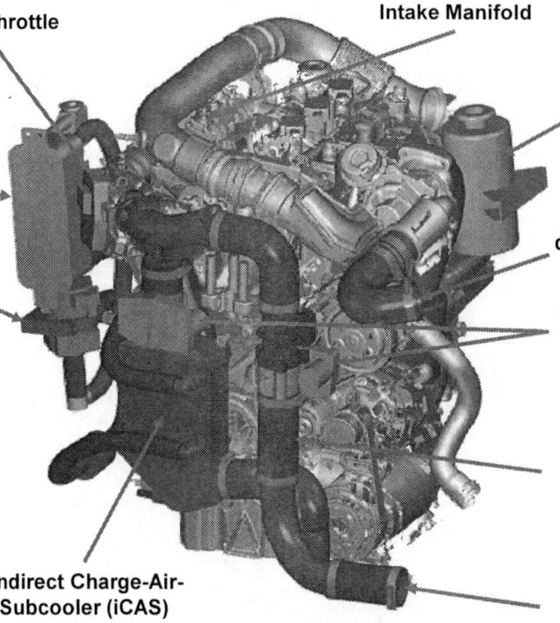

Figure 3: CAD picture of iCAS system mounted to engine for integration into the engine bay of a typical C-segment vehicle

While for the prototype phase conventional throttle flaps are used to realise the bypass functionality, a mass production solution will see this function integrated into a single actuator. The MAHLE Flexible ECU (MFE) controls the fluid pump speed, the actuation of the bypass valves and the A/C compressor load depending on the set value for the water out temperature of the chiller.

To integrate all of the required components into the vehicle is subject to package constraints and is not yet optimised in terms of a mass production solution. However, for a first demonstrator the components are packaged within the engine bay of a typical C-segment vehicle.

Figure 4 shows the increase in charge air density increase with turbocharging and charge air cooling, and illustrates the benefits achieved through charge-air sub-cooling. The aim is to achieve a high charge air density with the lowest pressure and temperature. Starting from ambient conditions of 1 bar and 25 °C, increasing the pressure to 2 bar (100% increase) only increases the density by about 50% due to the temperature rise in the compressor. Downstream of the compressor, a charge air cooler increases the density by an additional 40% via a reduction of the temperature to +15 °C above ambient conditions. Thus, both of these technologies contribute to the overall charge density increase by roughly the same amount. With conventional charge air cooler technology and heat rejection to ambient, charge air temperatures below ambient cannot be achieved. However, 'sub-cooling' the charge air for instance by 15 °C below ambient would further increase the air density by an additional 21%. Comparing this to more conventional measures, a 30 °C lower temperature roughly equates to 200 mbar of extra boost pressure.

Whereas this 200 mbar of extra boost pressure may be achieved with conventional T/C-CAC configurations by closing the wastegate in regions with excessive exhaust enthalpy available (torque plateau), this does not work in the run up line region, given that the wastegate is already closed. On the torque plateau, sub-cooling can be either used to achieve the same density with a 200 mbar lower boost pressure or improved performance. Thus, Turbocharging with charge air sub-cooling is effectively a two-stage charging approach using a 'thermal compressor' as the second stage.

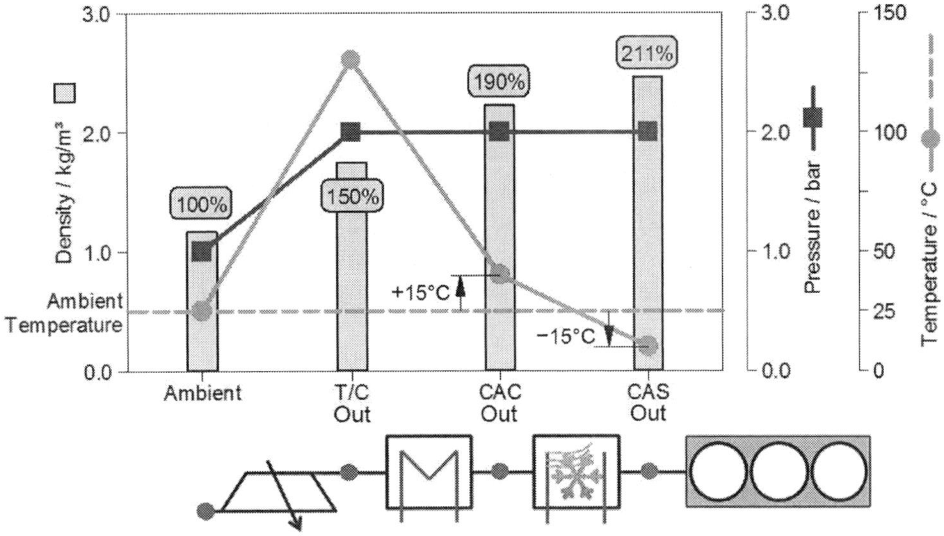

Figure 4: Overview of charge air density increase (bar chart) with Turbo-Charger (T/C), Charge Air Cooler (CAC) and Charge Air Sub-cooler (CAS), Ambient Temperature shown by dashed line

Figure 5 shows 1D simulation results of the charge air temperature in the intake port during a step change in engine load, at a constant engine speed of 1500 rev/min. It can be seen that a reduction in charge air temperature of 35 °C is achieved in less than 0.6 seconds.

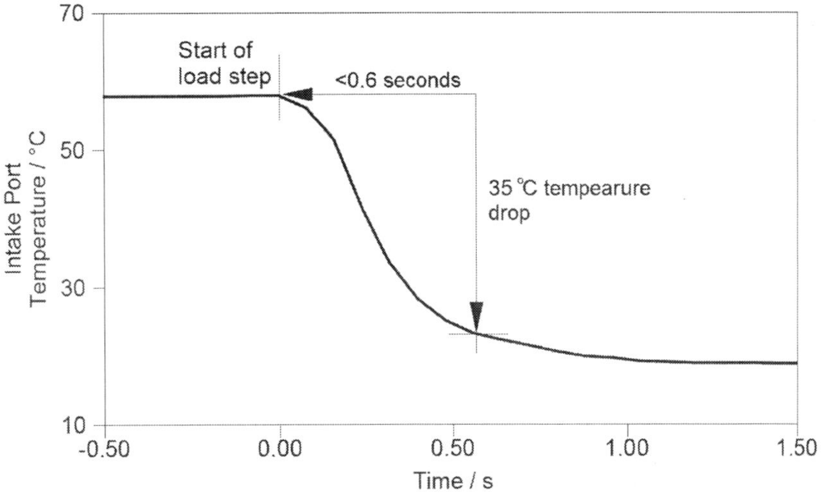

Figure 5: Temperature curve in the intake port for a 1D simulation of a constant-speed load-step at 1500 rev/min

3 ENGINE TEST RESULTS

3.1 Transient Engine Test Results

A current production, turbocharged, direct injection gasoline 3-cylinder engine is used for the first engine tests [14]. The engine was tested in steady state and transient conditions on the testbed, using the MFE as the engine controller. An air conditioning rig enables the temperature and humidity of the intake air to be tightly controlled over a wide range of conditions. The direct charge air cooler is replaced by an indirect testbed charge air cooler setup. The outlet temperature of the charge air cooler is set to provide air at a fixed temperature of 40 °C from the outlet. Likewise, the charge-air outlet temperature of the iCAS cooler is fixed at 10 °C. For measurements with sub-cooling and to adjust for the additional A/C compressor load, the A/C compressor power is calculated from the air mass flow, temperature and humidity (latent heat of condensation). Assuming a coefficient of performance (COP) of two for the A/C system, the brake specific fuel consumption (BSFC) and brake mean effective pressure (BMEP) values are adjusted accordingly. The baseline engine has a nominal torque plateau of 21.4 bar BMEP and an over-boost functionality up to 25.1 bar BMEP for a limited period of time.

Figure 6 shows the improvement in transient response with iCAS. The maximum performance benefit is achieved with the iCAS system being active during the entire load step (iCAS permanent). A reduction in the time to achieve 90% of full-load (T90) of almost 30% is achieved over the baseline case. Despite the significant transient response improvement, sub-cooling in part load regions has a negative impact on the fuel economy, but this can be avoided using the bypass system. Though the bypass system is not optimised in terms of volume reduction, an improvement of the T90 time of almost 20% is still possible. As the A/C compressor is typically decoupled during hard accelerations, the A/C compressor load for sub-cooling is not included in these load step results and it has been assumed that the thermal inertia of the system is large enough to enable sub-cooling to be achieved for the entire duration of the load step.

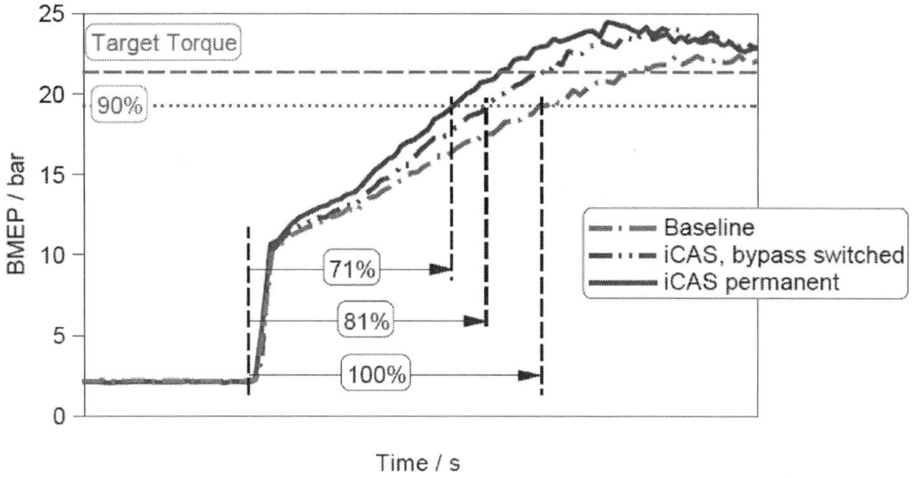

Figure 6: Transient response improvement with iCAS at 1300 rev/min

3.2 Steady State Engine Test Results

Figure 7 shows steady state engine test results at rated performance. Between 1100 and 1300 rev/min, a 16 % to 19 % low end torque improvement is possible with the lowered charge air temperature. Compared to the baseline at 40 °C, the boost pressure demand running at 10 °C charge air temperature is reduced by up to 200 mbar and the combustion phasing can be advanced by 3 °CA – 5 °CA. Whereas the 40 °C baseline calibration requires minor deviations from stoichiometric combustion, with sub-cooling stoichiometric operation over the entire power curve can be achieved, this gives rise to a corresponding reduction in brake specific fuel consumption (BSFC), which is also shown in Figure 7.

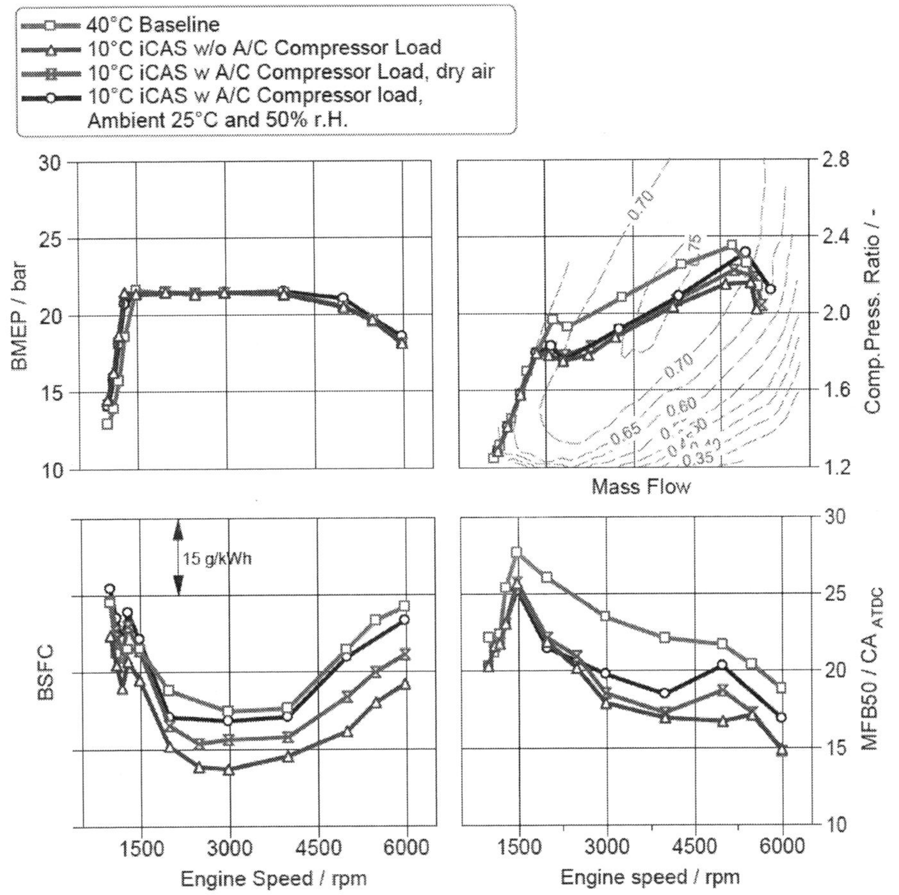

Figure 7: Engine test results at rated load with lowered charge air temperature from 40 °C (baseline) to 10 °C

Figure 8 shows BMEP (top) and BSFC (bottom) results for an intake and exhaust cam timing sweep (maximum valve overlap upper right, minimum valve overlap lower left corner in each plot) at 1200 rev/min, WOT, with 40 °C (left) and 10 °C (right) charge air temperature. The baseline BMEP of 17.3 bar is shown by a dashed line, the borderline of stoichiometric operation, i.e. relative air fuel ratio $\lambda = 1$, is represented by a dotted line. Running equal, or less, valve overlap leads to both in-cylinder and exhaust stoichiometric engine operation. Increasing valve overlap beyond the borderline, however, results in approximately stoichiometric in-cylinder (CO < 1.5%) conditions and lean exhaust gas due to scavenging.

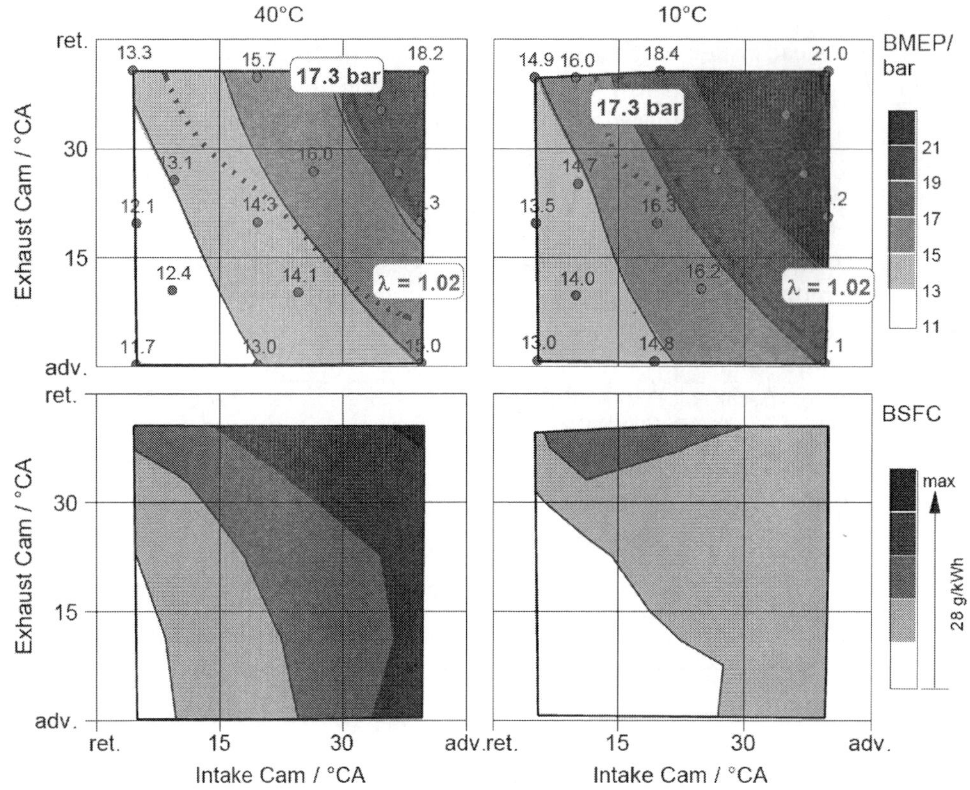

Figure 8: Cam timing optimization for 40 °C (left) and 10 °C (right) charge air temperature, 1200 rev/min, WOT; rated BMEP 17.3 bar (dashed line)

To achieve the rated BMEP of 17.3 bar at 1200 rev/min with 40 °C charge air temperature, significant valve overlap and scavenging is required, which leads to lean exhaust gas. New emission legislation like Real Driving Emissions (RDE) may require emission targets are achieved across the whole engine map [15]. For the baseline either in-cylinder enrichment or a reduction of the low end torque would be necessary to ensure proper three-way catalyst operation at all times. Though in-cylinder enrichment might be a solution to achieve stoichiometric conditions in the exhaust with scavenging, this normally causes both higher raw emissions (CO and HC), and fuel consumption penalties compared to in-cylinder stoichiometric operation. With a reduction of the charge air temperature to 10 °C, exhaust stoichiometric operation and rated BMEP can be achieved without in-cylinder enrichment. In addition, the fuel consumption values are generally better than the 40 °C baseline as well.

3.3 Condensation

With sub-cooling temperatures around 10 °C and pressures of approximately 2 bar absolute, the relative air humidity downstream the iCAS cooler, depending on the ambient conditions, is often 100 % and therefore condensation of moisture from the charge air will occur. An investigation into whether this condensation would become an issue in terms of droplet puddling, and/or combustion instability, was undertaken. A number of tests were performed at with the ambient conditions fixed at 20 °C and 90% relative humidity. In the vehicle the position of the iCAS cooler is constrained by the under bonnet package space available, however, on the testbed the position of the iCAS cooler can be easily varied between locations above and below the intake manifold.

To identify how much of the condensate goes directly through the engine with a mounting position of the iCAS cooler below the intake manifold, steady-state tests were performed at 18 bar BMEP and 3000 rev/min. The condensate was collected from a drain below the iCAS cooler and compared to the theoretical amount of condensate generated. This testing revealed that even for the low mounting position, more than 80 % of the condensate goes directly through the engine without causing any combustion instability.

4 OPERATING STRATEGY

As the steady state engine test results show, the BSFC benefit is dependent on the ambient conditions, i.e. air humidity, when taking the additional A/C compressor load into account. A charging strategy is necessary to achieve both a transient performance and a fuel consumption benefit simultaneously. Figure 9 shows simulation results of a C-segment vehicle over the US06 drive cycle at 25 °C and 50 % relative humidity with a thermal mass of the sub-cooling system of 3 kg water/glycol (chiller fluid). Note that the chiller fluid is pre-conditioned to a chilled state at the start of the test, i.e. the results shown in Figure 9 do not show the warm-up (or cool-down) period.

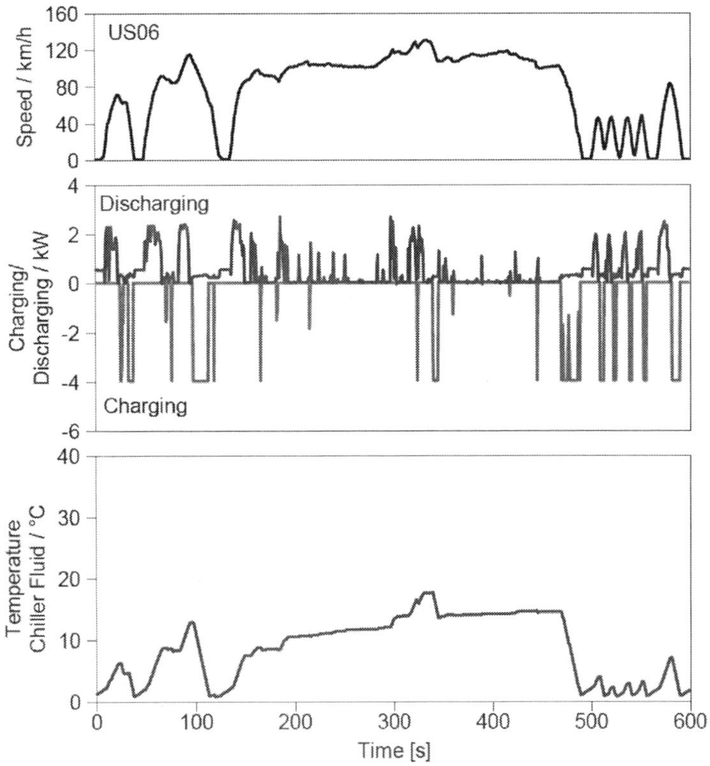

Figure 9: Simulation of the chiller fluid temperature for a C-segment vehicle using charge air sub-cooling; charging strategy focuses on deceleration phases

Discharging the thermal inertia (positive values) relates to sub-cooling the charge air, whereas (re)charging the thermal inertia is represented by negative values, see middle graph in Figure 9. The maximum charging rate is limited by the chiller and/or A/C compressor power to 4 kW. The resulting temperature of the chiller fluid, with a charging strategy that prioritises charging during braking, is shown in the bottom graph. While the maximum temperature of the chiller fluid is not limited, the lower temperature needs to stay above 1 °C in order to avoid freezing of the condensate within the inlet system.

The US06 drive cycle is representative of high load driving with harsh accelerations and thus the A/C compressor is decoupled frequently. However, the temperature of the chiller fluid stays below 18 °C at all times. The cumulative sub-cooling power demand including latent heat of condensation can be completely generated during engine over-run if the components are sized appropriately.

5 VEHICLE TEST RESULTS

Figure 10 shows the increase in driving dynamics that can be achieved in the vehicle by charge air sub-cooling of about 25 °C (measured just upstream of the intake port) for an example of acceleration from 30 km/h in fourth gear. The measurements were taken on a chassis dynamometer in a climatic wind tunnel at 30 °C ambient temperature, with 800 W/m² solar radiation and 50 % relative humidity. In addition to vehicle speed (top), the relative air mass (middle) and temperature profile upstream of the intake port (bottom) are shown. For the iCAS variant, the bypass valve is switched at the start of the load step, i.e., the charge air is then run through the sub-cooled path, and for both variants the air conditioning compressor is decoupled for ten seconds. During this time, charge air sub-cooling is provided by the thermal inertia of the low-temperature circuit. In both variants, the electronic control unit is calibrated such that the same target torque is maintained once the maximum torque has been achieved. The increased performance resulting from charge air sub-cooling is thus limited to the operating range at low engine speeds, whilst the wastegate is closed. At engine speeds above the knee-point the potential of the colder charge air temperature is converted to lower boost pressure demand and earlier ignition timing, which gives rise to earlier combustion phasing and improved fuel consumption in comparison with the baseline.

From Figure 10 it can be seen that the increased air charge density, and increased engine torque at engine speeds below the knee point, gives rise to an improved vehicle acceleration. The time taken to reach 50 km/h, from 30 km/h in 4th gear, is reduced by 0.7 seconds compared to the baseline. It can also be seen that the T90 time (time until 90 % of the final steady-state value is reached) is reduced by 1.1 seconds compared to the baseline. At this point in time, the relative air mass of the baseline is 168 %, while that of the iCAS is 207 %. The transient air charge gain of 39 % (absolute) is well above the theoretical (steady-state) expected value of 21 % for sub-cooling by 25 °C (see Figure 4).

Once maximum torque has been achieved, the acceleration is identical for both variants after this point in time. However, a comparison of the charge air temperature upstream of the intake port confirms the reduced charge air temperature with charge air sub-cooling (iCAS). It is believed that the temperature measurements shown in Figure 10 are severely delayed due to the thermal inertia of the thermocouple. It is believed that the sub-cooled air is available to the engine more quickly (as per the simulated result, shown in Figure 5).

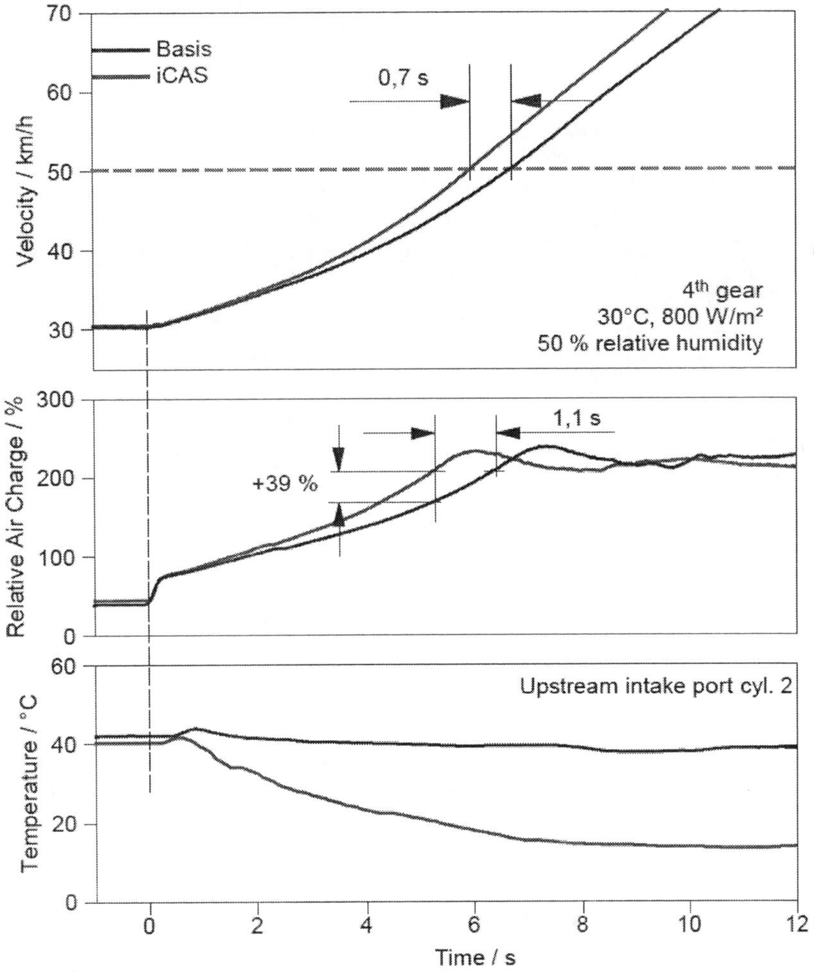

Figure 10: Acceleration benefit from charge air sub-cooling; measured on the chassis dynamometer in a climatic wind tunnel; 4th gear, start at 1080 rev/min, 30 km/h

6 SUMMARY AND OUTLOOK

The climatic wind tunnel tests confirm that typical A/C systems are designed for a rapid cabin cool down and are oversized for regular operation afterwards. Depending on the ambient conditions and cabin temperature in many situations excess cooling power is available which can be used to sub-cool engine charge air and act as a 'thermal compressor'. The promising engine and vehicle test results presented in this paper reveal that charge air sub-cooling allows for both transient response and fuel economy benefits using the right sized components and an appropriate charging strategy. Although the components (indirect charge air cooler, chiller, bypass) tested in this first demonstrator are 'off the shelf' and not yet optimised in terms of package size or functional integration, they do fit within the engine bay of a typical C-segment vehicle. The next step will be the development a performance optimised system that is more representative of a mass production solution along with the integration of the parallel operation of the cabin evaporator and iCAS chiller, as depicted in Figure 11.

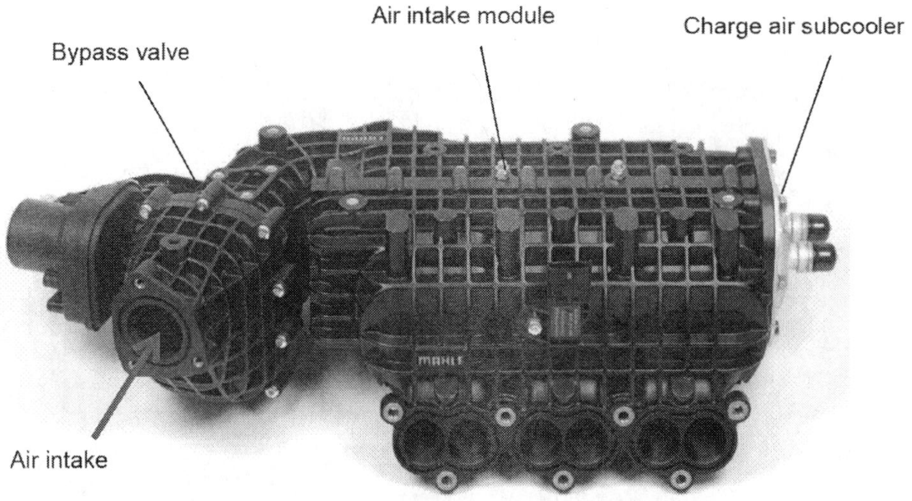

Figure 11: Air intake module with indirect, integrated charge air subcooling and bypass valving [16]

REFERENCES

1. Hancock, D., Fraser, N., Jeremy, M., Sykes, R. et al., "A New 3 Cylinder 1.2l Advanced Downsizing Technology Demonstrator Engine," SAE Technical Paper 2008-01-0611, 2008, doi:10.4271/2008-01-0611.
2. Lumsden, G., OudeNijeweme, D., Fraser, N., and Blaxill, H., "Development of a Turbocharged Direct Injection Downsizing Demonstrator Engine," SAE Int. J. Engines 2(1):1420-1432, 2009, doi:10.4271/2009-01-1503
3. Fraser, N., Blaxill, H., Lumsden, G., and Bassett, M., "Challenges for Increased Efficiency through Gasoline Engine Downsizing," SAE Int. J. Engines 2(1):991-1008, 2009, doi:10.4271/2009-01-1053.
4. Turner, J., Popplewell, A., Patel, R., Johnson, T. et al., "Ultra Boost for Economy: Extending the Limits of Extreme Engine Downsizing," SAE Int. J. Engines 7(1):387-417, 2014, doi:10.4271/2014-01-1185.
5. Turner, J., Popplewell, A., Marshall, D., Johnson T. et al., "SuperGen on Ultraboost: Variable-Speed Centrifugal Supercharging as an Enabling Technology for Extreme Engine Downsizing," SAE Technical Paper 2015-01-1282, 2015.
6. Bassett, M., Hall, J., Freeland, P., Servant, J., Gray, K., "E-supercharging for Heavily Downsized Gasoline Engines", SIA Powertrain Conference, Versailles, France, May 27-28, 2015.
7. Wieske, P.; Lüddecke, P.; Ewert, S.; Elsäßer, A.; Hoffmann, H.; Taylor, J.; Fraser, N. Optimierung von Dynamik und Verbrauch beim Ottomotor MTZ 11/2009.
8. Burkert, A. BMTS entwickelt Abgasturbolader mit geringem Trägheitsmoment Nachrichten aus Automobil- und Motorentechnik Springer for Professionals, 03.09.2012.
9. Martin, S.; Beidl, C.; Frank, A.; Rinderle, M.; Downsizing and electrification - potential analysis using the example of a modern charged 3-cylinder engine 13. Stuttgarter Symposium, 2013.

10. Guhr, C., Verbesserung von Effizienz und Dynamik eines hubraumkleinen turboaufgeladenen 3-Zylinder-DI-Ottomotor durch Abgasrückführung und ein neues Ladeluftkühlkonzept Dissertation TU Dresden Dresden, 2011.
11. Guhr, C.; Zellbeck, H. Turbocharging with Low Temperature Charge Air Cooling and EGR MTZ 10/2012 Volume 73.
12. Kadunic, S., Baar, R., Scherer, F., Zegenhagen, T., Ziegler, F. Heat2Cool-Engine Operation at Charge Air Cooling below Ambient Temperature 22. Aachen Colloquium, 2013.
13. André, M. "The ARTEMIS European driving cycles for measuring car pollutant emissions", Science of the Total Environment 334-335 pp.73-78, Published by Elsevier B.V., 2004.
14. Ernst, R.; Friedfeldt, R.; Lamb, S.; Lloyd-Thomas, D.; Philips, P.; Russel, R.; Zenner, T. The New 3 Cylinder 1.0L Gasoline Direct Injection Turbo Engine from Ford 20. Aachen Colloquium, 2011.
15. Vlachos, T., Bonnel, P., Perujo, A., Weiss, M. et al., "In-Use Emissions Testing with Portable Emissions Measurement Systems (PEMS) in the Current and Future European Vehicle Emissions Legislation: Overview, Underlying Principles and Expected Benefits," SAE Int. J. Commer. Veh. 7(1):199-215, 2014, doi:10.4271/2014-01-1549.
16. Eilemann, A., Wieske, P., Warth, M., Stehlig, J., "Charge Air Subcooling – Enabler for improved driveability and fuel economy", 36th International Vienna Motor Symposium, 7-8 May 2015.

Heavy duty vehicle cooling system auxiliary load management control: a comparison of advanced control strategies

S SERMENO, E BIDEAUX and **T MORGAN**
INSA de Lyon/Renault Trucks SAS, France

ABSTRACT

Vehicle Thermal Management covers the engineering field of solutions that maintain the complete vehicle in acceptable operating conditions regarding components and fluid temperatures in an engine. In the heavy duty truck industry, the engine's cooling system has been designed to overcome the worst case scenario, which is rarely the case in the life cycle of a truck. Moreover the current control laws are generally based on experience (trial and error). Recent studies have shown that by improving the control strategy of the cooling system the fuel consumption can be improved, however no real data is available since its implementation has been limited. Other goals can be exploited through the use of an advanced control and the improvement of control specifications such as a reduction of thermal shock, reduction of thermal fatigue, minimization of system overcooling (directly impacts fuel consumption but also the noise levels).

This paper presents and compares different control strategies that have been implemented in a simulation environment which aim at a more efficient regulation of the cooling system operation of a Heavy Duty Truck. In this study, the new control laws are able to fulfill the system's constraints; but also coordinate in a logical and efficient way the command routines. The control strategies presented are receding horizon control (RHC), Flatness control and Model Free Control (MFC). Said controllers have been tested on a Simulation Platform using a Matlab/Simulink and compared to the existing system control using a reference driving cycle. Additionally an optimal control has been determined using a dynamic programming algorithm.

NOTATION

α	Constant	[-]
q	Pump mass flow	[kg/s]
q_1	Radiator mass flow	[kg/s]
q_2	Air compressor mass flow	[kg/s]
C	Pressure Constant specific heat	[J/(kg.K)]
F	Estimated unknown process	[-]
K	Constant	[1/K]

Q	Heat Flux	[J]
T	Temperature	[K]
$\dot{y}$	Measure derivate	[K/s]
u	command	[-]

Suffixes

pf	Pumpfeed
oc	Oil cooler
eng	Engine
rad	Radiator
ac	Air compressor
oil	Oil circuit

1 INTRODUCTION

1.1 State of the Art

Nowadays, thermal management research is focused on the following topics: development of new components (electric valves, electric pumps…), system simulations and advanced control. A detailed list of Thermal Management concepts for heavy vehicles was given and reviewed by Wambsganss [15]. He also discussed some thermal management trends (computer control, heat storage…) but no tests were pursued. The work of Allen and Lasecki[1] completed this list. Electric valves, electric pumps, control of EGR cooler flow are some of the discussed concepts. The expected benefits of these optimizations are an increase of fuel economy (by decreasing the fan "On-Time" for example), decreasing engine warm-up time, increase oil life (optimized oil temperature) and decrease engine emissions (Optimized Combustion temperatures, fewer parasitic losses).

Nonlinear algorithms were first presented by Setlur, Wagner, Dawson and Chen [12]. They used a simplified cooling system model for real time computer algorithms to regulate the smart valve, electric water pump and radiator fan using nonlinear controls (Lyapunov). Nonetheless, it's not the only nonlinear control found. Bhat, Joshi, Shiozaki, Ogasawara, Yamada and Somu [2] used a different strategy: the adaptive control of an electromagnetic fan clutch (ECFD), and then added a PID controller to be used only during warm up. In order to improve the energy management, Salah, Mitchell, Wagner and Dawson [14] compared different control strategies for a simple circuit with electric fans, pump an electric thermostat valve. The results showed that an advanced control (in their case backstepping) led to energy savings of up to 6 W (compared to the normal operation).

Most of the research has been done for passenger vehicles and hybrids (patents and papers). However the heavy duty vehicle manufacturers are beginning to follow the example set by the car industry.

The cooling system is becoming more complex and, in most cases, it results in a modification of the demands made to the cooling system (oil cooler bypass, EGR cooler, air compressor, engine brakes, engine performance …). When several changes are made, a global view is needed and the complexity of said systems might hide interactions between controlled components that become evident in the final stages of vehicle validation (troubleshooting).

The review of the current literature indicates that this hasn't been studied yet, since most papers relate to a specific modification to the system. But future vehicles will be possibly using many different technologies.

In this paper only the modeling of the cooling system and the proposed control strategies that have been developed will be presented. The controllers will not be detailed.

The control strategies presented are receding horizon control (RHC), Flatness control and Model Free Control (MFC). Said controllers have been tested on a Simulation Platform using a Matlab/Simulink and compared to the existing system control using a reference driving cycle. The new approaches presented have been compared to a dynamic programming algorithm that was developed earlier by the authors [10].

The benefits of the advanced control strategies are presented, as well as the challenges to implement them in a vehicle. The control strategies are known in the literature, but other than the MPC, their implementation into a cooling system is novel and no comparison between the different methods is available for this particular application.

2 BOND GRAPH MODELING

Firstly, the model used will be described. The bond graph tool was chosen to be used in order to determine interactions between actuators and study the structure of the cooling system.

It needs to be noted that for this research pseudo bond graphs were used in order to couple the two existing phenomena (thermal and hydraulic). We have hydraulic energy (coolant, oil flows), electrical (control of the fan and pump clutch), thermal (heat exchange) and mechanical (pumps and valves). Some dynamics have been neglected when needed. The system's dynamics are nonlinear mainly due to the coupling of different domains.

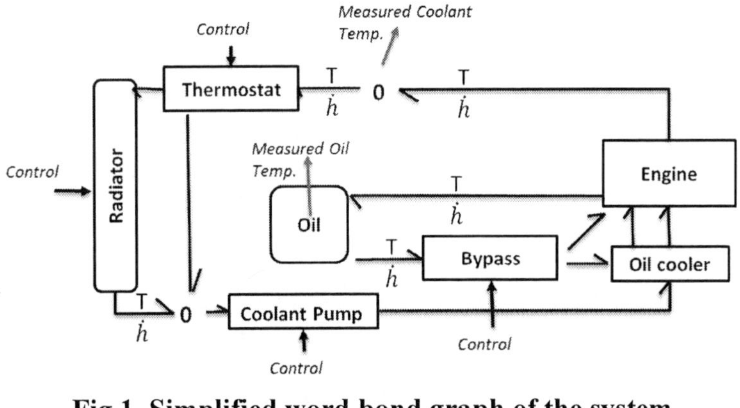

Fig 1. Simplified word bond graph of the system

The equations below describe the nonlinear model which we obtain for the Bond Graph representation in Figure 2. Where T_i is the temperature in a given node, C_i is the heat Capacity of a component, q is the coolant flow at the oil cooler, q_2 is the coolant flow through the air compressor, $(q-q_2)$ is the coolant flow through the engine, q_1 is the coolant flow through the radiator and Q_i are the respective heat inputs in each component when needed. Finally u_1 and u_4 are respectively the radiator heat exchange and the oil cooler heat exchange. It is related to our control specifications since u_1 depends on fan speed demand, and u_4 depends on the opening of the oil thermostat.

It has been assumed that heat capacity for each component is a function of the volume of coolant in the component and specific heat given by the mass of the material and is constant, as is the density of the fluid and the materials conductivity. Additionally, the dynamics of the actuator aren't included in the model presented; the goal of this paper is not the precise modeling of the system but the implementation of new control strategies.

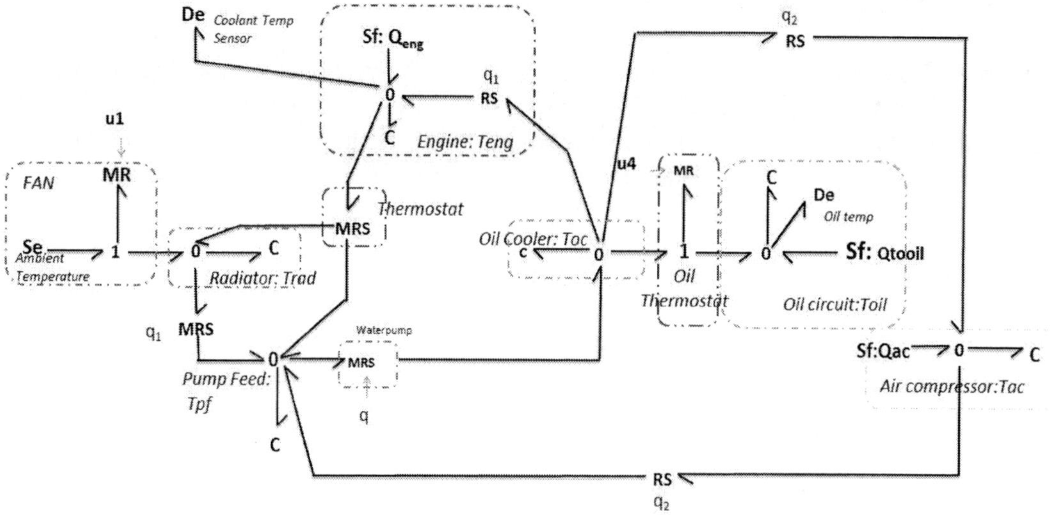

Fig 2. Simplified psuedo bond graph of the system

The resulting equations from such a system are

$$dT_{pf}/dt = \left[-cqT_{pf} + c(q - q_1 - q_2)T_{eng} + cq_1T_{rad} + cq_2T_{AC}\right] \times \frac{1}{C_{pf}} \tag{1}$$

$$dT_{oc}/dt = \left[cqT_{pf} - u_4\left(T_{oc} - T_{oil}\right) - cq_2T_{oc}\right] \times \frac{1}{C_{oc}} \tag{2}$$

$$dT_{eng}/dt = \left[c(q - q_2)T_{oc} - c(q - q_2)T_{eng} + \dot{Q}_{eng}\right] \times \frac{1}{C_{eng}} \tag{3}$$

$$dT_{rad}/dt = \left[cq_1T_{eng} - cq_1T_{rad} + u_1\left(T_{amb} - T_{rad}\right)\right] \times \frac{1}{C_{rad}} \tag{4}$$

$$dT_{AC}/dt = \left[cq_2 T_{oc} - cq_2 T_{AC} + \dot{Q}_{AC}\right] \times \frac{1}{C_{AC}} \quad (5)$$

$$dT_{oil}/dt = \left[u_4 (T_{oc} - T_{oil}) - \dot{Q}_{tooil}\right] \times \frac{1}{C_{oil}} \quad (6)$$

3 CONTROL LAWS DEVELOPMENT

In the study presented, the optimal control laws to be determined must be able to fulfill the system's constraints; but also coordinate (organize, class ...) in a logical or efficient way the command routines. Moreover, they can be exploited in order to reduce the energy used while still maintaining appropriate control. It needs to be noted that automatic control doesn't imply better fuel consumption just in the case it is specifically conceive for this purpose.

The following actuators have been assessed in this work; however their dynamics have not been modeled by the author:

- Water Pump: It consists on a two speed pump (low and high flow). A solenoid is used to engage or disengage the clutch (on/off).
- Fan: Running the fan is a large load for the engine (80 kW for high RPMs). The clutch allows a proportional regulation of the fan Speed.
- Coolant Thermostat.
- Oil Thermostat: An electric controlled thermostat is installed to improve fuel consumption.

As follows, the control strategies will be described.

3.1 Dynamic Programming (DP) Optimization

DP solves problems where decisions are made in succeeding stages. For each stage, the possible commands are evaluated using a cost function which implies the current and previous (expected) cost. The expected cost of a future or past stage is known.

A recursive function is used and it simplifies our optimization problem since it divides a global problem in different steps to be optimized.

The purpose of this step is to determine the actual gain of an advanced control strategy for the system. DP will determine the ultimate solution for a given cycle for instance. Currently we are not able to embed such a solution in a real time controller (calculation time, data needed, memory needed…).

A running model has been built and the results are showed. The used algorithm uses a simpler model than the one detailed in this paper. Due to the algorithm limitations an order reduction was made.

Further information can be found in [10].

3.2 Receding Horizon Optimization

A second approach will be to test the RHC method. It is a method that, at each sampling instant, calculates the current control input by solving an open-loop optimal control problem. This method has being introduced in the industry recently.

This method is also called Model Predictive Control since it uses a model of the system to predictive future states.

In particularly it has been developed by Ma et al [8] for the operation of building cooling systems, and a heavy duty cooling system application was studied by Güthler et al[6].

The following requirements need to be clearly stated:

- The prediction method
- The time horizon
- The constraints

One of the major drawbacks of the development of this technique to real world applications relies mainly on the needed computational time. So this approach has been mainly used for slow system with large sampling times (seconds, minutes, hours). New improvements in chips, ECUs… will expand its reach.

The optimization principle presented in this article it's a DP optimization for a shorter time horizon.

Concerning the prediction it has been assumed that we accurately know the future (we use the known values of the driving cycle). A prediction of the future can be made, but it is not presented for this paper. A less accurate estimation of the future will provide different results to the ones presented in this article, however it will be more accurate to what could be implemented currently in a vehicle.

The prediction horizon has been chosen so that the dynamics of the actuators are respected (activation time of the fan is of about 30s).

3.3 Flatness control

An approach based on the Flatness principle proposed by Fliess et al. [4] was used to develop nonlinear and linear controllers have been developed for multiple inputs multiple output system (MIMO) and single input single output system (SISO) systems. For the MIMO controllers the fan speed and oil thermostat opening is controlled while the coolant temperature at engine outlet and oil temperature are monitored. For the SISO controller, just the fan speed is controlled while monitoring the coolant temperature at engine outlet.

The linear controller has been introduced in [11]. The flat outputs have been determined using the canonical form presented by Fliess et al. [5]. No method exists to determine the flat outputs for a nonlinear system. If they can be determined, they are not unique.

As mentioned, it could be possible to control up to 4 actuators (coolant thermostat, oil thermostat, water pump and fan) while only 2 sensors (outputs) are used (Engine outlet temperature and oil feed temperature). As a result of such architecture, a full system decoupling is not possible, there's an input/output mismatch. In order to apply the flatness control directly to the system it would be necessary to define 4 outputs. Nevertheless, it is possible to consider different operating conditions which are common in every driving cycle, in order to select the adequate input set in each case. These driving conditions are presented in the following table and are explained in the following paragraphs.

The thermostat valve will not be controlled. It will remain the current wax thermostat that is temperature dependent. The coolant pump is a two speed pump (High and low speed). The fan speed is variable accordingly to the engine speed. And finally a variable bypass valve is used to control the flow through the oil cooler oil side.

In the first mode (wu), the primary objective consists in allowing the cooling system to achieve the set operating temperature (oil and coolant) In this case the fan action is unwanted (fan can only cool the coolant). The oil bypass is kept closed so the oil reaches its operating temperature as soon as possible (colder oil increases the engine friction losses).

For the second mode (nc), the system has reached its operating temperature (coolant and oil). In this mode's objective is to regulate the coolant temperature at a given range. According to table 1, two controllable components are available. However in this paper, a simple control law has been used for the coolant pump (similar to the current control).

Finally the last mode (ho) corresponds to a too high oil temperature. The oil quality and its properties could be deteriorated if the temperature keeps rising. In this mode it is assumed that the pump is fully engaged and two outputs can be controlled (fan and oil bypass).

Table 1. Operating conditions studied

Operating Conditions/ Component	Warm Up (wu)	Driving around an operating Temperature (nc)	Driving high oil temp (ho)
Thermostat	Opening according to coolant temperature	Completely Open	Completely Open
Coolant Pump (low/high speed)	Low/high (not controlled using flatness)	Low/high (not controlled using flatness)	High
Fan (% speed)	OFF	Variable	Variable
Oil Bypass (% valve)	OFF	OFF	ON

Then the SISO controller correspond to the nc mode and the MIMO linear controller correspond to the ho mode. The nonlinear system is also determined for the ho operating condition.

The flat outputs determined either by canonical form (only applies for linear SISO or MIMO systems) and the nonlinear methodology (based on experience mainly) are a combinations of different temperature nodes of the system and characteristics of the system (specific heat of components, flow distribution,..). This introduces an implementation challenge due to the lack of information in the current architecture.

3.4 Model Free Control

The previous assumptions are valid for the model free controller.

Model free control is a novel control strategy which does not need any precise mathematical model, thus offering some advantages for the control of complex systems, such as:

- Simplicity,
- Fast implementation
- Low computation power needed
- Robustness

It has been called a universal control, a plug and play control of complex systems.

Model free control, although novel, applications have been proven successful in several practical implementations in diverse fields (Transportation system, energy management Spal [13] Choi [3]).

Using the super twisting algorithm and an adaptive super twisting algorithm proposed by Levant [7] and algorithm was built in order to implement this solution.

The command is calculated as follows for a single output:

$$u = -\frac{F}{\alpha} + \frac{\dot{y}}{\alpha} + K_p e + K_i \int e + K_D \dot{e} \qquad (7)$$

Where F is the estimation of the unknown system, y is the system derivative. A PID has been added for robustness.

3.5 Reduced Order model

Its development is similar to the model free control, however some dynamics are introduced into equation (7) as follows:

$$u = -\frac{F}{\alpha} + \frac{\dot{y}}{\alpha} + \frac{c_p q T_{eng}}{\alpha c_{rad}} + K_p e + K_i \int e + K_D \dot{e} \qquad (8)$$

A simple dynamic is introduced. The knowledge of the system is expected to improve the solution provided by the controller.

Results will be compared in a later paragraph.

3.6 Implementation

In figure 3, the implementation of a linear SISO controller is presented. A similar representation is used for nonlinear (NL) and MIMO controllers.

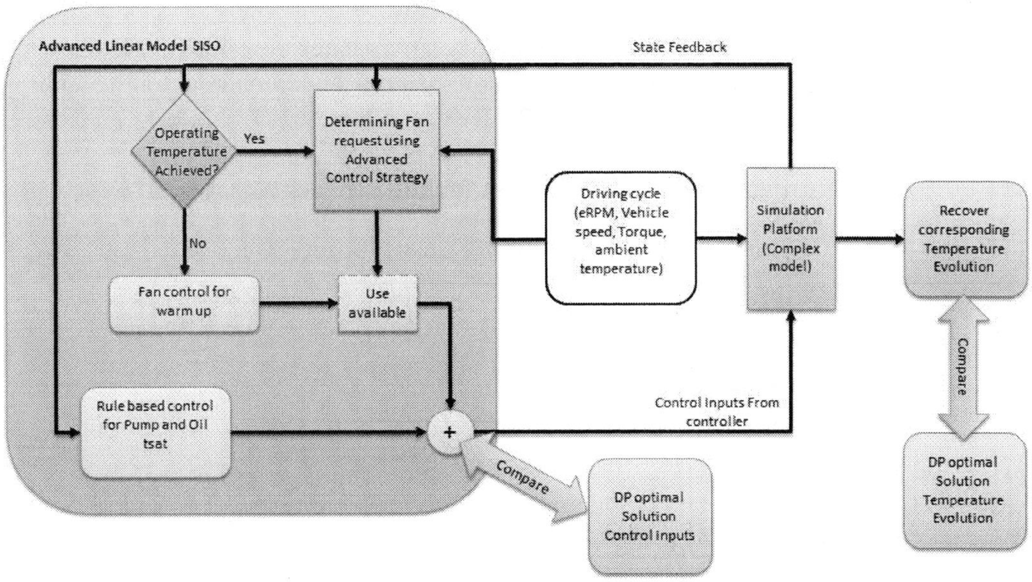

Fig 3. Linear SISO implementation

A closed loop is used in order to give robustness to the system. The correction is needed to deal with model uncertainties and better deal with unknown inputs. It has been set as a proportional or a PI gain.

The controllers have been implemented in a MATLAB/SIMULINK simulation environment, in which the cooling system model used is the one presented in this paper.

The systems characteristics (volumes, masses, heat input) were determined for a euro 6 compliant heavy duty vehicle with an 11 liters engine.

The same cycle has been used for all the simulations presented. The information used is the engine speed (RPM), load, vehicle speed and ambient temperature.

4 RESULTS

As previously stated the proposed controllers will be compared to the current control and an optimal control using the dynamic programming algorithm which will be assumed to be optimal. All controllers comply with the requisites of the current cooling system (temperature regulation) and prove to be somewhat beneficial (reduction in overall fan-on time, reduction of noise, fuel economy, and limitation of the engine derating time).

For simplicity only the energy consumptions gains will be showed based on early simulations.

However the plant model used is not thoroughly validated. The model was developed in order to test and build the control strategies. The main drawback of the simulation platform used is that the dynamics of the actuators are not accounted for in the model (e.g. fan drive, valve opening).

In table 1, a comparison of the control run and the approaches described previously are presented. They show the energy consumption of the fan compared to the energy consumed with the rule based controller currently in production.

Table 1. Fan energy consumption for the different strategies

Method	Energy Consumption (%)
Control run	100
DP	45
RHC	52
Flat SISO	72
Flat MIMO	70.4
Flat NL	66
MFC SISO	85
MFC MIMO	79

Some results are shown in this paper. For clarity, only the results for the DP, RHC SISO and NL flatness controller are presneted in figures 4 to 7. They are compared to the test results. In figure 4, the fan speed request for the same cycle is presented. Zoom ins have been included to identify behaviours of the fan request (figure 5 and 6).

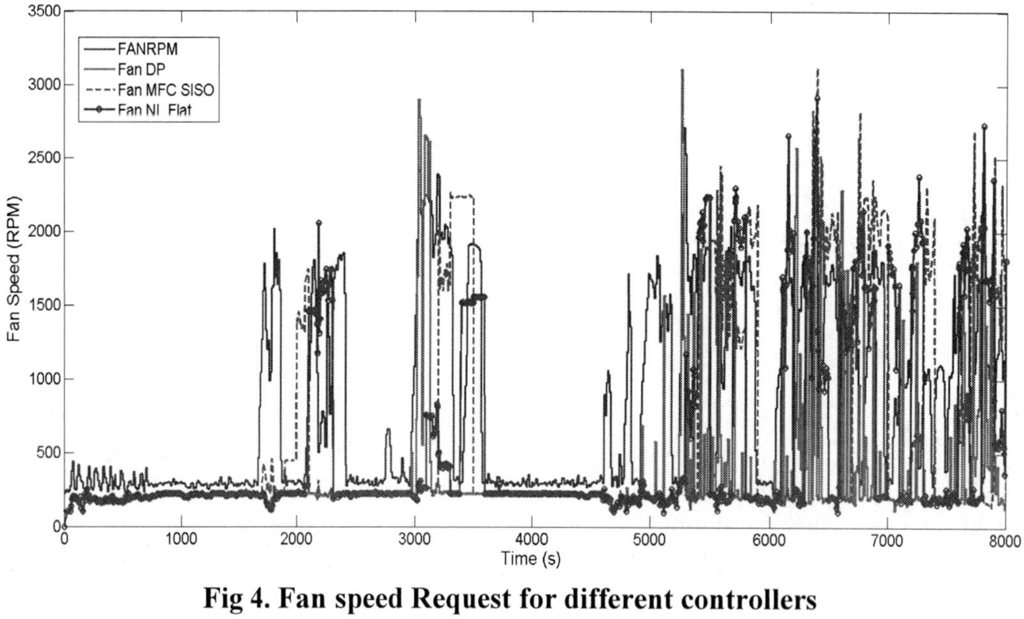

Fig 4. Fan speed Request for different controllers

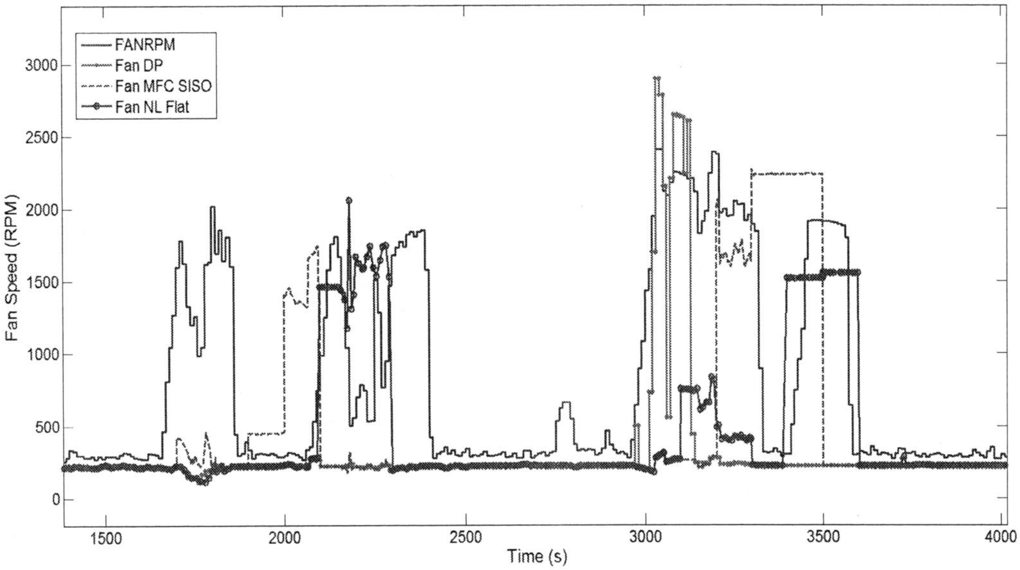

Fig 5. Zoom in of figure 4

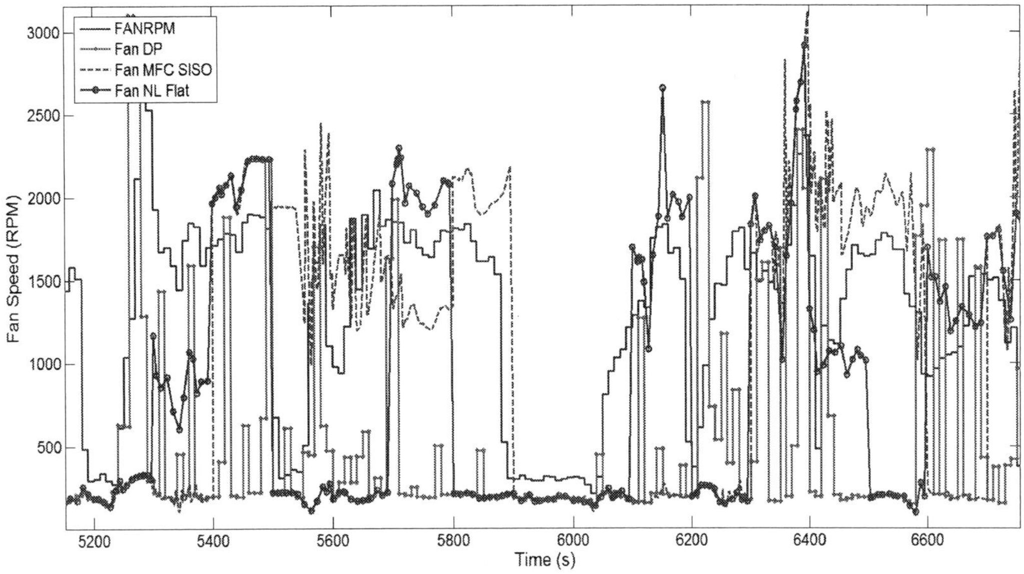

Fig 6. Zoom in of figure 4

In figure 7 the coolant temperature at engine outlet is presented. All of the controllers obtain allow a higher temperature for the coolant than the current control. The limit temperature currently used is 110 ° C. Enegy consumptions gain occur by allowing this higher temperatures. In figure 6, the fan speed is usually lower for the new advanced strategies compared to the current control. The fan on time is reduced as well. The DP results excells since the controller is able to predict the temperature behavior. It allows it to run freely, and to take advante of the airflow from the vehcile speed to cool the coolant.

A gain is also made since it seems, an early activation (figure 5) is determined by the rule based control laws, however it is not requiered by the controllers (or it is less penalising).

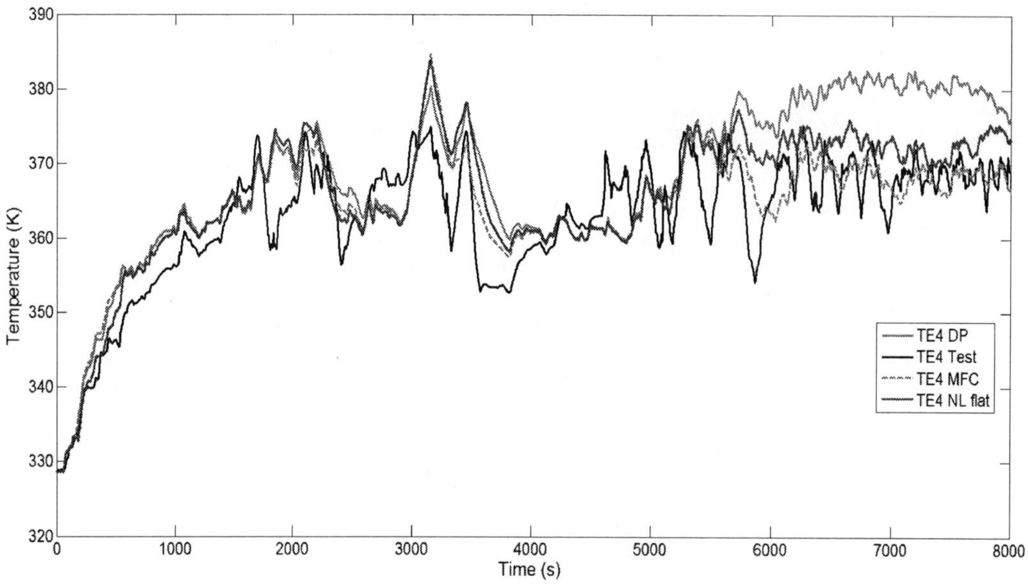

Fig 7. Coolant temperature dynamics for the differnet controllers

5 CONCLUSIONS

In this paper several new approaches for the control of the cooling system components were introduced. An implementation of said controllers is possible in a simulation environment.

Even though results are not conclusive, just an early implementation phase has been presented, the evaluation of the concepts remains of interest.

Significant energy gains (15-55 %) are shown. However uncertainties of the current model might hide other effects on the cooling system. Additionally, the dynamics of the actuators haven't been introduced into the models; this is a limit of the simulation platform. Further testing is planned using a more realistic model.

Conclusions presented in this paper are issued from a solely energy consumption point of view. This is the most interesting aspect for the manufacturer since it is a marketing opportunity.

The possible gains in energy consumption have been presented in percentage compare to the current control. No estimations have been made regarding fuel consumptions because of the assumptions made for this study and the limitations of the simulation platform. For example, the fan used is idealized (it is assumed that the requested speed is achieved immediately).

The proposed strategies remain promising; nevertheless, implementation in a vehicle architecture presents some challenges. They will be listed in the following paragraph.

Regarding the flatness method presented the implementation challenge is the availability of the information required to calculate the command. Currently only few states are measured in a normal vehicle configuration (coolant temperature at engine outlet and oil temperature in the sump). Either an estimator needs to be design to rebuild the states or new sensors need to be installed in order to implement the procedure in a vehicle. A second challenge is that the selected outputs do not correspond to existing sensors, and its physical representation is not easily understood. Nowadays, it is not possible to measure said outputs, so as stated previously they have to be calculated using the information from the system.

The MFC and ROM strategies implementation appear to be simple. However, these strategies rely on the implementation of a differentiator, which has been tested in the simulation platform presented. If a complete vehicle implementation is requested, special attention needs to be given to the performance of said differentiator.

The dynamic programming is an offline strategy and was never intended to be implemented as a real time controller. However, by comparing it to the performance of other strategies, we can assess the pertinence of a given strategy.

Finally, as stated briefly earlier, the greatest challenge to overcome for the MPC is the estimation of the future. In this paper and ideal estimation was used, a less accurate prediction will lead to a different result. For this paper the control horizon wasn't a constraint due to the assumptions made, however it should be properly studied for other cases. As stated in the literature, a compromise is generally made between accuracy, calculation time and overall gain of the strategy.

REFERENCES

[1] Allen, D. and Lasecki, M., "Thermal Management Evolution and Controlled Coolant Flow," SAE Technical Paper 2001-01-1732, 2001, doi:10.4271/2001-01-1732.

[2] Bhat, N., Joshi, S., Shiozaki, K., Ogasawara, M. et al., "Adaptive Control of an Externally Controlled Engine Cooling Fan- Drive," SAE Technical Paper 2006-01-1036, 2006, doi:10.4271/2006-01-1036.

[3] Choi. S, D'Andréa-Novel, Fliess, Mounier, Villagra. "Model-free control of automotive engine and brake for Stop-and-Go scenarios". European Control Conference (ECC'09), Aug 2009, Budapest, Hungary.

[4] Fleiss, M. Levine, J, Martin, P. and Rouchon, P. "Flatness and defect of nonlinear systems: Introductory theory and examples". Int. J. Control, Vol 61, p1327-1361; 1995.

[5] Fleiss, M. Levine "Generalised controller canonical form for linear and nonlinear dynamics". IEEE Transaction on Automatic Control, 35 No : 9 :994 – 1001, (1999).

[6] Güthler, D., Traub, M. and Wünsche, M., "Cooling system for diesel engine of lorry, has controller designed to determine operating parameter from group of operating parameters under consideration of measurement parameter, so as to lead to minimum fuel consumption of engine", Patent Number: DE102008014518A Company: DAIMLER AG, Filling date: 2009.

[7] Levant, A. "Full real time control of output variables via higher order sliding modes". European Control Conference, 1999, 6p.

[8] Ma, Y., Borrelli, F., Hencey, B.; Packard, A.; Bortoff, S., "Model Predictive Control of thermal energy storage in building cooling systems," Decision and Control, 2009 held jointly with the 2009 28th Chinese Control Conference. CDC/CCC 2009. Proceedings of the 48th IEEE Conference pp.392,397, 15-18 Dec. 2009. doi: 10.1109/CDC.2009.5400677.

[9] Salah, M., Mitchell, T., Wagner, J. and Dawson, D., "Nonlinear-Control Strategy for Advanced Vehicle Thermal-Management Systems", IEEE Transaction on Vehicular Technology, Vol. 57, no.1, 2008.

[10] Sermeno, S., Bideaux, E., Morgan, T., Khoa Nguyen, D. "Heavy duty vehicle cooling system auxiliary load management control: Evaluating the maximum gain of implementing an advanced control strategy", Proceedings of Commercial Vehicle Engineering Congress 2014, 2014-01-2341.

[11] Sermeno, S., Bideaux, E., Brun, X., Ameur, O., Morgan, T. "Heavy duty vehicle cooling system auxiliary load management control: An application of linear control strategy (MIMO and SISO)." Proceedings of the ASME 2014 International Mechanical Engineering Congress & Exposition, IMECE2014-39534.

[12] Setlur, P., Wagner, J., Dawson, D., and Chen, J., "Nonlinear Controller for Automotive Thermal Management Systems", proceedings of the 2003 American Control Conference, Denver, Colorado, 2003.

[13] Spall, J.C., and Cristion, J.A. (1998), 'Model-free control of nonlinear stochastic systems with discrete-time measurements', IEEE Trans. Automat. Control, 43, 1198–1210.

[14] Salah, M., Mitchell, T., Wagner, J. and Dawson, D., "Nonlinear-Control Strategy for Advanced Vehicle Thermal-Management Systems", IEEE Transaction on Vehicular Technology, Vol. 57, no.1, 2008.

[15] Wambsganss, M., "Thermal Management Concepts for Higher-Efficiency Heavy Vehicles," SAE Technical Paper 1999-01-2240, 1999, doi:10.4271/1999-01-2240.

An integrated system approach to thermal management on the thermal gasoline concept vehicle

A SUTHERLAND, E SHARPE and **X LIU**
BorgWarner, Corporate Advanced Engineering, USA

ABSTRACT

Vehicle thermal management has changed focus in recent times from the traditional worst-case cooling system performance to understanding and improving the thermal energy effects on fuel economy and emissions. As such, the most efficient operation of the entire system is an important consideration when applying thermal technologies to a vehicle.

BorgWarner initiated the Thermal Gasoline Concept Vehicle (TGCV) project and applied new BorgWarner thermal technologies to the vehicle in order to study the effects of these new technologies individually, as well as their overall system interaction and performance. The TGCV project was supported by a detailed simulation model to predict fuel economy performance and to drive the control system development.

This paper covers the new technologies applied to the TGCV, their operation and results from the initial simulation and testing phases to demonstrate improved capability and fuel economy.

NOMENCLATURE

AWD	All-Wheel Drive
CCV	Coolant Control Valve
DMCP	Dual Mode Coolant Pump
ECCV	Electric Coolant Control Valve
EOHX	Engine Oil Heat Exchanger
EPA	Environmental Protection Agency
FTP	Federal Test Procedure
GTW	Gross Trailer Weight
GVW	Gross Vehicle Weight
NEDC	New European Driving Cycle
TGCV	Thermal Gasoline Concept Vehicle
TOHX	Transmission Oil Heat Exchanger

1 INTRODUCTION

Historically thermal management has been focused on cooling systems and cabin comfort. For the cooling system, the focus was on ensuring a vehicle did not exceed the system or components maximum allowed temperatures and were signed off at worst-case ambient hill-climbs at vehicle GVW/GTW (Gross Vehicle Weight/Gross Train Weight) capacity.

With the global shift in the industry towards emissions reduction and improving fuel economy, thermal management in vehicles has changed from a focus on durability aspects to a focus on making the best use of the available energy in a vehicle such that it can operate as efficiently as possible. As such, successful thermal management comes from operating each component and system at the optimum temperature for the best performance, fuel economy and emissions.

Thermal management in any vehicle covers many different areas. In order to provide genuine emission and fuel economy benefits, it is clear that a systems based approach is required. As such, it was recognised that products supporting thermal management cannot be developed independently of the system and a test bench was required to understand the system interaction of these components. As a result the Thermal Gasoline Concept Vehicle (TGCV) was commissioned to support the development of several of these new thermal product technologies.

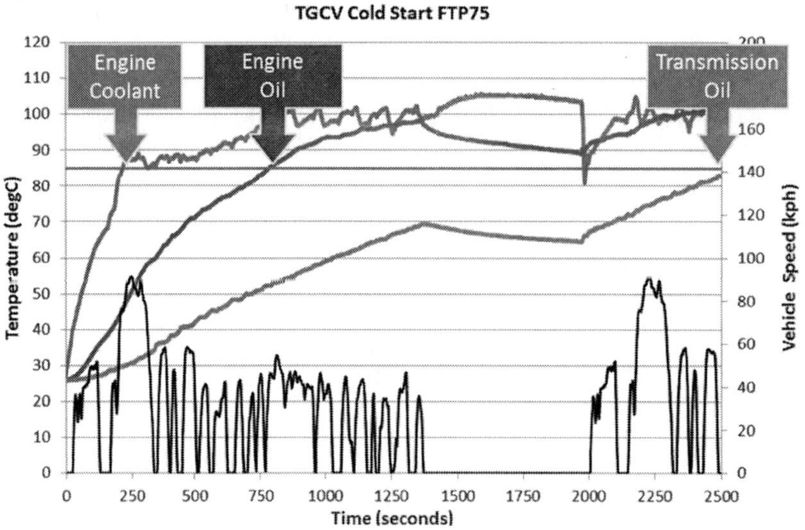

Figure 1: TGCV FTP75 fluid warm-up, baseline build condition

Emissions legislation in North America and Europe require the fuel economy and emissions to be assessed via the U.S. Federal Test Procedure (FTP) and New European Duty Cycle (NEDC) emissions. The NEDC and FTP test cycles are performed from an engine off cold soak condition, so there is a heavy influence on the warm up performance of the vehicle. Starting from a cold start condition, the high viscosities from engine and transmission oil can have a significant influence on the fuel economy or emissions of a vehicle. In many test conditions (Figure 1), the lubricating oils take a significant amount of time to reach operating temperature or do not reach it at all, meaning that the vehicle fluids are not operating at the optimum temperatures for the majority of the test cycle [2]. Low combustion efficiency is also associated with engine low temperature operation [10].

It was decided to develop the TGCV to enable faster warm-up using new product technologies already under development to achieve the following:

- Reduce engine and transmission lubrication based friction
- Reduce parasitic loses from components
- Take advantage of waste exhaust energy
- Provide intelligent control of the different systems and components

With the global trend for downsizing and boosting, a U.S. specification Ford Fusion (Ford Mondeo in Europe) with a 2.0l turbocharged EcoBoost® engine was selected as the base vehicle for the TGCV (Figure 2). The block and head are of cast aluminium. Performance figures for the engine are 231 hp at 5500 rpm, 270 lb-ft at 3000 rpm.

The vehicle is front-wheel drive and is fitted with a 6 speed automatic transmission, which was chosen to enable more repeatable vehicle testing by eliminating driver gear change effects.

Figure 2: Thermal Gasoline Concept Vehicle – Ford Fusion EcoBoost 2.0T
(Photo courtesy of Ford Motor Company)

Initially, the potential for fuel economy improvement was reviewed using the Environmental Protection Agency (EPA) publicly available data to understand the cold degradation effects from bag 1 to bag 3 of the FTP cycle (Figure 3). This was to ensure that a vehicle was selected with a realistic performance potential (instead of a vehicle with large cold degradation and therefore, an easy subject upon which to demonstrate improvement). However, it should be noted that the published EPA data only covered the all-wheel drive (AWD) version of the vehicle. During initial testing, the baseline complete cycle cold degradation performance was also tested and reviewed for the NEDC, World harmonised Light vehicle Test Procedure (WLTP) and FTP75 cycles. The differences between cold start and hot start, defined as complete cycle cold degradation, are shown in Table 1.

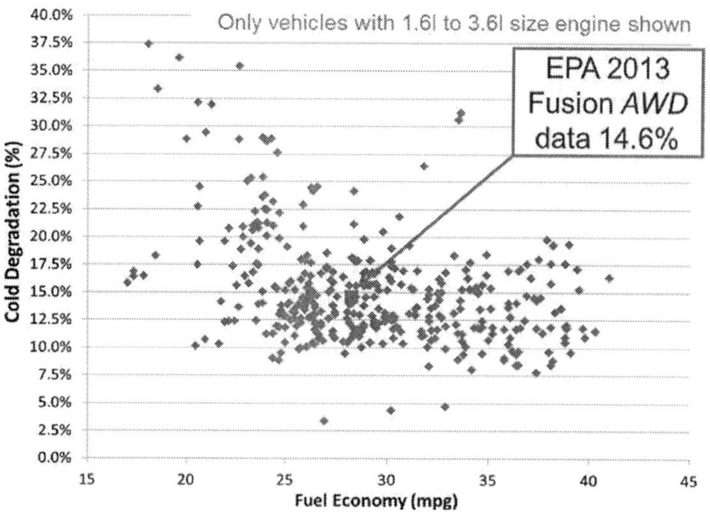

Figure 3: EPA FTP75 cold cycle degradation (bag 1 to bag 3 comparison)

Table 1: TGCV tested NEDC, WLTP and FTP75 cold start complete cycle fuel economy degradation

	FTP75	NEDC	WLTP
Maximum potential fuel economy improvement [%]	7.5	15.8	5.7

The TGCV focuses on using the main cooling system as a heat transfer medium to demonstrate how an OEM could take advantage of an existing cooling system for efficiency improvements using the products. The system is based around controlling how much coolant you flow, where you send the coolant and also when you control the coolant flow or direction.

1.1 Instrumentation

To enable the TGCV to be fully understood, a detailed set of instrumentation was added to the vehicle to include extensive fluid temperature, pressure and flow measurements as well as voltages and currents. Relevant CANBUS channels were also recorded.

Thermocouples were also machined into key critical metal temperatures in the exhaust valve bridges (2mm from gas face) and on the cylinder bore (1mm from gas face). This was done to help understand the critical friction surfaces in the engine and also to ensure that maximum metal temperatures were not exceeded under any conditions.

2 COMPONENTS FITTED TO THE THERMAL GASOLINE CONCEPT VEHICLE

A variety of thermal management products were installed on the TGCV for assessment individually and as a system. It should be noted that not all results are presented in this report.

2.1 Dual Mode Coolant Pump (DMCP)

Benefits offered by the control of coolant flow through controlled coolant pumps are well documented in the industry. For the TGCV, the Dual Mode Coolant Pump (DMCP) was chosen to provide the ability to control coolant flow accurately in the system.

The DMCP is a fully capable, fully controllable, main engine coolant pump that incorporates a 200W electric motor and a mechanical drive, connected through a clutch mechanism that is built into a single assembly. The DMCP is designed to mount onto the existing engine water pump location. The prototype level used on the TGCV is the EV1 level prototype as shown in Figure 4.

Figure 4: EV1 Level Dual Mode Coolant Pump (DMCP)

- The electric mode allows the water pump impeller to operate independently from the engine speed, allowing the impeller to underflow, overflow or zero-flow the coolant as required by the engine for optimized operation. The electrical mode also allows for engine off operation, for after-boil protection, stop/start enhancement and turbocharger protection.
- The mechanical mode allows for high speed and high load operations outside of the electric motor operating range and also for fail-safe operation in the event of electrical power loss to the DMCP.

The five modes of operation, and some of the system enablers this supports on the engine, are shown in Figure 5.

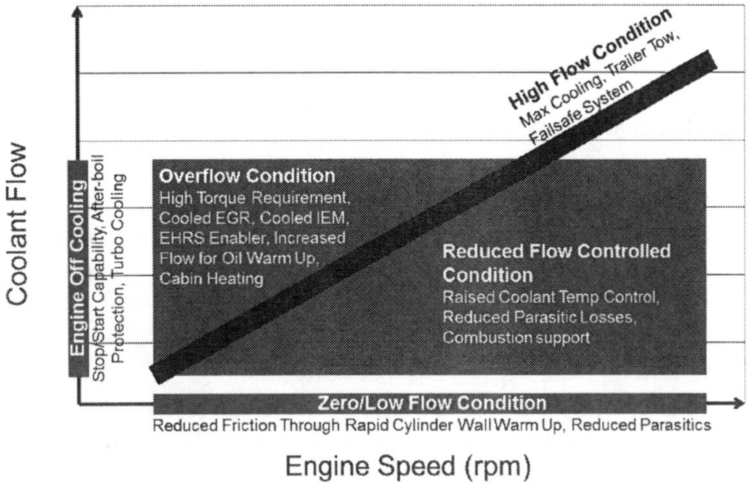

Figure 5: Coolant pump capability chart

To install the DMCP onto the 2.0l EcoBoost engine the decision was made to use the existing water pump location, although the engine block was modified to provide the inclusion of a new volute housing that allowed operation with a unique engine bypass routing (more details will be explained in section 2.7 under the system layout) and also allowed operation with the original wax thermostat in the baseline location. This meant that the inlet moulding could be re-used for packaging (Figure 6).

The base Ford stamped impeller was kept to ensure that the effect of the impeller design was eliminated from any comparisons, although a new high efficiency impeller design is available and is planned for testing on the TGCV to demonstrate further improvement.

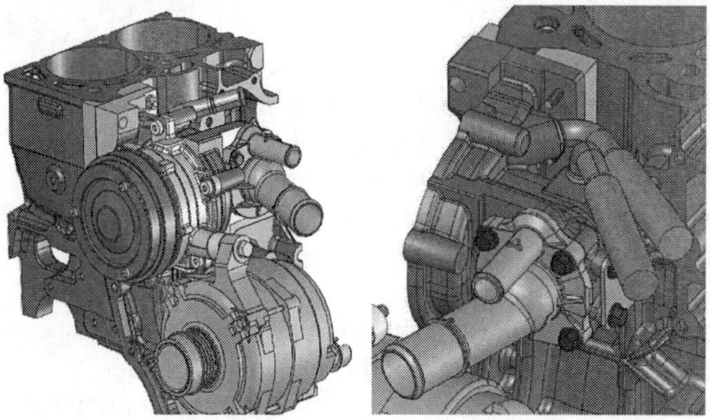

Figure 6: Dual Mode Coolant Pump installation

2.2 Electrical Coolant Control Valve (ECCV)

Alongside the ability to control the coolant mass flow that the DMCP offers, a coolant directional control device was required to provide coolant when and where required in the TGCV system. The Electrical Coolant Control Valve (ECCV) was chosen to meet this requirement.

The ECCV is a modular component with valve and actuator as separate components. The valve consists of one injection moulded housing with two spigots and a rotary, cylindrical sleeve with polymer based components (Figure 7). The coolant inlet for the ECCV enters axially into the main chamber. The rotary sleeve is mounted centrally on the main shaft in this chamber with bespoke size and shape openings.

The ECCV used on the TGCV is a prototype level rotary sleeve design (Figure 8), made from aluminium housing, with the ability to accurately control engine outlet flow based on angular position to the two main loops in the TGCV cooling system, the low pressure drop, low leakage rate and the rapid response time that an ECCV offers. The rotary sleeve was designed with outlet angles specific to support the TGCV system operation, allowing the two outlets to open as prescribed by the control requirements. The two ECCV outlets feed; 1) the auxiliary loop for all flow to all auxiliary heat exchangers focused on heating and cooling, and 2) the radiator loop for cooling requirements only (details of which are covered in section 2.7).

Figure 7: ECCV B1 prototype

The ECCV was mounted on the outlet of the cylinder head (figure 8) to allow the provision of zero-flow strategy from the ECCV alone and to take advantage of the Exhaust Heat Recovery System during zero-flow conditions, functions that the base EcoBoost 2.0l engine does not have. Since the base engine utilises a conventional wax thermostat at the inlet to the coolant pump, some modifications were required to accommodate the ECCV. The baseline plastic outlet manifold was replaced with a bespoke designed manifold which served as partial mounting for the ECCV and incorporated the cylinder head coolant outlet temperature sensor and the degas functions of the base vehicle.

To accommodate the additional multiple loops for the auxiliary coolant circuit, a bespoke outlet manifold was also produced. Initial discussions focused on using the ECCV to provide independent control for all loops, although the existing production vehicle package did not allow sufficient space.

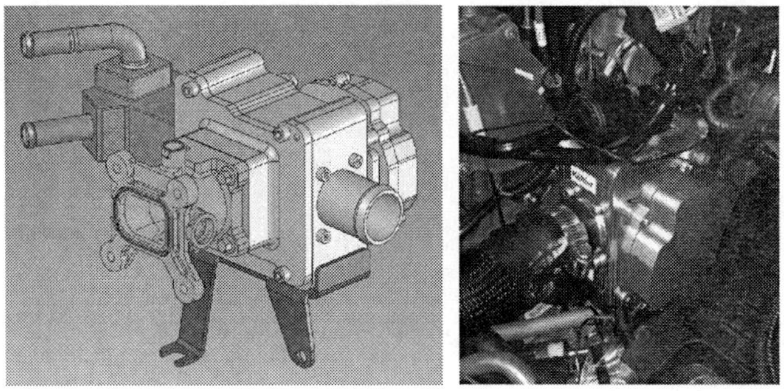

Figure 8: ECCV installation in TGCV

2.3 Exhaust Heat Recovery System (EHRS)

As well as controlling the coolant flow and direction, the opportunity to provide additional heat to the system was highlighted as a key requirement for the TGCV to enhance warm up. An Exhaust Heat Recovery System (EHRS) was selected for the TGCV and made of stainless steel construction (Figure 9) using BorgWarner's tube and shell design of heat exchanger, as used in EGR coolers. Pressure drop was considered a key concern for installation on a production vehicle, hence a design with minimal heat exchanger pressure drop on both the gas and coolant side with a straight through bypass option chosen. The bypass control is operated by a pneumatically controlled vacuum actuator. The design chosen also offers a high heat transfer capacity capable of ~11-16kW at 500°C gas temperature, depending on gas flow rate. 500°C was recorded at this location during cycle as an average exhaust temperature for the base EcoBoost 2.0l engine.

Figure 9: EHRS as installed in TGCV

The EHRS was installed underfloor on the Fusion post catalyst (Figure 10), to prevent any interference with the after-treatment.

Due to the existing vehicle design, a compromise in installation was made to place the EHRS after the catalyst outlet flex-pipe and exhaust hanger (Figure 10). This placement was chosen to prevent additional cantilevered mass on the turbocharger and catalyst assembly and to minimise the risk of failure of these components. The additional length of exposed exhaust pipe and associated heat loss meant a reduced exhaust temperature at the EHRS inlet, although the exhaust prior to the EHRS was insulated to reduce heat loss and to provide a realistic inlet gas temperature to the EHRS. This then required the location of the underfloor muffler to be moved rearwards. However, the increased coolant volume to reach the EHRS could not be compensated for and forced an increase in cooling system volume and inertia.

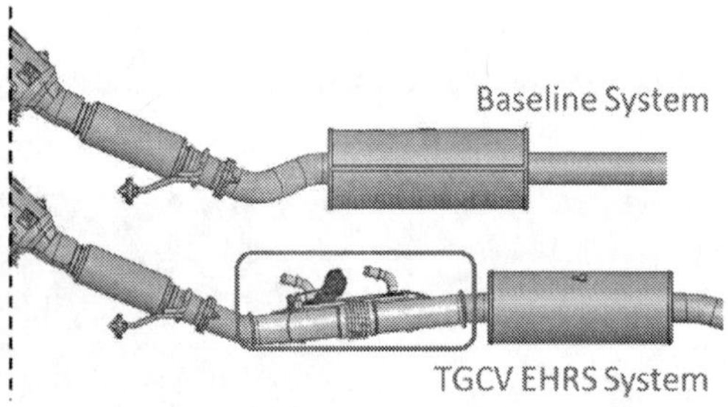

Figure 10: TGCV EHRS installation

2.4 Engine Oil Heat Exchanger (EOHX)

When using the cooling system as the base heat transfer medium, a method of transferring additional heat generated by the EHRS to the engine oil system through the coolant requires a heat exchanger. Since the production North American specification Ford Fusion 2.0l EcoBoost does not include an Engine Oil Heat Exchanger (EOHX), the EOHX and distribution casting from the Ford Explorer 2.0l EcoBoost installation was selected. The heat exchanger is a conventional circular plate heat exchanger mounted on the inlet side of the engine (Figure 11).

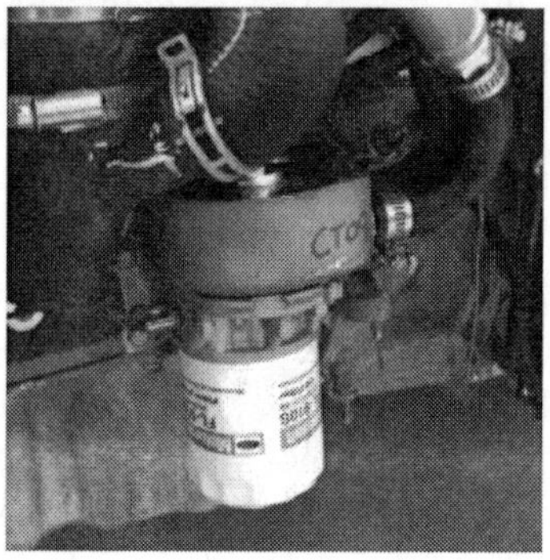

Figure 11: EOHX as installed on TGCV

2.5 Transmission Oil Heat Exchanger (TOHX)

As previously stated by Kunze et al. [6,7] and Janowski et al. [9], there are gains to be made from reducing the effect that the transmission slow warm up has on the vehicle fuel economy. Therefore, it was decided that the TGCV needed to optimize the transmission oil warm up by incorporating a Transmission Oil Heat Exchanger (TOHX). The production Fusion utilises a plate style heat exchanger mounted on the cooling pack. However, this is connected to the cold side of the radiator, so it does not flow until the thermostat opens. For the TGCV, the same heat exchanger was used (Figure 12), although re-connected to allow coolant flow and therefore transmission oil heating from any operating condition.

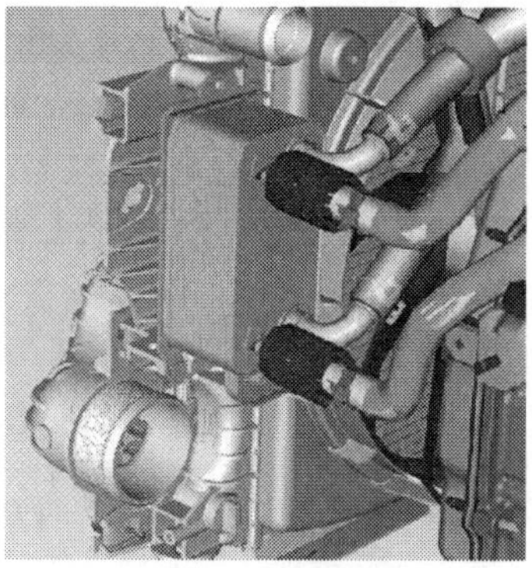

Figure 12: TOHX as installed on cooling pack

2.6 Auxiliary Coolant Control Valves

Once it was determined that the ECCV with the required outlets could not be packaged due to available space within the production vehicle, it was decided to install four small diameter coolant control valves (Figure 13) from an external source to control each of the auxiliary loops. Clearly this would not be an acceptable production solution, but was selected for research purposes to allow the independent control of coolant flow to key locations in the system. Unfortunately this introduced another compromise in the form of unrequired pressure drops to the system that would not exist if controlled through the ECCV.

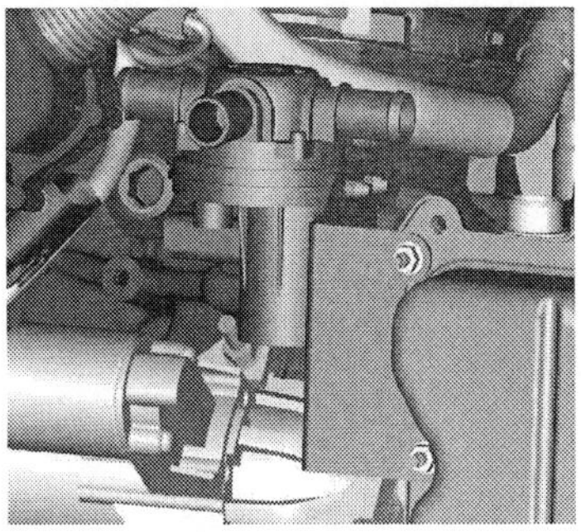

Figure 13: Small CCV installed to control EOHX coolant flow

2.7 System Integration

The production Ford Fusion 2.0l EcoBoost vehicle cooling system is a conventional system (Figure 14) with no intelligent control. It utilizes a mechanical water pump, belt driven from the crankshaft, an inlet wax thermostat controlling the bypass and the cabin heater as the only component on the auxiliary loop, with the transmission cooler fed from the radiator cold side.

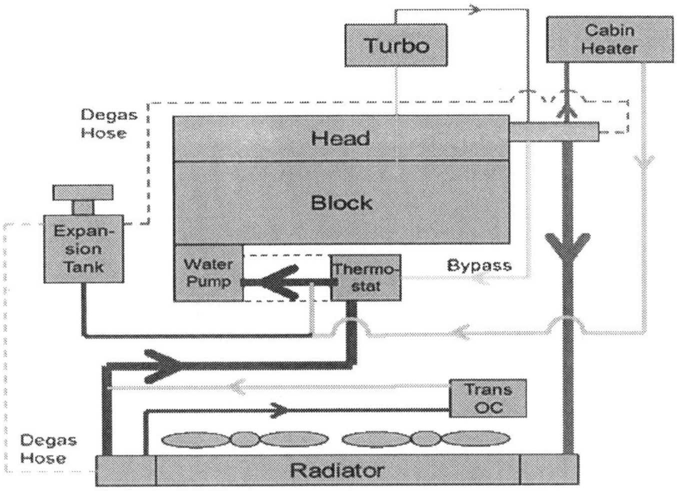

Figure 14: Ford Fusion 2.0l EcoBoost production cooling system

The TGCV updated cooling system (Figure 15) has attempted to incorporate new features alongside the components installed while considering the system effects, all within the constraints of the production vehicle package.

An early requirement was to assess the benefit of zero-flow on the engine with the DMCP and ECCV. However, the zero-flow strategy works in competition with the additional heat source from the EHRS. In order to take advantage of zero-flow benefits and the additional heat available through the EHRS, a new engine bypass on the outlet of the engine was installed, which feeds directly to the EHRS. So, during a cold start the ECCV can close to provide zero-flow to the engine, yet the DMCP can still operate to provide coolant flow to the EHRS to heat the coolant in the auxiliary circuit, which would otherwise be stagnant. This enhances heat transfer to the EOHX and TOHX for reduced parasitics and also to the cabin heater for enhanced comfort in the form of faster cabin warm up. The DMCP can also run at maximum electrical speed to enhance the heat transfer to these systems.

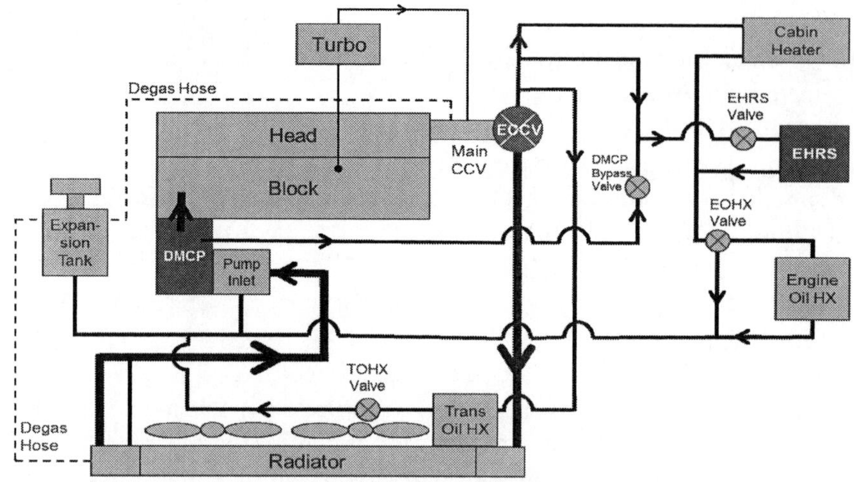

Figure 15: TGCV enhanced cooling system

The EOHX incorporated a coolant side bypass to allow for pressure drop and flow control to and from the EOHX as required.

The TOHX is now connected to the engine outlet for maximum engine outlet temperature and also can be flowed during zero-flow conditions.

With the increase in system complexity, an increase in coolant volume was expected. This of course has a negative effect on the cooling system thermal inertia, which will have a detrimental effect on the warm up rate. However, care was taken within production vehicle package limitations to ensure that the volume to each loop was minimized during warm up conditions.

3 TGCV CONTROL STRATEGY AND SIMULATION MODEL

3.1 TGCV Control Strategy

A detailed control strategy was developed to operate the TGCV cooling system which focused on achieving the best fuel economy under all operating conditions. There are seven different operating modes in the control strategy designed to cover all operating conditions. There are five warm up enhanced modes and two temperature control modes.

Warm up enhanced modes:

- Cold start zero engine flow
- Engine oil warm up
- Transmission oil warm up

- Hot operation with thermostat closed
- Engine off cold soak operation (for stop/start)

Temperature control modes
- Hot operation with thermostat open
- Engine off hot soak operation

To achieve the operation, the control system operates nine different actuators, shown in blue in Figure 16. These include:
- DMCP clutch
- DMCP electric motor
- ECCV rotational position
- EHRS bypass valve actuator (gas side)
- 4x auxiliary coolant control valves (TOHX valve, EHRS valve, DMCP bypass valve, EOHX valve)
- Electric cooling fan override

A control target established for the system is to assess the three main electrical devices (DMCP, ECCV and cooling fans) and select the lowest electrical load device that can control the system operation as required, since electrical load results in reduced fuel economy.

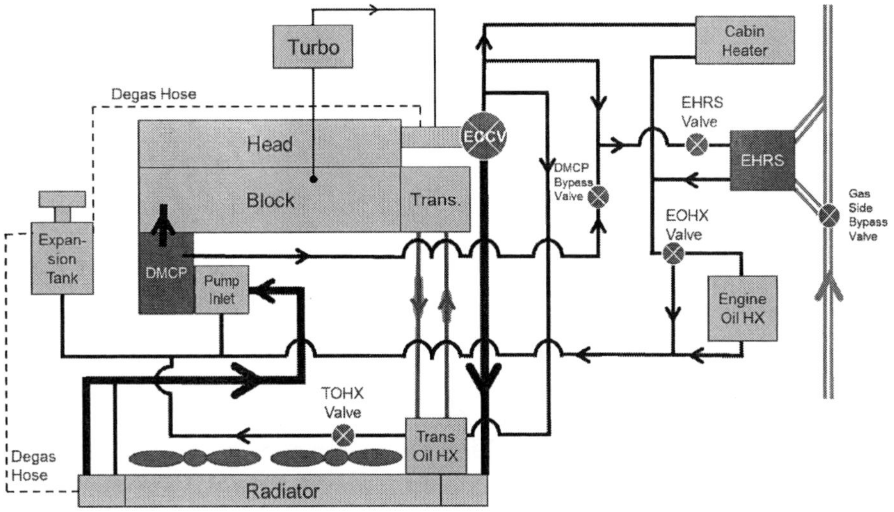

Figure 16: TGCV control actuators

More details of the control strategy and operating modes are shown in the Liu et al. paper [1], also presented at this conference.

3.2 TGCV Simulation Model

A detailed model of the production vehicle and the TGCV update was constructed in GT Suite to include engine system, vehicle system, hydraulic circuits and a neural net for intelligent control (Figure 17). The production and initial update build models were correlated to test data acquired from the test vehicle.

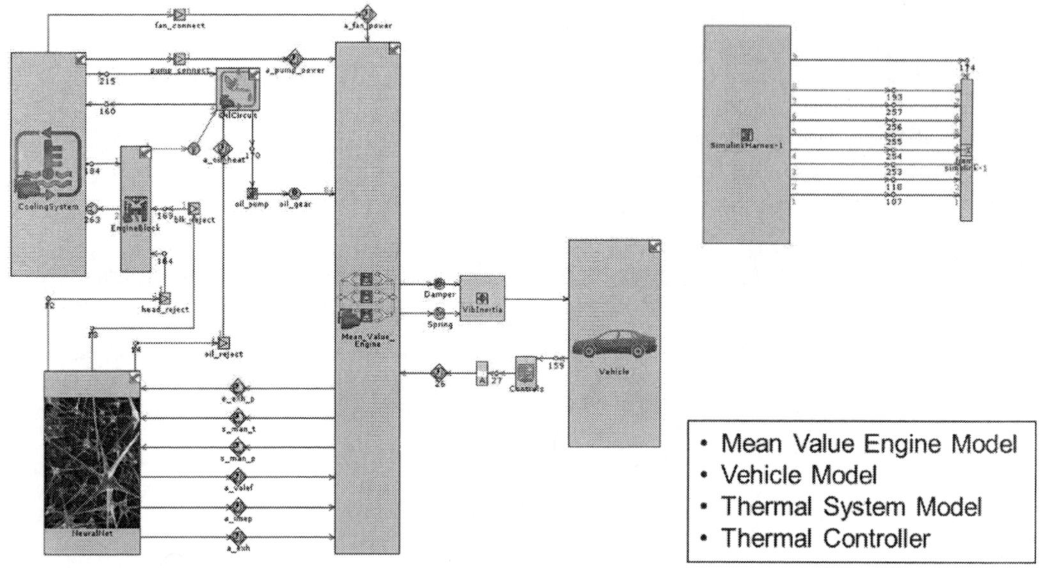

Figure 17: TGCV high level simulation model

There were two main targets for the simulation model:

- To support the cooling system schematic layout selection. With limited knowledge of the base engine and system operation, coolant flows were initially matched to production simulation model data where appropriate to ensure coolant flow to critical components was acceptable and allowed for optimized flow where required.
- To support the control system operation and assess strategy changes on the fluid system performance and vehicle fuel economy.

More details of the simulation model and correlation are shown in the Liu et al. paper [1], also presented at this VTMS12 conference.

4 TGCV SIMULATION RESULTS

With the flexibility that the simulation model offers, many of the capabilities of the products in the system were assessed in the simulation model to develop and understand the potential and strategy that could be supported.

Zero-flow was seen as a key benefit to engine and system warm up, with both the DMCP and ECCV able to offer this benefit. The effect of zero-flow on the engine was studied to assess the potential for improvement during the FTP75 condition. The potential was assessed over the bag 1 and bag 1+2 sections of the test cycle. The data in Figure 18 shows the results for various zero-flow times to study the effect. The simulation was run using the ECCV for zero-flow control, with the DMCP operating mechanically to simulate a conventional water pump and no additional heat exchangers connected (EHRS, EOHX, TOHX). The results suggested that there is an optimum time to operate the zero-flow strategy for, in this case around 180-200 seconds, whereas the common belief is that zero-flow strategies should operate for as long as possible until the valve bridge temperature or another factor limits it.

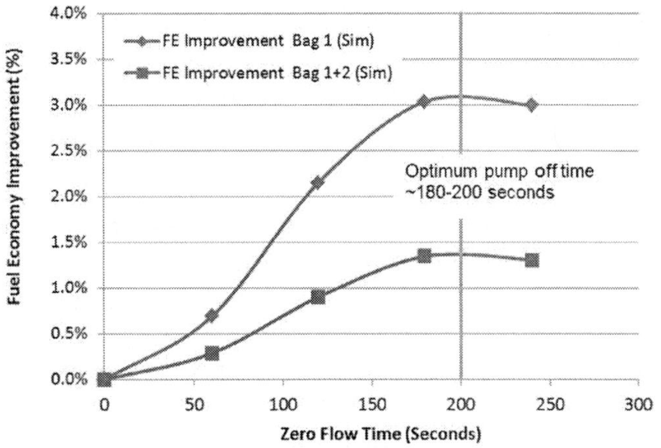

Figure 18: Effect of zero-flow time on fuel economy for FTP75 bag 1 & bag 1+2

The effect of the EHRS performance on the system was reviewed in addition to the zero-flow strategy. For this condition, the system was run as per the zero-flow condition, however, flow was allowed to bypass the engine and flow to the EHRS and onto the EOHX. DMCP was operating mechanically and there was no flow to the TOHX. Results for performance over bag 1 of the FTP75 cycle are shown in Figure 19. Using the EHRS to heat the coolant and engine oil through the EOHX offered additional improvement in fuel economy of approximately 2.0%, although it was accepted that losses in the oil gallery to the metal mass may have limitations in the model affecting results.

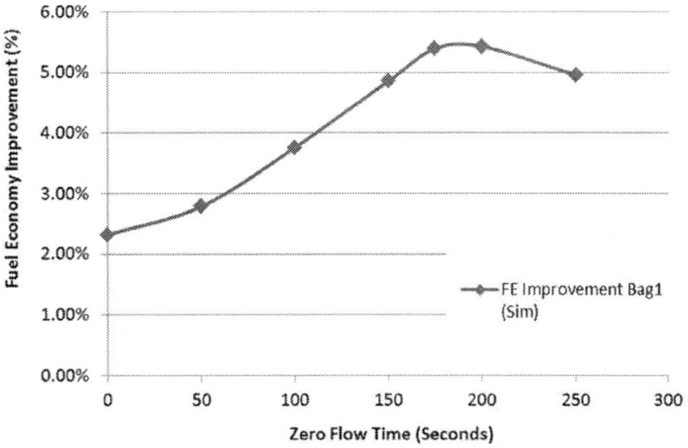

Figure 19: Effect of zero-flow duration with EHRS heat input on fuel economy during FTP75 bag 1

Many other effects were studied, but have not been included in this report.

5 TGCV TESTING RESULTS

To ensure successful operation of the technologies and the overall system, the TGCV was tested on North American and European emissions cycles in a certified emissions test facility to the drive cycle standards.

At the time this report was prepared, the TGCV vehicle testing and development was still in progress. Results presented here are initial system results focusing on developing a system that can provide overall performance benefits.

Figure 20 shows initial system test data results using only the DMCP, ECCV and cooling fans to control the system. Initial focus was on controls development of the DMCP and the ECCV as critical components to control the volume of flow and direction of flow in the system. EHRS was not operational and there was no coolant flow directed to the EOHX or TOHX. For this test, zero-flow performance was assessed using both the DMCP and ECCV as can be seen over the initial 180 seconds. This resulted in an initial rapid rise in coolant temperature and cylinder wall metal temperature. However, opening of the ECCV to the auxiliary loop and initiation of the DMCP caused a sudden drop in coolant temperature as the cold coolant remaining in the external coolant loop entered the engine. This is intended to be covered by the EHRS operation during a zero-flow condition. Once the ECCV was opened, focus was put on minimizing flow fluctuations to remove coolant temperature hysteresis as associated with wax thermostats. DMCP operation was controlled to run at minimum required speed, although this was hindered by the requirement to limit temperature rise across the engine and to maintain a trickle flow through the EHRS even though it was not operating. During the higher load bag 3 conditions focus was put on using the ECCV to control the coolant temperature as a lower electrical load than the DMCP, hence the reduced pump operation towards the end of the cycle.

The initial results showed a slightly lower oil and transmission oil temperature, which were concluded to be due to the changes to the cooling side layout, with increased inertia from larger coolant volume and increased metal mass from the added EOHX.

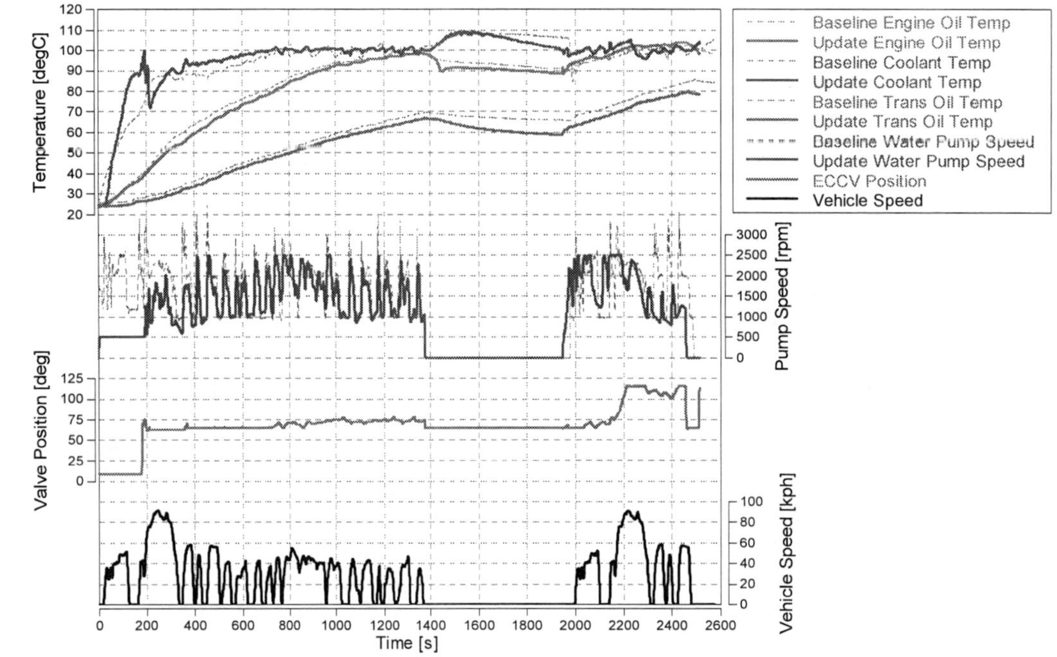

Figure 20: TGCV initial FTP75 non-optimized performance comparison

Figure 21 shows a further development of the system operation during bag 1 and 2 of an FTP75 cycle, once again using only the DMCP and ECCV for control, with no flow to EOHX or TOHX and the EHRS bypassed. In this test, the zero-flow condition was reduced to 80 seconds, before moving into the different operating modes.

Once the ECCV was opened to the radiator, post zero-flow, the coolant out of radiator was at a much lower temperature. As a result, opening sensitivity of the ECCV based on engine outlet temperature was initially quite large. However, fine adjustment of the ECCV opening resulted in a very stable coolant temperature out of the engine. DMCP speed was controlled to regulate delta temperature across engine block.

As engine oil and coolant warm up, the focus on heat transfer from coolant to oil is reduced and the ECCV further opens to allow more flow to the radiator. The DMCP speed can be reduced to save energy consumption due to lower flow restriction in the coolant circuit. The ECCV and DMCP work together to keep the electric cooling fans switched off, since the fans have the higher electrical load.

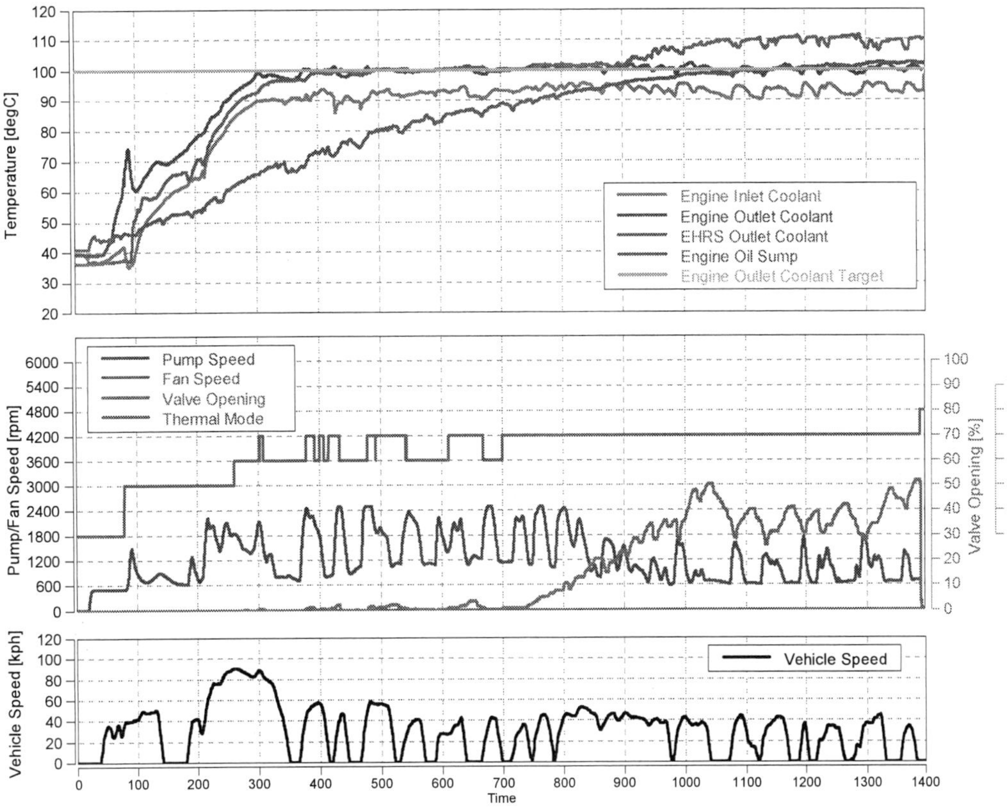

Figure 21: Example of TGCV initial FTP75 non-optimized performance control

NEDC tests focused on increasing cylinder bore temperatures for reduced friction by using the ECCV to control zero-flow strategy. The potential for increasing the zero-flow duration over the FTP cycle due to the lighter load NEDC cycle was assessed. The DMCP was operated mechanically during these NEDC tests.

Figure 22 shows the effect of zero-flow on the coolant and engine oil temperatures at four different zero-flow durations and compares the effects to baseline. Initially there is a slight delay in coolant temperature increase due to non-flowing coolant to the temperature sensor, but the temperature quickly rises over baseline levels.

The effect of zero-flow allows a faster warm up of engine oil at the main gallery, which is assisted by the increased coolant temperature once the zero-flow period ends. Overall benefit of zero-flow above 450 seconds was limited by the ECU strategy switching on the cooling fans at 104°C coolant temperature, which would be adjusted for further benefit.

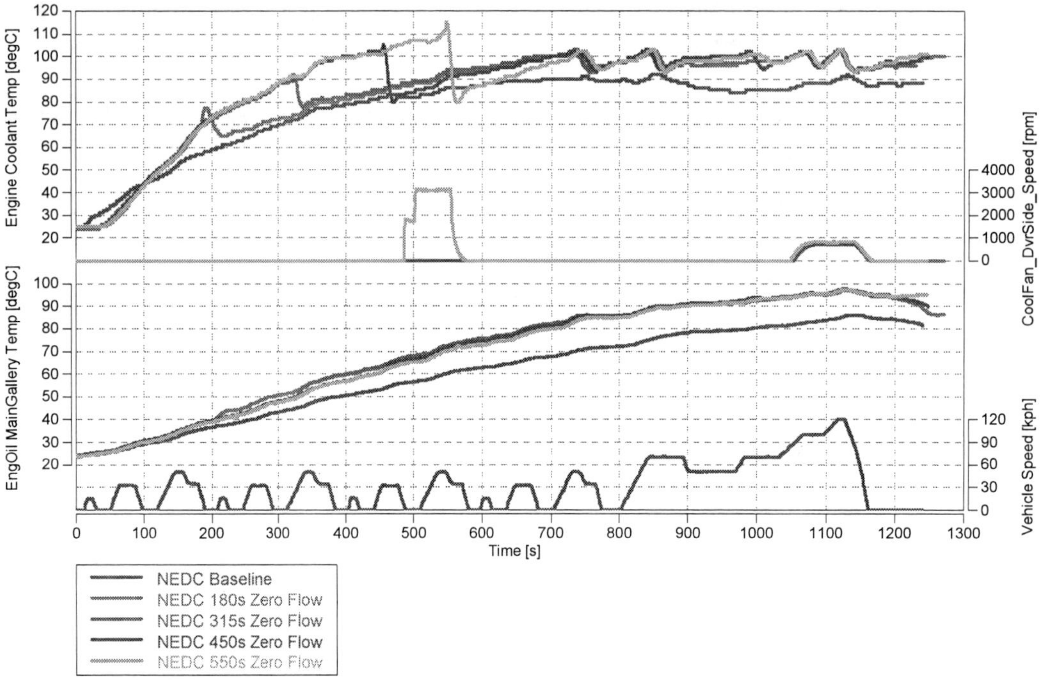

Figure 22: TGCV initial NEDC zero-flow test fluid temperature assessment

Figure 23 shows the effect of the same zero-flow tests as Figure 22 on engine metal temperatures. The results show a significant increase in bore metal temperatures during the zero-flow conditions, allowing for increased surface oil temperature and reduced friction, although some variation in increase is observed around the block caused by coolant jacket design. The durability critical valve bridge temperatures increase over baseline, although are still below guideline for peak temperature. Once the ECCV opens and coolant flow starts, all metal temperatures quickly return to match baseline.

Results from the more developed control system are shown in the Liu et al. paper [1].

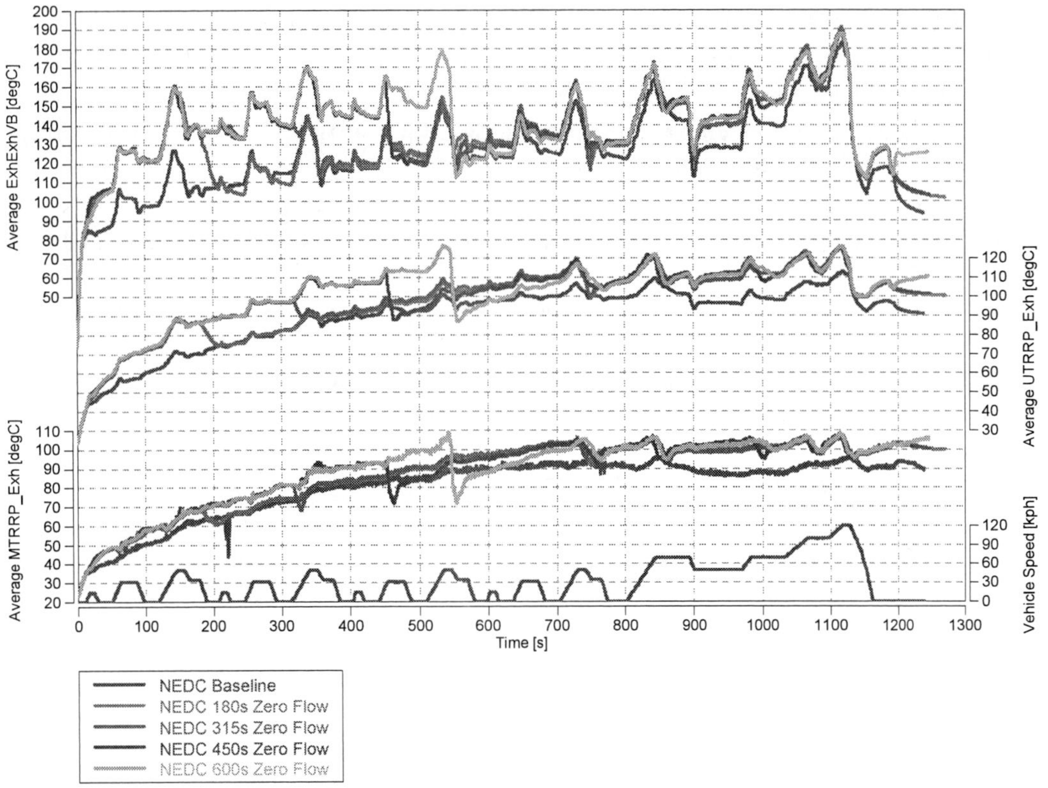

Figure 23: TGCV initial NEDC zero-flow test metal temperature assessment

6 CONCLUSIONS

BorgWarner has successfully integrated the Dual Mode Coolant Pump, Electric Coolant Control Valve and Exhaust Heat Recovery System into the existing vehicle thermal system on a production Ford Fusion 2.0l EcoBoost. The updated TGCV system allows for significant control opportunities of the thermal system and potential efficiency gains that can come with the enhanced control.

Fuel economy improvements have been demonstrated in simulation from the operation of components in the system.

The TGCV system operation has been demonstrated on vehicle emissions dyno, with the initial control strategy demonstrating system efficiency improvements. The data presented in this paper represents initial operating performance of the TGCV vehicle. As such, further improvement is targeted in the future as cooling layout and thermal components are further optimized and the control strategy is refined.

7 NEXT STEPS

The TGCV project was not yet complete at the time of preparing this paper. Further opportunities for improvement and confirmation of fuel economy and emissions data is yet to be finalised. Next steps planned for the TGCV are:

- Completion of initial cycle testing at other vehicle conditions.
- Assessment of vehicle efficiency potential with individual product installation on the vehicle.
- Updates to components for improved operational efficiency based on learnings from initial development.
- Inclusion of new fuel savings technologies for further assessment.
- Build changes to key components in the systems that have been noted as bottlenecks in the performance.
- Further development of the system control strategy for rapid warm up and maximum fuel economy improvement over baseline.
- Demonstration of the system and products to the industry.

ACKNOWLEDGEMENTS

The authors would like to acknowledge the following for their work on and support of the TGCV project:

- The many participants at BorgWarner who were involved in the project globally:
  - BorgWarner Thermal Systems, Marshall, Michigan, USA
  - BorgWarner Emissions Systems, Vigo, Spain
  - BorgWarner Emissions Systems, Esslingen, Germany
  - BorgWarner Transmission Systems, Auburn Hills, Michigan, USA
  - BorgWarner Advanced Engineering, Auburn Hills, Michigan, USA
  - BorgWarner Advanced Engineering, Ludwigsburg, Germany
  - BorgWarner Support Services, Auburn Hills, Michigan, USA
- The Engineering team at Ford Motor Company, Dearborn Michigan, USA

REFERENCES

1. **X Liu, R Rastelli,** Thermal gasoline concept vehicle, simulation and control. VTMS 12.
2. **E Sharpe, A Sutherland, D DeLong.** Cold Start vs. Hot Start Fuel Economy of Main Drive Cycles and Real World Condition to Show Potential Improvement from Thermal Management Technologies. SAE TMSS 2014.
3. **J Shutty, W Wenzel, M Becker, S Bohan, et al.,** Advanced Thermal Management of a Light Duty Diesel Vehicle, SAE Technical Paper 2013-01-0546.
4. **W Wenzel, T Buchholz, J Shutty, J Tsai.** Thermal management for light-duty-vehicle with diesel engine: Evaluation of an optimized cooling system with variable cooling components. 14th Stuttgart International Symposium, Automotive and Engine Technology, Volume 1, p 61, 2014.
5. **P Keller, W Wenzel, M Becker, J Roby.** Hybrid coolant pump with electric and mechanical drive. ATZ, Volume 10, p52, 2010.
6. **K Kunze, S Wolff, I Lade, J Tonhauser.** A Systematic Analysis of CO2-Reduction by an Optimized Heat Supply during Vehicle Warm-up. SAE Technical Paper 2006-01-1450.

7. **K Kunze, I Lade, S Wolff.** Customer Related CO2 Reduction By Selective Heat Supply During Vehicle Warm Up. VTMS8.
8. **J-P Zammit, P Shayler, I Pegg.** Thermal Coupling And Energy Flows Between Coolant, Engine Structure And Lubricating Oil During Engine Warm Up.
9. **P Janowski, P Shayler, S Robinson M Goodman**, The Effectiveness Of Heating Parts Of The Powertrain To Improve Vehicle Fuel Economy During Warm Up. C1305/054/2011. VTMS10.
10. **E Bent, P Shayler, A LaRocca, C Rouaud.** The Effectiveness Of Stop/Start And Thermal Management Measures To Improve Fuel Economy. C1365/027/2013 VTMS11.

Heat flow rate based thermal management for electric vehicles using a secondary loop heating and cooling system

T A WEUSTENFELD, W BAUER-KUGELMANN, J C MENKEN and **K STRASSER**
AUDI AG, Germany

J KOEHLER
University of Braunschweig, Germany

Thermal management energy consumption has crucial impact on the range of an electric vehicle [1]. This requires optimal use of available thermal energy while meeting cabin comfort and component temperature requirements. Secondary loop systems are a promising approach in order to meet these requirements while reducing refrigerant charge and simplifying the vehicle front end. Based on a simplified drivetrain model heat losses are estimated. Using first order thermal mass models and typical component temperature constraints, energy surpluses and energy deficits are calculated. Based on temperature levels and energy balance a central refrigeration unit is then used to provide additional heating power (heat pump) or cooling power (AC mode). An ambient radiator is used as heat sink if required. Simulation results for two different drive cycles show necessary heat flow rates between components and three cooling cycles for typical thermal management use cases. The energy flow rate based thermal management approach remains without further specification of a heating and cooling system (including pumps, valves, heat exchangers and electric heaters). Therefore presented results can be used for system design or lay the foundation for development of control strategies.

NOTATION

EM	Electric machine
PE	Power electronics
BAT	Battery
CHRG	Charging device
AMBX	Ambient heat exchanger
COP	Coefficient of performance
CC	Cooling cycle
HC	Heating cycle
LC	Low temperature cycle
CAB+	Cabin heating
CAB-	Cabin cooling

1 INTRODUCTION

Traditionally the main goal of electric vehicle thermal management is to protect powertrain subsystems from overheating. A more general view would be to consider climate control (including safety functions such as defrosting and defogging) as part of the thermal management system. So that the term "thermal management", in this case, refers to the total number of all cooling and heating systems including component cooling, component heating and the HVAC (heating, ventilation, and air conditioning) system. A schematic overview of a typical thermal management system for an electric vehicle is shown in figure 1.

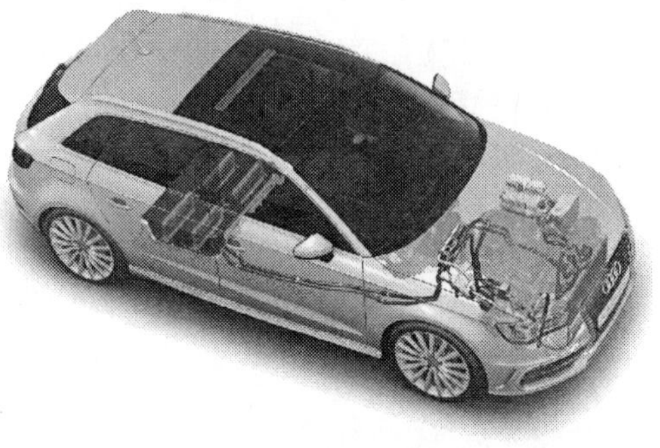

Figure 1 – Typical thermal management system for an electric vehicle

Recent attempts to find energy-efficient thermal management systems for electric and plug-in hybrid electric vehicles have led to secondary loop systems as an alternative approach to meet dynamic heating and cooling demands. Recently, much research has been performed regarding automotive applications of secondary loop systems. Ghodbane et al. [2] present a secondary loop system with the HFC refrigerant R152a. Kowsky et al. [3] demonstrate a hermetic encapsulated central thermal management unit for the use in electric and hybrid vehicles using the battery waste heat as energy source during heat pump mode.

Not always thermal management is considered as an interconnected system in vehicle thermal management, but instead dedicated cooling loops for each subsystem are developed. Much focus has therefore been drawn on isolated systems, instead of a holistic approach, that can potentially benefit from synergetic effects by combining cooling and heating loops [4]. Secondary loop systems hence focus on these benefits.

Figure 2 shows a secondary loop system with a compact refrigeration cycle and a secondary loop flow distribution unit. This work is intended to study thermal energy flow on an abstract level assuming a similar secondary loop system.

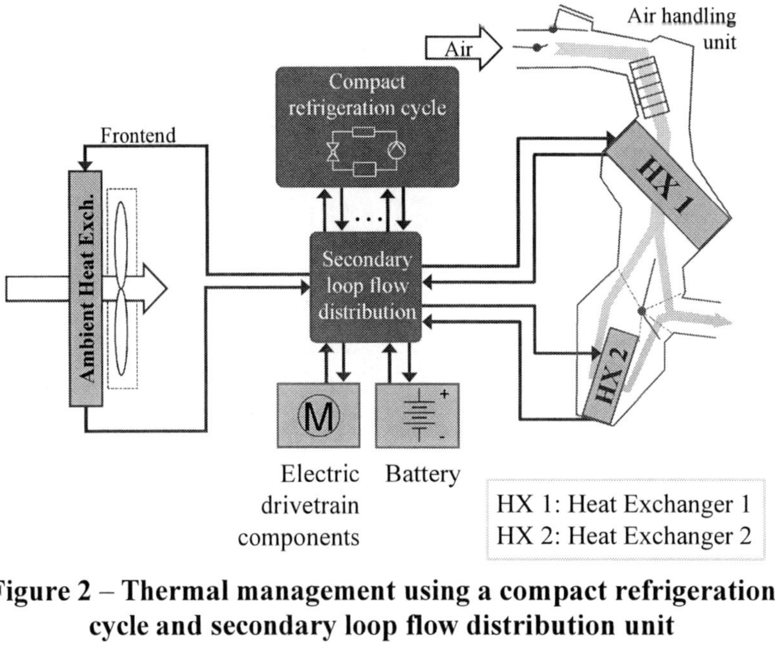

Figure 2 – Thermal management using a compact refrigeration cycle and secondary loop flow distribution unit

1.1 Electric Drivetrain

Thermal losses of electric drivetrain components result in heating of components. This could potentially lead to overheating. To avoid overheating a cooling system ensures that the maximum component temperatures are not exceeded. In case of the battery a given minimum component temperature can also require heating. Figure 3 illustrates a simple scenario for common temperature limits: While power electronics and electric machine usually tolerate a wide temperature range, the battery temperature must be maintained within narrow temperature bounds. This can necessitate to heat the battery if its temperature is too low. In general, typical batteries operate best between 10°C and 30°C [5].

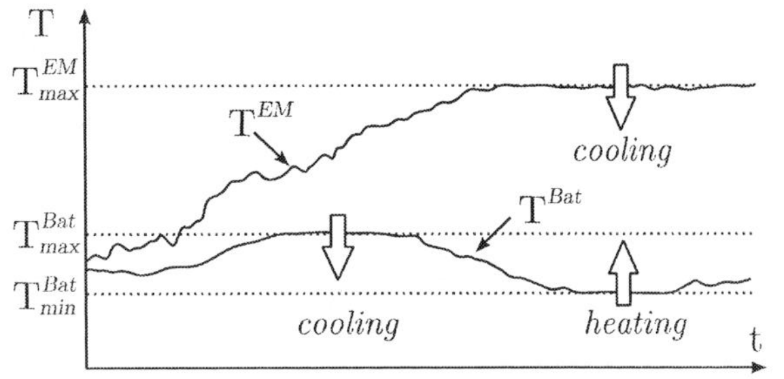

Figure 3 – Component cooling and heating

1.2 Air conditioning components

Usually the air handling unit is composed of a heat exchanger dedicated for cooling and a heat exchanger dedicated for heating, a blower and various air flaps. This allows heating and cooling of the passenger compartment as well as dehumidification of inflow air in a process called reheat. During reheat air is first cooled close to 0°C in order to expel saturated water and afterwards reheated to comfort temperature. Contrary to conventional A/C systems the described system uses a secondary loop heat exchanger as cooling device. Therefore, no refrigerant is entering the interior of the vehicle. This is done by providing cold and hot coolant for cabin heating and cooling.

1.3 Primary loop cycle

A refrigeration cycle is able to move heat from a lower temperature level to a higher temperature level using mechanical work. It can therefore be used as a refrigeration device for cooling or a heat pump device for heating depending on the perspective. In automotive application the refrigeration cycle is mainly used for air conditioning. However, for electric vehicles in many cases waste heat is only available at lower temperature level because the overall drivetrain efficiency is much higher compared to conventional vehicles. Therefore, both heating and cooling are required applications for a refrigeration cycle.

The vapor compression refrigeration cycle contains an electrical driven compressor, an (indirect) condenser, an expansion valve, an (indirect) evaporator and a refrigerant accumulator. Using a simple and compact refrigeration cycle allows reduction of refrigerant [6] and hermetic encapsulation. "Indirect" refers to the fact that the compact refrigeration cycle only provides hot and cold coolant as secondary fluid. This means that all connection logic has to be implemented by a coolant distribution system. In order to identify the necessary topological configurations, this work focuses on simulation of an electric vehicle thermal management system based on thermal energy flow rate between subsystems using simplified models.

2 METHODOLOGY

All simulations are carried out using a simplified powertrain model, similar to [7], to calculate powertrain losses and an energy flow rate based thermal management model based on first order thermal masses and time-dependent solution of the energy balance equations in MATLAB/Simulink. HVAC requirements are calculated by assuming heating and cooling loads of a standard size passenger car with comparable comfort requirements as a conventional vehicle heated by heat losses from a combustion engine. Simulation of (temperature dependent) drivetrain losses is decoupled from the simulation of the thermal management system. Some models might be unsuited for general statements on energy consumption due to oversimplification. The main focus here is to describe required heat flow on a fundamental level. This justifies the use of such simplified models.

The simulation is performed for two different scenarios with different thermal management requirements: A winter scenario at average temperature of -3°C and a summer scenario at average temperature of 31°C. In both cases the battery is charged for 45 min. After 60 min a Motorway-150 CADC (Common Artemis Driving Cycle) is repeated three times resulting in a drive cycle of 53 min at 100 km/h on average.

3 SIMULATION MODEL

A compact electric vehicle with a range of 150 km and an average energy consumption of 15 kWh/100 km is assumed. Basic vehicle parameters are summarized in Table 1.

Table 1 – Assumed vehicle and environment parameters

Parameter	*Unit*	*Description*	*Value*
		Vehicle	
m_v	[kg]	vehicle mass	1750
f	[-]	friction coefficient	0.01
r_{tire}	[m]	tire radius	0.3125
c_w	[-]	air drag coefficient	0.27
A	[m^2]	frontal area	2.23
n_{gear}	[-]	transmission rate	5.66
		Environment	
ρ	[kg/m^3]	air density	1.0133
v_w	[m/s]	wind speed	0
g	[m/s^2]	standard gravity	9.81

The simulation model is broken down into three parts: Simulation of drivetrain losses, simulation of HVAC requirements and simulation of the thermal management system.

3.1 Drivetrain model and thermal losses

Figure 4 shows the powertrain components of an electric vehicle, including one or more electric machines (EM), power electronics (PE) and a high-voltage battery. All of these components have in common that their efficiency characteristics and therefore thermal losses vary depending on component temperatures and power requirements.

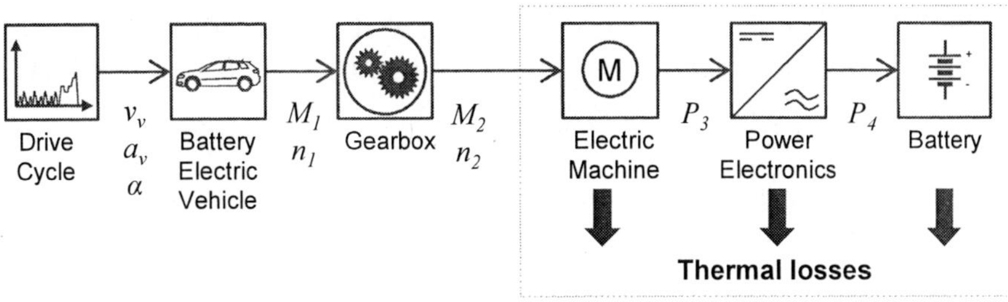

Figure 4 – Overview drivetrain model

The power requirements for each component are calculated based on the power requirements of the previous component. Feedback between components, that ensures that power requirements by one component can be met by the next component, is not considered. Plausibility checks show that for the considered drive cycles each component is able to deliver the required power. Vehicle speed v_v, acceleration a_v and slope α are given by the drive cycle. The overall drive resistance force is calculated by the sum of drive resistance forces [8]:

$$F_{drivetrain} = \underbrace{f \cdot m_V \cdot g \cdot \cos(\alpha)}_{F_{roll}} + \underbrace{m_V \cdot g \cdot \sin(\alpha)}_{F_{elev}} + \underbrace{c_w A \tfrac{\rho}{2} (v_V + v_W)^2}_{F_{air}} + \underbrace{m_V a_V}_{F_{acc}}$$

where F_{roll} is rolling resistance, F_{elev} is resistance caused by road slope, F_{air} is air resistance and F_{acc} is resistance due to acceleration or deacceleration. Mechanical power requirements are given by $P = F_{drivetrain} \cdot v_v$. For tire radius r_{tire} torque $M = F_{drivetrain} \cdot r_{tire}$ and rotational speed $n = 2\pi \cdot r_{tire}$ are then calculated. A fixed transmission n_{gear} of 5.66 is assumed resulting in a higher rotational speed and reduced torque. Based on efficiency maps $\varepsilon = f(n, M)$ for a synchronous electric machine and power electronics of a 100 kW drivetrain the required power drawn from the battery is estimated (Figure 5 a/b). Values for vehicle parameters m_v, f, c_w, A and environment assumptions g, ρ, v_w are given in table 1.

Based on the power each component provides $P_{provided}$ and the efficiency ε, (thermal) power loss P_{loss} (Q_{loss}) and required power $P_{required}$ are calculated.

$$P_{loss} = (1 - \varepsilon)/\varepsilon \cdot P_{provided} = Q_{loss}$$

$$P_{required} = 1/\varepsilon \cdot P_{provided}$$

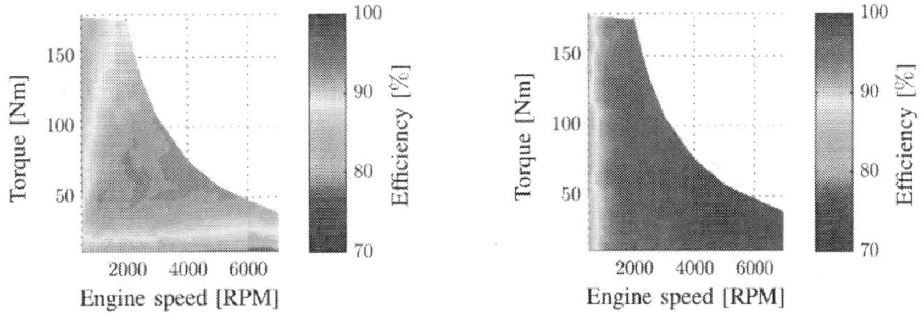

(a) Electric Machine (b) Power Electronics

Figure 5 – Efficiency maps and thermal power losses within the powertrain [9]

A 26.5 kWh lithium-ion battery pack with 3 parallel stacks of 84 series-connected cells is assumed (table 2). For the battery an equivalent circuit-based battery model is used

$$V_o = V_{oc} - I_d R_{int} = f(SOC) - I_d R_{int}(T_{Bat}, SOC)$$

where output voltage V_o depends on open circuit voltage V_{oc} and an equivalent internal resistance R_{int}.

Table 2 – Assumed battery parameters

		Battery	
Parameter	*Unit*	*Description*	*Value*
Type		Lithium-ion 3p84s	
C	[kWh]	Capacity	26.5
SOC_0	[-]	Initial state of charge	0.65

While internal resistance depends on average cell temperature T_{Bat} and state of charge SOC, cell temperature in turn, varies with battery internal state and external stress [10]. Any battery operation generates heat due to internal resistance when the cell delivers power to electric loads or is charged. These losses can be calculated from

$$P_{loss} = {I_d}^2 R_{int}.$$

Moreover, additional auxiliary loads of 450W are assumed for e. g. ECUs, lighting, entertainment. The charging device efficiency ε_{chrg} is assumed to be constant at 0.92.

3.2 Thermal Component Model

A simplified lumped first order model is used. The component temperature $T_m(t)$ is given by the differential equation

$$mc_{p,m}\, dT_m(t)/dt = Q = Q_{dis} + Q_{add/rem} + Q_{loss}$$

where $mc_{p,m}$ is the thermal mass of the component, $Q_{add/rem}$ is added or removed heat by the thermal management system and Q_{loss} is thermal loss of the component caused by non ideal operation. Dissipation losses to the environment Q_{dis} are estimated with overall heat transfer coefficient k and area A from

$$Q_{dis} = kA \cdot (T_\infty - T_m).$$

The temperature of the thermal mass T_m has to stay within given boundaries:

$$\mathrm{T}_{min} \leq \mathrm{T}_m \leq \mathrm{T}_{max}$$

Temperature boundaries, thermal masses mc_p and heat transfer coefficients kA are given in table 3.

Table 3 – Allowed temperature range, thermal mass and heat transfer coefficient

Component	T_{min} [°C]	T_{max} [°C]	mc_p [J/K]	kA [W/K]
Electric machine	-30	65	60000	1
Power electronics	-30	85	2000	3
Battery	-5	35	130000	3
Charging device	-25	70	2000	4

3.3 Cabin heating and cooling demand

Cabin heating and cooling demands are calculated based on a first order cabin model. Only loss to the environment Q_{loss} and heating or cooling effort Q_{hvac} are considered.

$$Q_{loss}(t) + z \cdot Q_{hvac}(t) = mc_{p,cab} \frac{dT_{cab}(t)}{dt}$$

$$Q_{loss}(t) = kA_{cab}(T_{cab}(t) - T_{amb})$$

Equivalent thermal capacity of the cabin $mc_{p,cab} = 123\,kJ/K$ and heat transfer coefficient between cabin and environment $kA_{cab} = 90\,W/K$ are estimated based on simulation data from a higher order simulation model. Mass flow in and out of the cabin is captured implicitly. During "ignition on" (z = 1) the heating and cooling power Q_{hvac} that will drive T_{cab} towards the comfort temperature $T_{comfort}$ is modeled by a PI controller

$$Q_{hvac}(t) = K_p(e(t) + \frac{1}{T_i}\int_0^t e(\tau) \cdot d\tau)$$

with error $e(t)$ defined as

$$e(t) = T_{cab}(t) - T_{comfort}(T_{amb}, \Phi_{amb}).$$

Controller parameters are chosen to meet typical dynamic heating and cooling requirements. Different gains K_p for heating and cooling ($K_{p,heating} = 286$ and $K_{p,cooling} = 212$) are chosen to consider different dynamic behaviors during cooling and heating. The integral time is set to $T_i = 1400\,s$. Based on ambient temperature, humidity and solar radiation the comfort target temperature $T_{comfort}$ and the coolant side temperature level T_{cab} at which heating or cooling is provided are calculated using a model of an air conditioning control unit.

$$T_{comfort} = f_1(T_{amb}, \Phi_{amb})$$

$$T_{cab} = f_2(T_{amb}, \varphi_{amb}, \Phi_{amb})$$

3.4 Thermal management strategy

The thermal management concept is based on thermal energy flow. At each time step the optimal energy flow is calculated assuming steady state. Heat is transferred between components and coolant cycles. Each coolant cycle is balanced to net power of 0. Coolant cycles are divided into three sub cycles: A **heating cycle (HC)**, corresponding to the hottest system temperature – usually the cabin heater, a **cooling cycle (CC)** which is connected to the ambient heat exchanger and a **low temperature cycle (LC)** corresponding to the lowest system temperature. All powertrain components (on board charging device, battery, power electronics, electric machine) can be included in the **cooling cycle** for "passive cooling" or in the **low temperature cycle** for "active cooling". Electric machine and power electronics can also be included in the **heating cycle** as heat source for cabin heating depending on the current temperature level and heating/cooling requirements. It is assumed that the thermal management will only actively interfere when maximum or minimum allowed temperature are reached.

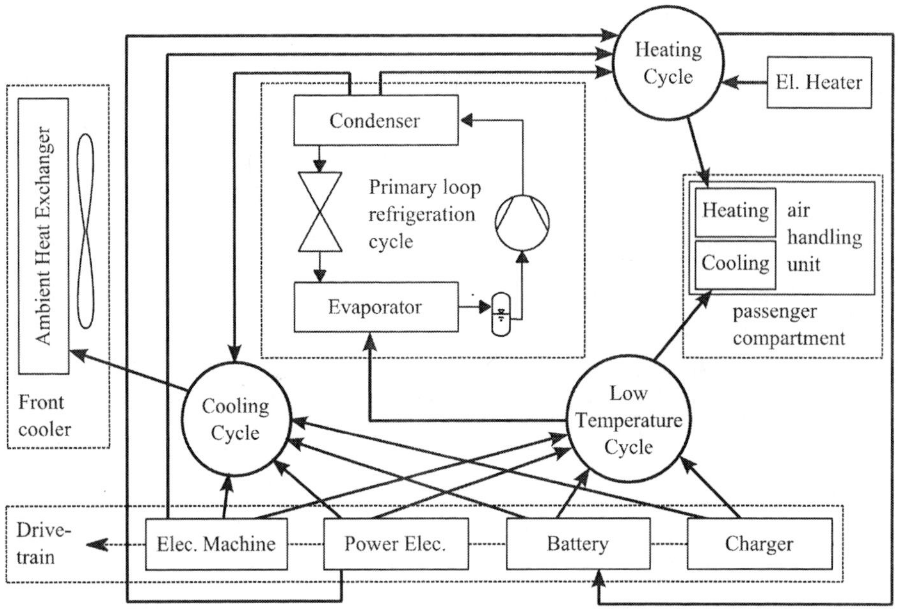

Figure 6 – Possible energy flows between components and coolant cycles

During passive cooling (not to be confused with "passive" air cooling) coolant circulates between components that must be cooled and an outside air heat exchanger. During active cooling the refrigeration cycle is used as heat sink. Active cooling allows cooling of components below ambient temperature T_{amb}. Active cooling is also required for air conditioning by providing coolant below ambient temperature for the cabin cooler. In heat pump mode the low temperature cycle serves as heat source for heat at low temperature, which is then transferred to a higher temperature level.

The thermal management system remains without further specification of a technical realization which would consist of pumps, valves, heat exchangers, heaters, expansion devices and a compressor. A purely energy flow based approach allows to derive basic understanding of the thermal demands of an electric vehicle being true for a wide range of systems in contrast to an approach solely suited for a specific system. In order to capture essential effects of thermal management, the following assumptions are made:

- Each component can only be associated with one or no coolant cycle.
- The cooling cycle is always associated with the ambient heat exchanger.
- The condenser of the compact refrigeration cycle (HOT) is either connected to the heating cycle or to the cooling cycle.
- The evaporator (COLD) is always connected to the low temperature cycle.
- Cabin heater (CAB+) and cabin cooler (CAB-) are always connected to heating and cooling cycle respectively.
- When sufficient heat is available by thermal losses of components inside the powertrain, heat from the hottest component is used for cabin heating.
- A refrigeration cycle (A/C) is used for cabin cooling or (active compressor) battery cooling. Abundant waste heat is dissipated to the environment by the front cooler.
- When component temperature is high enough, waste heat is directly used for cabin heating if necessary.
- No heating/cooling is done until min./max. allowed temperature is reached.

Table 4 summarizes the criteria.

Table 4 – Criteria for assigning a component to a certain coolant cycle

	BAT Battery	**EM** El. Machine	**PE** Power. Elec.	**CHRG** Charger	**CAB+** Cabin heat.	**CAB-** Cabin cool.	**HOT**	**COLD** ← COP →	**AMBX** Cooler
Heating Cycle (HC)	Active battery heating $T_{Bat} < T_{min}$	Heating with EM $T_{HC} < T_{EM}$ *deficiency HC*	Heating with PE $T_{HC} < T_{PE}$ *deficiency HC*	Heating with CHRG $T_{HC} < T_{CHG}$ *deficiency HC*	Heating (reheat) $Q_{heat} > 0$		heat source (heat pump)		
Cooling Cycle (CC)	Passive Cooling $T_{Bat} > T_{max}$ $T_{Bat} > T_{amb}$	Passive cooling $T_{EM} > T_{max}$	Passive cooling $T_{PE} > T_{max}$	Passive cooling $T_{CHG} > T_{max}$			heat source (A/C)		heat sink
Low Temp. Cycle (LC)	Active cooling $T_{Bat} > T_{max}$ $T_{Bat} < T_{amb}$	Heat pump *deficiency LC*	Heat pump *deficiency LC*	Heat pump *deficiency LC*		Cooling (reheat) $Q_{cool} > 0$		heat sink	
None	Circulation $T_{min} < T_{Bat} < T_{max}$	Circulation $T_{EM} < T_{max}$	Circulation $T_{PE} < T_{max}$	Circulation $T_{CHG} < T_{max}$	Cooling only $Q_{heat} = 0$	Heating only $Q_{cool} = 0$			

Required heating and cooling demand for each component is calculated at every time step from

$$Q_{rem} = mc_{p,m}(T_m - T_{max}) \text{ when } T_m > T_{max}$$

$$Q_{add} = mc_{p,m}(T_m - T_{min}) \text{ when } T_m < T_{min}$$

Dynamic behavior is derived from the steady state thermal management model by only transferring 10% of the required heating or cooling demand. This can results in violation of the temperature constraints (figure 7), but gives results closer to a real system. The primary loop cycle is used for balancing missing heating and cooling demand. The mutual relationship between heating power Q_{hot}, cooling power Q_{cold} and compressor work P_{comp} for the refrigeration cycle at steady state is given by the steady state coefficient of perfomance (COP):

$$Q_{hot} = COP \cdot P_{comp} \text{ and } Q_{cold} = (1 + COP) \cdot P_{comp}$$

$$Q_{hot} = COP/(1 + COP) \cdot Q_{cold}.$$

Therefore, a given cooling Q_{cold} or heating load Q_{hot} (depending on demand) will result in a certain heating or cooling power that must be considered. Primarily the COP depends on the temperature levels between which the refrigeration cycle is operating. For this work a simplified linear relationship is used

$$COP = 3.509 - 0.0494 \cdot T_c + 0.0461 \cdot T_e$$

where T_c is the inlet temperature of the condenser and T_e is the inlet temperature of the evaporator. Coefficients are derived from steady state measurements performed with a comparable refrigeration cycle.

4 RESULTS

Simulation results for the hot summer scenario are shown in figure 7. The first graph shows the vehicle velocity with time frame for charging and driving. While driving the "ignition on" signal indicates the start and end of the drive cycle. The next three graphs show the ambient conditions with 31°C temperature, 36% relative humidity and 329 W/m^2 sun radiation.

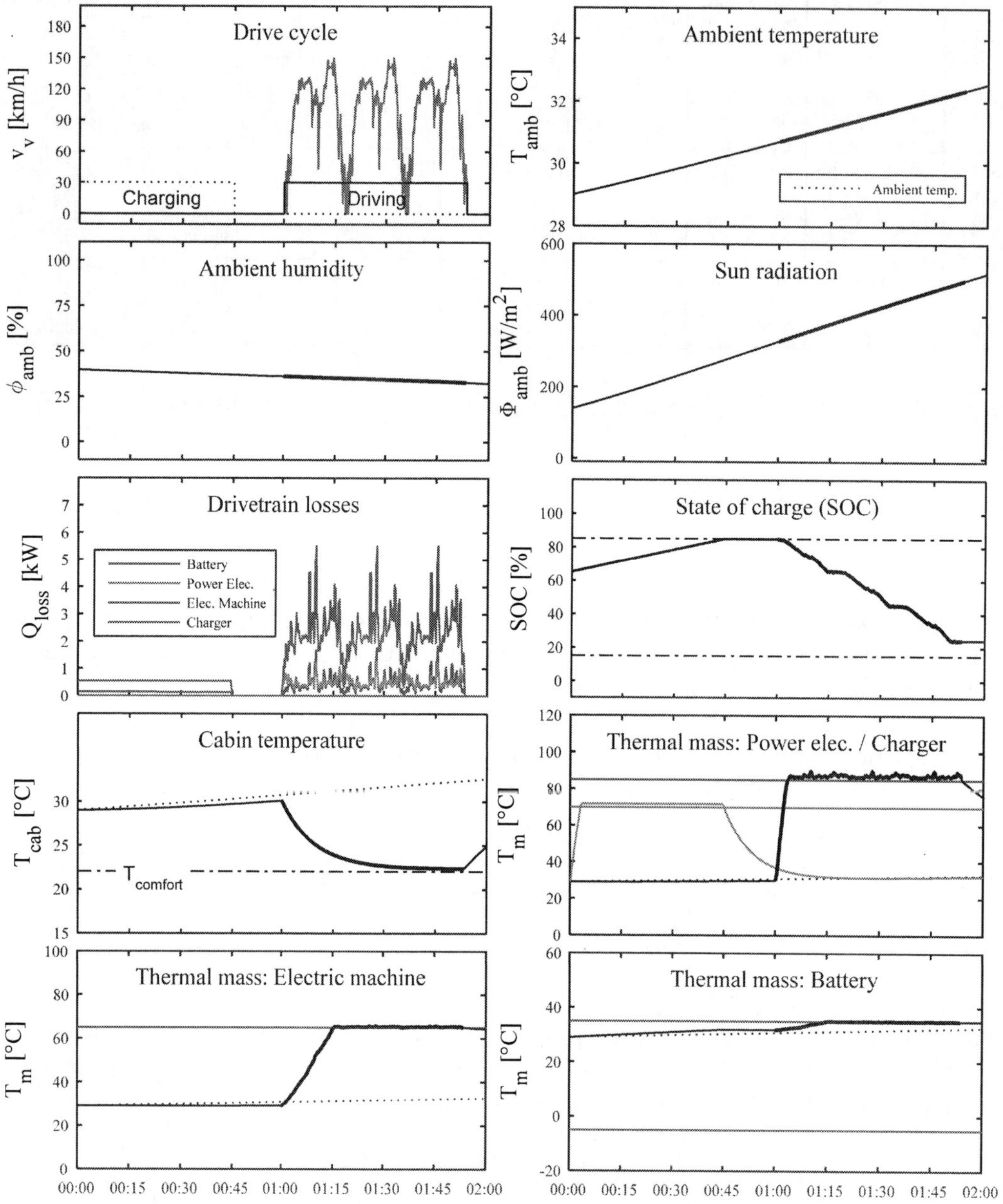

Figure 7 – Simulation results for the summer scenario

Thermal power train losses for battery, power electronics, electric machine and charging device are shown in the next graphs. Next the state of charge (SOC) is shown. After charging the maximum allowed amount of energy is stored in the battery at a SOC of

85%. During the drive cycle the battery is discharged to approximately 24%. Subsequently cabin temperature (together with ambient temperature and comfort temperature) is shown. The last three graphs comprise the temperature of the drivetrain components. Maximum/minimum allowed temperatures are indicated as well. Starting at approximately 25°C all components reach their maximum allowed temperature with power electronics first, then electric machine, then battery. The charging device only heats up during the 45 min charging period. During that period the charging device is cooled most of the time.

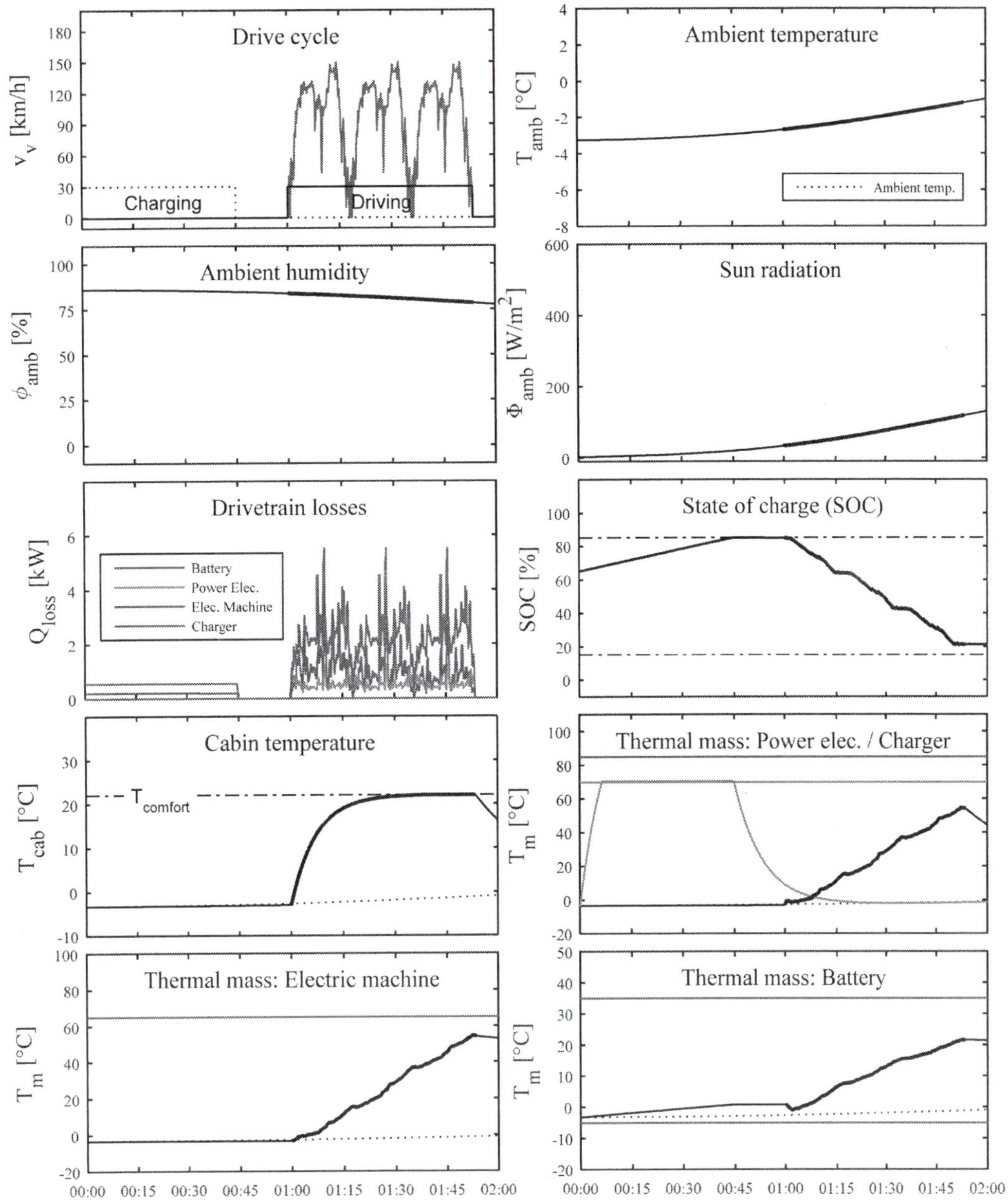

Figure 8 – Simulation results for the winter scenario

When the maximum allowed temperature is reached, the thermal management system draws heat from each component so that component temperature stays at maximum temperature. Simulation results for the cold winter scenario are presented in the same order (Figure 8). It can be seen that component heat up is delayed. This is mainly due to cabin heating requirements that result in a utilization of waste heat either as low temperature heat source for the heat pump or as heat source that can be used directly at high temperature level.

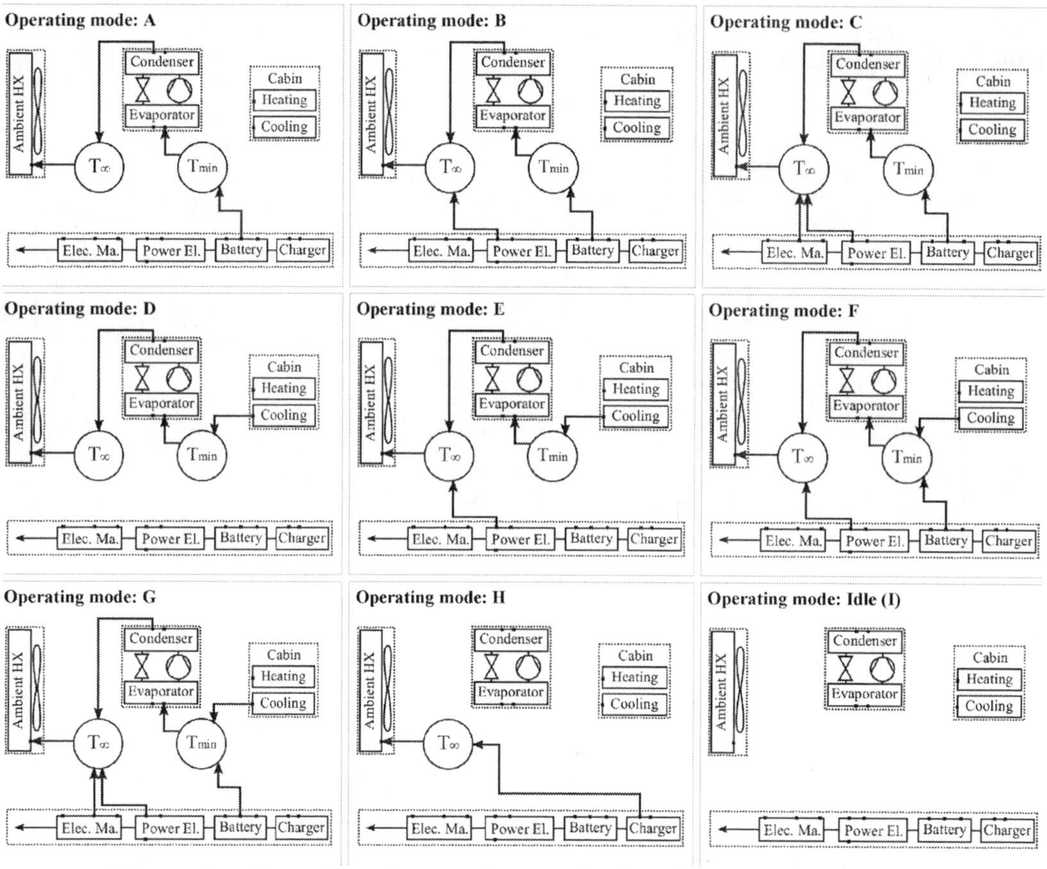

Figure 9 – Overview of operating modes for the summer scenario

Different cooling requirements and component conditions result in different thermal energy flow requirements during the drive cycle. These operating modes are shown in figure 9. Figure 10 shows the active operating mode including cooling of the battery charger between 0:00 and 0:45. The drive cycle itself is highlighted as "ignition on".

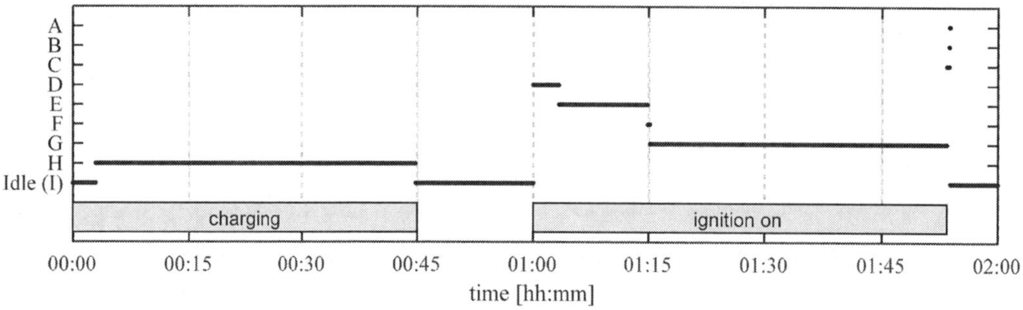

Figure 10 – Operating modes over time summer scenario

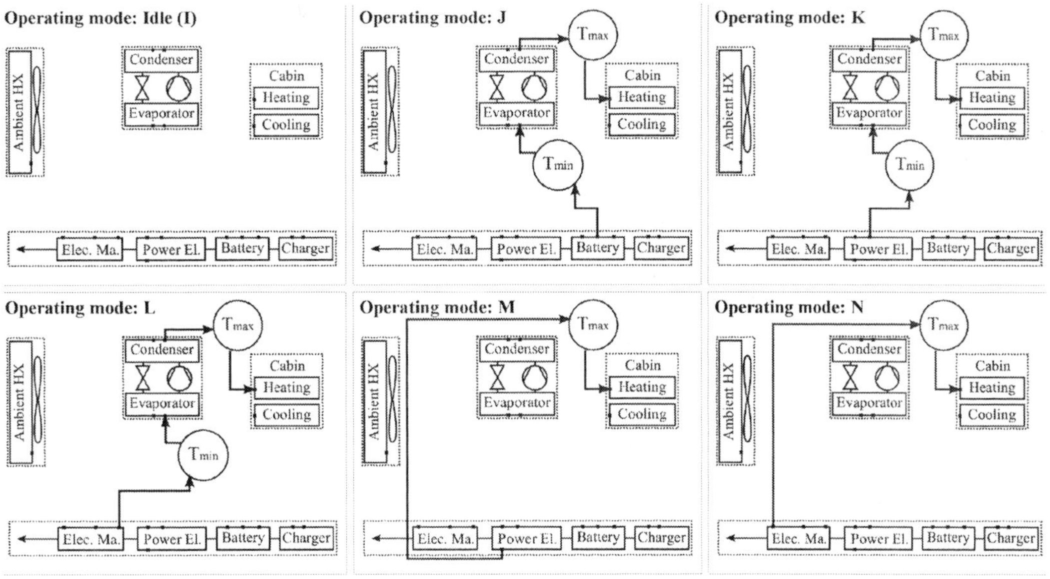

Figure 11 – Overview of operation modes for the winter scenario

Accordingly figure 11 and 12 show required thermal energy flows and operating modes for the winter scenario. Cooling of the battery charging device is included as well (operating mode H). Operating mode I for both scenarios represents an idle case where no thermal management is required.

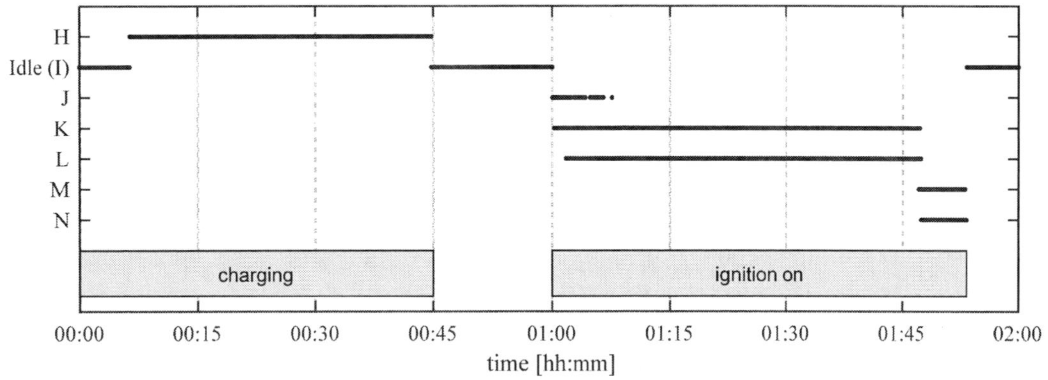

Figure 12 – Operating modes over time winter scenario

The percentage share of each operating mode is given in figure 13 and 14 for both scenarios. As well the total sum of active time for each operating mode is indicated. Idle mode (I) is excluded from the calculation of percentage share.

During the summer drive cycle operating mode A, B, C and F are relatively rare with accumulated active time below 1 min each. Operating mode G is mainly observed. It represents cooling at very high load. Beside active battery and cabin cooling, cooling of power electronics and electric machine is required. During the winter drive cycle different usage of components can be observed. For example power electronics and electric machine are used as low temperature heat source first (K, L) and later as heat source at high temperature (M, N). For both scenarios charger cooling is the most common operating mode due to a long charging period.

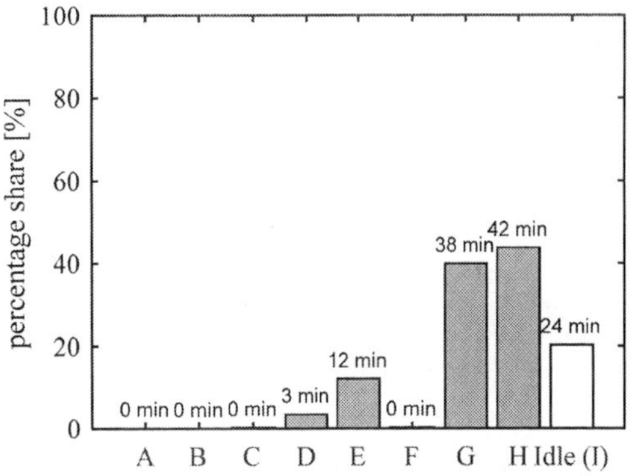

Figure 13 – Histogram summer scenario

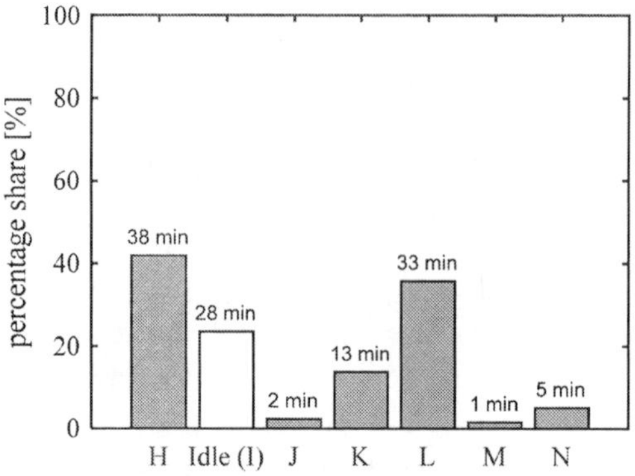

Figure 14 – Histogram winter scenario

5 DISCUSSION

The results indicate that many topological configurations can be useful and are practical use cases for vehicle thermal management. However, reduction of complexity should be an important goal. Therefore, not every operating mode must or should be implemented. The results clearly show that some operating modes are of higher importance compared to others. Especially operating modes with low frequency of occurrence (e.g. A-C, F, J and M) must be checked for necessity. Also not every energy flow rate configuration must result in an individual piping. For example in most cases (e.g. heat pump mode) power electronics and electric machine can be encapsulated in one reconcilable unit.

6 CONCLUSION

Based on simplified models drivetrain heat losses and thermal requirements for cabin comfort are estimated. Using first order thermal mass models and typical component temperature constraints, energy surpluses and energy deficits are calculated. A heat flow rate based thermal management strategy then transfers heat between components and three coolant cycles. Simulation results for a realistic winter and summer scenario are presented. Up to 14 useful operating modes are identified and importance is prioritized by calculating the total active time of each operating mode. The results remain without further specification of a heating and cooling system design (including pumps, valves, heat exchangers and electric heaters). Obviously, the provided results depend on the assumptions that are made. Thermal capacity of components, different driving behavior and drive cycles will result in new operating modes.

The main goal of a heating and cooling system should be low cost and low complexity as more topological states result in higher switching time, more complex control schemes and finally more development and testing time. Therefore, more advanced studies should focus on overall system performance when selected operating modes are prohibited in order to reduce system complexity. Furthermore focus of further consideration will be a holistic evaluation including statistical profound driving behavior and weather conditions. Especially the number of operating modes will increase when reheating and more driving scenarios are included. However this will reflect the overall distribution of operating modes more accurately and result in more representative result for overall system energy consumption.

REFERENCES

[1] R. Farrington and J. Rugh, "Impact of vehicle air-conditioning on fuel economy, tailpipe emissions, and electric vehicle range," in Earth Technologies Forum, 2000.

[2] Ghodbane, M., Craig, T. D., and Baker, J. A. "Demonstration of an energy-efficient secondary loop HFC-152a moile air conditioning system," Final Report for the U.S. Environmental Protection Agency, 2007.

[3] Kowsky, C., Wolfe, E., Leitzel, L., and Oddi, F. "Unitary HPAC system, " SAE International Journal of Passenger Cars - Mechanical Systems, 5(2)., 2012.

[4] J. P. Rugh et. al., "PHEV/EV integrated vehicle thermal management – development of a KULI model to assess combined cooling loops," VTMS, 2011.

[5] A. A. Pesaran, "Battery thermal management in EV and HEVs: Issues and solutions," Advanced Automotive Battery Conference, vol. 43, no. 5, 2001.

[6] J. C. Menken, T. A. Weustenfeld, and J. Koehler, "Experimental comparison of the refrigerant reservoir position in a primary loop refrigerant cycle with optimal operation, " in 15th International Refrigeration and Air Conditioning Conference at Purdue, 2014.

[7] K. Wipke, M. Cuddy, and S. Burch, "Advisor 2.1: a user-friendly advanced powertrain simulation using a combined backward/forward approach, " Vehicular Technology, IEEE Transactions on, vol. 48, no. 6, pp. 1751–1761, 1999.

[8] Konz, M.; Lemke, N.; Försterling, S.; Eghtessad, M.: Spezifische Anforderungen an das Heiz-Klimasystem elektromotorisch angetriebener Fahrzeuge. In: Forschungsvereinigung Automobiltechnik e.V. (Hrsg.): FAT-Schriftenreihe 233, 2011.

[9] A. Enthaler, T. A. Weustenfeld, F. Gauterin and J. Koehler, "Thermal management consumption and its effect on remaining range estimation of electric vehicles," IEEE 3rd International Conference on Connected Vehicles & Expo (ICCVE), 2014.

[10] Eugene Kim, Jinkyu Lee and Kang G. Shin, "Real-Time Battery Thermal Management for Electric Vehicles," ICCPS'14, 2014.

A process for battery thermal design

E TATE[1], W JANSEN[2], I HUGHES[2], J HAN[1], A VELIVELLI[1] and Z YANG[1]
[1] Exa Corporation, USA
[2] Jaguar Land Rover, UK

ABSTRACT

Traction battery packs are one of the largest and most expensive components in an electrified powertrain. Their durability is reduced by hot spots and high average temperatures. Meeting durability targets and integrating a battery pack into a vehicle involves trade-offs. A process that starts with electrical characterization and progresses to vehicle thermal models is used. At each step, design issues are resolved, simplifying the next stage in the process. This process starts with cell models and concludes with a mix of system and 3D models at the vehicle level. The process allows exploration of alternatives and trade-offs in limited time.

1 INTRODUCTION

Vehicle electrification is one of the strategies automakers are pursing to decrease fuel consumption and CO2 emissions. Electrification can provide all of the energy to move a vehicle as in an EV, or be used to make a conventional powertrain more efficient as in a start-stop HEV. Between these extremes are a large range of powertrain options. While the powertrains in these vehicles may vary significantly, they all share a need to design and package a battery pack which provides the electrical energy. Aggressive electrification leads to greater reductions in fuel consumption, but presents greater design challenges.

One of the largest challenges with batteries is making sure they are durable. A battery pack may be one of the most expensive components in an advanced powertrain. Customers and regulators expect traction batteries to be as durable as other powertrain components[1][2]. However, unlike typical powertrain components, batteries have failure mechanisms which can be greatly accelerated by small differences in duty cycles and mildly high ambient temperatures[3][4]. Battery thermal protection schemes enhance pack durability, but typically reduce fuel savings, reduce performance, increase vehicle weight, and increase vehicle complexity.

Balancing these tradeoffs in a design requires the ability to predict electrical performance and temperatures in a pack over complex drive cycles. Prototyping and physical testing are used in a development cycle, but several factors make this undesirable. The primary challenge is that battery electrical and thermal behavior changes with charge, temperature, and age. Any physical test needs to control these factors to ensure repeatable and trustworthy results. Since batteries take appreciable time to reach a given charge and temperature, a significant fraction of test time can be consumed conditioning a pack prior to a test. Battery aging also affects test results. Battery aging is a complex phenomenon [3] that affects all batteries and cannot be reversed or directly measured in situ. This makes it difficult to consistently reproduce results with a test fleet. This problem is even greater in a development fleet where vehicles are used for multiple purposes. Additionally, early physical prototypes may contain immature control systems which do not properly guard batteries. This can lead to early pack failures while testing. Finally, evaluation of extreme operating conditions can stress and permanently damage a pack. Fortunately, using virtual batteries in simulation, these challenges are surmounted. In a simulation based process, setting the charge, temperature, and age for a test is simply a matter of selecting initial conditions for a simulation. This eliminates the conditioning time required in a physical test. Permanent damage to a pack is not an issue with virtual batteries so extensive testing in extreme conditions and with immature calibrations occurs without consequence. Additionally, if a virtual battery model is able to execute at real time or faster, it can be used in physical testing. Such a virtual battery, created by combining a power supply with a simulation, allows bench testing powertrains and thermal components using hardware-in-the-loop (HIL) physical test apparatus.

With these considerations in mind, a battery design process was developed based on 3D models in PowerFLOW, PowerTHERM and system models in Simulink®. The outputs of this process are thermal data and fast running simulations. The thermal data is generated at cell, module, and pack levels. It includes temperatures and heat flows. The fast running models are useful for both long duration system simulations and HIL physical testing.

2 PROCESS GOALS

The goals of this battery design process are to ensure a vehicle meets fuel economy and performance targets, while meeting durability and cost targets. The process needs to support evaluation of a battery pack at different vehicle locations, with different internal geometries, different electrical configurations, and different cooling mechanisms. Additionally, the process needs to generate fast models which can be used for HIL and long duration drive cycle simulation. These targets and needs are satisfied by breaking the process into multiple steps.

3 DESIGN PROCESS

The design process is illustrated in Figure 1. This process consists of three major steps. The process progresses from high level durability, fuel economy, and performance targets into simulations which verify that the pack design is capable of meeting these targets.

The first step in the design process is to set targets for the pack electrical and thermal behaviour. Durability targets are converted into peak temperature and temperature distribution targets at the pack level. Additional thermal targets may also be set at the module and cell level. Fuel economy and performance targets are converted into electrical targets and cooling power targets on drive cycles. By converting from vehicle and durability targets into pack electrical and thermal targets, the pack analysis task is simplified. By defining the design targets in terms of thermal and electrical behaviors, the pack can be evaluated without a need for a full powertrain model.

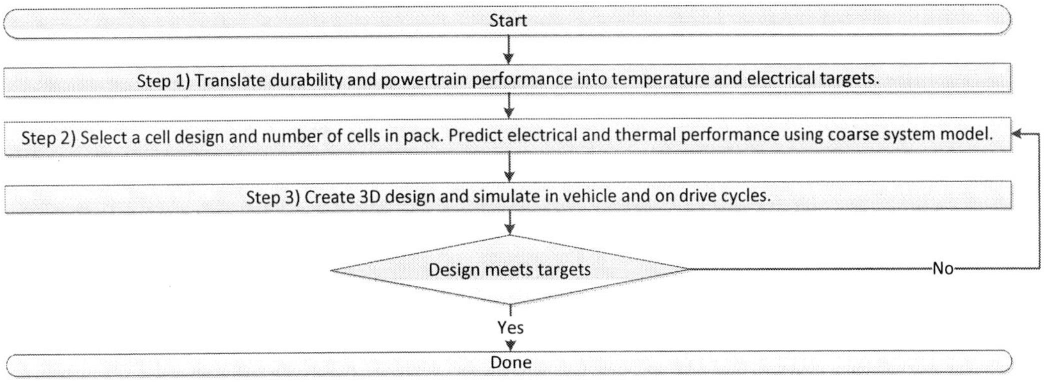

Figure 1 – Design process flow

The second step in the design process is to size the parts of a battery pack. This is done by starting with the electrical and thermal targets for the pack. The cell chemistry, the number of cells, and the module configuration are selected. These elements are combined with a model of the cooling system to create a system model which describes the behaviour of the battery pack electrically and thermally. A key point is that detailed geometry is not generated at this point. This behavioural model is generated to enable coarse predictions of electrical and thermal behaviour. This system model behaves like a final design is expected to behave. This model is sufficient to predict behaviour on multiple drive cycles when integrated with a vehicle model. It can establish if the decisions made at this point will likely achieve the electrical and thermal targets. A key aspect of this process is that this coarse model will be replaced later by a battery model generated from the 3D geometry.

One of the challenges at this step is determining if the behaviours in this coarse model can be realized. It is possible to generate a system model which cannot be built as a pack. For example, the heat transfer from the pack to the coolant may be overestimated. When modules are placed in the pack and integrated with the rest of the cooling system, there may be no combination of materials and geometry which is able to remove the required heat. If this is discovered after design options are exhausted, a high level specification for the vehicle performance or the pack must change.

In the third step of the design process, the pack geometry is generated. The geometry of the pack includes the cell, the module, cooling plates, and enclosure design. It also includes the placement of the pack on the vehicle. Since the key aspects of the pack were determined using a coarse model, it is important to verify that the detailed geometry satisfies the electrical and thermal targets. To verify electrical and thermal performance, analysis is performed in several steps involving 3D and system simulations. This outline of the analysis steps are illustrated in Figure 2.

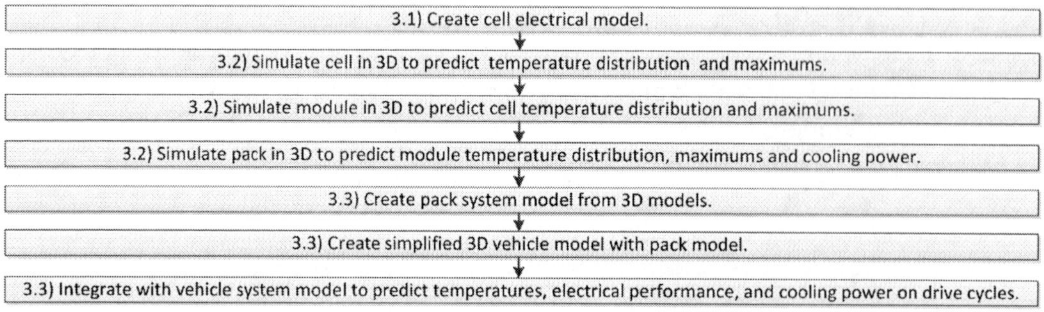

Figure 2 – Analysis and simulation steps

The analysis steps involve electrical characterization of the cell from physical test data, thermal analysis and characterization using PowerFLOW and PowerTHERM, and drive cycle simulation using Simulink. Additionally, the models which are generated for drive cycle simulation are also reusable for HIL testing. If analysis shows that the pack does not meet design targets, then alternatives such as geometry changes, operational changes, battery changes, and design target changes are considered. If the geometry is changed, then analyses are rerun.

3.1 Cell Electrical Characterization

In order to analyse the pack's thermal behaviour, an electrical model is required. An electrical model is used to predict the heat generation and the power capability of the pack. For this process, an electrical circuit model is used. This process uses a model which is compatible with the circuit model supported in PowerTHERM[5]. The losses in this circuit model are used to predict the heat generation. The heat generation due to an electrical circuit is illustrated in Figure 3. In this figure, heat is generated by the ohmic heating in the resistors and by entropic heat associated with the open circuit voltage. The capacitors are assumed lossless. Furthermore, the values associated with the elements in this model vary with the charge, temperature, and age of the battery.

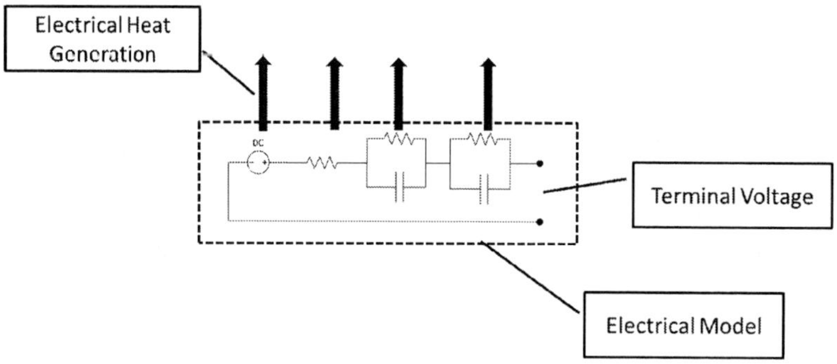

Figure 3 – Battery electrical model and heat rejection

This electrical model is developed from test data. If possible, the test data is structured to simplify identification of the equivalent circuit parameters. For example, measurements are taken of voltage and current, while pulses of various amplitudes and durations are applied to a battery. During pulses, the charge and temperature are maintained over a tight range.

For this process, the pulse amplitude and timing is selected to satisfy a couple of goals. The peak pulse amplitude is selected to be about as large as the maximum amplitude expected in operation. The pulses are in groups of positive and negative currents. This grouping allows better measurement of time constants in the battery. When the SOC is changed to prepare for another pulse train, there is a large period of time provided so the battery can settle into equilibrium. Ideally this rest period should be long enough to nearly reach equilibrium. However, in practice there is a tradeoff between ideal testing and the time to complete a test. Additionally, the data needs to be collected with minimal delay between each voltage and current sample. Large skews between voltage and current samples cause errors in estimation of resistance and time constants.

Figure 4 and Figure 5 illustrate structured current pulses and the resulting voltage measurements from this kind of test. Given this data, multiple methodologies are available to convert the data into a circuit model [6][7]. One unique aspect of batteries compared to traditional powertrain components is the sensitivity to temperature. Where a conventional powertrain will be slightly derated at extremes of temperatures, this is typically only for a short duration. In contrast, a battery may suffer significant reductions in electrical performance for long periods. These extended periods are due to the large thermal time constants which limit how quickly a pack can be brought into an optimal temperature range. Since battery performance typically changes with SOC and temperature, model parameters are obtained for a range of states of charge and temperatures. Because of the changes with battery age, this characterization may be applied to batteries of different ages.

One of the key assumptions in fitting an electrical model is that the measured behaviour can be predicted accurately by selection of available model parameters. Depending on cell chemistry and microstructure, a linear circuit like the one shown in Figure 3 may have unacceptable differences between measured and predicted voltages. If this occurs, alternative circuit models are used. Common modifications to circuit models include asymmetric resistance, nonlinear resistance, and hysteresis voltages[8]. The best alternative is chosen by selecting the simplest model with least error in matching test results.

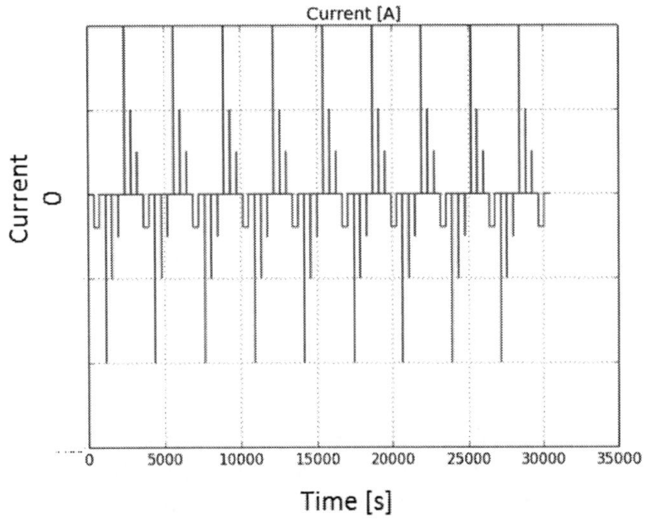

Figure 4 – Structured pulse data

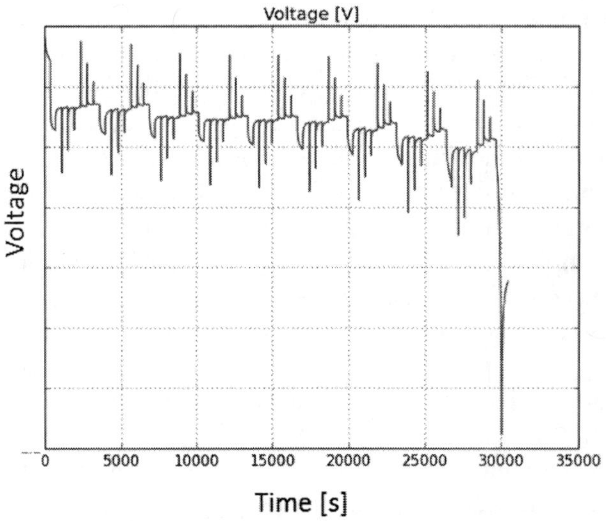

Figure 5 – Battery voltage response

3.2 3D Thermal Simulation

Once electrical behaviour is characterized, it is combined with the geometry and material properties to generate cell, module, and pack thermal simulations. The module model is created from the cell models. The pack model is created by combining modules with a cold plate, enclosure, wiring, and support hardware. This process is illustrated in Figure 6. At each level of model complexity, different questions are answered.

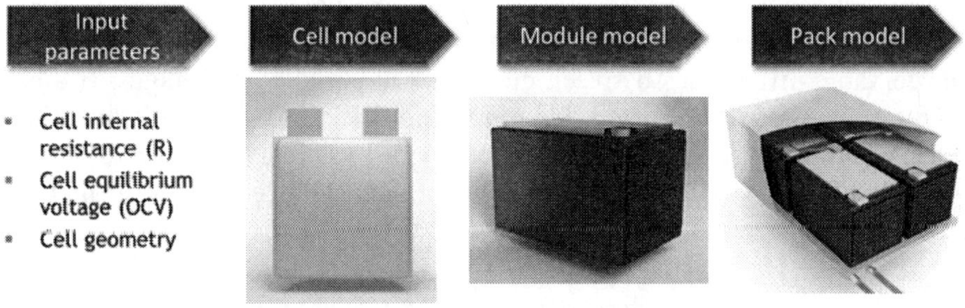

Figure 6 – 3D modeling progression

With a cell model, the uniformity of temperatures within a cell is evaluated. If parts of a cell are significantly hotter than the rest of the cell, those areas may fail earlier due to the higher temperatures. Since a cell failure leads to a pack repair, it is important to minimize these types of failure mechanisms. Hot spots can be reduced by design changes or usage changes. Design changes include selection of different materials and changes to cell geometry. Alternatively, the heating in the cell can be altered by changing electrical targets. This may be done by revising the number of cells in a pack or changing how the vehicle uses electrical power.

With a module model, the effect of heat spreaders on the cell temperature distribution is evaluated. Additionally, the distribution of temperatures among the cells is evaluated. A good module design exhibits nearly uniform cell temperatures from end to end. This is important because if cells at the ends or center of a module tend to operate at higher average temperatures, those cells will degrade faster and require premature pack repairs. If this is discovered, it can be mitigated by changing the heat spreaders within the module, modifying the electrical interconnect, adding heat spreaders external to the module or by changing the average electrical load on the module.

At both the cell and module levels, the behaviour is evaluated through simulation of multiple electrical duty cycles. One such cycle is illustrated in Figure 7. This is an example of a duty cycle derived from a vehicle level test. Figure 8 shows temperatures measured on a module and the simulation results from PowerTHERM. The maximum error between the measured and predicted values is less than one degree C. This shows the level of accuracy that can be achieved using electrical circuit models fit to physical tests and thermal models generated from geometry and material properties. Additional information on 3D battery modelling is found in [9].

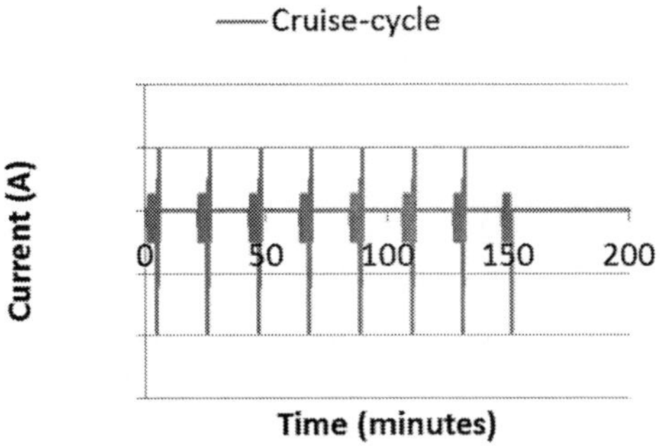

Figure 7 – An example electrical duty cycle

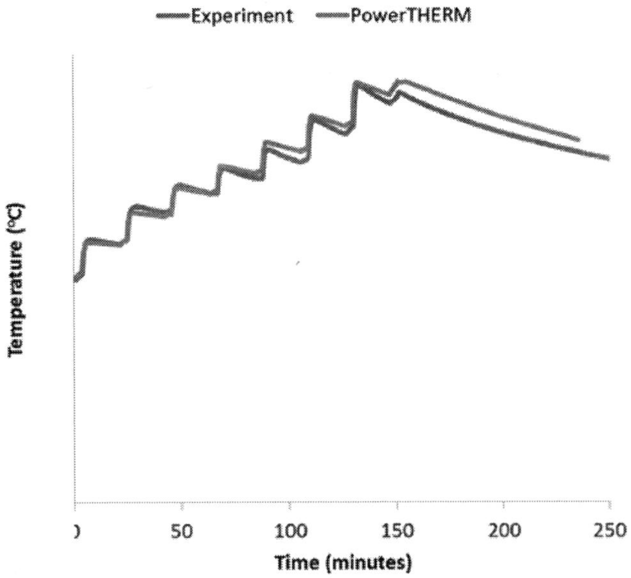

Figure 8 – Temperature prediction results

At a pack level, there are two approaches to generating a 3D model. One approach uses fully detailed modules. This provides high fidelity in a simulation and access to the most information about how the pack will behave. However, this level of detail can be a challenge because the complexity of a pack model can approach the complexity the rest of a model of the rest of a vehicle. Pack models can be complex because the each module is built from layers of cells and heat spreaders with electrical interconnect between each set of tabs. As an alternative to simulating all of the detail in a module, a reduced order model can be used.

In a reduced order model, the boundary conditions on the module are simplified and the heat equations for the internal geometry and materials are replaced by simplified systems of equations. These equations are developed to match the observed behavior of the boundary conditions and temperatures. Using a reduced order model, faster simulation results are obtained. A key advantage of this approach is that for the same amount of time and resources, more duty cycles, pack designs, and pack locations can be evaluated.

3.3 System Model Generation

System models are generated for three purposes. One is to provide reduced order models which can be combined with a 3D model for faster execution as discussed in section 3.2. A second purpose is to generate a model of the pack on the vehicle to combine with other system models and to predict the behaviour of the entire vehicle on a drive cycle. A third purpose is to generate models which run at real time so they can be combined with HIL test stands to replace physical battery packs. In all of these cases, the key is simplification of 3D model into a system of ordinary differential equations (ODEs). Once a model is in this form, it can be simulated in a system modelling tool like Simulink®.

Reducing the number of equations typically involves changing models in three ways[7]. First, the degrees of freedom in boundary conditions are reduced. For example, in a 3D model, the surface temperature on the side of a battery may have 10's or 100's of values which represent the temperature at different points on the surface. In a system model, this boundary may be reduced to a single value. Second, the system model only predicts specific temperatures. These specific temperatures may correspond to thermocouples or other key temperatures. Third, a trade-off is usually applied. Some additional error is accepted in order to eliminate equations. When these conditions are applied, a simplified model which executes 1000's of times faster with very small errors can be generated. To execute a reduced model in a system simulation tool like Simulink®, the equations are changed from a temporally and spatially discretized partial differential equation (PDE) into a continuous time ODE.

For example, in one test case, a battery model in PowerTHERM was reduced from more than ten-thousand equations to less than 10 equations. This reduction in the number of equations and variables increased the execution speed by a factor of ten-thousand. This speed up was accomplished while keeping the difference in the predicted temperatures between the 3D and system model to 0.1C on a 20C step change in temperature. This example shows that fast and accurate models can be generated from a 3D model. However, this performance improvement is obtained by simplifying boundary conditions and only simulating specific measurements. Fortunately, in a typical battery pack, the design goal of nearly uniform temperature distribution leads to relatively uniform boundary conditions. In fact, if a 3D cell analysis established that hot spots were within limits, and a 3D module analysis established the cell temperature variations were acceptable, then the information lost to these system model simplifications will not result in missing important pack issues.

3.4 Combined 3D and System Models for Improved Performance

For some attribute evaluations, neither a system model nor a 3D model provides satisfactory results. A system model offers very fast execution, but since details are omitted, some attributes of a design are not available in this form. A 3D model offers high spatial resolution, but is often not practical to use to simulate long duration tests. In these two cases, a 3D model and a system model are combined. This involves replacing a part of a 3D model with a system model which maps boundary conditions to system model equations. By a combining simplified equations and boundary conditions, a faster 3D model is created. If the original 3D model has relatively uniform boundary conditions where the system model is substituted, then the loss in fidelity is minor.

To see how this works, consider a pack model with several modules as illustrated in Figure 9. In this model, the modules represent approximately 90% of the elements in the pack model. Replacement of the modules with system models reduces the number of elements proportionally. The equations in the system model are trivial compared to the number of equations they replace.

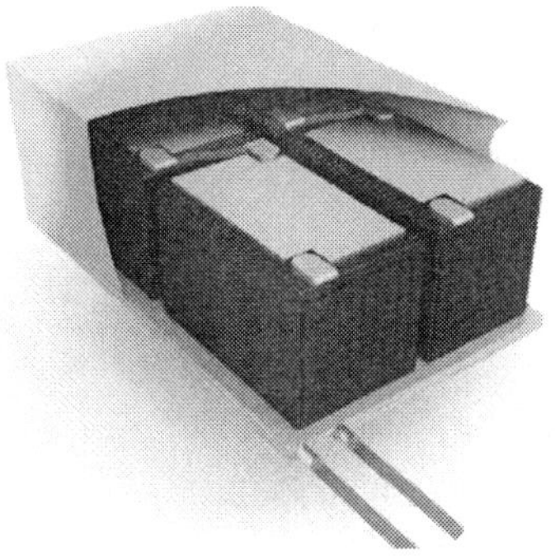

Figure 9 – Pack 3D model

3.5 Vehicle Level Simulation

Once the pack design is selected, one of the remaining questions is how the placement within the vehicle affects battery durability and the cooling power requirements. In general, there are several regions on a vehicle where a pack can be located. Depending on the vehicle, a pack may be located in the passenger compartment, in the trunk, or under the body. Each of these regions has unique challenges because of different sources of heat that affect the pack. If the pack is located in the cabin, solar loads will affect it. If the pack is located in the trunk, hot ambients in the cabin will affect it. If the pack is located under the vehicle, it may be subject to hot spots from the vehicle exhaust, heat rejection from the engine during thermal soaks, and hot road surfaces. Figure 10 illustrates some common locations on a vehicle for a pack.

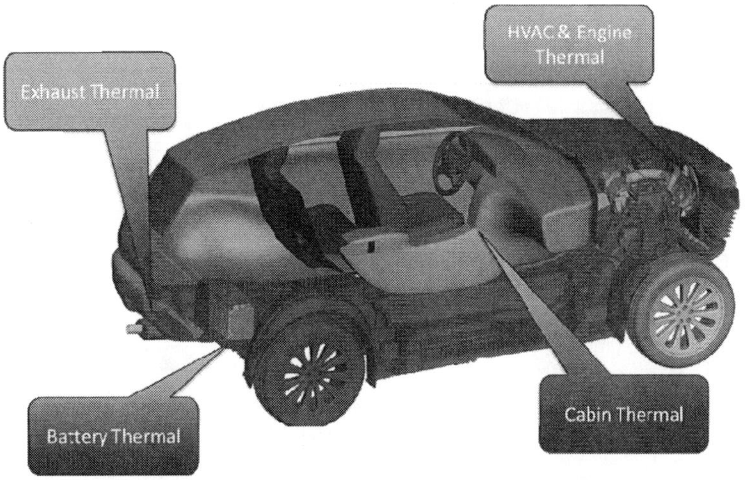

Figure 10 – Potential pack thermal interactions

Accurate 3D simulation involves extending established best practices for simulating thermal protection problems in areas where a pack may be located. For example, if a pack is located in a cabin, then the best practices for cabin soak and cool down provide a basis for the pack simulation in a vehicle. If a pack is located under body, then exhaust thermal protection best practices provide a basis for pack simulation in a vehicle. These best practices help ensure that the most representative boundary conditions are applied to the pack enclosure, while executing a full vehicle simulation.

3.6 Drive Cycles

The final check on a pack design is how it performs during a drive cycle. This performance includes the peak temperatures within the pack and the energy consumption for cooling. Since the time constants for batteries are large, transient evaluation requires long drive simulations. In this process, a system model for the pack is generated from the 3D models. This system model is then integrated into a full vehicle system model. The full vehicle model establishes enclosure, electrical and cooling system boundary conditions. The pack system model simulates heat transfer within the pack, heat generation from electrical operation, and heat rejection into the cooling system. Because of the speed of this fast model, it can be used to evaluate many drive cycles in a very short time frame.

Real world usage of a battery typically involves very complex electrical excitations. Figure 11 illustrates this kind of signal. In this example, the signal involves many short duration charge and discharge pulses. Unlike laboratory testing, the amplitudes and durations of pulses while driving are irregular. However, using an equivalent circuit with different values for positive and negative currents, an electrical voltage prediction error of less than 1% was achieved on battery chemistries of interest. This voltage prediction error was almost unbiased with a mean error of less than 0.001% on a drive cycles. Since the predicted heat generation in the battery is calculated from the losses in the equivalent circuit model, the accuracy in this equivalent model contributed to the accuracy of thermal predictions.

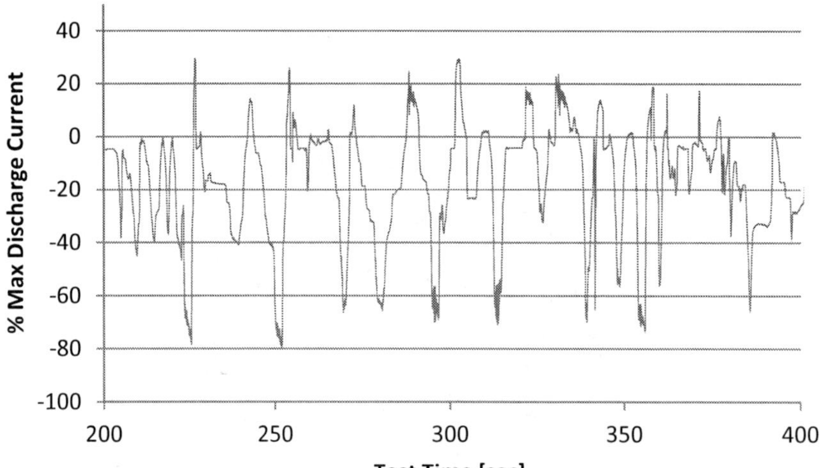

Figure 11 – Example of current during drive cycle

4 CONCLUSIONS

Designing battery packs for use in an electrified powertrain is a challenging problem. It requires consideration of complex real-world conditions and interactions from the entire vehicle. Additionally, the changes in battery behavior with charge, temperature and age, requires special attention in both physical testing. The need to evaluate a vehicle over a large range of conditions leads to significant periods of time dedicated to conditioning a pack for a physical test.

The use of simulation eliminates many of these issues. The design and analysis process outlined here, using PowerFLOW and PowerTHERM, demonstrate an ability to accurately predict battery electro-thermal behavior. However, because of the long time periods that batteries need to be evaluated over, the exclusive use of 3D simulation can be prohibitive. Therefore, a mix of system modeling and 3D modeling is used. The 3D modeling is used to generate comprehensive and highly accurate results, which are then used to generate reduced order models. The reduced order models predict a subset of the temperatures predicted by the 3D tools, but execute thousands of times faster. This faster performance enables the use of the system models on very long duration simulations and in HIL test benches.

Overall, a combination of 3D modelling in PowerFLOW, PowerTHERM, along with system simulation in Simulink® enables an efficient design process for a battery pack and prediction of its real-world performance.

REFERENCE LIST

[1] http://www.epa.gov/obd/pubs/420f09048.pdf

[2] http://www.nrel.gov/vehiclesandfuels/energystorage/pdfs/42469.pdf

[3] Pesaran, Battery Thermal Management in EVs and HEVs: Issues and Solutions, Advanced Automotive Battery Conference, Las Vegas, Nevada February 6-8, 2001, http://www.nrel.gov/vehiclesandfuels/energystorage/pdfs/aabc_lv.pdf

[4] Groot, J., State-of-Health Estimation of Li-ion Batteries: Cycle Life Test Methods, Division of Electric Power Engineering, Chalmers University of Technology, 2012 http://komar.bitcheese.net/files/JensGroot.pdf

[5] ThermoAnalytics, RadTHERM 11.0 Battery Model User Manual.

[6] PNGV Battery Test Manual, DOE/ID-10597, Rev 2, Feb 2001, http://avt.inel.gov/battery/pdf/pngv_manual_rev3b.pdf accessed on Dec 18, 2014.

[7] Bechtold, T., Rudnyi, E., Model Order Reduction for Dynamic Electro-Thermal Simulation of Microsystems, Communications to SIMAI Congress, ISSN 1827-9015, Vol 1, 2006.

[8] Verbrugge, M., Tate, E., Adaptive state of charge algorithm for nickel metal hydride batteries including hysteresis phenomena. Journal of Power Sources 2004; 126:236–249.

[9] Velivelli, A., Han, J., Jansen, W., Alajbegovic, A., Tate, E., Thermal Modeling of 48-Cell Module For Realistic Load Cycles, SAE 2014 Thermal Management Systems Symposium, Sept 22-24, 2014, Denver, Colorado, USA.

Effects of variable accelerations on a PHP-based cooling system

M MANZONI[1,5], M MAMELI[1], C DE FALCO[2], L ARANEO[3], S FILIPPESCHI[4] and M MARENGO[1,5]
[1] University of Bergamo, Department of Engineering and Applied Science, Italy
[2] Politecnico di Milano, MOX Mathematics Department, Italy
[3] Politecnico di Milano, Energy Department, Italy
[4] University of Pisa, DESTEC, Italy
[5] University of Brighton, School of Computing, Engineering and Mathematics, UK

ABSTRACT

A PHP is a passive, wickless heat transfer device consisting of a capillary loop, evacuated and partially filled with a working fluid, with alternated heating and cooling zones. The PHP light structure (4kg/m^2), flexibility and moderately high heat transfer capability (30W/cm^2) are interesting for automotive cooling applications. During their motion, vehicles withstand different speeding-up, which may influence the PHP behavior. The braking and the turning accelerations are the most critical reaching over ±1g for passenger cars, more for sport cars. Therefore, numerical simulations have been carried out to analyze the performances of a stand-alone PHP-based cooling system under transient accelerations.

Keywords: Pulsating heat pipe, Two-phase flow, Acceleration effects, Cooling system

NOMENCLATURE

Latin symbols

A	Cross flow area, [m^2]	m	Mass, [kg]
A_{ex}	External lateral area, [m^2]	m_{LV}	Evaporated/Condensed mass, [kg]
A_{w-f}	Internal lateral area, [m^2]	p	Pressure, [Pa]
a	Acceleration, [ms^{-2}]	Pr	Prandtl number, []
c_v	Specific heat constant volume, [Jkg^{-1}K^{-1}]	$\dot{Q}$	Heat power, [W]
		q_{ex}	External heat flux, [Wm^{-2}]
f_τ	Friction coefficient, []	q_{w-f}	Heat flux between wall and fluid, [Wm^{-2}]
g	Gravity acceleration, [ms^{-2}]		
h	Convection coefficient, [Wm^{-2}K^{-1}]	r	Tube internal radius, [m]
h_{LV}	Latent heat of vaporization, [Jkg^{-1}]	R_{eq}	Equivalent Thermal Resistance
i	Vapor plugs index, []	Re	Reynolds number, []
j	Liquid slugs index, []	T	Temperature, [K]
k	Thermal conductivity, [Wm^{-1}K^{-1}]	t	Time, [s]
K	Wall slices index, []	u	Velocity, [ms^{-1}]
L	Length, [m]	x	Axial coordinates, [m]

Greek symbols			
ρ	Density, [kg/m^3]	hom	Homogenous phase change
ε	Surface roughness, [m]	het	Heterogeneous phase change
ϑ	Inclination to horizontal, [rad]	int	Internal
Subscripts		l	Liquid
$cond$	Condenser zone	sat	Saturated conditions
$heat$	Heater zone	v	Vapor
ex	External	w	Wall
		∞	Environmental

1 INTRODUCTION

In the past decades, the need of the automotive industry to become environmental friendly has intensified. The more and more stringent targets for CO_2 production, the request for particulate reduction and the lack of fossil resources led to the development of electric vehicles (e.g. Hybrid Electric Vehicles, Fuel Cell Electric Vehicles) able to combine different power sources. In addition, the great advance in electronics and micro-electronics induced even tradition cars to get "electronics addicted".

However, batteries and electronic components generate heat; optimum thermal performance and high reliability throughout the product life cycle are required key factors, even if the thermal environment of a motor compartment is extreme. In order to maximize the electronics performances and lifetimes, a precise temperature and temperature gradient control is requested.

Many different technologies are available for the thermal management of such systems: air cooling, forced liquid loops and phase change materials. In case of very high heat input fluxes, a possible solution lies in two-phase systems which make use of the high efficient thermal transport process of evaporation and condensation to maximize the thermal conductance and reduce the temperature gradient between a heat source and a heat sink. As relatively new and promising members of the wickless heat pipes family, Pulsating Heat Pipes (PHPs) have recently reached the attention of the automotive industry as a practical way to remove heat from the electronic components, [1], [2].

Patented by Akachi [3], [4] in the earlier '90s, a PHP consists of a capillary tube closed end to end in a loop, evacuated and partially filled with a working fluid. Since the tube diameter is small enough, the fluid usually distributes itself in a constraint pattern of vapor plugs and liquid slugs (Fig. 1). When power is provided at the heater section, evaporation occurs within the liquid phase, vapor pressure raises inducing oscillation and circulation of the fluid along the channel. Thus, the train of fluidic elements is pushed toward the condenser zone where condensation occurs and both sensible and latent heat are rejected to an external cold source. Even if slug and plug is the most common hydraulic regime within a PHP, other possible fluidic paths characterized by separation of vapor and liquid can appear (e.g. annular, stratification) [5]. In addition, if too much power is provided at the evaporator section evaporating all the working fluid, the PHP reaches the unwanted condition of dry-out showing very poor thermal performances.

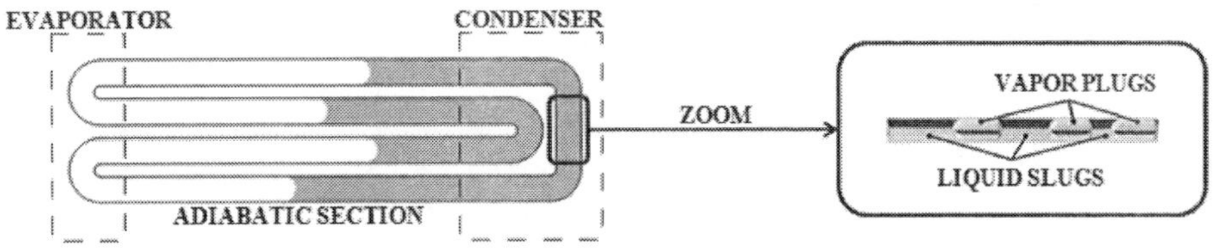

Figure 1 - Schematic of a generic PHP.

During their motion, vehicles withstand different accelerations, which must be accounted for in the design of their sub-systems. Speed and acceleration are very intrinsic to vehicle classes and traffic conditions. Usual accelerations for cars and heavy vehicles are fractions of g, but even more high values can be reached during particular sudden maneuvers (e.g. breaking and cornering): over ±1g for a passenger car and even more for sport cars.

The complex PHP thermal-hydraulic behaviour mainly depends on the interplay between phase change phenomena, capillary forces, and possible acting accelerations. The coupled effect of such phenomena, however, has not been thoroughly examined yet. The very few studies performed on planar PHPs operating under different acceleration loads [6], [7], [8], [9] have shown a strong relationship between PHP performances and applied accelerations. Most of these studies, however, are focused only on the steady-state conditions. Mameli et al. [9] presents also transient results underling a rapid response of the PHP thermal behavior to variation of the acceleration field for a very large range of g-levels. However, no tests have been performed at negative g-conditions (i.e. accelerations from the heater to the cooler zone). Thus, a validated numerical analysis appears as a useful tool in order to predict the transient response of a hypothetic PHP-based cooling system to sudden manoeuvres, which implies both positive and negative levels of acceleration.

Therefore, in the present work, numerical simulations have been carried out to analyze the performances of a PHP-based cooling system under variables accelerations from 2g to -2g (worst scenario). The employed one-dimensional lumped parameter numerical tool has been developed improving Mameli et al. [10], [11] previous code and consists of a two-phase separated flow model where capillary slug flow is assumed a priori. A complete set of balance differential equations accounts for thermal and fluid-dynamic phenomena. The originality of this numerical tool is the suppression of the standard assumption of saturated vapor plugs as well as the consequent embedding of heterogeneous and homogeneous phase changes. Moreover, vapor is not treated as an ideal gas anymore, but Van der Waals equation has been implemented. This tool has already been validated against experimental data under different gravity steady conditions [12], [13].

2 NUMERICAL MODEL

The proposed model is applicable to PHPs in confined operating regimes (i.e. slug flow). Its main assumptions are:

a. The model is one-dimensional. Mass, momentum and heat transfer are calculated along the axial direction. Heat transfer in radial direction is lumped.

b. Liquid slug momentum equation is lumped. Friction between vapor and wall elements is neglected.
c. Liquid menisci maintain spherical shape with negligible contact angle at the wall.
d. Vapor may exist in saturated or not saturated conditions.
e. Vapor is treated as real gas. Density is calculated as mass over volume. Vapor pressure is defined as function of temperature and density (Van der Waals equation).
f. All other fluidic thermo-physical properties are standardly calculated as polynomial functions of temperature only. Liquid slugs pressure is set equal to the average value of the adjacent vapor plugs pressure.
g. The liquid film around each vapor plug is constant in space and time.

2.1 Fluidic model

Liquid slugs and vapor plugs are spatially traceable during the whole simulation time. A Lagrangian approach is applied in order to follow each fluidic element in its chaotic motion within the channel. In addition liquid slugs are subdivided into smaller domains characterized by the same velocity and pressure but, eventually, different temperatures. Fig. 2 shows a schematic of the numerical domains.

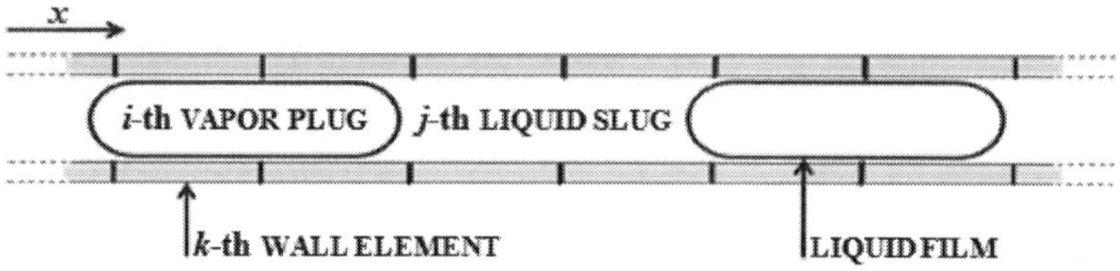

Figure 2 - Numerical domains. Liquid slugs sub-domains are not indicated.

2.1.1 Phase changes

Since the PHP tube is a closed domain, the total mass of fluid remains constant during the simulation time. However, the mass of single fluidic elements could vary due to phase changes. Heterogeneous and homogenous evaporation/condensation must be taken into account because of assumption *d*.

Heterogeneous condensation/evaporation can occur when a vapor/liquid comes into contact with a surface that has a temperature below/higher than the fluid one. This involves heat transfer to/from the solid wall.

Otherwise, if vapor/liquid pressure is higher/lower than the saturation value at its actual temperature, homogeneous condensation/evaporation occurs in the fluidic bulk. In order to evaluate the heat exchange between liquid slugs and vapor plugs due to this kind of transition the equivalent homogeneous heat is evaluated a posteriori as in Eq. 4.

Heterogeneous evaporated or condensed mass, $m_{LV,het}$, could be calculated as in Eq. 1. According to Fig. 3, a double check on pressure and temperature must be performed to assert if this kind of phase change may take place: if liquid pressure is lower than the saturation value or its temperature overcomes the saturation point, then evaporation occurs; if vapor pressure exceeds the saturation value or its temperature is lower than the saturation point, then condensation occurs. In both cases, the wall temperature T_w must satisfy a proper condition.

$$m_{LV,het} = \int_0^{\Delta t^*} \frac{A_{w-f} q_{w-f}}{|h_{LV}|} dt \quad if \begin{cases} evaporation: \ T_w \geq T_l + \Delta T_{superheat} \\ condensation: T_w \leq T_v - \Delta T_{supercool} \end{cases} \tag{1}$$

where $\Delta T_{superheat}$ and $\Delta T_{\text{supercool}}$ are fixed (see Table 2); A_{w-f} is the contact surface area between each solid domain and the fluidic elements, q_{w-f} is the heat exchanged between the wall and the fluidic elements (section 2.3), h_{LV} is the heat of vaporization and Δt^* is a characteristic time associated to new fluidic elements generation:

$$\Delta t^* = m_{min} \left(\frac{A_{w-f} q_{w-f}}{|h_{LV}|} \right)^{-1} \tag{2}$$

A new element is, indeed, produced only if $m_{LV,het} \geq m_{min}$, otherwise the mass exchange occurs only between adjacent elements; Δt^* is proportional to the evaporation/condensation ratio at steady state, while m_{min} is calculated assuming that every new element must have at least the volume of a sphere whose diameter is equal to the tube internal one ($m_{min,eva(cond)} = {}^4/_3 \pi r^3 \rho_{v(l)}$). For the present operative conditions, the evaporative and condensed Δt^* are respectively 5×10^{-3} s and 1.5 s.

According to Fig. 3, if vapor pressure exceeds the saturation value at the actual temperature, homogeneous condensation occurs in the vapor bulk; if vapor pressure is lower than the saturation point, then homogeneous evaporation is induced in the adjacent liquid elements. Homogeneous evaporated or condensed mass, $m_{LV,hom}$, could be calculated at each time step by solving the simple linear systems of Eq. 3.

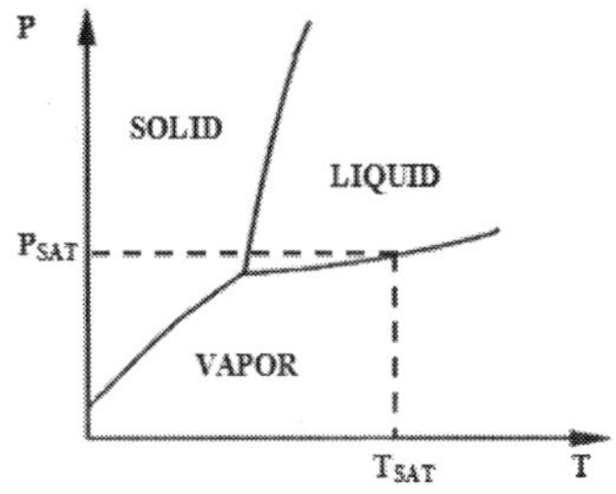

Figure 3 - Phase change diagram of a generic working fluid.

$$\begin{array}{ll} Evaporation \quad p < p_{sat} & Condensation \quad p > p_{sat} \\ \begin{cases} m_{LV,hom} = A\rho_{sat}L_{sat} - A\rho_v L_v \\ L_{sat} = L_v + \dfrac{m_{LV,hom}}{A\rho_l} \end{cases} & \begin{cases} m_{LV,hom} = A\rho_v L_v - A\rho_{sat}L_{sat} \\ L_{sat} = L_v - \dfrac{m_{LV,hom}}{A\rho_l} \end{cases} \end{array} \tag{3}$$

where A and L are the fluidic element cross section area and length, while p and ρ are its pressure and density. The sub-scripts v refers to vapor plugs, l to liquid slugs and sat to saturated conditions.

The heat released/requested during homogeneous phase changes at each time step Δt is calculated as:

$$Q_{hom} = \pm \frac{m_{LV,hom}|h_{LV}|}{\Delta t} \begin{cases} - \ \ evaporation \\ + \ \ condensation \end{cases} \tag{4}$$

For both kind of phase changes, if $m_{LV,het(hom)}$ exceeds the mass of the corresponding fluidic element, then this vanishes and the adjacent slugs/plugs merge.

2.1.2 Momentum equation

The momentum equation is calculated for liquid slugs only. Vapor plugs are, therefore, drag along the tube by the liquid motion. Accounting for assumption *b*, the momentum equation of the *j*-th liquid slug integrated along its length is:

$$\frac{d(mu)}{dt} = mgsin(\vartheta) + A(p_i - p_{i+1}) - 0.5 f_\tau \frac{m}{2r} u^2 \tag{5}$$

where u is the velocity and r the tube internal radius. The first term on the right side is the gravity force: ϑ is the local angle between the gravity vector and the flow direction. The second and the third terms are the forces due to adjacent vapor (i and $i+1$) expansion/compression and to viscous losses. Pressure drop due to capillary forces has been neglected since the cross section along the tube length is constant and because of assumption *c*.

The friction coefficient f_τ is evaluated by means of semi-empirical correlations, either for fully developed laminar (i.e. Hagen-Poiseuille) [14] or turbulent flow (i.e Haaland) [15]. Minor losses are evaluated only if the j-th liquid slug passes through a bend. The friction coefficient is evaluated according to Darby 3K method [16], [17].

The solution of Eq. 5 provides the velocity of each liquid slug. Their new position for each time step $x_{j,t}$ can be estimated as in Eq. 6 where a is the *j*-th liquid slug acceleration defined as the ratio between the forces applied on the fluidic element and its mass.

$$x_t = x_{t-1} + u\Delta t + 0.5 a \Delta t^2 \tag{6}$$

If the spacing between two adjacent fluidic elements is zero or if there is an overlap, they merge and the total number is accordingly reduced.

2.1.3 Energy balance

Fluidic temperatures are calculated by solving the energy balance applied to each lagrangian sub-domain.

$$c_v \frac{d(Tm)}{dt} = \left(q_{w-f} A_{w-f} - m_{LV.het} h_{LV}\right) - \left(kA \frac{\partial T}{\partial x}\bigg|_i^j + kA \frac{\partial T}{\partial x}\bigg|_{i+1}^j \right) - p\frac{dV}{dt} + Q_{hom} \tag{7}$$

where c_v and k are the specific heat and the thermal conductivity. These equations apply for each of the phases. The first term on the right side of Eq. 7 represents the heat exchanged between the wall and the fluidic elements, the second is the axial conduction within the fluid, the third term is the compression work calculated only for vapor plugs and the last term is the heat associated to homogenous phase changes.

2.2 Solid model

The PHP tube is sub-divided into smaller Eulerian domains with constant mass and fixed position over time. Therefore, only the energy equation must be implemented in order to evaluate the PHP wall temperature.

$$m_w c_{v,w} \frac{dT_w}{dt} = q_{ex} A_{ex} - k_w A_w \left(\left. \frac{\partial T_w}{\partial x} \right|_{K-1}^{K} + \left. \frac{\partial T_w}{\partial x} \right|_{K+1}^{K} \right) - q_{w-f} A_{w-f} \tag{8}$$

where A_{ex} is the contact area between the wall and the external environment and A_w the tube wall cross section area.

The second term on the right side of Eq. 8 accounts for the heat conduced within the wall, the third term is the heat exchanged between the wall and the fluidic elements (section 2.3), while the first term is the heat exchanged between the wall and the external environment.

Constant heat input power Q_{ex} is supplied to the evaporator zone; forced convection is applied at the condenser; no heat exchange occurs between the wall and the environment in the adiabatic region.

$$q_{ex} = \begin{cases} Q_{ex}/A_{ex} & evaporator \\ 0 & adiabatic\ zone \\ h_\infty (T_w - T_\infty) & condenser \end{cases} \tag{9}$$

where h_∞ is a constant external heat transfer coefficient (Tab. 1). External air is assumed at constant temperature T_∞. The radiative heat loss has been neglected since it has been estimated as ~0.1% of q_{ex}.

2.3 Solid/fluid coupling

Solid and fluidic domains are routinely coupled by the heat exchange between wall and slugs (plugs):

$$q_{w-f} = h(T_w - T) \tag{10}$$

where h is the chosen heat transfer coefficient.

If no phase changes occur, h can be estimated by means of different semi-empirical formula: Shah and London correlation [18] is used for the laminar flow thermally developing region ($Re \leq 2000$), the Gnielinski correlation [14] is implemented for the transient/turbulent flow ($2000 < Re < 10000$), while for the fully developed turbulent flow ($Re \geq 10000$) Dittus-Boelter correlation [14] is applied.

If phase changes occur, constant heat transfer coefficients have been presently assumed: 2000 W/m^2K for condensation and 4000 W/m^2K for evaporation. The values are chosen accounting for the coefficients usually identified in literature [5].

3 NUMERICAL RESULTS

The numerical simulations have been performed for an ideal planar PHP, meaning that the thermal inertia of the peripheral elements has been neglected. The physical and numerical characteristics of the tested design are listed in Tab. 1 and Fig. 4. The device dimensions and the heat input fluxes have been chosen in feasible ranges for electronics cooling. The provided power is kept constant during the whole simulation time to test the device in the worst load condition. The temperature of the external environment has been set thinking of a PHP cooled down by forced convection from a radiator in contact to the ambient outside the car motor hood.

Since fixed flux at the evaporator (Q_{ex}) and fixed temperature at the condenser (T_∞) have been imposed as boundaries conditions, the heat rejection rate is an output of the model and it equals the heat input rate when stationary conditions are reached.

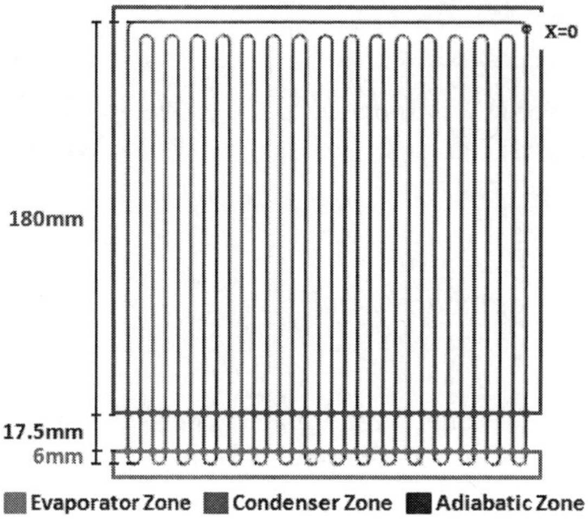

Figure 4 - PHP dimensions.

Table 1 - PHP physical and numerical characteristics.

Input parameter	Value	Input parameter	Value
Working fluid	FC-72	Lateral acceleration	-2g / 2g
Volumetric filling ratio	0.5	External temperature	20 °C
Tube material	Copper	External HTC	300 W/m^{2}K
Internal diameter	1.1 mm	Evaporator heat flux	3.23 W/cm^2
External diameter	2.0 mm	ΔT superheat	2.5 °C
Surface roughness	50 μm	ΔT cooling	0.1 °C
Total length	6.62 m	N° of wall grids	430
N° of *U* bends	31	N° of slug/plug (t=0)	20
Gravity acceleration	1g	Time step	0.001 sec

The results are presented in form of temperature responses to accelerations and decelerations since temperature control is the key parameter for electronic cooling once the heat input is known.

The model is one-dimensional, thus only accelerations parallel to the fluidic path can be accounted for. Since the cornering and the breaking acceleration of a car lay in the plane perpendicular to the gravity direction, in order to study their effect on the PHP performance, a horizontal orientation has first been tested (Fig. 5). The employed coordinate system is also displayed in Fig. 5.

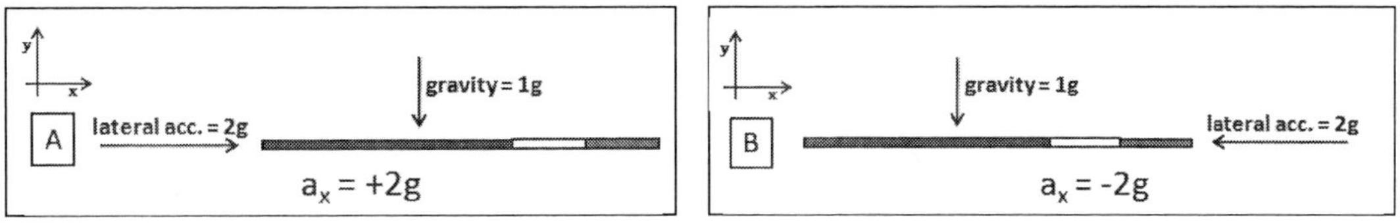

Figure 5 - Schematic of the horizontal simulated cases. Condenser zone in blue, evaporator zone in red.

The simulations have been carried out providing a step variation (10s) of the acceleration acting on the system, which previously has reached steady state. This tries to reproduce sudden breaking or cornering maneuver ($\pm 2g$) of a car in uniform motion (e.g. a car on a highway). The results in term of PHP averaged wall temperature are presented in Fig. 6.

The ideal capillary PHP shows a quick response to sudden modifications of the acceleration acting along the fluid path direction improving its performances, because the new associated force increases the fluid circulation helping the liquid to reflow from the condenser to the evaporator zone (Case A). When the disturbance disappears, the PHP returns to its normal steady state. A negative acceleration, otherwise, does not provide any variation on the PHP performances (Case B).

The absence of acceleration, as well as the application of an acceleration directed from the evaporator to the condenser zone, indeed, damps the fluid motion and thus the phase changes within the system. The device works like a tube full of liquid and vapor, which transports heat only by axial conduction with, consequently, higher temperatures.

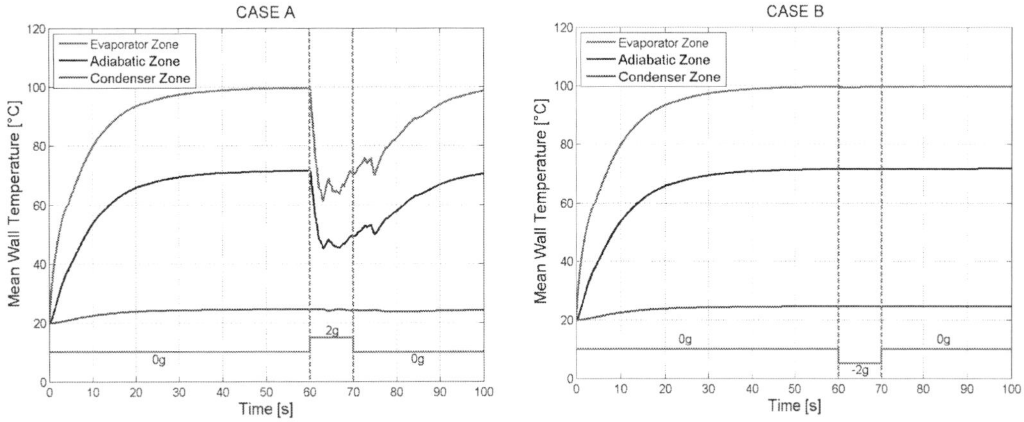

Figure 6 - Horizontal PHP thermal response to acceleration step variation in the PHP plane. Normal gravity constant.

This explanation is also supported by Fig. 7, which shows a particular of the liquid velocity during the acceleration step. It is clearly visible that the velocity assumes not zero values only when +2g is applied on the system; the application of -2g induces a first oscillation on the velocity signal that, however, is quickly recovered.

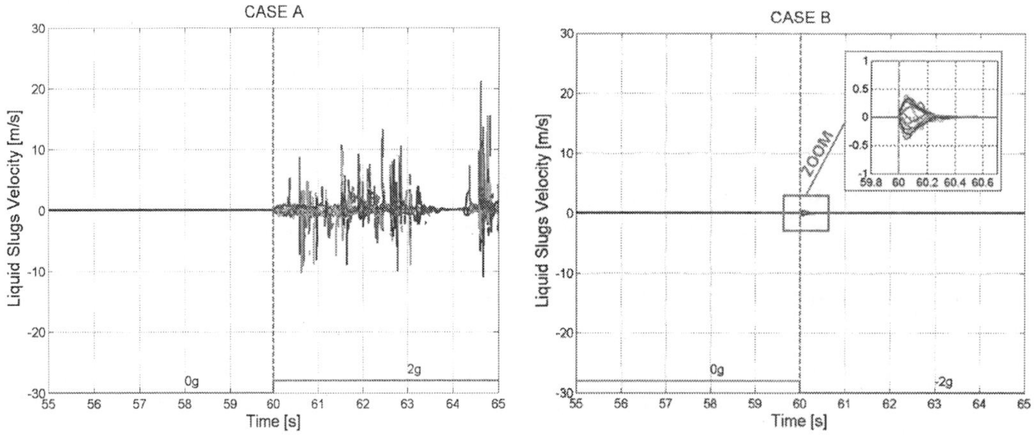

Figure 7 - Particular of the liquid velocity during the acceleration step. Each color is associated to its own slug.

3.1 Inclined PHP

Since the PHP performances are better when a force is applied continuously from the condenser to the evaporator zone, simulations have been carried out for a 45° inclined PHP (Fig. 8) in order to reduce the PHP thermal resistance during vehicles uniform motion.

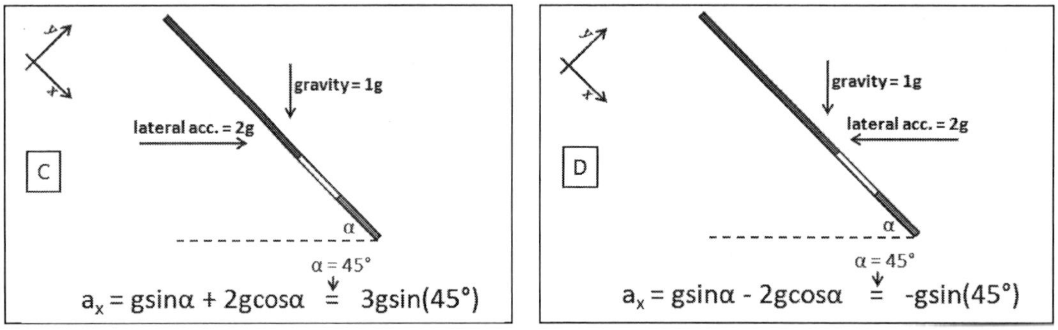

Figure 8 - Schematic of the 45° inclined simulated cases. Condenser zone in blue, evaporator zone in red.

The results in term of PHP mean wall temperature are presented in Fig. 9. Again, like in the case of the horizontal PHP, the system shows a quick response to acceleration step improving its performances when the acceleration is increased (Case C), worsening them otherwise (Case D). This last simulation also shows that the recovery after the acceleration step is very rapid, meaning that the PHP is a robust system with respect to variation of the acceleration level. Moreover, it should be notice that, since the simulated PHP is stand-alone, its transient response is quicker with respect to a real device due to peripheral elements thermal inertia. This implies that the temperature enhancement shown for the Case D is hereby amplified (conservative situation).

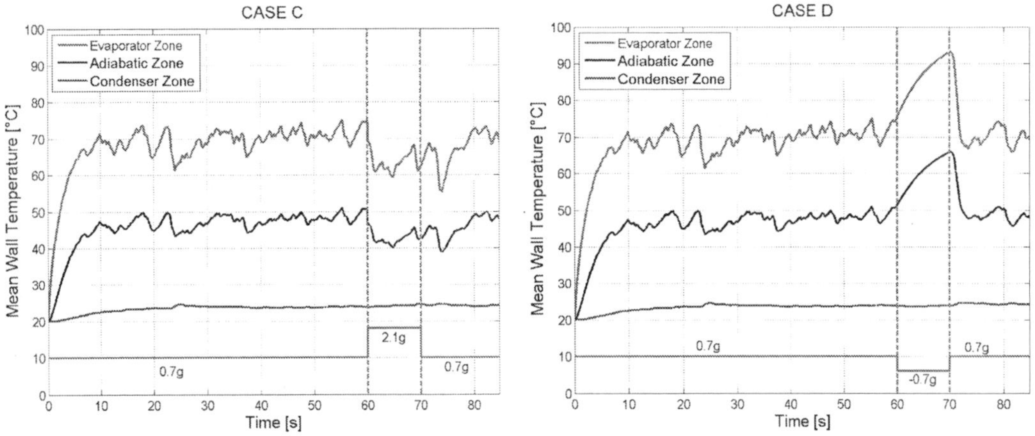

Figure 9 - Inclined PHP thermal response to acceleration step variation.

Case C underlines, also, that the PHP temperature characteristics are not so different for such positive variations of the acceleration level. This can also be deduced from Fig. 10, which shows the equivalent thermal resistance of the device, as:

$$R_{eq} = \frac{T_{ev} - T_{cond}}{\dot{Q}} \tag{11}$$

where the numerator represents the difference between the evaporator and the condenser average temperatures in the pseudo-steady state, while $\dot{Q}$ is the external heat power.

The calculation of the PHP equivalent thermal resistance displays great improvements of the PHP performances only when passing from negative/zero to positive accelerations. This is an interesting aspect meaning that if the device is designed/ oriented such as it senses only positive accelerations for all the vehicle maneuvers, the performances of the PHP remain quite constant over time.

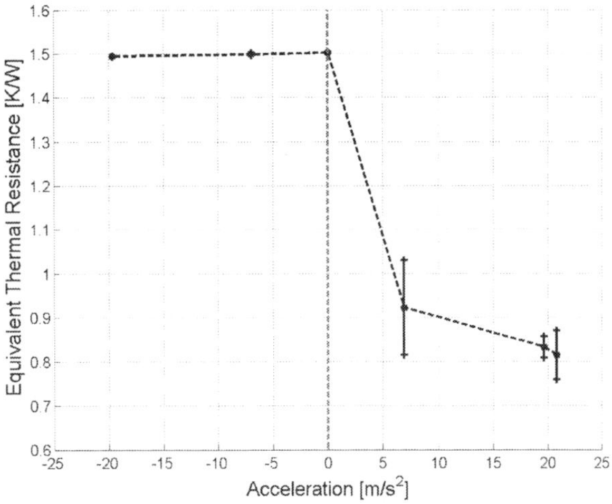

Figure 10 - PHP equivalent thermal resistance for different accelerations.

4 CONCLUSIONS

The thermal performances of an ideal PHP-based cooling system have been simulated for transient step variation (10s) of the applied acceleration from 2g to -2g. An already validated one-dimensional lumped parameter numerical tool has been employed for such analyses. Since the model is one-dimensional, only accelerations parallel to the fluidic path can be accounted for. Thus two different orientations ($\alpha = 0°, \alpha = 45°$) have been considered in order to study the combined effect of gravity and sudden cornering/breaking accelerations. The results can be summarized in the following points:

- If no acceleration is applied on the system, the PHP behaves like a tube full of liquid and vapor that transmits heat only by conduction with associated relative high temperatures.
- The PHP shows a quick response to positive accelerations improving its performances since the associated force helps the liquid to flow from the condenser to the evaporator zone.
- Negative accelerations damp the fluid motion worsening the PHP performances. The recovery from negative to positive accelerations is very rapid, meaning that the system is robust to variation of the acceleration levels.
- The equivalent thermal resistance is not so influenced by variation of positive accelerations, meaning that if the PHP-based cooling system is designed/oriented such as it senses only positive accelerations during all the vehicle maneuvers, its performances remain quite constant over time.

Thus, a PHP represents an innovative, valid candidate for automotive electronics cooling, not only for its light structure, flexibility and reasonable heat transfer capability, but also for its robust behavior with respect to variation of the acceleration levels.

REFERENCE LIST

[1] Polášek, F., & Štule, P. (1997). Thermal control of IGBT modules in traction drives by pulsating heat pipes. *Proc. of 10th International Heat Pipe Conference.*

[2] Akachi, H. & Miyazaki, Y. (1997). Stereo-type heat lane heat sink. *Preprints 10th Internat. Heat Pipe Conf.*

[3] Akachi, H. (1990). *US Patent* No. 4921041.

[4] Akachi, H. (1993). *US Patent* No. 5219020.

[5] Mameli, M., Marengo, M., & Khandekar, S. (2014). Local heat transfer measurement and thermo-fluid characterization of a pulsating heat pipe. *Int. J. of Thermal Sciences*, **45**, 140-152.

[6] Kiseev, V.M., & Zolkin, K.A. (1999). The Influence of Acceleration on the Performance of Oscillating Heat Pipe. *Proc. of the 11th International Heat Pipe Conference.*

[7] Van Es, J., & Woering, A.A. (2000). High-acceleration performance of the Flat Swinging Heat Pipe. *NLR-TP-2000-265.*

[8] Gu, J., Kawaji, M., & Futamaca, R. (2004). Effect of gravity on the performance on pulsating heating pipe. *J. of Thermophysics and Heat Transfer*, **18**, 370-378.

[9] Mameli, M., Manzoni, M., Araneo, L., Filippeschi, S., & Marengo, M. (2014). Experimental investigation on a closed loop pulsating heat pipe in hyper-gravity conditions. *Proc. of the 15th IHTC Conference.*

[10] Mameli, M., Marengo, M., & Zinna, S. (2012). Thermal simulation of a pulsating heat pipe: effects of different liquid properties on a simple geometry. *Heat Transfer Engineering*, **33**, 1177-1187.

[11] Mameli, M., Marengo, M., & Zinna S. (2012). Numerical model of a multi-turn closed loop pulsating heat pipe: effect of the local pressure losses due to meanderings. *Int. J. Heat Mass Transfer*, **55**, 1036-1047.

[12] Manzoni, M., Mameli, M., DeFalco, C., Araneo, L., Filippeschi, S., & Marengo, M. (2014). Towards a numerical simulation of a closed loop pulsating heat pipe in different gravity levels. *Proc. of the 32nd UIT Conference.*

[13] Manzoni, M., Mameli, M., DeFalco, C., Araneo, L., Filippeschi, S., & Marengo, M. (2014). Toward a design of a micro pulsating heat pipe. *Proc. of 4th European Conference on Microfluidics.*

[14] Incropera, F. P., & DeWitt, D. P. (2007). Chapter 8: Internal Flow. *Fundamentals of Heat and Mass Transfer*, 6th Ed.

[15] Haaland, S. (1983). Simple and Explicit Formulas for the Friction Factor in Turbulent flow. *Transactions of ASME, Journal of Fluids Engineering*, **103**, 89-90. Darby, R. (1999). Correlate pressure drop through fittings. *Chemical Engineering*,**106**,101-104.

[16] Darby, R. (2001). Correlate pressure drop through fittings. *Chemical Engineering*, **104**,127-130.

[17] Shah, R. K., & London, A. L. (1978). Laminar Flow Forced Convection in Ducts. *Advanced in Heat Transfer*, Supple. 1.

Waste Heat Recovery

Modelling and optimization of thermoelectric generator systems for automobiles

A AGURTO-GOYA
Jaguar Land Rover Ltd, UK

S CHIWANGA
European Thermodynamics Ltd, UK

P SHAYLER
University of Nottingham, Faculty of Engineering, Thermofluids Research Group, UK

ABSTRACT

Although significant progress has been made in the field of vehicle waste heat recovery, the business case for thermo-electric power generation in automotive applications remains very challenging. Reducing the cost and increasing the efficiency of these devices is a key enabler for the adoption of the technology. This paper describes new design practices and a novel CAE approach aimed at minimising cost, whilst maintaining good performance, to enhance the commercial viability of this technology. A model of a thermoelectric generator (TEG) waste heat recovery system for passenger car is developed based on electro-thermal analogies through an equivalent lumped parametric circuit. The model is employed to optimize a baseline design to minimize cost, weight and exhaust back-pressure without sacrificing performance

DEFINITIONS, ACRONYMS, ABREVIATIONS

CAE	Computer Aided Engineering
CAT	Catalytic Converter
GTDI	Gasoline Turbo Direct Injection
NEDC	New European Driving Cycle
OEM	Original Equipment Manufacturer
P&O	Perturb and Observe
TEG	Thermo Electric Generator
TEM	Thermo Electric Module
WHR	Waste Heat Recovery
VIPER	Vehicle Integrated Powertrain Energy Recovery
WLTP	Worldwide harmonized Light vehicles Test Procedures
ZT	Figure of Merit

1 INTRODUCTION

With the increasing environmental effects of fossil fuels and stricter emission legislation it has become the utmost priority of automobile manufacturers and OEMs to improve the efficiency of Internal Combustion (IC) engines by investing on carbon reduction technologies. Jaguar Land Rover (JLR) endeavours to meet these demands through significant investment on R&D activities to ensure JLR products remain competitive in this respect.

One such project is VIPER 2 (follow up of VIPER). This three year collaborative project is part funded by Innovate UK, and follows on from the successful VIPER IDP5 project. The project started in June 2014 and aims to accelerate the implementation of new, cost-effective powertrain thermal management and energy recovery systems for on-cycle and real world CO_2 improvements. The target is 5.5% WLTP CO2 reduction via accelerated powertrain warm-up and waste heat recovery.

One of the waste heat recovery technologies under development in VIPER 2 is thermoelectric generation which relies on the Seebeck effect to produce useful electrical power from a temperature gradient. The electricity generated by the TEG can then be utilized to charge batteries, energize ancillaries or store heat (thermal batteries) unloading the engine in one case and accelerating engine warm up in others leading to a reduced fuel consumption.

Although power conversion efficiency of existing thermoelectric materials is still low (currently ~5% [1]) there are other valuable attributes that make thermoelectric generation attractive to automobile industry. Compactness, lack of moving parts and low noise match the attribute requirements for passenger vehicle applications well, and there are further synergies for hybrid vehicle applications, where increased electrical energy storage capacity is available.

This paper describes advanced CAE methods and design practices targeting cost reduction of TEG devices. The paper starts giving an overview of thermoelectric WHR for vehicles following with a description of the TEG system modelling based on electro-thermal analogies through an equivalent lumped parametric circuit. The model is then correlated against prototype component tests. At the end of the paper a global optimization of a baseline TEG system design to minimize size and cost, without compromising performance is described.

2 THERMOELECTRIC WASTE HEAT RECOVERY FOR PASSENGER VEHICLES

2.1 TEG System configuration

Heat to Power technologies in a vehicle are being heavily researched. Heat to Power technologies based on thermoelectric (TE) effect in comparison with other related technologies like electric turbo compound and Rankine cycle has the advantage of having no moving parts, being silent and very reliable. The thermoelectric effects are the direct conversion heat differentials to electric voltage and vice versa [2]. The Seebeck effect is the conversion of temperature differences directly into electricity. This effect was first discovered by Thomas Johann Seebeck in 1821, who found that a voltage existed between two ends of a metal bar when a temperature gradient existed in the bar. In contrast, the Peltier effect was first observed in 1834 by Jean Peltier, who found the reverse of the Seebeck effect. For passenger vehicles (conventional and non-conventional), the waste exhaust heat can be recovered directly to electrical energy for battery charging, thereby increasing the overall vehicle fuel efficiency.

In figure 1a a typical temperature distribution of the exhaust gas system is shown, which it is composed of the exhaust manifold, exhaust pipe, catalytic converter, center muffler and rear muffler [3]. It is a natural choice to install the heat exchanger at the location with the highest temperature, namely at the front-end location however the corresponding temperature may exceed the allowable working temperatures of the selected thermoelectric materials leading to a need to bypass the excess of energy. A typical layout of TEG for waste energy recovery is shown in figure 1b where a TEG system is placed after CAT to have minimum disruption on exhaust gas after treatment.

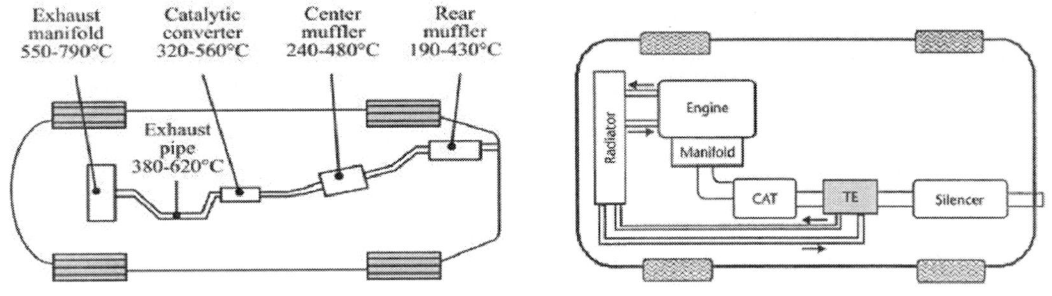

(a) (b)

Figure 1. (a) Temperature distribution of exhaust gas and (b) typical TEG layout for waste energy recovery [4]

The heat energy available from exhaust gas needs to be extracted by the heat exchanger. The heat exchanger is combined with the thermoelectric devices to form a united TEG system structure (figure 2). The heat exchanger in figure 2 works with the thermoelectric devices by means of a pump which circulates a working fluid around them. The use of a pump allows for a wide variation of exhaust gas flows which is normally the case on duty cycles of passenger cars. It is worth noting that the design of heat exchanger involves a trade-off between the thermal exchange efficiency and the exhaust gas flow rate. The thermal exchange efficiency gained by the restriction of the flow path may result in exhaust backpressure which demands more engine pumping work leading to an offset of the overall energy efficiency.

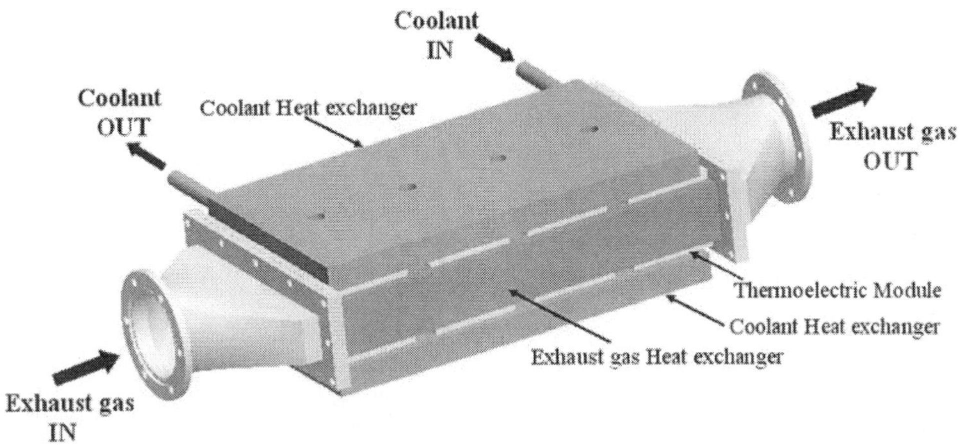

Figure 2. United TEG system structure [5]

2.2 TEG material

The suitable thermoelectric material needs to offer high energy conversion efficiency which is measured by the figure of merit, ZT:

$$ZT = \frac{S^2T}{\kappa\rho} \quad (1)$$

Where S is the TE power, T is the absolute temperature, κ is the total thermal conductivity and ρ is the electrical resistance [6]. Figure 3 shows ZT characteristics of typical thermoelectric materials [7, 8]. It can be seen that different materials offer optimal ZT at different ranges of temperatures: p-type and n-type Bi_2Te_3 at lower temperature range (<200°C) of exhaust gas; p-type Te-Ag-Ge-Sb alloy (TAGS) at the medium temperature range (200-500°C); and the Skutterudites (p-type $CeFe^4Sb^{12}$ and n-type $CoSb^3$) at the higher temperature range (500-800°C).

Despite the recent improvement in TE material ZTs considerable effort is needed to make this technology economically viable. It is estimated that ZT of around 2 is necessary in order to obtain a system efficiency of ~10% (which correspond to a material efficiency of ~15%) [4]. Commercially available modules are made of Bi_2Te_3 based materials with a relatively low operating temperature (lower than 450K). As exhaust gasses from vehicle can be up to 850K materials with a large ZT at high temperature are required. In order to achieve the highest average *ZT* over the entire operating temperature range of exhaust gas, a module should consist of a combination of different materials with different ZT values in cascade. The concept has been implemented through advanced structures adding different materials with different thickness to build the TE couple seen in figure 4 [9, 10].

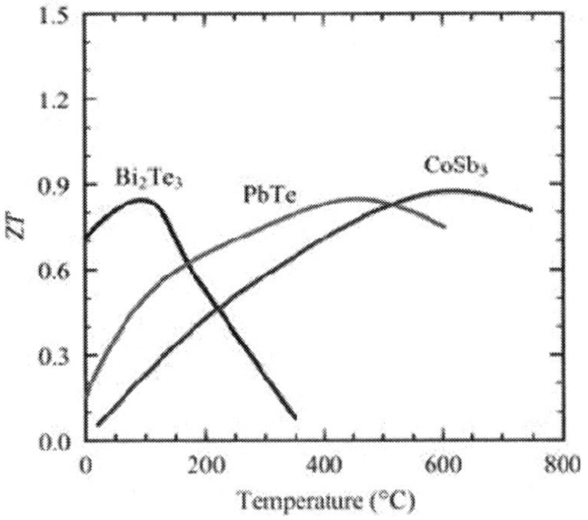

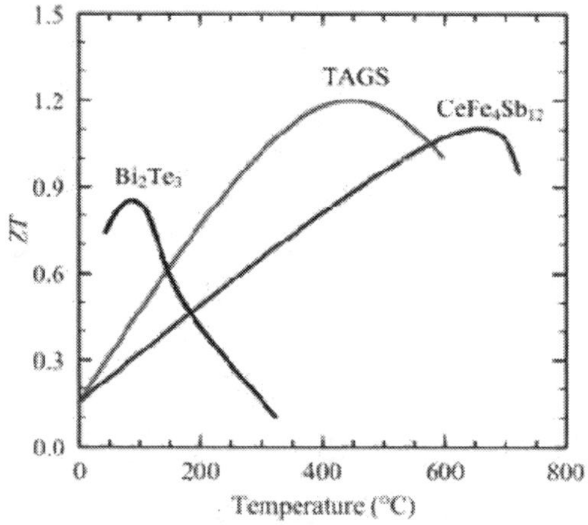

Figure 3. Comparison of TEG p-type and n-type materials [11]

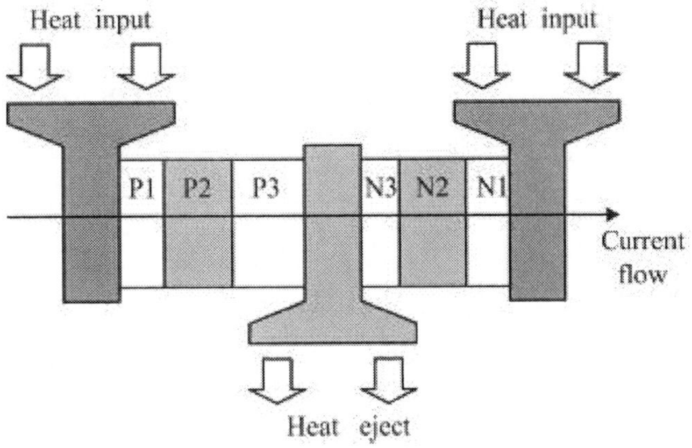

Figure 4. Segmented and multilayer TE couple configuration [11]

2.3 TEG Power Conversion

The TEG system needs to charge the battery or supply electrical energy to the vehicle power network. Since the TEG output voltage changes dynamically with the temperature of exhaust gas and hence varies in a wide range, power conditioning between the TEG and the load is necessary as shown in Figure 5. The output voltage range of a typical TEG is 0-25 V, which needs to be converted to 12.3-16.5 V for battery charging [12]. Therefore, a dc-dc converter, having step-up and step-down characteristics, is required to serve for power conditioning.

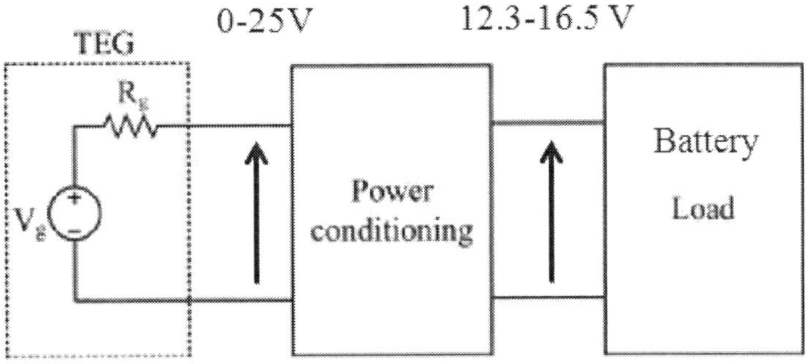

Figure 5. TEG power conditioning [11]

In order to fully utilize the thermoelectric material and maximize the recovery of waste heat energy, the TEG needs to adopt the maximum power point tracking (MPPT). The MPPT methods that have been applied to TEG are mostly developed for photovoltaic generation, such as the load matching method [13], the curve-fitting technique [14], the incremental conductance method [15], the ripple correlation method [16] and the perturbation and observation (P&O) method [17]. Moreover, the fuzzy logic [18] and neural network [19] have also been adopted to improve the MPPT for photovoltaic generation. Among the aforementioned MPPT methods, the P&O method is most commonly adopted [11].

3 MODELLING TEG SYSTEM

A generic TEG system consists of a large number of thermoelectric modules sandwiched between heat exchanger pipes or ducts carrying hot input and cold output flow of a working fluid [1] as seen in figure 6.

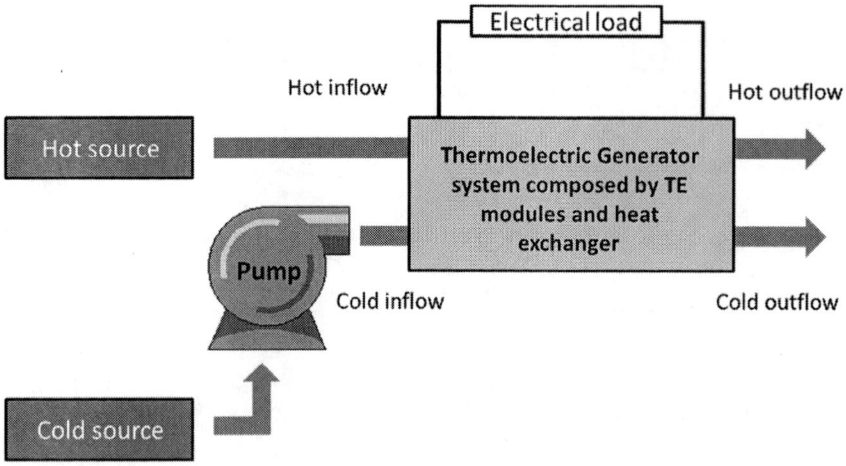

Figure 6. Basic arrangement of a complete TEG system

The design of a TEG system therefore involves optimizing thermoelectric modules and a heat transfer device. The thermoelectric generator system was modelled and simulated in the MATLAB/Simulink environment utilising a well-known thermal network approach in which an electrical circuit equivalent of thermoelectric modules is built based on electro-thermal analogies.

In this paper the heat exchanger device chosen was a parallel flat plate with offset strip-fin. Offset strip fins are widely used for industries that require lightweight high-performance exchangers due to their high heat transfer rate relative to heat exchanger volume [20], which is also suitable for the challenging space claim in automotive industry. The fins create a greater pressure drop but also allow greater heat transfer coefficient. Based on latest correlation data available of flow friction factor and Colburn modulus data [21] an equivalent thermal network resistance was performed to model the heat exchanger. The model parameterises fin geometry (figure 7) to adjust their dimensions without compromising back pressure created when the TEG system is sitting on vehicle exhaust system.

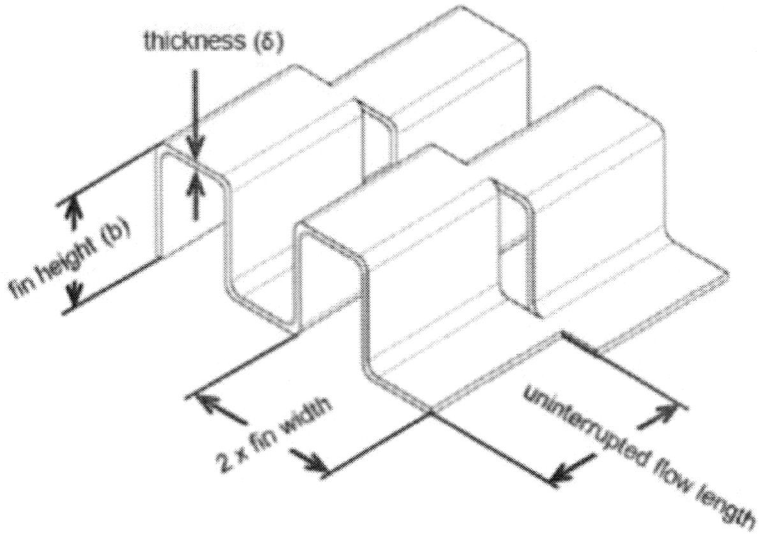

Figure 7. Offset strip fin heat exchanger parameters [22]

There are a large amount of books quoting comprehensive formulas to explain thermoelectric effects at all sorts of detail so it is not the purpose of the paper to repeat those details. Instead it is more important to show the electro thermal equivalent circuit derived from the energy balance happening in the junctions of a thermoelectric module. The bottom of figure 8 shows the energy balance over the thermoelectric device on the top. The resultant electro thermal circuit model is shown in figure 9 and is the basic block used to build the whole TEG system model.

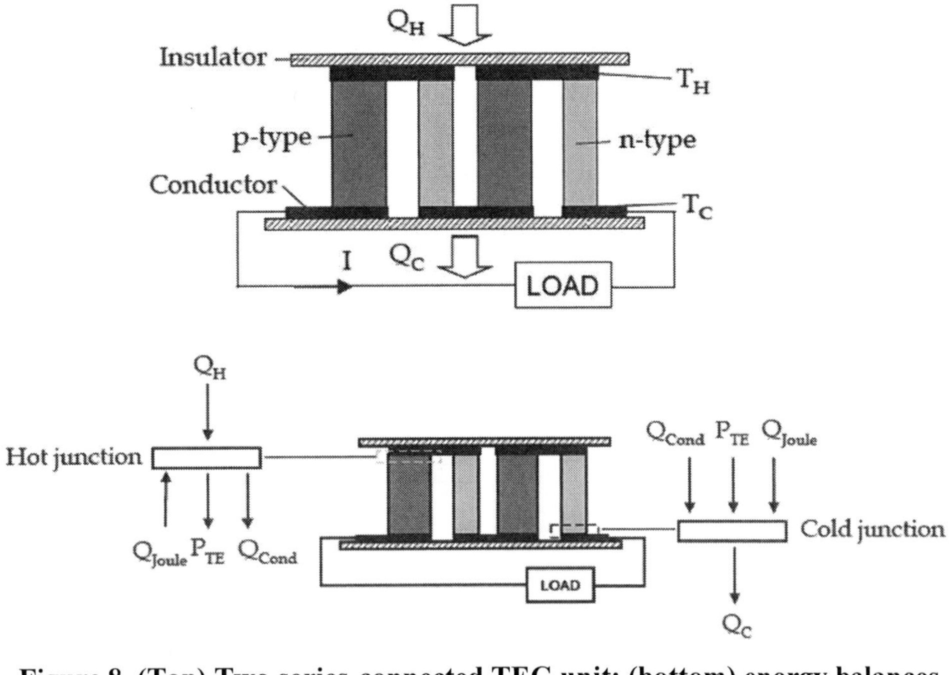

Figure 8. (Top) Two series-connected TEG unit; (bottom) energy balances on thermoelectric junctions [23]

For the hot junction the energy contributors are:

- Thermal input (Q_H) from an outside heat source;
- Internal input from resistive heating (Q_{Joule});
- Electrical energy output from thermoelectric effect (P_{TE}); and
- Thermal output by conduction (Q_{Cond}) away from the hot junction toward the cold one.

For the cold junction, the contributions are:

- Thermal output (Q_C) to the cold source;
- Internal input from resistive heating (Q_{Joule});
- Electrical energy input of thermoelectric power (P_{TE}), and
- Thermal input from conduction (Q_{Cond}) from the hot junction

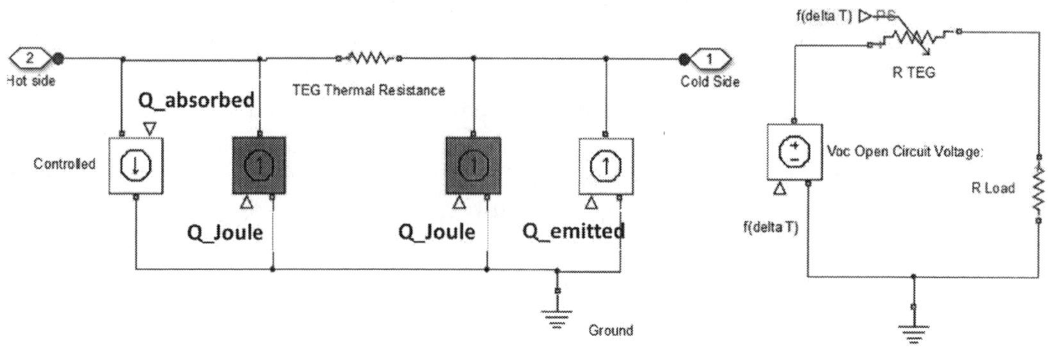

Figure 9. Lumped electrothermal equivalent circuit model

3.1 1-D Model

Figure 10 describes how the heat exchanger and thermoelectric module can be described in one-dimension by using the controlled volume approach and under conditions of the working fluid flow approximating to a constant across the width of the heat exchanger. A sector is considered to be isothermal for the plane perpendicular to the heat flow through the thermoelectric modules. The TEG system is divided up into discrete volumes (sectors) and an energy balance is applied between adjacent sectors resulting in a progression of energy from one sector to the next one. The axial heat conduction within pellets is ignored, as transverse conduction will be dominant. The number of discrete volumes or sectors adds resolution to the model but at the expense of computation time.

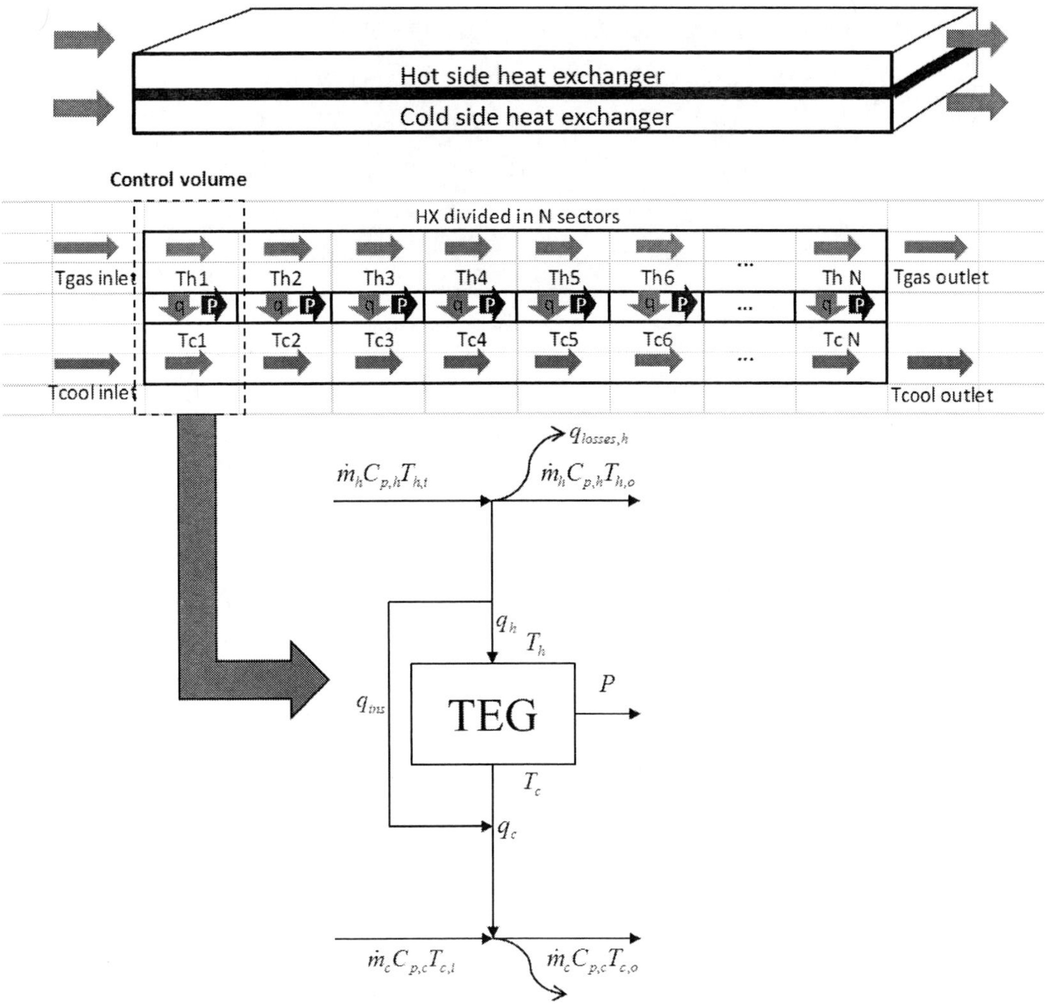

Figure 10. TEG system 1D model

As the sectors are stepped through, some of the inputs of the next sector are provided from the previous one such as working fluid temperatures. The performance of each sector is added together to give a resultant for the whole TEG system.

Following the modelling approach explained in this section a TEG system model having a flat plate with offset-strip-fins as a heat exchanger and a variable number of thermoelectric modules was proposed to meet space demand and heat available from the vehicle's exhaust system. One of the variants of this model consists of 4x4x8 thermoelectric modules shown in figure 11. This model was used to evaluate the TEG system performance under boundary conditions from a vehicle exhaust system after the catalytic converter (figure 1b). Having as inputs mass flow and temperature of heat source and cold sink the model can be customised with parameters corresponding to physical properties and geometry dimensions of the components in order to perform the optimisation studies detailed in section 6.

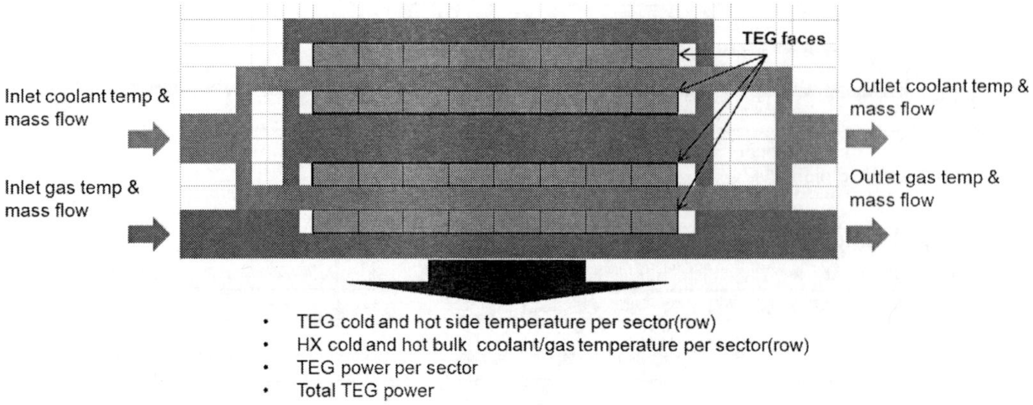

Figure 11. TEG system with 4x4x8 thermoelectric modules

4 MODEL VALIDATION

In this section a validation of the results from the TEG system modelling is compared to the results from the physical testing performed at the University of Nottingham. The TEG system prototype, built by European Thermodynamics (ETL) was placed after the CAT of a 2.0L GTDI engine exhaust system. Validation was performed for (1) hot and cold surface temperatures of the thermoelectric modules along the axial direction of the heat exchangers, and (2) outlet temperature of the coolant. The electrical power generated is still pendent as power conversion unit was under development by the time of this paper was written.

4.1 Experimental Setup

Figure 12 shows the TEG system prototype consisting of 4x4x8 TE modules and a dual split rectangular heat exchanger and bypass pipe [24]. The coolant is provided via rectangular box that covers the cold side temperature slab. Two arrays of TE modules are placed on top and bottom of each rectangular hot slab. Both coolant and hot the slab have offset strip fins designed not to exceed the 2kPa back pressure limit. TEG module specifications in the prototype were MFG P/N GM250-241-10-16 and MFG P/N GM250-241-10-12 from ETL with dimensions of 40mmx40mmx3.4mm.

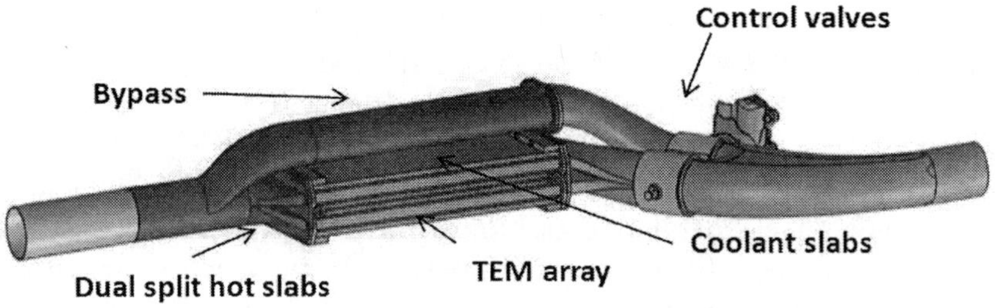

Figure 12. ETL TEG system prototype

4.1.1 Testing

The ETL TEG prototype was tested at the test rig facilities of University of Nottingham (figure 13) under several steady state driving conditions and over one emission driving cycle (NEDC).

- Steady State: six road load conditions: idle, 33, 50, 72, 100 and 120 kph
- Transient: NEDC

Figure 13. Test rig at University of Nottingham

4.2 Validation for temperature across TEG arrays

Half of the whole assembly TEG system was instrumented due to the symmetry of the unit. As can be seen in figure 14, thermocouples were distributed at three sections along the direction of the gas and coolant flow (sector 1, 4 and 7). The quality of prediction of the model at steady state condition is shown in figure 15 with a temperature survey of the TEG system at the aforementioned road load conditions and a good level of correlation is observed. The model prediction validity at transient condition is proven over the NEDC and the temperature survey is shown in figure 16.

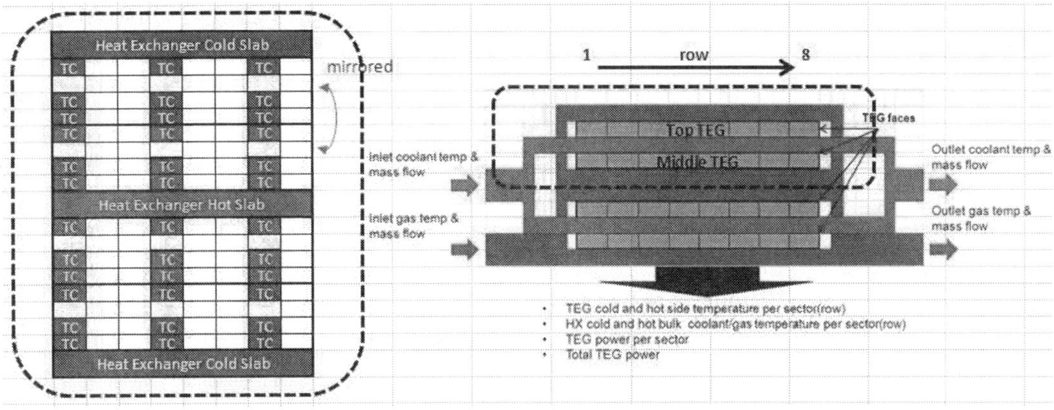

Figure 14. Thermocouple designation map

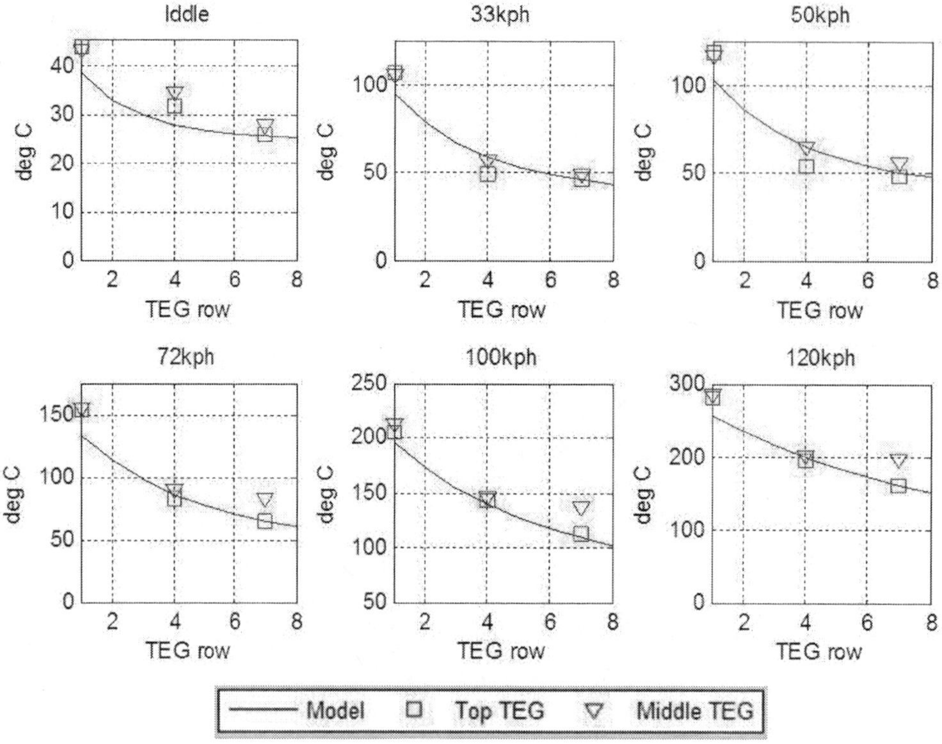

Figure 15. Correlation of TEG hot surface temperature at steady state (six road load conditions)

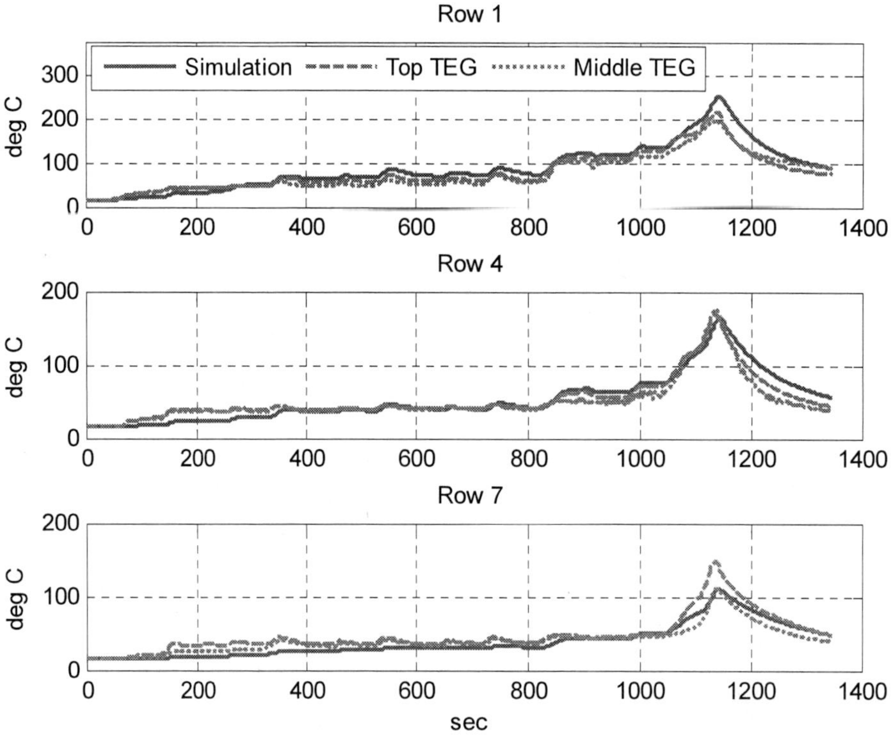

Figure 16. Correlation of TEG hot surface temperature at NEDC

4.3 Validation for temperature of coolant outlet

Figure 17 shows that the model predicts with reasonable accuracy the coolant outlet temperature for six road load running conditions and one NEDC run.

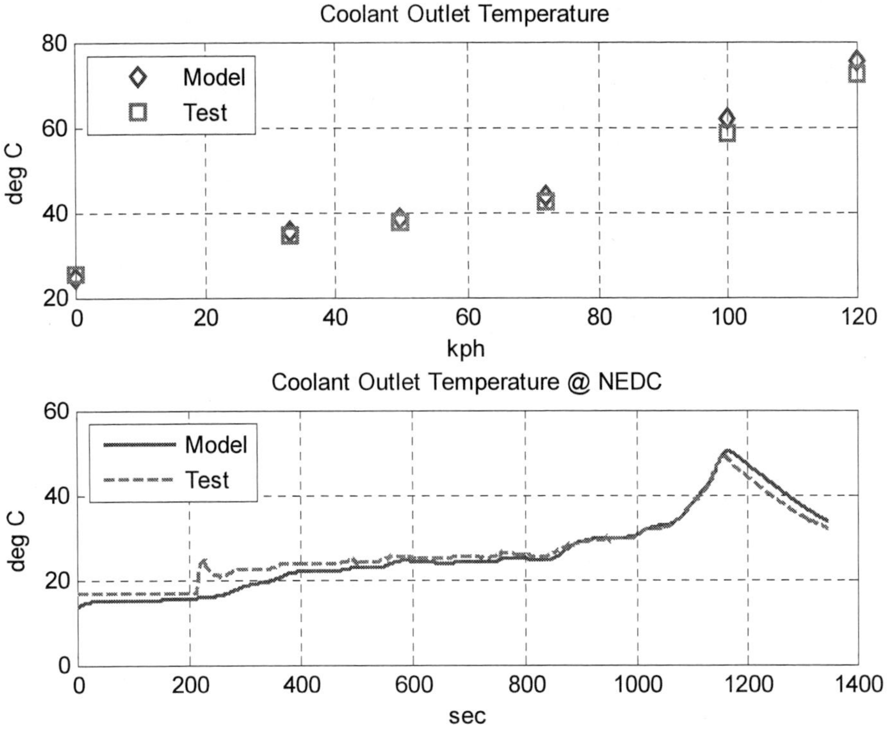

Figure 17. Correlation of coolant outlet temperature for six road load conditions cases (top) and NEDC (bottom)

4.4 Further work for validation

Based on prototype of figure 12, a conveniently simplified version was created by ETL in order to perform sensitivity studies to speed up further development of the TEG for VIPER. The prototype consists of 2x4x4 (2 arrays, 4 TEM per sector, 4 sectors or rows) TEMs of same specification than the first prototype. The model was modified to the simplified version and to support the sensitivity studies.

The assembly was tested on NEDC and deliberately overload in the last part of the cycle to understand the achievable peak temperature differential across the TEGs arrays that suits higher temperature rate thermoelectric materials. The model follows the test data (figure 18 left and bottom right) and predicts the expected power generation when the TEG is protected to reach the rated temperature (figure 18 top right). Prediction of electrical power is expected to be in agreement with test data (once power converter become available) as the characterization of individual TEMs has been proven.

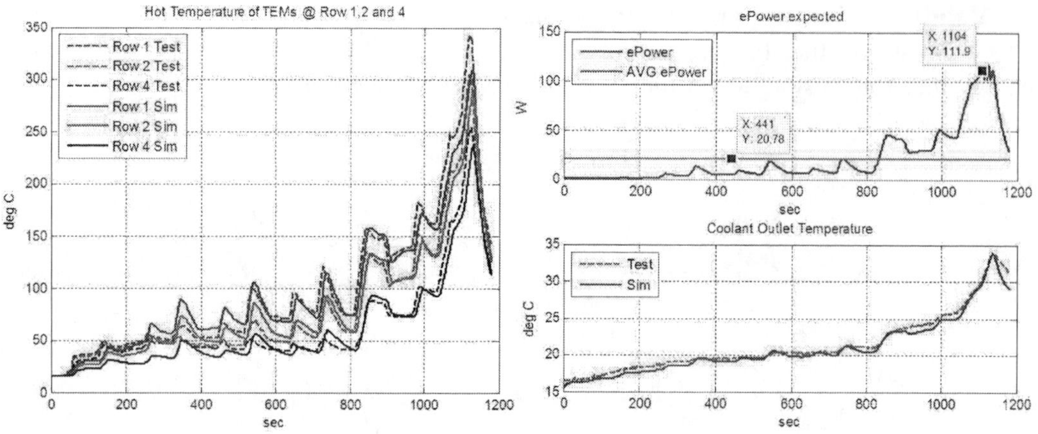

Figure 18. Validation work of 2x4x4 (72 TEMs) prototype

5 TEG SYSTEM OPTIMIZATION

Thermoelectric generation as a vehicle waste heat recovery technology has been on the OEMs portfolio for years, but different factors have prevented a move forward into commercialisation. Basically the cost and inadequate thermo-electric conversion efficiency are the limiting factors to achieving a cost effective technology with an attractive cost per gram of CO2 reduction. One of the main TEG design aims must be minimize the amount of TEMs along with maximizing heat transfer. The model proposed in this paper is used to find a cost effective solution.

5.1 Defining Cost Functions

To perform the optimization we have to decide what criteria of the design need to be taken into consideration. To help with this decision the design objective (minimum cost per CO2 gram reduction) is decomposed into a hierarchical criteria tree (figure 19) and, following a Analysis Hierarchical Process [25], the critical aspects to optimise, apart from cost, were identified. From the hierarchical criteria tree a further parameter tree was derived (as seen in figure 20) that defines the cost functions for the optimization process.

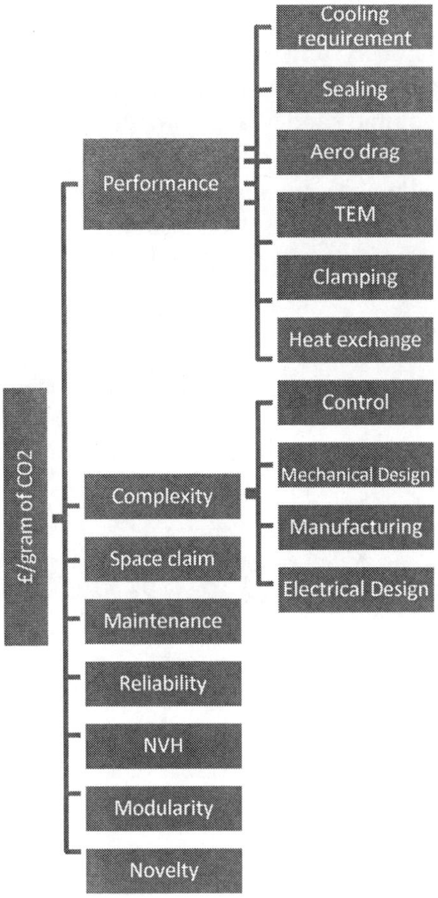

Figure 19. Design criteria of TEG design

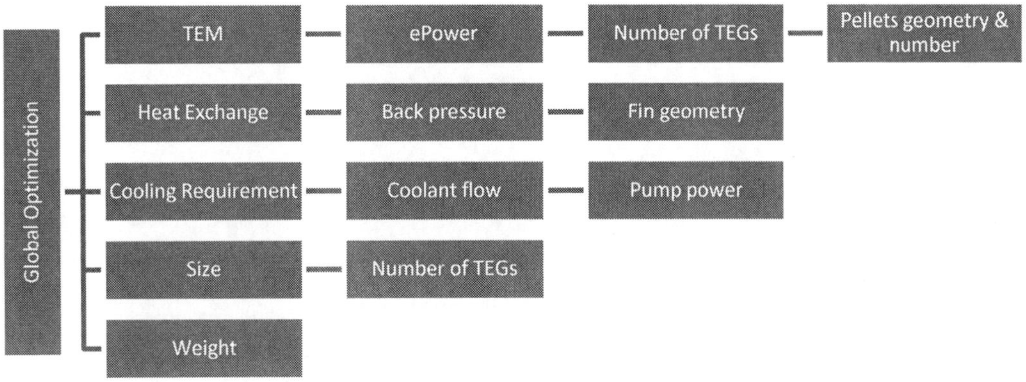

Figure 20. Critical parameter tree

In the higher levels of the tree there are parameters that capture most of the critical attributes of the design; these are chosen as cost functions (electrical power, back pressure and number of TEMs in this paper). The tree can be expanded further down to reveal parameters that can be used as degrees of freedom parameters (e.g. the length of pellet sensitivity to electrical power generation; fin thickness sensitivity to heat exchange performance, etc.).

5.2 Optimization Problem

The generic parametric TEG system model described in section two was used to optimise an existing design in terms of cost, size and weight without compromising performance for given volume and back pressure limits. As mentioned in the introduction of this section the main driver for cost is the thermoelectric material, therefore the degree of freedom parameters for the optimisation problem are related to the amount of thermoelectric material, as seen in figure 21.

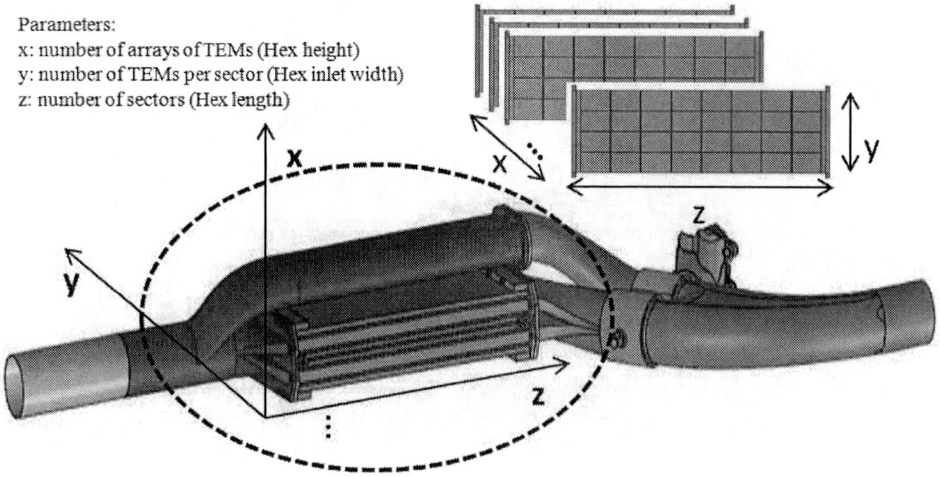

Figure 21. Optimization parameters

The optimization problem is summarized as this:

$$Max\{F(x,y,z); Min\{G(x,y,z)\}; Min\{H(x,y,z)\}\} \quad (2)$$

Such that:

$1 \leq x \leq 4;$

$2 \leq y \leq 8;$

$2 \leq z \leq 8;$

Where: F: electric power;
G: back pressure;
H: number of TEMs;
x: number of arrays of TEMs (heat exchanger height)
y: number of TEMs per sector (heat exchanger inlet width)
z: number of sectors (heat exchanger length)

Note: For simplicity of the computation the width of a single sector is assumed to be the width of a single TEM.

5.3 Optimization Process

The optimization problem was solved using a Design of Experiment (DoE) approach; completing each experiment (through virtual simulation in this case) by running the parametric TEG system model in Matlab&Simulink 2013b.The DoE was analysed using the recognised quality tool Minitab 17 and the global optimisation of all cost functions (electric power, number of TEMs and back pressure) using the response optimiser tool. The optimum design is then compared with the baseline model variant described in figure 11.

The following boundary conditions were used in the study: exhaust gas at vehicle speed of 120 kph and co-flow heat exchanger with constant pump flow of 8 l/min. Temperature limit of TEM in the study is 250 deg C.

5.4 Optimisation Results

5.4.1 DoE Results

The DoE results for power generation, exhaust back pressure and number of TEMs were used to produce the contour plot shown in figure 22. The levels of contour plot correspond to the number of TEMs which link directly to cost, mass and weight incurring on the designs. Looking into the baseline it belongs to a region of high performance but expensive designs. There are regions of less expensive designs with same or even lower back pressure if a moderate 20% of power generation is accepted. The saving in weight and cost could offset the reduction in power generation therefore there are opportunities of cost reduction.

The design cannot be bias to maximize power generation as the penalty to introduce back pressure on CO_2 emission may be significant (e.g. penalty of 6kPa increase is ~1% increase of CO_2 emission). Given the compromise between power generation and back pressure this DOE will show main effects and interactions of the parameters associated before performing the global optimization.

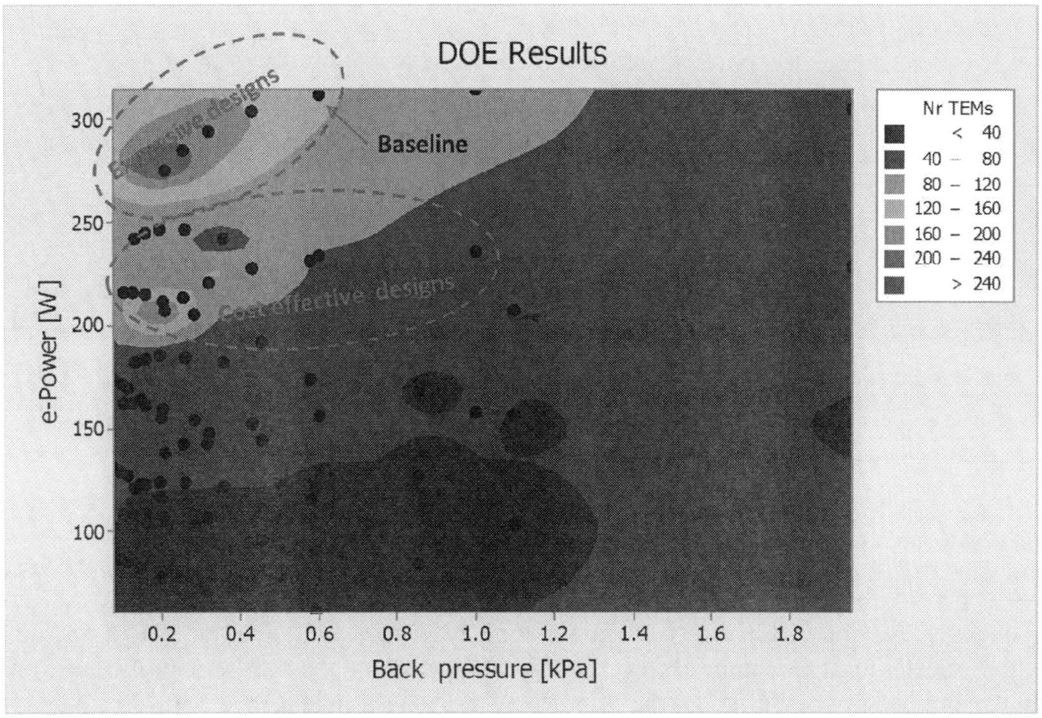

Figure 22. Opportunities for cost reduction

5.4.2 *Parameter sensitivity for e-Power*

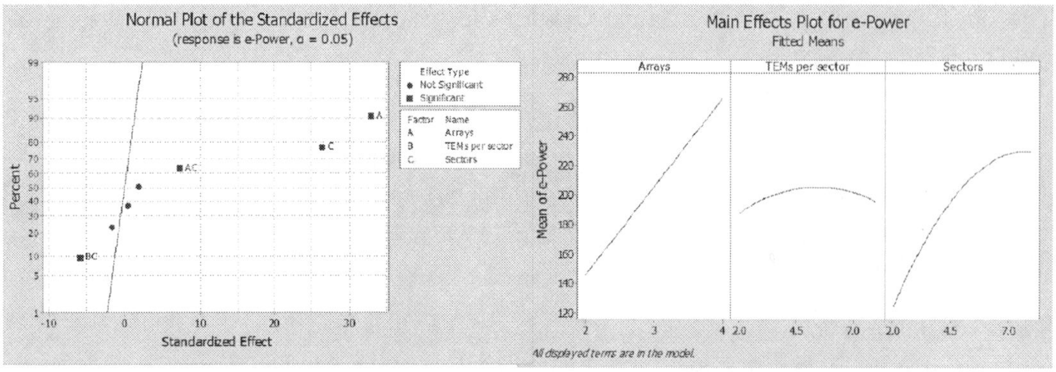

Figure 23. (a) Parameter significance for power generation; (b) Parameter main effect for power generation

Figure 23a exhibits parameters with a more significant effect on power generation. The number of arrays and sectors of the design have the larger effect on maximizing power generation. It is also shown that a choice of TEMs per sector and number of sectors can negatively interact to reduce power generation. The main effect plot, figure 23b, tells basically the same information of figure 23a, but also reveals an interesting insensitivity of power generation to number of TEMs per sector. From the parameter sensitivity information of figure 23 we can infer that for a local optimisation (power generation only) the number of TEMs per sector (i.e. the width of the TEG system assembly) is the parameter to exploit in order to minimize the number of TEMs without sacrificing power generation.

5.4.3 *Parameter sensitivity for Back pressure*

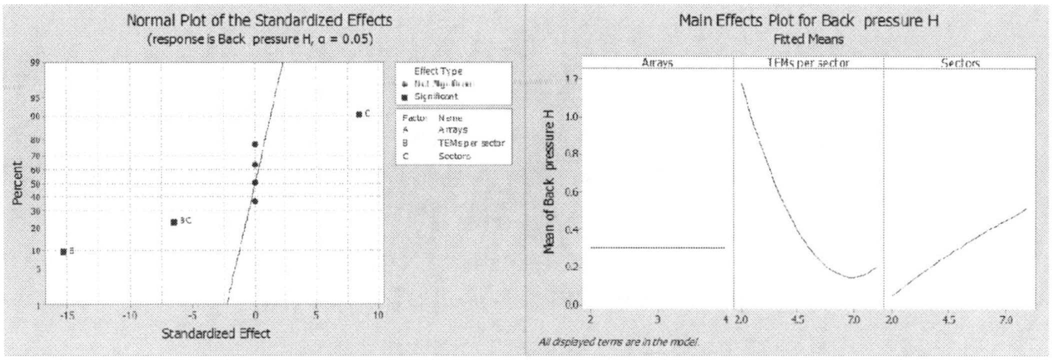

Figure 24. (a) Parameter significance for back pressure; (b) Parameter main effect for back pressure

Figure 24a shows parameters that have more significant effect on exhaust gas pressure. As expected the number of TEMs per sector, which relates to the width of heat exchanger, has the larger effect on minimizing back pressure. The number of sectors which relate to the length of the heat exchanger, has the larger significance on increasing back pressure whilst the number of arrays (i.e. height of the heat exchanger has negligible effect. The interaction between the width and length of the ducts also plays a part on the back pressure. The main effect plot of figure 24b confirms what the previous plot says. From the information of figure 24 we can infer that for a local optimisation (back pressure only) the number of TEMs per sector (width of the TEG system assembly) and the number of sectors (length of the TEG system assembly) have to be compromised to achieve minimum back pressure and number of TEMs.

5.4.4 Global Optimization

Having understood the local parameter sensitivity to the cost functions of the optimization problem it is important to understand the interactions of the parameters when multiple functions are optimized. Figures 23b and 24b compare the main effects plots for power generation and back pressure to understand these interactions. The number of TEMs per sector and number of arrays are the parameters which compromise maximum power and minimum back pressure whilst minimising the number of TEMs. In figure 25 all cost functions (back pressure and electrical power) are optimized at the same time using Minitab®17 via Response Optimizer tool. The tool computes the number of TEMs to fit in the design in order to satisfy a composite desirability score (weighted number based on cost function objectives. The optimum design consists of a TEG system assembly with 4 arrays; 6 TEMs wide and 3 TEMs length (total 72 TEMs).

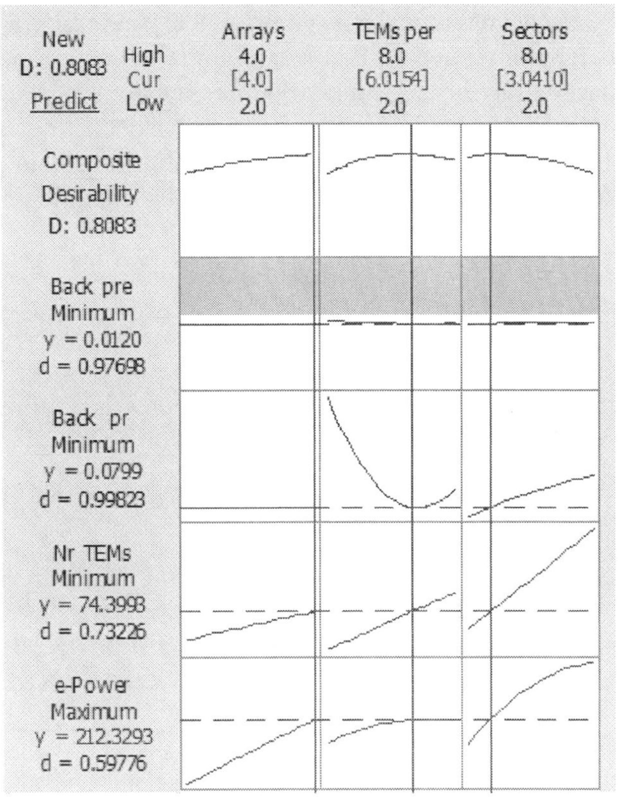

Figure 25. Global optimization

6 FURTHER WORK

The next step of this work is to complete the correlation work of the TEG system model for power generation and extend the modelling approach and optimization process to the second phase of VIPER 2.

7 CONCLUSIONS

- A parametric and modular TEG system model with compact offset-strip fin has been developed. The model can be used to scale the TEG system to the requirements of different vehicles in terms of fuel consumption and space available.
- A global optimization approach has been described using two widely available commercial tools. The importance of extending the parameter sensitivity knowledge from a local optimization to a global optimization was highlighted to help identify conflicting factors to multiple cost functions.
- The baseline design was optimised to use 43.5% less TE modules (72 rather than 128) at the expense of reducing electrical power generation by 20%. It represents about 30% weight reduction when accounting for cooling and exhaust components and close to 50% in cost reduction. The grams of CO_2 savings will not be affected as power reduction will be recovered by the substantial reduction on weight and back pressure when accounting for fuel economy improvement.
- Although details are not shown in this paper, the existing design can be further downsized by using less than 50% of the TE modules without sacrificing power generation performance if the reliability of the TE module is improved up to nominal temperature of 315 deg C.

8 ACKNOWLEDGEMENT

The authors are grateful for the support received from industrial partners in the VIPER 2 project including European Thermodynamics Ltd, University of Nottingham, JLR INNOVISTA and INNOVATE UK.

9 REFERENCES

[1] K. Smith and M. Thornton. *Feasibility of Thermoelectrics* for*Waste Heat Recovery* in. *Conventional Vehicles*. Technical Report. NREL/TP-540-44247. April 2009.

[2] Rowe, D. M., *Thermoelectrics Handbook: Macro to Nano*, CRC/Taylor & Francis, 2006.

[3] LaGrandeur, J., D. Crane, and A. Eder, *Vehicle Fuel Economy Improvement Through Thermoelectric Waste Heat Recovery*, Proceedings of Diesel Engine Emissions Reduction Conference, 1-7, 2005.

[4] Stephen Beeby, Neil White., *Energy harvesting for autonomous systems*, Chapter 5, 2010.

[5] M. A. Karri, B. T. Helenbrook, and E. F. Thacher. *Exhaust energy conversion by thermo- electric generation: Two case studies. Energy Conversion and Management,* 52(3):1596 – 1611, 2011.
[6] Yang, J., *Potential Applications of Thermoelectric Waste Heat Recovery in the Automotive Industry*, Proceedings of International Conference on Thermoelectrics, 170-174, 2005.
[7] Crane, D., and L. E. Bell, *Progress Towards Maximizing the Performance of a Thermoelectric Power Generator*, Proceedings of International Conference on Thermoelectrics, No. II-1, 1-6, 2006.
[8] Caillat, T., J. P. Fleurial, G. J. Snyder, and A. Borshchevsky, *Development of High Efficiency Segmented Thermoelectric Unicouples,* Proceedings of International Conference on Thermoelectrics, 282-285, 2001.
[9] LaGrandeur, J., D. Crane, S. Hung, B. Mazar, and A. Eder, *Automotive Waste Heat Conversion to Electric Power Using Skutterudite, TAGS, PbTe and BiTe,* Proceedings of International Conference on Thermoelectrics, No. B10-2, 1-6, 2006.
[10] Bell, L. E., High Power Density Thermoelectric Systems, Proceedings of International Conference on Thermoelectrics, 1-5, 2004.
[11] Xiaodong Zhang, K. T. Chau, C. C. Chan, *Overview of Thermoelectric Generation for Hybrid Vehicles, Journal of Asian Electric Vehicles* , vol. 6, no. 2, pp. 1119-1124, 2008.
[12] Kim, R. Y., and J. S. Lai, *Aggregated Modeling and Control of a Boost - Buck Cascade Converter for Maximum Power Point Tracking of a Thermoelectric Generator*, Proceedings of IEEE Applied Power Electronics Conference and Exposition, 1754 -1760, 2008.
[13] Khouzam, K., *Optimum Load Matching in Direct-Coupled Photovoltaic Power Systems - Application to Resistive Loads*, IEEE Transactions on Energy Conversion, Vol. 5, No. 2, 265-271, 1990.
[14] Kislovski, A., and R. Redl, *Maximum Power Tracking Using Positive Feedback*, Proceedings of IEEE Power Electronics Specialists Conference, 1065-1068, 1994.
[15] Kuo, Y. C., T. J. Liang, and J. F. Chen, *Novel Maximum Power Point Tracking Controller for Photovoltaic Energy Conversion System*, IEEE Transactions on Industrial Electronics, Vol. 48, No. 3, 594-601, 2001.
[16] Shmilovitz, D., *On the Control of Photovoltaic Maximum Power Point Tracker via Output Parameters*, IEE Proceedings - Electric Power Applications, Vol. 152, No. 2, 239-248, 2005.
[17] Femia, N., G. Petrone, G. Spagnuolo, and M. Vitelli, *Optimization of Perturb and Observe Maximum Power Point Tracking Method*, IEEE Transactions on Power Electronics, Vol. 20, No. 4, 963-973, 2005.
[18] Veerachary, M., T. Senjyu, and K. Uezato, *Neural Network Based Maximum - Power - Point Tracking of Coupled - Inductor Interleaved - Boost Converter - Supplied PV System Using Fuzzy Controller*, IEEE Transactions on Industrial Electronics, Vol. 50, No. 4, 749-758, 2003.
[19] D'Souza, N. S., L. A. C. Lopes, and X. Liu, *An Intelligent Maximum Power Point Tracker Using Peak Current Control*, Proceedings of IEEE Power Electron-ics Specialists Conference, 172-177, 2005.
[20] M.H. Teruel, C.Y. Nakashima, P. Paglione, *Rectangular offset strip-fin* plate *heat exchanger lumped parameters* dynamic *model*, Brazilian Symposium on Aerospace Eng. & Applications, 2009.

[21] Manglik, R. M., Bergles, A. E., 1990, *"The Thermal-Hydraulic Design of the Rectangular Offset-Strip-Fin Compact Heat Exchanger"*, Compact Heat Exchangers, Hemisphere Publishing, New York, pp. 123-149.

[22] M AKarri, *MODELING OF AN AUTOMOTIVE EXHAUST THERMOELECTRIC GENERATOR*, 2005.

[23] Bitschi, Andreas, *Modelling of thermoelectric* devices for *electric power generation*, ETH, 2009.

[24] D Brennan, R Gilchrist, B Wicksteed, *Heat exchanger design and analysis for an automotive thermoelectric generator*, Vehicle Thermal Management Systems Conference, Coventry TechnoCentre, UK, May 15-16, 2013.

[25] Omkarprasad S. Vaidya, Sushil Kumar, *Analytic hierarchy process: An overview of applications*, European Journal of Operational Research, Volume 169, Issue 1, 16 February 2006, Pages 1-29.

A new approach for predicting the maximum fuel saving potential of an automotive thermoelectric generator in its early development stage

A BAUKNECHT, M REXEIS and **R A ALMBAUER**
Institute of Internal Combustion Engines and Thermodynamics,
Graz University of Technology, Austria

ABSTRACT

This paper presents a novel methodology for predicting the fuel saving potential of an automotive thermoelectric generator (ATEG) in its early development stage. The methodology incorporates a newly developed model describing a power optimised ATEG using two main variables: the number of transfer units of the exhaust gas heat exchanger and the ZT value of the thermoelectric material. Using the presented methodology various ATEG concepts, mainly differing in their integration position along the vehicle's exhaust system, are described and studied. The maximum fuel saving potential is calculated for legal and customer driving cycles.

NOTATIONS

Latin and Greek symbols

A	Area [m²]	P	Power [W]
c_p	Specific heat capacity [J/kgK]	$\dot{Q}$	Heat flux [W]
c_x	Air resistance coefficient [-]	R	Electrical resistance [Ω]
FC	Fuel consumption [kg]	t	Time [s]
h	Specific enthalpy [J/kg]	T	Temperature [K]
$\dot{H}$	Enthalpy stream [W]	ZT	Figure of merit [-]
k	Heat transfer coefficient [W/m²K]		
K	Thermal conductance [W/K]	α	Seebeck coefficient [V/K]
m	Mass [kg]	η	Efficiency [-]
$\dot{m}$	Mass flow rate [kg/s]	θ	Temperature ratio [-]
M	Torque [Nm]	Θ	Temperature difference ratio [-]
NTU	Number of Transfer Units [-]	τ	Optimisation parameter [-]
p	Pressure [N/m²]	ω	Angular velocity [rad/s]

Subscripts

amb	Ambient conditions	ICE	Internal Combustion Engine
ATEG	Automotive TEG	in	Incoming
c	Cold side	int	Integrated scenario
cool	Coolant	m	Mean value
e	Effective	max	Maximum value
el	Electrical	opt	Optimum value
exh	Exhaust gas	out	Out-going
gen	Electric generator	TE	Thermoelectric
h	Hot side	TEG	Thermoelectric Generator
HX	Heat exchanger	veh	Vehicle

1 INTRODUCTION

Thermoelectricity is well known for means of waste heat recovery (WHR), especially in the automotive world. In the past few years several prototypes of thermoelectric generators have been built and tested [1]. However, the transition to series production of such a component has not been made yet. The reason for this are poor cost to benefit ratios in comparison to other CO_2 saving technologies, but also technological burdens such as designing long-time stable and high-temperature resistant TE modules [2,3].

The integration of a thermoelectric generator (TEG) is usually considered for the first time when a new vehicle is in the early stage of the development process. In this phase the main question is whether the utilisation of a TEG would lead to a positive cost to benefit ratio or not. It is well known that any changes made later in the development process will lead to a substantial amount of effort and will also require relatively large financial expenses. This means there is a need for knowledge and tools to describe the TEG's characteristics when only little details about its actual design are already known.

One way to predict the positive or negative effects of a new vehicle component is to use a retrofit approach. This task aims for blending in the new subsystem into a given vehicle architecture and involves dealing with so-called integration effects. In general this term denotes any consequences that alter the original vehicle characteristic profile in regard to the vehicle's entire life cycle (manufacturing, operating, servicing and recycling). In terms of a WHR component, such as a TEG, the most relevant integration effects are those which affect the vehicle's energy management systems and have an impact on fuel consumption. A significant reduction in fuel consumption can result from lowering the load of the electric generator and by minimising the side effects, such as the possible increase in exhaust gas back pressure, vehicle weight and air resistance at the same time [4]. When combined the integration effects need to have a positive balance in order to qualify the thermoelectric generator as a useful measure to reduce CO_2 emissions. For predicting these relationships a wide range of simulation tools is (commercially) available, but they mostly require the TEG component to be specified in an advanced way. In an early development stage the required level of detail can usually only be provided by relying on empirical values. Since this practice is limited in covering all possible solutions, it does not necessarily lead to a stable answer.

This paper presents a novel approach for predicting the fuel saving potential of an automotive thermoelectric generator (ATEG). The methodology is based on a modular simulation system comprising a well-established model for calculating vehicle longitudinal dynamics and a newly developed system-level model of the ATEG component. In this paper the simulation system is used to analyse system behaviour in legal and customer driving cycles.

2 THE FUEL SAVING POTENTIAL OF AN AUTOMOTIVE TEG

The fuel saving potential of an ATEG is not only determined by its electrical power output, but also by other possibly occurring integration effects such as increased exhaust gas back pressure, higher vehicle weight or larger air resistance. The resulting difference in fuel consumption can be calculated by integrating the modified torque profile of the engine as shown in eq. (1) and eq. (2) [4]. The subscript "int" denotes the scenario of an ATEG-equipped vehicle.

$$\Delta FC_{\mathrm{int}} = \int FC(\omega_{\mathrm{ICE}}(t), M_{\mathrm{ICE}}(t))\mathrm{d}t - \int FC(\omega_{\mathrm{ICE}}(t), M_{\mathrm{ICE,int}}(t))\mathrm{d}t \tag{1}$$

$$M_{\mathrm{ICE,int}}(t) = M_{\mathrm{ICE}}(t) - \Delta M_{\Delta P_{\mathrm{gen}}}(t) + \Delta M_{\Delta p_{\mathrm{exh}}}(t) + \Delta M_{\Delta m_{\mathrm{veh}}}(t) + \Delta M_{\Delta c_{\mathrm{x}}}(t) \tag{2}$$

The electrical power output of an ATEG generally depends on two parameters: the system's conversion efficiency and the available exhaust gas energy. The former can be formulated as a product of the exhaust gas heat exchanger efficiency and the thermoelectric (TE) efficiency [5]. The available exhaust gas energy can be determined in a simplified way by neglecting the kinetic and potential energy of the fluid since these parts are usually low in comparison to the enthalpy stream. The latter is then calculated using the exhaust gas mass flow rate and the specific enthalpy of the exhaust gas entering the ATEG and at ambient conditions respectively. The resulting physical correlation is shown in eq. (3).

$$P_{\mathrm{el}} = \eta_{\mathrm{TEG,veh}} \dot{H}_{\mathrm{exh,in}} = \eta_{\mathrm{HX,h}} \eta_{\mathrm{TE}} \dot{m}_{\mathrm{exh}} \left(h_{\mathrm{exh,in}}(T_{\mathrm{exh,in}}) - h_{\mathrm{exh,amb}}(T_{\mathrm{amb}})\right) \tag{3}$$

In order to determine the TE efficiency numerous models have been proposed [1,6]. As is usual for models they can be classified by the position of their system boundaries. Table 1 gives an overview of some commonly known models that differ in the modelled irreversibility (i.e. thermal and electrical resistances) and thus also in the way the system's energy conversion efficiency is calculated. All four models have in common that this quantity is determined by using the ratio θ between the temperature of the TE material's hot and cold side, see eq. (4).

$$\theta = \frac{T_{\mathrm{TE,h}}}{T_{\mathrm{TE,c}}} \tag{4}$$

Table 1: System characteristics of selected TEG models.

Position of system boundaries	Type of system modelling	Thermal / electrical resistances	System's energy conversion efficiency	
Entire TEG component	Reversible heat engine (Carnot engine)	No / No	$\eta_{\mathrm{TEG}} = 1 - \theta$	(5)
	Endoreversible heat engine (Curzon-Ahlborn engine)	Yes / No	$\eta_{\mathrm{TEG}} = 1 - \sqrt{\theta}$	(6)
TE module	Real system operated at "maximum power point"	Yes / Yes (for all functional layers)	$\eta_{\mathrm{TE}} = (1-\theta)\left(\frac{\sqrt{1+Z_{\mathrm{e}}T_{\mathrm{m}}}-1}{\sqrt{1+Z_{\mathrm{e}}T_{\mathrm{m}}}+\theta}\right)$	(7)
Thermo-couple	Ideal system operated at "maximum power point"	Yes / Yes (but only within TE material)	$\eta_{\mathrm{TE}} = (1-\theta)\left(\frac{2(1+\theta)}{ZT_{\mathrm{m}}}+\frac{\theta}{2}+\frac{3}{2}\right)^{-1}$	(8)

The first two approaches model the TEG component as a (endo-)reversible heat engine [7-9]. They are capable of providing results without specifying TEG characteristics and can therefore be rated as theoretical reference models. The second set of models treat the TEG on the level of a TE module or a thermocouple and are commonly known in literature [10-13]. For the present study only the operation at the so-called "maximum power point" is of interest since this use case is mainly relevant for an ATEG. In contrast to the first set of models information about the (effective) $ZT_{(e)}$ value is required which is defined as shown in eq. (9) and eq. (10) respectively [10,14].

$$ZT = \frac{\alpha^2}{KR}T \tag{9}$$

$$Z_{\mathrm{e}}T = \frac{\alpha^2}{K_{\mathrm{e}}R_{\mathrm{e}}}T \tag{10}$$

The parameters K and R denote the thermal conductance and electrical resistance of the TE materials. The corresponding effective values, K_{e} and R_{e}, additionally account for the functional layers of the TE module, such as metal conductors or isolation layers, as well as contact resistances at the interfaces in-between. When modelling the TEG as a real system these quantities can only be determined when actual design considerations or at least some well-founded assumptions are made. In case of the ideal system it is sufficient specifying the TE material in terms of its ZT values.

3 METHODOLOGY

The novel methodology proposed in this paper aims to provide a practical way for predicting the maximum fuel saving potential of an ATEG in a wide range of typical driving scenarios. For this purpose the ATEG is modelled as a power optimised component which can be installed at any position along the flow path of the exhaust gas, i.e. starting at the manifold and ending with the rear muffler. Heat rejection is realised by connecting the heat sink of the ATEG to the vehicle's cooling system. A further physical connection is made with the on-board electrical system using a maximum power point tracking (MPPT) converter [15].

The methodology is assisted by a computational simulation system which is of modular design and comprises two stand-alone computational models as well as several databases. An overview of its hierarchical structure is given in figure 1.

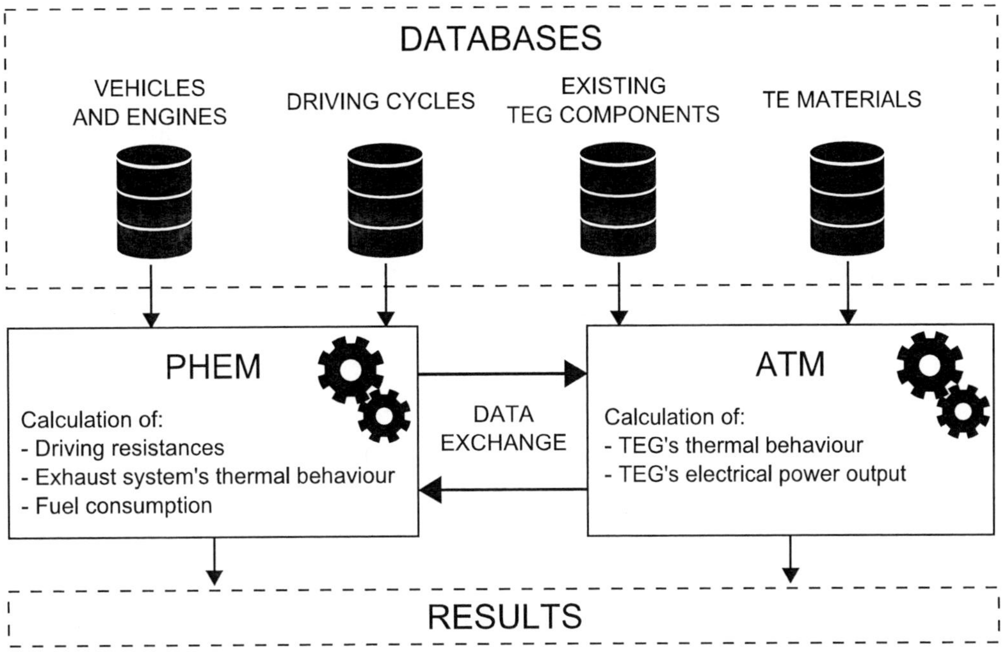

Figure 1: Hierarchical structure of the simulation system.

The basic module of the simulation system is the Passenger Car and Heavy Duty Emission Model (PHEM) which has been designed to describe vehicle longitudinal dynamics and to calculate emission characteristics as well as fuel consumption [16]. Its information structure is shown in figure 2.

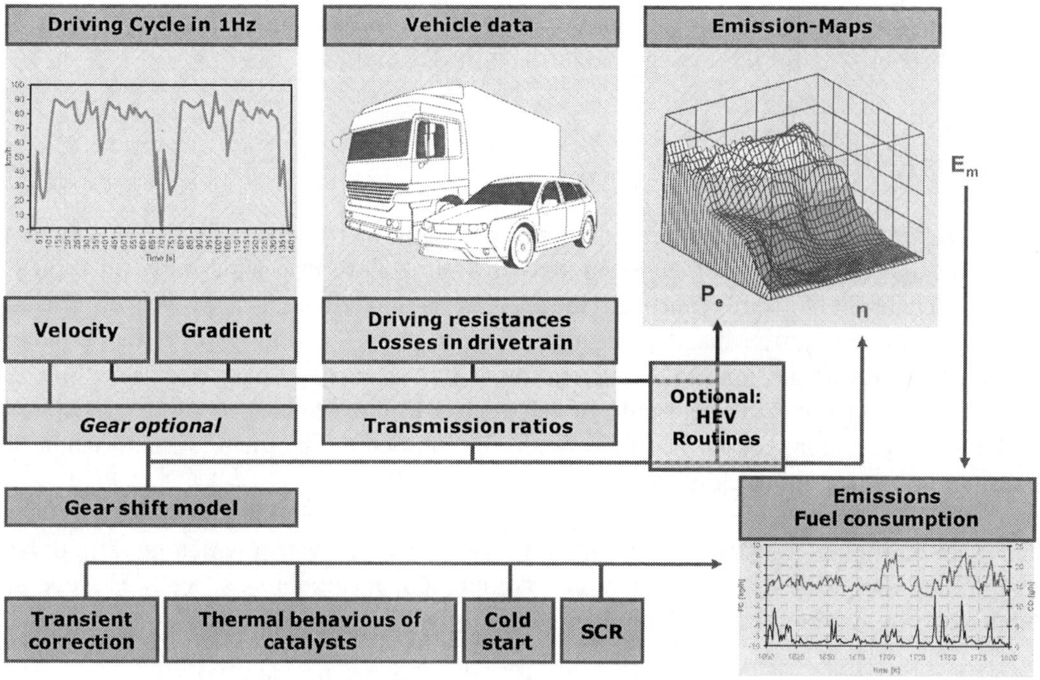

Figure 2: Schematic layout of the information structure of the Passenger Car and Heavy Duty Emission Model (PHEM) [16].

PHEM offers the functionality to specify arbitrary velocity and shifting profiles representing either steady-state operating conditions or dynamic driving cycles. The model also accounts for the mechanical and electrical energy consumption of auxiliary components, such as the air-conditioning compressor or the alternator. Additionally PHEM is capable of modelling the vehicle's exhaust system parametrically so that mass flow rates and exhaust gas temperatures at any position can be obtained.

Within the simulation system the Passenger Car and Heavy Duty Emission Model utilises a two-way data interface for sharing interim results with the second main module, the Automotive TEG Model (ATM). It is designed using several MATLAB® routines and describes the quasi-stationary behaviour of a power optimised ATEG. For this purpose it is based on the model of the ideal system (cf. chapter 2, table 1) and additionally incorporates special boundary conditions that result from the automotive application. The ATM also accesses databases storing characteristics of already realised TEG components (e.g. heat transfer coefficients) and data of selected TE materials which are suitable for automotive use. The ATM itself is presented in the following paragraphs on the basis of the ATEG's energy balance shown in figure 3.

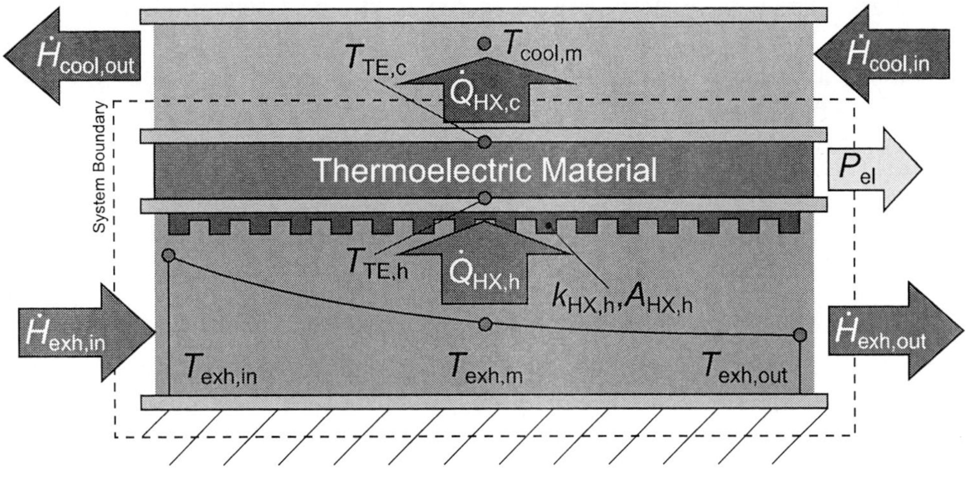

Figure 3: Energy balance of the automotive thermoelectric generator.

As shown in eq. (3) the TEG component's efficiency can be formulated as the product of the exhaust gas heat exchanger's efficiency and the thermoelectric efficiency. The former is generally defined as the ratio of the actually transferred heat flux on the hot side and the enthalpy stream of the exhaust gas entering the flow channel of the exhaust gas heat exchanger. This ratio can also be expressed by using the Number of Transfer Units (NTU_h) and the temperature difference ratio Θ_h as shown in eq. (11).

$$\eta_{HX,h} = \frac{\dot{Q}_{HX,h}}{\dot{H}_{exh,in}} = \frac{k_{HX,h} A_{HX,h} \left(T_{exh,m} - T_{TE,h}\right)}{\dot{m}_{exh} c_{p,exh} \left(T_{exh,in} - T_{cool,m}\right)} = NTU_h \Theta_h \tag{11}$$

In order to determine the temperature difference ratio the following quantities have to be described. The thermodynamic mean temperature $T_{exh,m}$ accounts for the fact that the heat transfer occurs at the gliding temperature which is defined as given in eq. (12).

$$T_{exh,m} = \left(T_{exh,in} - T_{exh,out}\right) \left[\ln\left(\frac{T_{exh,in}}{T_{exh,out}} \right) \right]^{-1} \tag{12}$$

The hot side temperature $T_{TE,h}$ can be formulated as shown in eq. (13), where the parameter τ (value range 0 to 1) describes the ratio of the temperature drop across the hot side of the TEG component and the total available temperature difference.

$$T_{TE,h} = T_{exh,m} - \tau \left(T_{exh,m} - T_{cool,m}\right) = T_{exh,m} - \tau \Delta T_m \tag{13}$$

For a power optimised ATEG the ideal value of τ can be derived analytically with the result that half of the total available temperature difference should drop across the hot side of the TEG component [17]. Correspondingly a "maximum power condition" in regard to the exhaust gas heat exchanger can be defined, see eq. (14).

$$\tau_{ATEG,opt} = 0.5 \tag{14}$$

A further assumption is made in regard to the cold side temperature $T_{TE,c}$. For the present model it shall equal the temperature of the coolant which is assumed to be constant along the flow channel of the coolant heat exchanger. This practice is justified by the fact that the coolant's specific heat capacity and mass flow rates are much higher in comparison to the exhaust gas [17].

$$T_{TE,c} = T_{cool,m} = T_{cool,in} \tag{15}$$

The electrical power output can then be calculated using the ideal model (cf. chapter 2, table 1) and equations (11) to (15) which account for the irreversibility that occur outside of the TE material. Using this new approach allows to solve eq. (3) as long as suitable values for the two unknown quantities, i.e. the ZT value of the TE Material and the Number of Transfer Units of the exhaust gas heat exchanger (NTU_h), are chosen. These values are usually found in literature or can be derived from measurements of already existing TEG prototypes. An example is given in the next chapter, where the application of the model is shown.

4 APPLICATION OF THE MODEL

4.1 Model validation

The model presented in the previous chapter is initially validated against test data of an ATEG prototype which shall be selected from a group of three existing hardware variants. In this regard the most important eligibility criterion is the compliance with the "maximum power condition" (cf. chapter 3) which is analysed for the three available ATEG prototypes. The results are shown in figure 4.

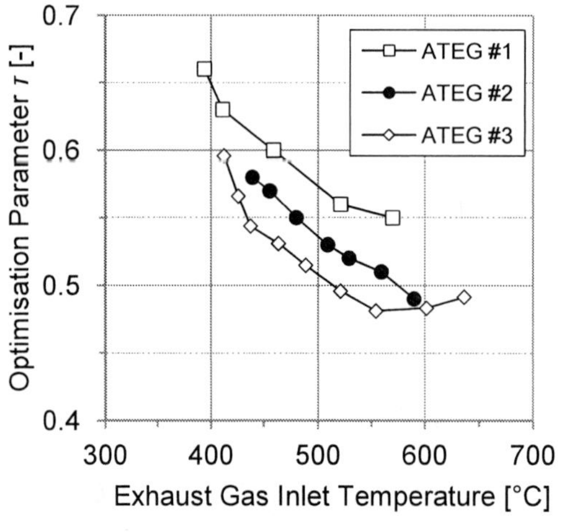

Figure 4: Comparison of three different ATEG prototypes on the basis of the optimisation parameter τ.

The comparison of the three ATEG prototypes shows that ATEG #1 is not suited for validation since the value $\tau_{ATEG,opt} = 0.5$ is never reached within the investigated operating points. In case of ATEG #2 and ATEG #3 a much better characteristic can be observed. The mean values are $\tau_{ATEG,2,m} = 0.54$ and $\tau_{ATEG,3,m} = 0.52$ respectively. As a conclusion ATEG#3 is selected for the validation of the ATM.

The model validation itself is based on an on-road test drive which reflects the characteristics of the relevant legal and customer driving cycles. In detail the test drive is a cold-started and rather dynamic driving cycle that also includes several engine idling phases. The vehicle speed on average is $v_{veh,m}$ = 55 km/h, while the maximum value is $v_{veh,max}$ = 132 km/h. Figure 5 shows the measured quantities that are used as input data for the ATM, while figure 6 presents a comparison of the measured values and the corresponding simulation results.

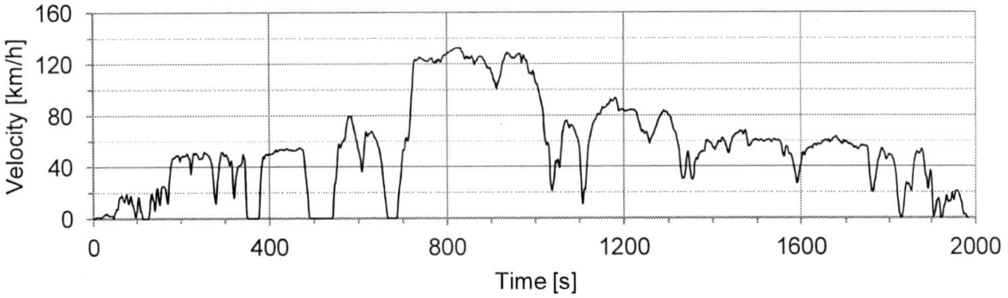

(a) Vehicle speed.

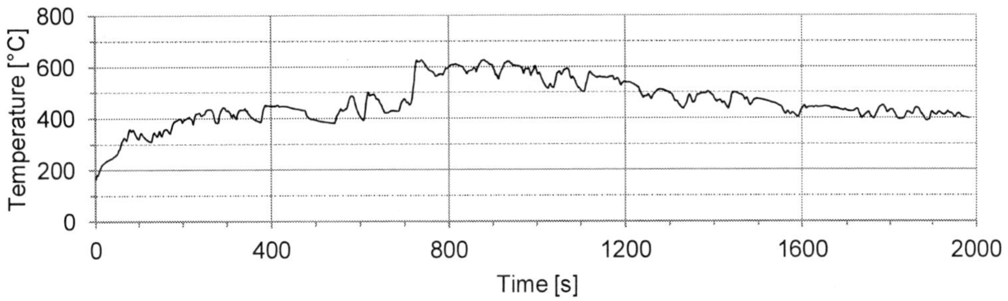

(b) Exhaust gas temperature at inlet of ATEG component.

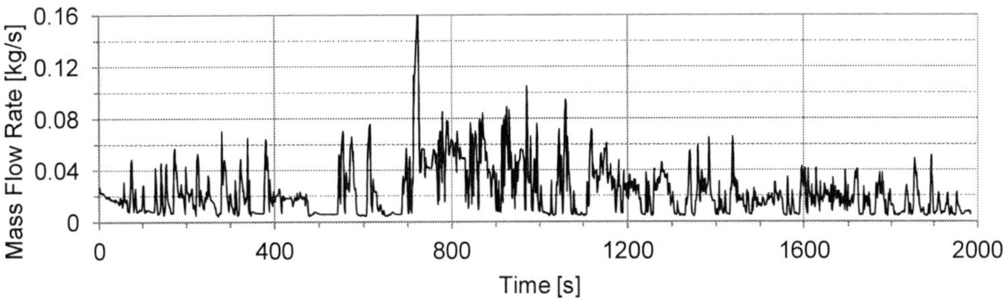

(c) Exhaust gas mass flow rate.

Figure 5: Simulation input data recorded during a dynamic on-road test drive.

Figure 6 (a) indicates that the exhaust gas outlet temperature can be modelled very well. On average the simulated values are $\Delta T_{exh,out,m}$ = 8.8 K higher than the measured values. Highly dynamic phases such as full load accelerations cannot be captured properly due to the quasi-stationary approach of the ATM. A similar picture emerges when analysing the comparison of the TE hot side temperature, shown in figure 6 (b). In this case the mean difference between simulation and measurement is $\Delta T_{TE,h,m}$ = 14.5 K, while the biggest differences occur during heat up phases of the ATEG component. Figure 6 (c) shows the comparison of the measured and simulated electrical power output. Since the latter is calculated using the measured mass flow rate, its dynamic is also reflected in the simulation results. On average a good quantitative agreement with a mean difference of $\Delta P_{el,m} = 6.2$ W can be achieved.

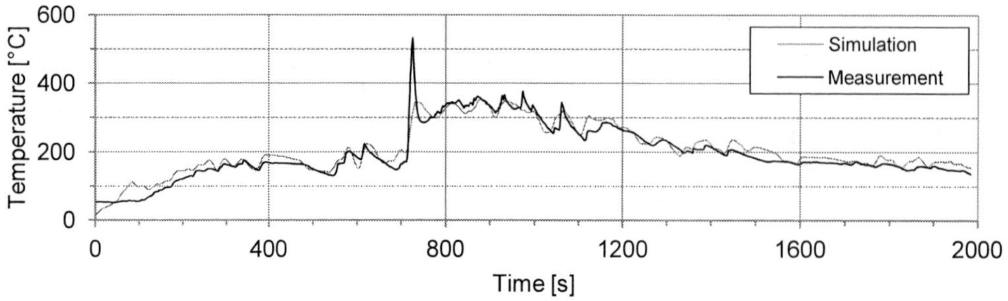

(a) Exhaust gas temperature at outlet of ATEG component.

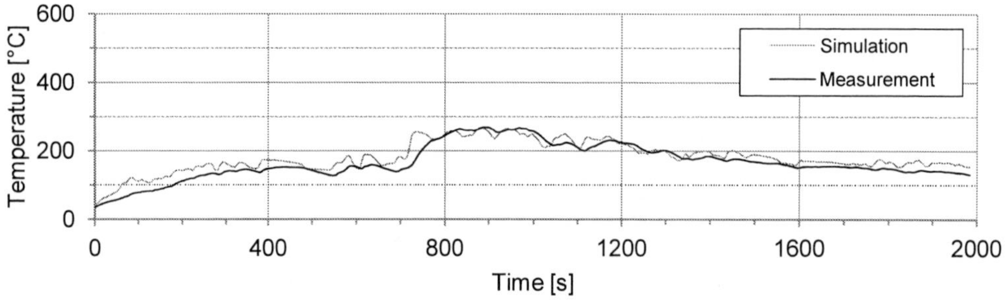

(b) Material temperature at TE hot side.

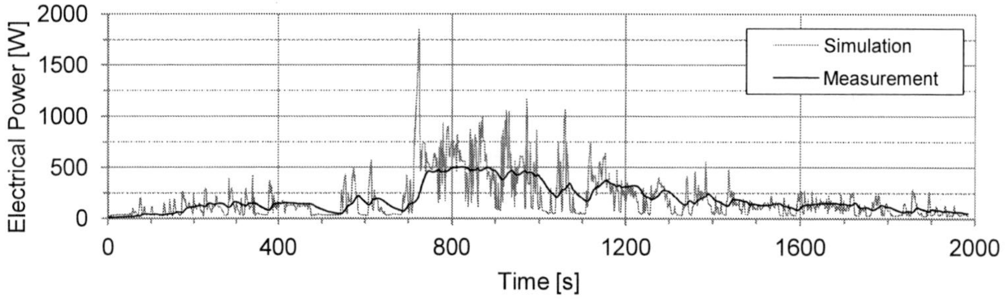

(c) Electrical power output.

Figure 6: Comparison of measured and simulated results.

In summary the simulation results show a good qualitative and quantitative agreement in regard to the measured values. The discussed discrepancies mainly occur due to the fact that the ATM is based on a quasi-stationary approach. Additionally the ATEG prototype does not exactly fulfil the "maximum power condition" in all relevant operating points. Taking into account these aspects the model can be considered as validated.

4.2 Scope of the simulation study

The model is now used to predict the fuel saving potential of a power optimised ATEG installed in a reference vehicle which reflects the current state of automotive technology. The choice for the present study is a mid-sized sedan equipped with a 4 cylinder petrol engine (direct fuel injection, turbo-charger, variable valve actuation) and manual six-speed transmission. The vehicle has been analysed at the Institute's chassis dynamometer and the obtained data, in particular in terms of driving resistances and exhaust system characteristics, is used as modelling parameters in PHEM (cf. chapter 3).

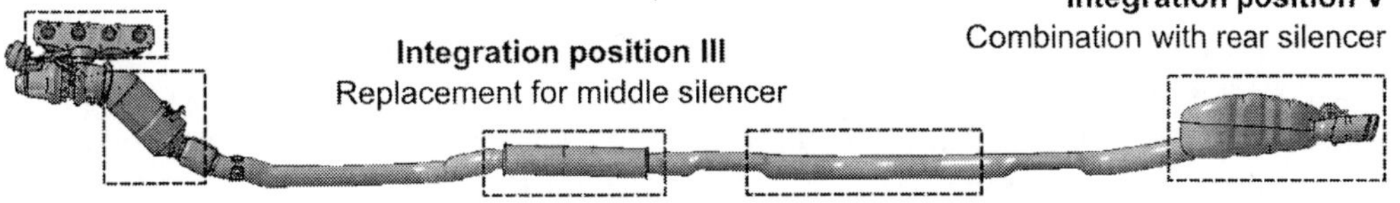

Figure 7: Overview of the possible integration positions along the exhaust system of the reference vehicle.

The performance of the ATEG will be studied for five different integration positions (as shown in figure 7) which are located along the single-entry exhaust system of the reference vehicle. In the first two scenarios the ATEG is combined with the manifold and the catalytic converter respectively. The third integration position is based on the assumption that the ATEG replaces the middle silencer. A further option is to use free installation space which would only be needed in case the vehicle is equipped with a Diesel engine and therefore employs additional emission control components. For the fifth scenario the ATEG is combined with the rear silencer.

The TEG itself is described using two modelling parameters. The first one is the heat transfer characteristic of the exhaust gas heat exchanger, i.e. the correlation between the exhaust gas temperature at the inlet of the ATEG component and the Number of Transfer Units. Figure 8 shows the characteristic curves of the three existing ATEG prototypes which will be used for this study.

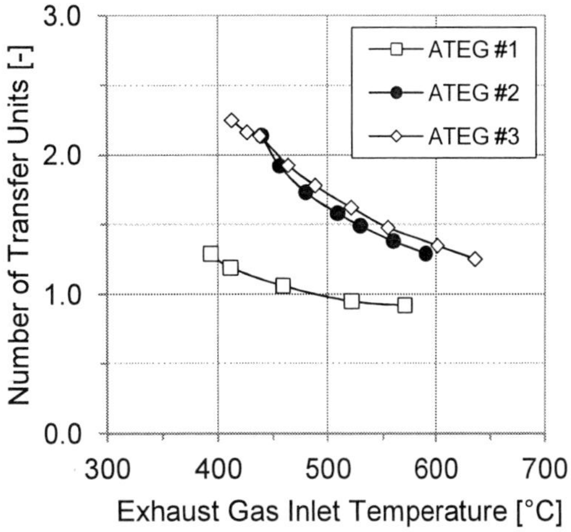

Figure 8: Comparison of three ATEG prototypes on the basis of the exhaust gas heat exchanger performance expressed by the Number of Transfer Units.

The second modelling parameter is the ZT value of the deployed TE materials. The present study is made on the basis of skutterudites which currently seem to be one of the best choices for automotive applications, especially in terms of ZT values, high temperature stability and costs [18].

In order to determine the maximum fuel saving potential only the inevitable integration effects shall be considered. The first integration effect is the additional weight of the TEG component which is assumed to be m_{ATEG} = 15kg. The second integration effect is the increase in exhaust gas pressure which is considered in correspondence with the assumed heat exchanger characteristics.

In terms of the cooling system it is assumed that no additional radiator is needed which would increase vehicle weight as well as air resistance. Any warm-up effects are neglected and the electrical power demand is supposed to be unlimited. The possible influence of these additional integration effects will be considered in future studies.

The fuel saving potential is determined for three different driving cycles, namely the New European Driving Cycle (NEDC), the Worldwide Harmonized Light Vehicles Test Cycle (WLTC) and the Common Artemis Driving Cycle (CADC). The corresponding vehicle speed profiles are shown in figure 9. The NEDC and WLTC are legal driving cycles, while the CADC is supposed to reflect customers' real driving behaviour [19]. The three cycles are not only different in duration and distance, but also in terms of maximal and average vehicle speed, as shown on the right hand side of figure 9.

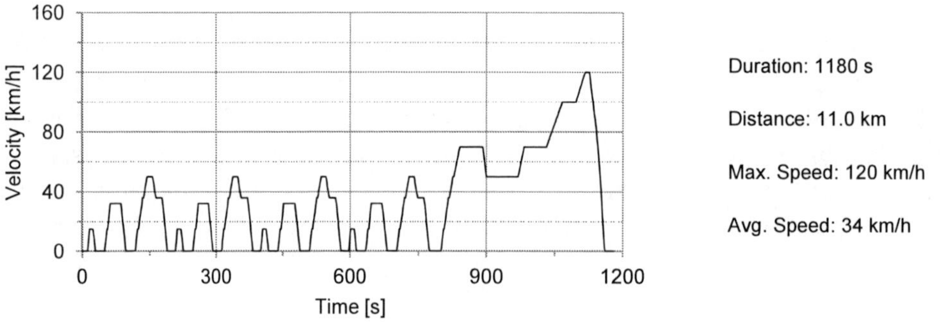

(a) New European Driving Cycle (NEDC).

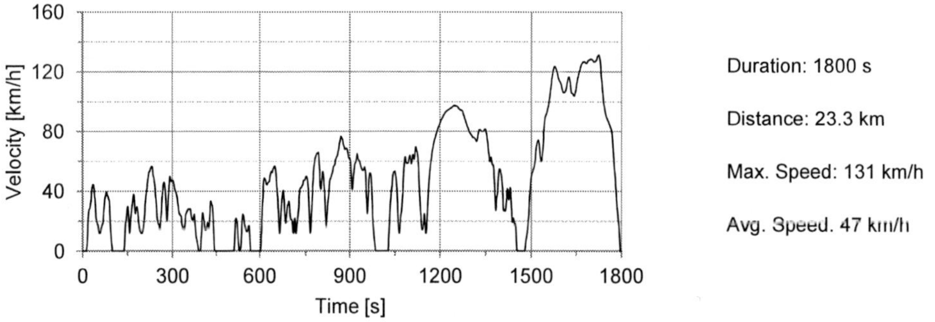

(b) Worldwide Harmonized Light Vehicles Test Cycle (WLTC).

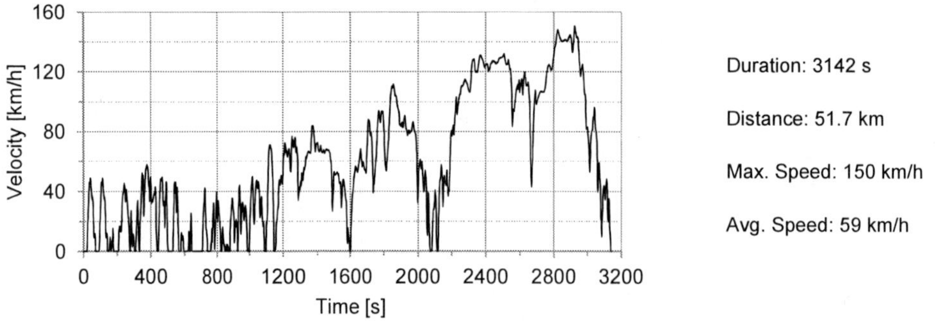

(c) Common Artemis Driving Cycle (CADC).

Figure 9: Vehicle speed profiles and characteristics of the simulated driving cycles.

4.3 Simulation results

Figure 10 shows the average electrical power output of the simulated ATEGs for the three driving cycles. It can be seen that the integration position is the major effect in terms of device performance and has a much higher influence than the characteristics of the exhaust gas heat exchanger. Especially in case of the NEDC the variation of the NTU characteristics has only little consequences. The highest power output can be achieved for the second integration position, i.e. the combination with the catalytic converter, in all driving cycles. This is due to the exothermic reactions occurring in this component that lead to higher exhaust gas temperatures in comparison to the engine's manifold [20].

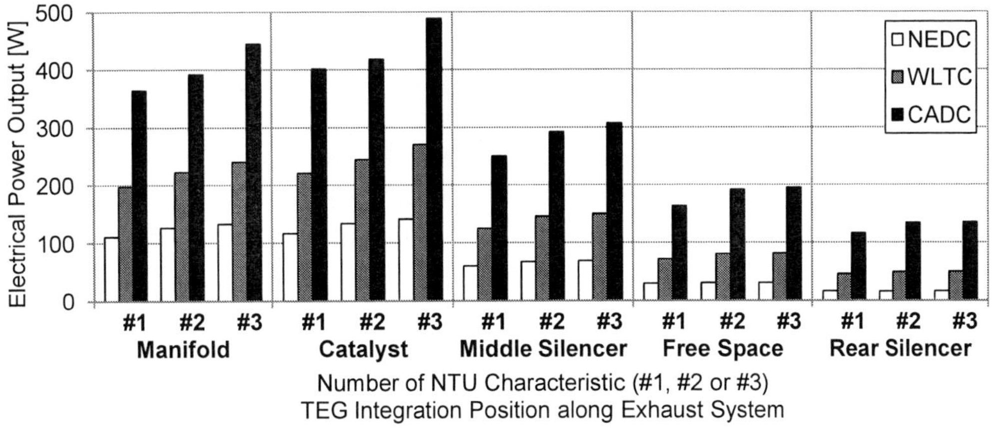

Figure 10: Average electrical power output in different driving cycles as a function of integration position and NTU characteristics.

The following results are presented on the basis of the best NTU characteristic (i.e. of ATEG #3) and for three different scenarios: In scenario A no integration side effects are taken into account. Scenario B also includes the increased exhaust gas back pressure, while scenario C additionally incorporates the ATEG's component weight. Figure 11 shows the fuel saving potential for these scenarios.

In scenario A a reduction in fuel consumption can be achieved for every integration position with values up to -2.7% in case of integration position II and the CADC. If the described integration effects are also taken into account the fuel saving potential decreases for all options investigated. The results of scenario B illustrate the influence of the increased back pressure which can be quantified with values of about 0.2%. Scenario C shows the additional influence of the increased vehicle weight which decreases the potential further by values of about 0.4 %. Both numbers are in line with what has already been reported in literature and sum up to a total reduction in fuel saving potential of about 0.6% [4,21,22].

As a consequence two of the simulated integration positions, i.e. position IV (free installation space) and position V (rear silencer), only lead to a minor or to no reduction of fuel consumption at all. Integration position III (middle silencer) shows only minor fuel saving potential in the NEDC with a value of -0.1 %. In case of the other two driving cycles values of -0.6% and -1.1% can be reached. The highest potential in a legal driving cycle can be generated if the ATEG is combined with the catalytic converter (-0.8% in the NEDC and -1.5% in the WLTC) or, as a second choice, with the manifold (-0.7% and -1.3% respectively). In regard to customer driving situations, which are reflected by the CADC, a maximum reduction of fuel consumption of -1.8% (integration position I) and -2.0% (integration position II) can be achieved.

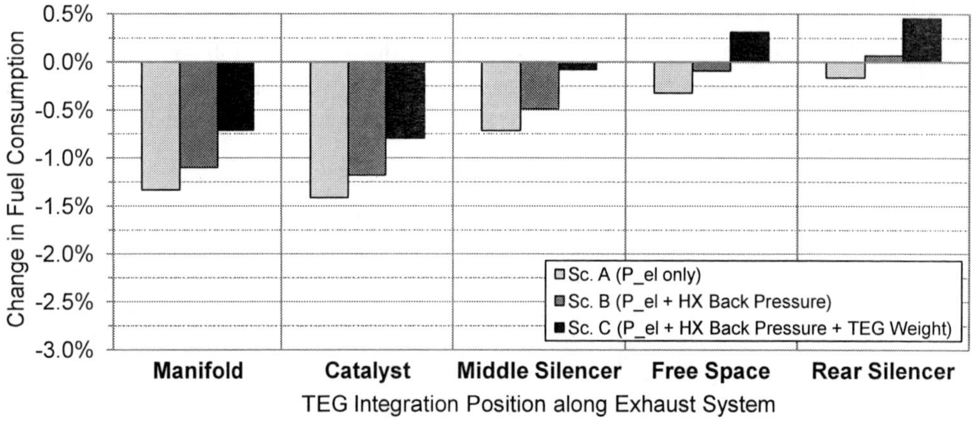

(a) NEDC.

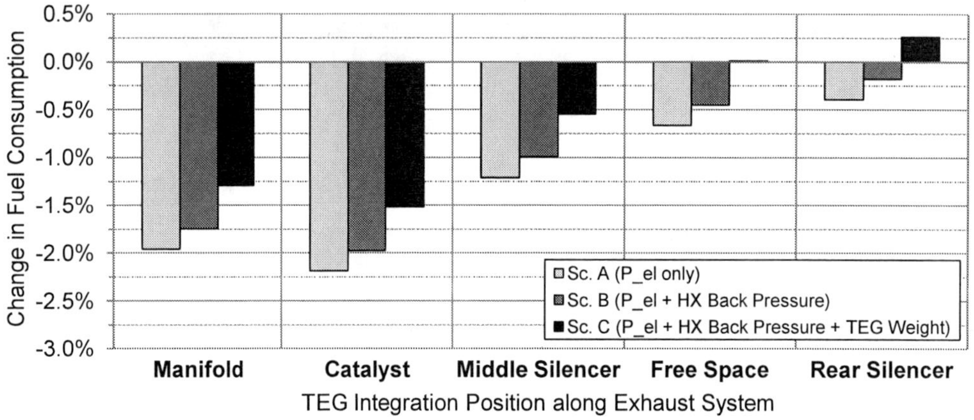

(b) WLTC.

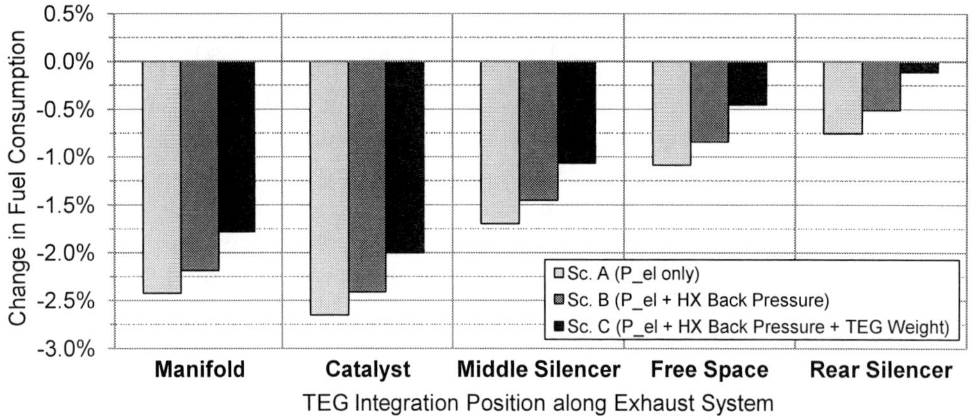

(c) CADC.

Figure 11: Fuel saving potential in different driving cycles (assumptions made: HX characteristics of ATEG #3 and skutterudites as TE material).

5 CONCLUSIONS

The methodology presented in this paper was used to determine the fuel saving potential of a power optimised automotive generator in legal and customer driving cycles. The results of the simulation study lead to the following findings. The integration position of the ATEG has a higher influence on power output than the variation of the exhaust gas heat exchanger characteristics in case the power optimisation criteria are fulfilled. The evaluation of the simulated scenarios illustrates that integration effects, such as increased exhaust gas back pressure and additional vehicle weight, have to be taken into account since they reduce the fuel reduction potential by up to 0.6%. Considering all influences included in this model, a significant fuel reduction can only be achieved if the ATEG is combined with the manifold, the catalytic converter or, as a much less attractive choice, is used as a replacement for the middle silencer. Fuel reductions which are better than -1.0% cannot be achieved in the NEDC for any of the simulated integration positions. For the other two driving cycles, i.e. the WLTC and the CADC, maximum values up to -1.5% and -2.0% can be achieved if the ATEG is combined with the catalytic converter.

In regard to future studies and developments the following recommendation is made. The integration positions located very close to the internal combustion engine require much more detailed investigations. As shown in this study suitable concepts promise fuel saving potentials which are high enough to reach attractive cost to benefit ratios of automotive thermoelectric generators.

REFERENCE LIST

[1] S. Kumar, S. D. Heister, X. Xu, J. R. Salvador and G. P. Meisner, "Thermoelectric generators for automotive waste heat recovery systems part I: Numerical modeling and baseline model analysis" *Journal of Electronic Materials*, vol. 42, no. 4, pp. 665–674, 2013.

[2] C.-S. Ernst, I. Olschewski and L. Eckstein, "CO2 reduction potentials for passenger cars until 2020", Study carried out on behalf of the Federal Ministry of Economic Affairs and Energy, Institute for Automotive Engineering (ika), RWTH Aachen University, Aachen, 2012.

[3] M. Ranalli, M. Adldinger, D. Kossakovski and M. Womann, "Thermoelectric generators - From aerospace to automotive" *ATZ Worldwide*, vol. 115, pp. 60–65, 2013.

[4] T. A. Horst, W. Tegethoff, P. Eilts and J. Koehler, "Prediction of dynamic Rankine cycle waste heat recovery performance and fuel saving potential in passenger car applications considering interactions with vehicles' energy management" *Energy Conversion and Management*, vol. 78, pp. 438–451, 2014.

[5] N. D. Love, J. P. Szybist and C. S. Sluder, "Effect of heat exchanger material and fouling on thermoelectric exhaust heat recovery" *Applied Energy*, vol. 89, no. 1, pp. 322–328, 2012.

[6] A. Bitschi, "Modelling of thermoelectric devices for electric power generation", Dissertation, ETH Zurich, 2009.

[7] D. C. Agrawal and V. J. Menon, "The thermoelectric generator as an endoreversible Carnot engine" *Journal of Physics D: Applied Physics*, vol. 30, no. 3, pp. 357–359, 1997.
[8] H. T. Chua, K. C. Ng, X. C. Xuan, C. Yap and J. M. Gordon, "Temperature-entropy formulation of thermoelectric thermodynamic cycles" *Physical Review E*, vol. 65, no. 5, p. 056111, 2002.
[9] J. M. Gordon, "Generalized power versus efficiency characteristics of heat engines: The thermoelectric generator as an instructive illustration" *American Journal of Physics*, vol. 59, no. 6, pp. 551–555, 1991.
[10] D. M. Rowe, Ed., *Thermoelectrics handbook: Macro to nano*. Boca Raton: CRC Press, 2006.
[11] A. Z. Sahin and B. S. Yilbas, "The influence of operating and device parameters on the maximum efficiency and the maximum output power of thermoelectric generator" *Int. Journal of Energy Research*, vol. 36, no. 1, pp. 111–119, 2012.
[12] K. Yazawa and A. Shakouri, "Energy payback optimization of thermoelectric power generator systems" in *Proceedings of the ASME 2010 International Mechanical Engineering Congress & Exposition (IMECE2010)*, Vancouver, 2010.
[13] K. Yazawa and A. Shakouri, "Optimization of power and efficiency of thermoelectric devices with asymmetric thermal contacts" *Journal of Applied Physics*, vol. 111, no. 2, p. 024509, 2012.
[14] A. Miner, "The compatibility of thin films and nanostructures in thermoelectric cooling systems" *Journal of Heat Transfer*, vol. 129, no. 7, pp. 805–812, 2007.
[15] C. Yu and K. T. Chau, "Thermoelectric automotive waste heat energy recovery using maximum power point tracking" *Energy Conversion and Management*, vol. 50, no. 6, pp. 1506–1512, 2009.
[16] S. Hausberger, R. Luz and M. Rexeis, "New emission factors for EURO 5 & 6 vehicles" in *Proceedings of the 19th International Transport and Air Pollution Conference 2012*, Thessaloniki, 2012.
[17] P. M. Mayer and R. J. Ram, "Optimization of heat sink–limited thermoelectric generators" *Nanoscale and Microscale Thermophysical Engineering*, vol. 10, no. 2, pp. 143–155, 2006.
[18] K. Salzgeber, P. Prenninger, A. Grytsiv, P. Rogl and E. Bauer, "Skutterudites: Thermoelectric materials for automotive applications?" *Journal of Electronic Materials*, vol. 39, no. 9, pp. 2074–2078, 2010.
[19] T. Woll, "Verbrauch und Fahrleistungen" in *Hucho - Aerodynamik des Automobils*, 6. Aufl., T. Schuetz, Ed. Wiesbaden: Springer, 2013, pp. 137–176.
[20] R. Parise, "Electrical energy production in an automobile catalytic converter" in *Proceedings of the 2nd International Energy Conversion Engineering Conference (IECEC)*, Providence, Rhode Island, 2004.
[21] B. Mazar, "Gesamtsystemoptimierung eines thermoelektrischen Generators fuer eine Fahrzeuggruppe" Dissertation, Technische Universitaet Dresden, 2011.
[22] C. Koffler and K. Rohde-Brandenburger, "On the calculation of fuel savings through lightweight design in automotive life cycle assessments" *Int. Journal of Life Cycle Assessment*, vol. 15, no. 1, pp. 128–135, 2010.

A novel and efficient thermoelectric-generating (TEG) system of energy harvesting from exhaust gases of passenger vehicles

B LI, V AKDOGAN, Y LI and **Y YAN**
HVACR & Heat Transfer Research Group, Faculty of Engineering, University of Nottingham, UK

J LI and **J WANG**
FAW Co. Ltd. com, China

ABSTRACT

The exhaust gas from an internal combustion engine carries away about 30% of the heat of combustion. The major technical constraint that prevents successful implementation of waste heat recovery is due to its intermittent and time mismatched demand and availability of energy. In the present work, a thermoelectric generator (TEG) prototype with advanced heatpipe heat exchanger has been developed for capturing waste exhaust heat in vehicles. The completed assembly consists of 60 TEG modules with full electric system component. The overall TEG prototype was tested in a hot–air environment. Improvements in the performance of TEG system have been achieved by facilitating higher heat flux in radial direction of exhaust flow. This high heat flux flow is achieved by means of sandwiching heat pipe based heat exchanger between multi-stage TEG modules. The fuel efficiency is improved by 5% optimally under verifying TEG device and heat pipe exchanger system performance at varies operating conditions. The performance parameters pertaining to the heat pipe heat exchanger and TEG modules such as amount of heat recovered, heat lost, flow rate, conversion efficiency and percentage energy saved are evaluated and reported in this paper.

Keywords: Waste heat recovery; Heat recovery heat exchanger; Heat Pipe, thermoelectric generation.

1. INTRODUCTION

Over the last decades, the increasing concern regarding the issues of global warming and environmental pollution has enhanced the research towards novel and environmentally friendly technologies for power generation purposes. Within this stimulus, thermoelectric generators (TEGs) have continually attracted great attention as a promising alternative to other green technologies although the efficiency of thermoelectric conversion is still relatively low [1, 2]. Nevertheless, TEGs are solid state devices of heat engine type, where the flow of charge carriers serves as the working fluid. They are naturally silent in operation, do not have mechanical moving parts and require much less maintenance than other energy recovery technologies. Their overall size and weight is small and light, with respect to the power delivered and depending on the materials used for manufacturing they can easily operate at elevated temperatures. As mentioned, the efficiency of thermal to electric conversion of practical TEGs is lower than 5% that be constrained to a relatively low efficient level since the discovery of Seeback effect in semiconductors [3]. Car manufacturers, such as BMW, General Motors, Jaguar Land Rover, have reported to develop several prototypes with power output ranging from 80W to 600W over the last few years despite the overall performance of power supply is anticipated sufficient to offer to passenger car electric system. Practically, the existing TEGs prototypes installation is inevitably challenged by limited space and weight requirements under or in the chassis [4-6]. Therefore, a compact and higher denser TEG system is needed to both fetch exhaust waste heat maximally and integrated with existing exhaust pipeline easily. Many investigators has developed various TEGs heat exchanger structures such as counter cross flow patterns, multi-stacked channel, pin-fin heat sinks in order to enhanced the heat gain from exhaust gas.[2, 7, 8]

Stobart et al. suggested [9] that valuable systematic data to emerge the viability of this technology has to be obtained but many serious issues remain to be solved technically especially for optimising the heat transfer performance of the TEGs and, optimising the TE materials. They carried out the initial investigation to optimise the TEG heat exchanger geometry and suggested that the heat exchanger geometry needs to be carefully designed to maximise the heat transfer while minimising the pressure drop across the device. Some other researchers also attempt to enhance the rate of heat transfer by a special heat exchanger in the exhaust of diesel engine for energy recovery purpose. Wang et al. [10] and Feru et al. [11] used plate heat exchangers to change the exhaust flow pattern and enhance the heat transfer for extracting the heat from the exhaust but resulted in higher pressure drop through TEGs test section. Pandit et al. [12] analysed the pin fin geometry effect on heat transfer performance in rectangular channels. Amaral et al. [13] also conducted TEGs power output assessment using inner channel geometries with alternating flow impeding panels. Traditional shell and tube heat exchanger were proposed by Bari et al. [14]. Similar TEG performance researches were investigated by Lesage et al. [15], Brennan et al. [16].

Therefore, many prototypes are designed to enlarge its heat exchanger surface coarsely due to extracting heat from the exhaust gas is difficult resulting from the low heat transfer coefficient of gas flow, typically around 90W/m2-K. This makes heat conduction through TEG uneven and lower efficient. Furthermore, the ratio of the extracted heat to the total heat from exhaust in terms of energy density utilisation in gas flow direction is normally much lower than the expected value which could be transferred to the hot side of thermoelectric materials. Moreover, some or most prototypes designed changed the exhaust pipeline by enlarging the contacting surface area for heat transfer and splitting gas flow into many small channels due to common squared plate-shaped thermoelectric modules. Such heat exchanger arrangements will exert negative pressure change on the engine gas exhausting [16, 17]. In the present study, we have designed a novel and efficient TEG system of energy harvesting from car exhaust gases. The innovative system can generate electricity with power output density as high as 2.93 W/mm. The results of prove the concept of the design are presented in this paper.

Table 1 shows a comparison of recent developments of TEGs in exhaust heat recovery from major car manufacturers. The declared power output, improved fuel economy and utilised energy density are also given in the table. To compare the performance of power output, the term 'Power output density' here is defined as declared power output divided by prototype actual length considering other dimensions of prototype are similar in size.

Table 1. Comparisons of TEG system performance and energy density

Automotive Manufacture	**Type of Vehicle**	**Thermoelectric material**	**Expected power output**	**Drive cycle**	**Fuel economy Improved**	Power output density
Ford Motor	Ford Fusion 3.0litre V-6 engine	Bi_2Te_3	500w	EPA highway	2%	0.42
General Motors		Bi_2Te_3 skutterudite (Future)	35W	FTP city	5%	0.04375
BMW	4.0litre V8 diesel engine	Ring-type PbTe Lead-free TE by 2018	300W	Highway	7.5%	0.5
Jaguar Landrover	2.0 GDI Sport saloon	Bi_2Te_3	150W 600W	NEDC ARTEMIS	2.5% 3.2%	0.33 1.33
Renault Trucks	11litre Renault Truck	Bi_2Te_3	300W for diesel, 500W for gasoline 1kW for diesel truck		0.35%	0.25 0.412 0.833
Our results in Nottingham	**Passenger car**	$\mathbf{Bi_2Te_3}$	**440W**	**N/A**	**N/A**	**2.93**

2. EXPERIMENT

2.1 SYSTEM configuration

As shown in Figure 1. A 20cm L X 30 cm W X 20cm H (20 layers with 120 TEG modules capacity) squared shape co-axial TEG heat exchanger is built up. The design was proposed that can prominently enhance the heat transfer performance in radial direction of gas flow which in turn keeps the overall TEGs heat exchanger compact and clean. The completed assembly consists of segmented TEG modules which are elaborately configured together in flow direction. The heat pipes at each segment, which replaces the fins-stack structure, provide a super thermal conducting pathway into each TEGs hot side surface. Therefore high heat flux flow is achieved by means of sandwiching heat pipes between these multi-stage TEG modules, and also such heat flux can be conducted away by heat pipes into cooled water loop.

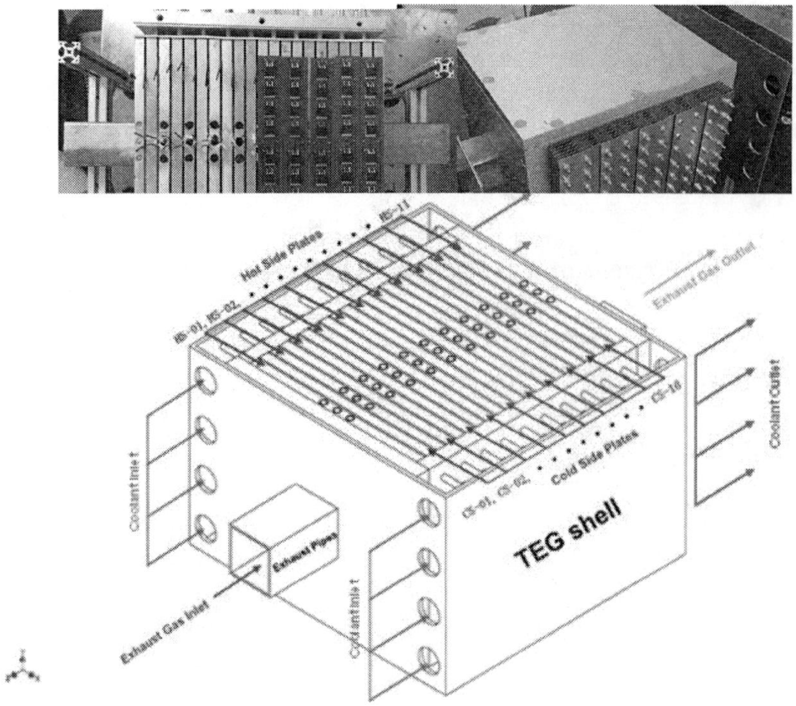

Figure 1. Proof-of-Concept design with integrating high performance heat pipes in radial direction of exhaust flow

In the experiment, only 60 TEGs (63.5mm L x63.5mm W x 4mm H) are installed with 5 smart bypass diodes PCB connected on top of each layers. Water coolant jacket are shaped by the outside shell, and the coolant will flow through finned heatpipe heat sinks in order to take heat away from cores of TEGs. The experiment is tested at boundary conditions below:

Table 2. Experiment on proof-of-concept prototypes

Operation condition	Ranges	Notes
Heated Air Temperature	300°C ~ 550°C	Varied by PID control
Air flow rate	650 L/min	Varied by frequency control
Coolant Inlet Temperature	20°C ~ 50°C	Coolant circulated by water chiller

3. RESULTS

3.1 Transient Heat transfer Enhancement by heat pipes

The transient state of square shape heating plates and cooling plates were shot by FLIR infra camera which depicted in Figure 2. The temperature contour can be seen from camera, and the cooling plates are constantly cooled by cold side heatpipes and maintained in a relatively cold operational condition for TEGs cold side when system in start-mode. Meanwhile, the heating plates were heated up to 65.6°C in 5 mins, it is shown hot side heat pipes conduct hot air energy into heating plates rapidly in radial direction of air flow.

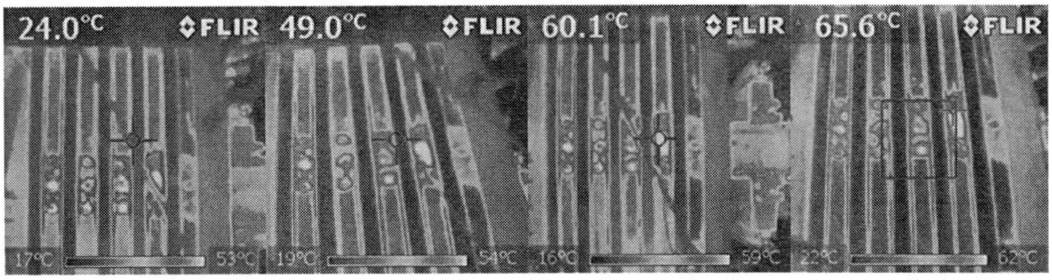

Figure 2. The temperature changes at starting mode of exhaust heating

The proof-of-concept design here has also proved a faster thermal response performance than conventional flat fins heat sink at inner gas wall surface. The overall thermal resistance at both sides of TEGs modules are minimised by means of integrating heat pipes in such thermal flow. Therefore, the performance of thermal response is expected higher and hot side temperature of TEGs can be reached at higher level in order to integrate high thermoelectric figure of merit TEGs material such as PbTe, PbSe, etc. In addition, the prototype has proof that this segmented and repeated pattern of heat exchanger works and shows a promising results in our next stage improvements.

3.2 Temperature difference measurements at each TEGs layer

The transient state temperature measurement was carried out in this test by separating the whole testing into 4 procedures:

1. The water chiller is switched on at the beginning of the test in order to maintain a constant temperature in the TEGs heat exchanger;
2. The air blower is switched on after 20mins after system stabilising period, air flow rate with 650L/min is maintained by air blower into the squared exhaust tube;
3. The air heater is switched on with hot air blowing into exhaust pipe inlet at temperature of 500°C, the heating process lasts for 30 mins;
4. Then the air heater is switched off with coolant circulating in the whole loop until system temperature decrease to ambient temperature.

It is shown in figure 3 that each layers of TEG modules produce different level of temperature gradients. The major reason of such various temperature gradients is due to effective heat pipe heat conducting performance, and the hot air temperature decreased from 500°C at inlet to 212.2°C at outlet. The heat energy is sufficiently drawn into water loop after conducting into TEGs modules. The highest one in the figure was monitored at $\Delta T = 220°C$ at 12th layer where no TEGs modules installed between hot plates and cold plates. This also approved that the exhaust heat will be transferred away to coolant through TEGs surface, however, the temperature difference will be built up and higher than those layers with TEGs installed. It is further advised that transient overheating may cause problems when the temperature limits reached at some critical insulation material.

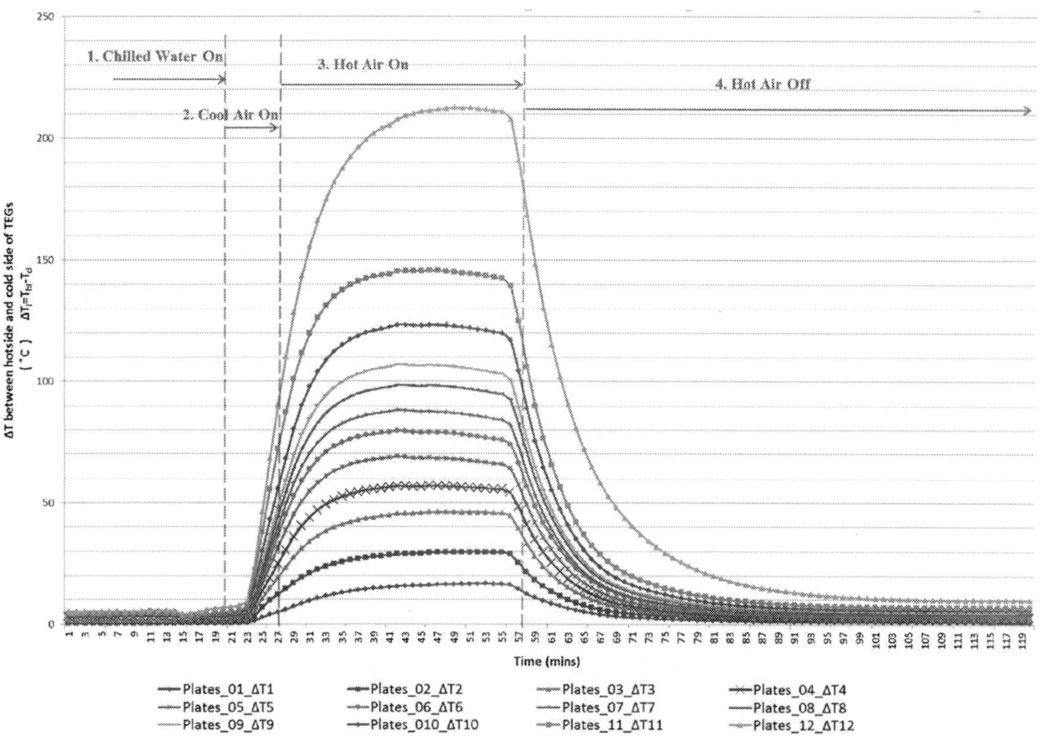

Figure 3. Transient thermal performance of each TEG modules

In the direction of air flow, it is to be clearly seen from figure 3 that each layer will generate different level of power into bypass diodes PCBs. There are two solutions to solve this problem:

1. Install bypass diodes individually for each layer rather than combing them together into MPPT modules. However, significant cost, space of PCB installation, electronics cooling will incur inevitably. Further optimisations of such parameters need to be investigated;
2. Ongoing optimisation of heat pipe positioning in the exhaust pipe is carried out in order to achieve better heat fetching and distribution into the heating plates. Therefore, the high temperature heatpipes will be varied by size and the gas wall surface will be adjusted by attaching localised small fin stacks. It is targeted that all the heat in different TEGs layers with be distributed evenly in the direction of air flow as long as the high temperature heat pipe is capable of conducting the energy at such high temperature range.

From the initial results, it is concluded that the generated temperature difference ΔT are still not adequate for all-time operation condition. The lesson we learned from this prototype is the problem of TEGs modules misalignment in each layer due to the surface contact. Each layers consist of 6 series-linked TEG modules, therefore, unevenness of the surface will be created at each cross section which result in numerous surface contact resistance between metal plates and TEGs ceramic surface. Precision manufacturing is then required in next stage to minimise such problem as concerned in most of our cases.

3.3 CFD simulation of Concentric TEGs heat exchanger

As mentioned above, an ongoing configuration of concentric TEG heat exchanger is being prototyped. There are serval major improvements from our previous proof-of-concept design as shown in figure 4: Single concentric TE module will be made with higher density of thermoelectric couples in each TE. All the heatpipes are reflow soldered after installing into the metal plates and wavy fins stacks in order to reduce the surface contact resistance among different materials interfaces. Meanwhile, the surface unevenness has been solved by means of higher one dimensional pressure (flow direction) pressed on whole segmented TEGs modules without structural constrains from radial direction of heat exchanger. In addition, water coolant jacket are designed that all high temperature components are fully immersed the in the coolant bath with a solid to solid heat conduction by heatpipes. The advantages of such design is to divert the exhaust heat into radial direction and enhance the heat transfer into the heating plates.

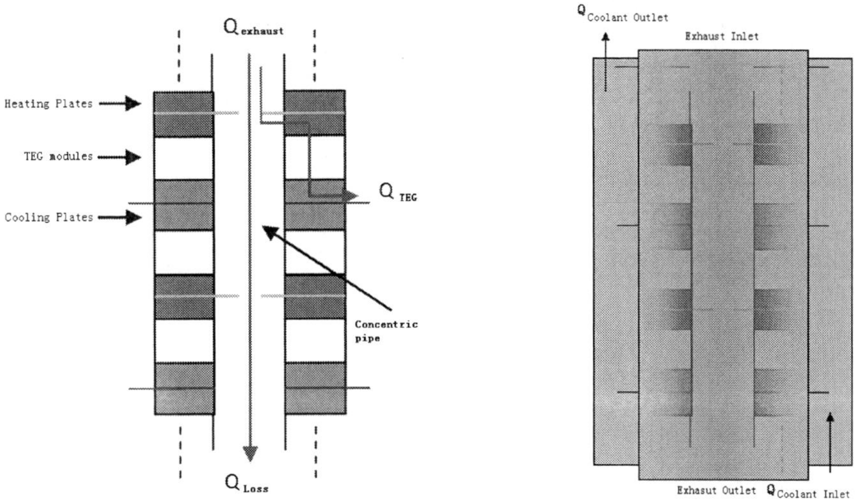

Figure 4. Schematic of concentric TEG heat exchanger with heat pipes

For optimisation purpose, CFD simulation of model validation has been carried out before prototyping process. Here we have choose a typical condition for interpreting our CFD results. The Nitrogen gas property is used as ideal gas model in Fluent, turbulent model and energy model are enabled, counter cross flow pattern was presented below. The major boundary conditions are listed in table 3:

Table 3. Boundary conditions for CFD simulation

Hot air temperature	**Coolant temperature**	**Heatpipes (Radial)**	**Copper**	**TE material**	**Insulation**	**Aluminium**
773 K (500°C)	300 K (27°C)	8000 (W/m.K)	387 (W/m.K)	1.2 (W/m.K)	0.03 (W/m.K)	180 (W/m.K)
Mass flow rate of gas	**Velocity of coolant**	**Heatpipes (Axis and Tangential)**	**Overall Geometry**	Concentric TEGs	**Water hydraulic diameter**	**Gas hydraulic diameter**
0.04kg/s	1.5m/s	387(W/m.K)	200mm OD x 500mm L	40 pairs	100mm	48mm

As seen from figure 5a, the heat absorbed by heat pipe is distributed into TEGs by heating up same concentric Aluminium plates. The temperature difference between TEGs hot side and cold side are about 80°C. And staggered high temperature heat pipe shaped a repeated pattern with 20 layers external tube flow within inner channel. Each layer consist of 6 heat pipes in a circle, and transfer from 100W to 40W degrading in the direction of gas flow as shown in figure 5b.

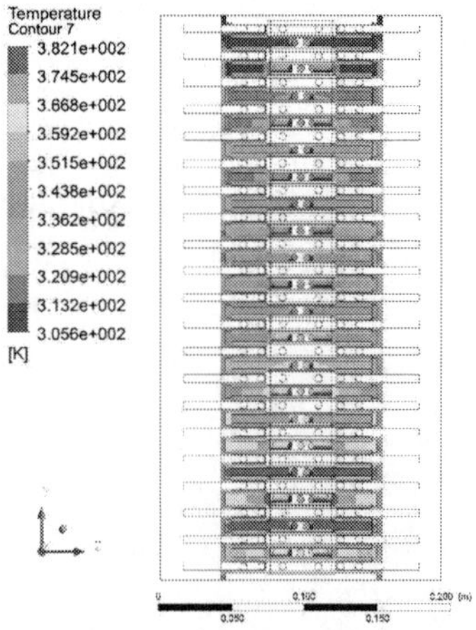

Figure 5a. Temperature contour of heatpipes and TEGs

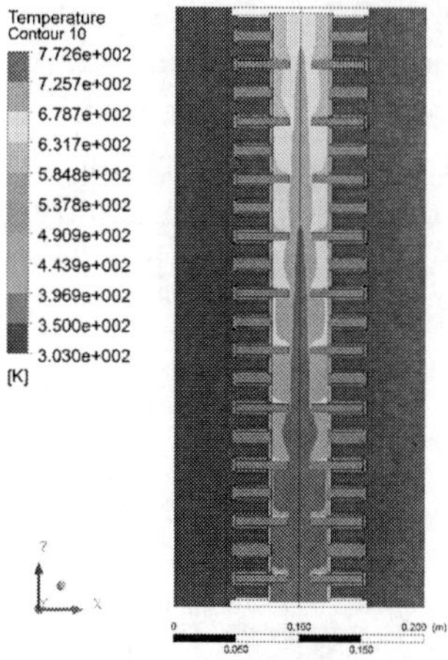

Figure 5b. Temperature contour of gas metal plates

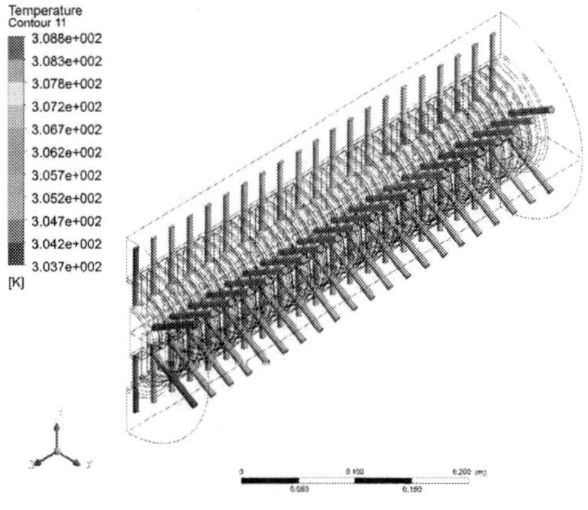

Figure 5c. Temperature contour of cooling heatpipes

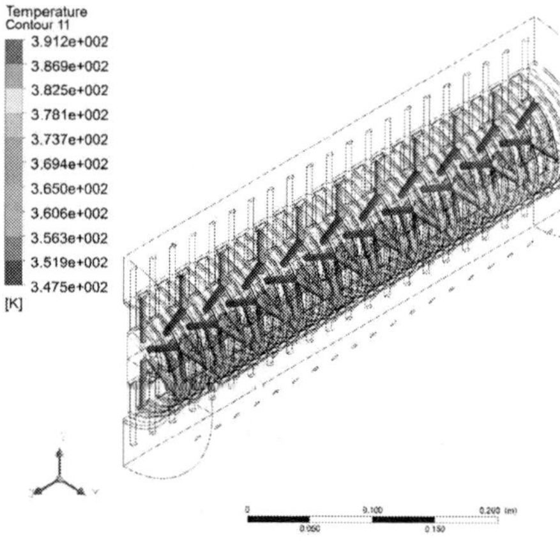

Figure 5d. Temperature contour of heating heatpipes

In Figure 5c, the cooling heat pipes perform quite well regarding to only 5°C temperature increasing at such circulating velocity. Therefore, it is concluded that the water pumping power can be reduced as long as the coolant are not overheated through the TEG heat exchanger. Otherwise, a water side bypass loop may needed in order to maintain the constant coolant temperature into the engine. Nevertheless, the adjustment of cooling heat pipes are needed in further development. For lightening weight awareness of heat exchanger, advanced cooling structures such as pin fin, spiral coils or metal foam heat sinks are in the scope of consideration.

Same as cooling heat pipes, the heating heat pipes as depicted in figure 5d needed to be further investigated regarding to its performance and durability at temperature range from 300°C to 600°C. Normal coper-water heat pipes will reach its upper limits at condensation section. We have addressed to introduce Synthetic Organic Fluids DOWTHERM™ A as high temperature heat transfer fluid in the heat pipes. Then it is expected that the actual performance will be improved from our current prototype experimental results.

4. CONCLUSION

In this paper, we have presented a novel design of TEGs for energy harvesting from car exhaust gas. In the design, the heat pipe technology has been effectively applied to heat exchangers. The prototype designed and the results have proved that the durability of copper–water heat pipe in its mid-high temperature range in the exhaust gas flow. In addition, the cooling capacity of heat pipe has been approved through the testing, the immersed heat pipes in the coolant are capable of removing counterpart heat flux from TEGs cold side ceramics. The experimental results show that the TEGs performance decreased in the direction of exhaust gas flow under reducing temperature difference between TEGs modules. Therefore, the configuration of hot side heatpipe needs be adjusted and optimised by means of reducing the contacting surface with gas, or cross stacking heatpipes in the exhaust in our ongoing design process.

5. REFERENCES

1. Hatami M., Ganji D. D. & Gorji-Bandpy M. *A review of different heat exchangers designs for increasing the diesel exhaust waste heat recovery*. Renewable and Sustainable Energy Reviews, (2014). 37: 168-81.
2. Deng Y. D., Liu X., Chen S. & Tong N. Q. *Thermal optimization of the heat exchanger in an automotive exhaust-based thermoelectric generator*. Journal of Electronic Materials, (2013). 42: 1634-40.
3. Stevens R. J., Weinstein S. J. & Koppula K. S. *Theoretical limits of thermoelectric power generation from exhaust gases*. Applied Energy, (2014). 133: 80-88.
4. Wang W. S. & Liu X. *Experiments and analysis on thermoelectric generators of automotive exhaust under the multi-field coupling*. Advances in Applied Sciences and Manufacturing, Pts 1 and 2, (2014). 850-851: 217-20.
5. Wang Z. R., Yao J. H., Chen S. & Yin S. *Research on the affection of thermoelectric module performance related to thermal deformation in the automotive exhaust-based thermoelectric generator*. Advances in Applied Sciences and Manufacturing, Pts 1 and 2, (2014). 850-851: 188-91.
6. Love N. D., Szybist J. P. & Sluder C. S. *Effect of heat exchanger material and fouling on thermoelectric exhaust heat recovery*. Applied Energy, (2012). 89: 322-28.
7. Liu X., Deng Y. D., Chen S., Wang W. S., Xu Y. & Su C. Q. *A case study on compatibility of automotive exhaust thermoelectric generation system, catalytic converter and muffler*. Case Studies in Thermal Engineering, (2014). 2: 62-66.
8. Crane D. T. & Jackson G. S. *Optimization of cross flow heat exchangers for thermoelectric waste heat recovery*. Energy Conversion and Management, (2004). 45: 1565-82.

9. Stobart R. & Wijewardane A. Exhaust system heat exchanger design for thermal energy recovery in passenger vehicle applications. In I. o. M. Engineers (Ed.), Vehicle thermal management systems conference and exhibition (vtms10) (609-22). Woodhead Publishing:2011.
10. Wang T., Luan W., Wang W. & Tu S.-T. *Waste heat recovery through plate heat exchanger based thermoelectric generator system*. Applied Energy, (2014). 136: 860-65.
11. Feru E., de Jager B., Willems F. & Steinbuch M. *Two-phase plate-fin heat exchanger modeling for waste heat recovery systems in diesel engines*. Applied Energy, (2014). 133: 183-96.
12. Pandit J., Thompson M., Ekkad S. V. & Huxtable S. T. *Effect of pin fin to channel height ratio and pin fin geometry on heat transfer performance for flow in rectangular channels*. International Journal of Heat and Mass Transfer, (2014). 77: 359-68.
13. Amaral C., Brandão C., Sempels É. V. & Lesage F. J. *Net thermoelectric generator power output using inner channel geometries with alternating flow impeding panels*. Applied Thermal Engineering, (2014). 65: 94-101.
14. Bari S. & Hossain S. N. *Waste heat recovery from a diesel engine using shell and tube heat exchanger*. Applied Thermal Engineering, (2013). 61: 355-63.
15. Lesage F. J., Sempels É. V. & Lalande-Bertrand N. *A study on heat transfer enhancement using flow channel inserts for thermoelectric power generation*. Energy Conversion and Management, (2013). 75: 532-41.
16. Brennan D., Gilchrist R. & Wicksteed B. Heat exchanger design and analysis for an automotive thermoelectric generator. In ImechE (Ed.), Vehicle thermal management systems conference proceedings (vtms11) (115-26). Woodhead Publishing: 2013.
17. Lu H. L., Wu T., Bai S. Q., Xu K. C., Huang Y. J., Gao W. M., Yin X. L. & Chen L. D. *Experiment on thermal uniformity and pressure drop of exhaust heat exchanger for automotive thermoelectric generator*. Energy, (2013). 54: 372-77.

Electrical heat assisted Lean NOx Trap system to improve NOx emissions in low load engine operating conditions

Z AHMED, G CAIRNIE, M NDIAYE, J RODRIGUEZ and **A SANKAR**
AVL Powertrain UK Ltd., UK

ABSTRACT

The future introduction of Real World Driving Emissions (RDE) standards increases the challenge for Diesel passenger cars to meet the main regulated emissions. In particular, NOx emissions control remains a challenging task and requires the use of expensive exhaust aftertreatment devices such as Selective Catalyst Reduction (SCR) or Lean NOx Traps (LNT).

In the case of LNT systems, although the calibration and control of their performance is achievable without major difficulties on the standard homologation cycle, NOx removal performance issues appear when it comes to some specific real world driving conditions. For passenger cars applications in city driving cycles, the exhaust gas temperatures are usually low and limit the effective operation of the LNT, mainly due to the low NOx storage and the rich purge conversion efficiency.

In this study, a LNT catalyst is combined with an Electrically Heated Catalyst (EHC) to form an E-LNT system. The EHC is powered by the vehicle electrical system (battery and alternator) and used to thermally manage the exhaust conditions at the LNT. With a model-based approach, different E-heater and rich purging strategies have been developed to improve the performance of the LNT in real world low load driving conditions.

1 INTRODUCTION

NOx emissions reduction is one of the most challenging Exhaust Aftertreatment Systems (EAS) engineering tasks for automotive catalysts applications. Lean NOx Trap (LNT) and Selective Catalyst Reduction (SCR) systems are the two main NOx emissions reduction technologies developed in order to meet the increasingly stringent tailpipe emissions standards for lean-burn Internal Combustion Engines (ICE) applications such as Diesel or Gasoline Direct Injection. The choice of a particular NOx reduction solution, for a given application, depends on different considerations such as cost (catalyst formulation, precious metals loading), vehicle class (inertia weight), and most importantly targeted reduction performance with regards to the specific legislation market (raw engine-out NOx range, NOx tailpipe limits).

The LNT catalyst basic operation principle consists of a discontinuous NOx removal from the exhaust gas with two distinct phases. NOx is first oxidized (NO part) into NO_2, and then stored on adsorption sites contained within the catalyst washcoat during the normal lean operation of the engine. In a second phase, the engine is operated, during a short pulse, into a forced rich mode (purge) that desorbs the NOx from the storage sites and reduces them into nitrogen. The overall LNT NOx conversion efficiency is the combined result of the NOx storage efficiency and the rich purge conversion efficiency, which are highly dependent on the LNT inlet or substrate temperature. Typically for most LNT catalyst formulations, there is low conversion efficiency at higher exhaust temperatures due to the instability of adsorbent nitrate compounds leading to the thermal desorption of NOx during the storage phase. On the other hand, at low temperatures, the low conversion efficiency is mainly due to low NO to NO_2 oxidation activity and poor conversion efficiency [1].

The calibration of a LNT system can usually be optimized to achieve good NOx conversion and meet emissions limits on homologation cycles like the NEDC. However the forthcoming European legislation, with the progressive introduction of Real World Driving Emissions (RDE) [2] for Euro 6c and beyond, is likely to increase the challenge for using LNT systems for NOx emissions reduction. For example with light-duty diesel passenger cars applications, the relatively low exhaust gas temperatures during low load conditions like urban or city driving are critical for an effective purging of the LNT catalyst. Also the current engine downsizing trend in the automotive industry [3], towards more fuel efficient engines and CO_2 reduction, has an adverse effect on the EAS with the subsequent decrease of exhaust temperatures.

To maintain the status of the LNT catalyst as one of the preferred NOx emissions reduction solutions, it needs to have as wide as possible temperature operating range, especially covering the low end region, e.g. using more efficient formulations at lower temperature [4, 5] usually with a cost increase compared to conventional LNT formulations and a reduction in peak conversion efficiency. One alternative solution is to use the LNT with the combination of an upstream Electrically Heated Catalyst (EHC) to help achieve its low temperature boundary condition when needed in certain low load-duty cycles.

EHC have been used in the past and more recently, in a variety of studies, mostly to improve the cold start emissions performance of EAS [6, 7, 8] in order to achieve stringent emissions limits such as US ULEV standards.

In this study an EHC is positioned directly upstream of a LNT catalyst to form an E-LNT system for an I5 Diesel engine. The aim is to study with a model-based approach the performance enhancement of a LNT catalyst, especially under low load operating conditions of the engine, where the exhaust gas temperatures are too low to achieve a good NOx conversion efficiency. From an existing Diesel I5 engine model and EAS model including the EHC and the LNT, an integrated electrical heating and LNT purge strategy has been developed to improve the performance of the LNT catalyst in the low exhaust temperature range for city like driving conditions.

In this paper, we present the model of the engine and the E-LNT, the improved electrical heating and purge strategy and the system performance in terms of NOx emissions reduction and CO_2 penalty for a range of city and low load duty cycles.

2 MODEL DESCRIPTION

MoBEO (Model Based Engine Optimization) is a custom AVL tool which can be used for engine and Aftertreatment calibration optimization in a simulation environment.

A global MoBEO model is built from a library of models such as engine models, EAS models and Electronic Control Unit (ECU) models. The core model is the engine model, which is specific to a given engine definition and contains the cylinder and the combustion models built from physical equations and with real engine data from testing. MoBEO is typically used for cycle or steady state optimization during the engine calibration development process.

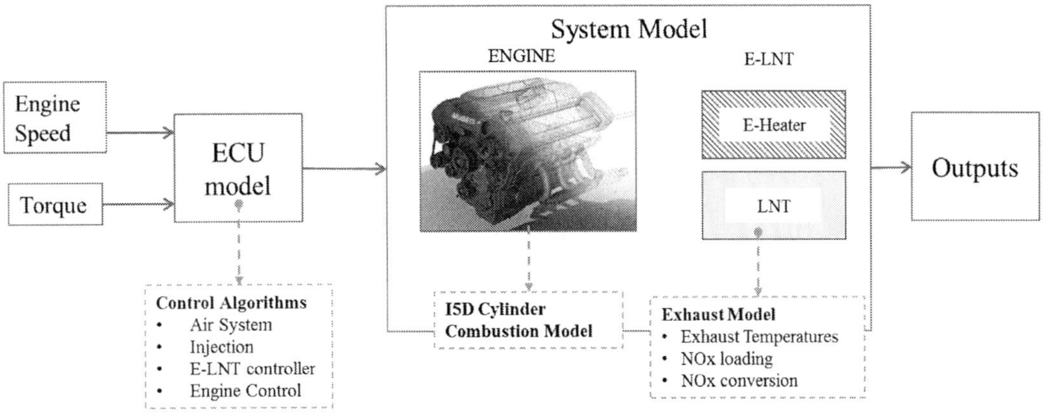

Figure 1: Overview of the global I5D MoBEO model set-up for the study

2.1 Engine model

The engine model used in this study is built from an I5D 2.4L Volvo Diesel engine with the characteristics listed in Table 1 below:

Table 1: Volvo I5D engine characteristics

Engine Type	Inline 5-Cylinder Turbo Charged Diesel
Power	151 kW at 4000 rpm
Torque	420 Nm at 1500 to 3250 rpm
Max. Engine Speed	4800 rpm
Displacement	2400 cm^3
Stroke : Bore	93 mm : 81 mm

2.2 LNT model

A simple parametric LNT model is implemented in the global I5D MoBEO model. It is based on inputs such as exhaust mass flow, exhaust gas temperature, LNT substrate temperature and exhaust gas lambda. The LNT model simulates the NOx storage in the lean operation mode and the NOx release and conversion during the rich operating mode of the engine (purge mode).

The LNT model has been built based on a specific LNT with an optimized formulation for RDE emissions. The LNT formulation is given in Table 2:

Table 2: LNT specification used for the study

Matrix	Cordierite
Ageing	160000 [Km]
Volume	1.9 [l]
Diameter	143.8 [mm]
Cpsi	400
Wall thickness	4.3 [mm]
PGM loading	118gm/ft^3
PGM Ratio	90:25:3

One important LNT characteristic is the NOx storage capacity which is the maximum amount of NOx that can be stored. It is a function of substrate temperature, exhaust space velocity and NOx slip ratio. Figure 2 shows the NOx storage capacity for the LNT used in this study.

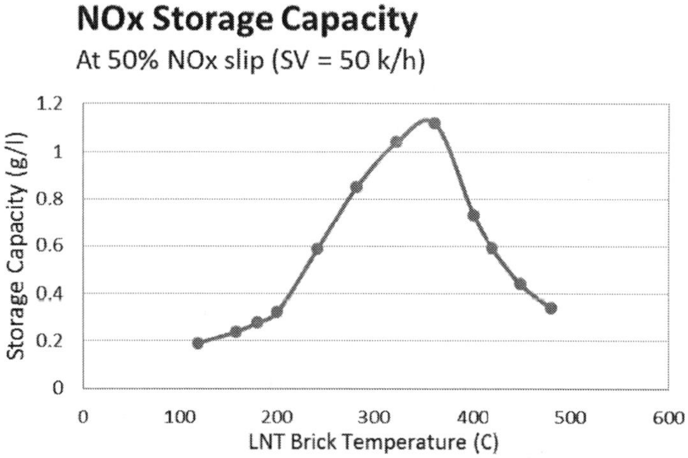

Figure 2: NOx storage characteristics for the LNT used in the study at 50 kph Space Velocity

2.3 E-Heater model

In the EAS layout, an EMITEC EHC system (EMICAT) is located immediately upstream of the LNT in order to control the substrate temperature. The E-heater is powered by the vehicle battery via an extra loading on the alternator. The device operation is limited to a maximum power of 2kW.

A simple E-heater model has been implemented into the EAS subsystem of the global I5D model according to the specific EMICAT hardware characteristics. The E-heater model is a function of the exhaust conditions:

• Exhaust mass flow rate
• LNT inlet temperature
• Desired LNT upstream temperature
• Electrical heater power

2.4 ECU model

The ECU model replicates the real Engine Control Unit of the I5D engine. It contains the main algorithms and the calibration parameters necessary for the control of the engine and the EAS:

• Airpath system
• Injection system
• LNT purge control logic
• E-heater control logic
• System switches between the different engine combustion modes

The default LNT purge and E-heater logics implemented in the ECU model are discussed in the following sub-sections.

2.4.1 LNT purge control logic

The LNT rich purge logic controls the request for the rich engine operating mode based on the following boundary conditions.

- LNT substrate temperature – There is a temperature window in which the LNT rich conversion can be effective. This temperature typically depends on the LNT formulation. For this study the range is as follows:
  $$200°C < \text{LNT substrate temperature} < 500°C$$
- NOx storage – The rich purge is requested based on the NOx storage level of the LNT. The filling ratio of the LNT is defined as the ratio of the current NOx stored over the LNT capacity:
  $$R\text{-}theta = \frac{NOx\ stored}{LNT\ capacity}$$
  The minimum R-theta value to request a purge is set by default to a fixed value of 0.58 in the LNT purge control logic.
- Exhaust mass flow –To achieve an effective LNT emptying, the rich mode is also carried out in a certain exhaust mass flow window. For this study the range is as follows:
  $$70\ kg/h < \text{Exhaust mass flow} < 250\ kg/h$$
- Engine operating point – Rich operation mode is not achievable across the entire engine operating range, mainly due to limitations such as combustion stability, smoke limit and hardware limits such as pre-turbine temperature.
  The boundary conditions for this I5D engine are:
  - 1000 rpm < engine speed < 3250 rpm
  - 45 Nm < torque < 250 Nm

2.4.2 E-Heater control logic

The E-heater logic implemented in the ECU model controls the request for the activation of the EHC and the delivery of the electrical power to the device. The activation boundary conditions are as follows:

- 50°C < LNT substrate temperature < 450°C.
- 70 Kg/h < Exhaust mass flow < 250 Kg/h.

In the default E-heater logic, the activation is simply requested manually via a switch.

3 MODEL VALIDATION

In this section, the accuracy of the global model (engine, EAS and ECU) is assessed by comparing simulation results with physical measurements obtained from existing tests on transient dynamometer. The same I5D engine and exact EAS layout, with the EHC upstream of the LNT, is used, as outlined in the model description above. The aim is to assess the model accuracy in normal and rich combustion mode operation, with and without E-heater activation.

Different model simulation outputs such as airpath and injection variables, exhaust conditions and main emissions are checked against the physical measurements.

Table 3 and 4 give a summary of the emissions results for a Paris city driving cycle test in two different configurations:

- No E-heater activation
- E-heater manually activated 40 s before every rich purge with a demanded power of 950 W. The heater is switched off at the end of the rich purge event.

Table 3: Simulation vs measurements - emissions comparison, Paris cycle no E-heater

Without E-heater	Simulation	Experimental
CO_2 (g/km)	182.5	188.4
Pre LNT NOx (g/km)	0.280	0.251
Post LNT NOx (g/km)	0.121	0.128
Tail pipe CO (g/km)	2.053	1.643
Tail pipe HC (g/km)	0.101	0.164

Table 4: Simulation vs measurements - emissions comparison, Paris cycle with E-heater (P=950 W, activation 40 s before rich purge)

Test with E-heater	Simulation	Experimental
CO_2 (g/km)	192.4	194
Pre LNT NOx (g/km)	0.256	0.265
Post LNT NOx (g/km)	0.105	0.09
Tail pipe CO (g/km)	1.342	1.207
Tail pipe HC (g/km)	0.055	0.146

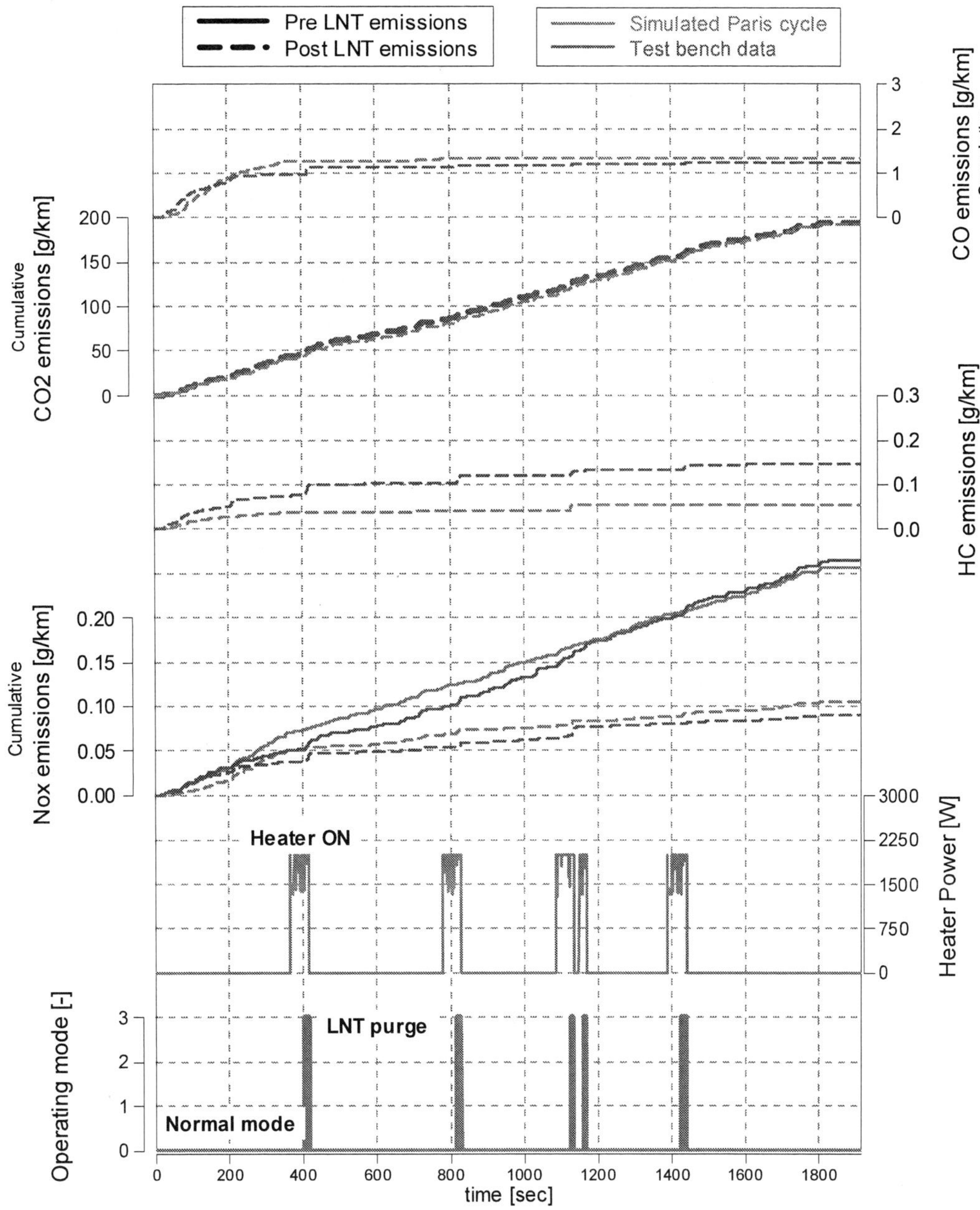

Figure 3: Simulation vs Test bench cumulative emissions for Paris cycle with E-heater

From the summary data in Table 4 and the dynamic traces in Figure 3, the initial observations for the default purging and e-heater logic are:

There is good correlation between simulation and experimental results for both CO_2 and NOx emissions in the two configurations tested. The activation of the E-heater just before the LNT purge incurs an additional CO_2 penalty due to the extra load on the alternator. The improvement in the NOx conversion efficiency shown in the post LNT NOx emissions can be explained by the increase in LNT temperature induced by the periodic E-heater activation.

HC and CO emissions show a deviation between the simulation and physical measurements. The model prediction inaccuracy is mainly due here, especially for HC, to the transient nature of the Paris cycle. However, the same trend can be observed in HC and CO emissions reduction with the E-heater activation due to the improved LNT HC/CO conversion efficiency with the increase in temperature. The HC and CO model accuracy improvement has not been investigated further, as the main emissions analysed in this specific study are NOx and CO_2.

4 INTEGRATED E-LNT SYSTEM CONTROL STRATEGY

The purge and E-heater logics are decoupled in the default ECU model described above.

In this section we present the development of an improved E-LNT system control strategy that combines the E-heater and the LNT purge logics. The aim is to allow a more effective control of the E-LNT system, especially in the critical low exhaust temperatures conditions.

Two different approaches for using the E-LNT system were primarily considered:

The first option is to use the E-heater to maximise both the storage capacity and the rich conversion efficiency of the LNT. With this specific LNT formulation, this would be achieved by trying to keep a constant temperature of 350°C throughout any given cycle (as we can see in Figure 2). This option was discarded due to the predictable high impact in engine out NOx and most importantly the CO_2 penalty with the constant extra load on the alternator. Any gain in NOx emissions would be negated by a significant fuel penalty.

The second option is to use the E-heater as a purge enabling device for the LNT. This meant it was necessary to identify operating conditions where the LNT would reach its maximum storage limit and still remain outside of the rich mode activation temperature window. In this case the E-heater would be activated to increase the LNT substrate temperature above the minimum purge activation threshold. This concept is shown in Figure 4:

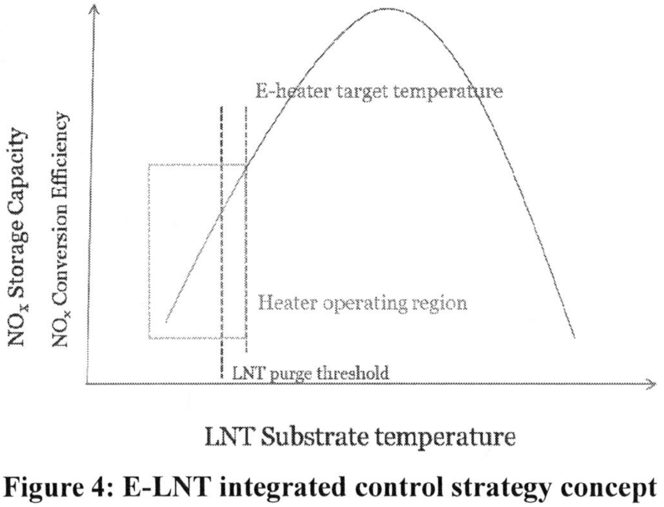

Figure 4: E-LNT integrated control strategy concept

The NOx storage ratio threshold (R-theta) for requesting a purge in the original purge logic is set to a fixed limit of 0.58. However, there is a possibility for very low load driving conditions and very low exhaust gas temperatures, that this threshold will not be reached at any time during the cycle and consequently the LNT purge trigger would not be achieved. The NOx storage limit to request a purge has then to be modelled as a function of the LNT substrate temperature as shown in Figure 5:

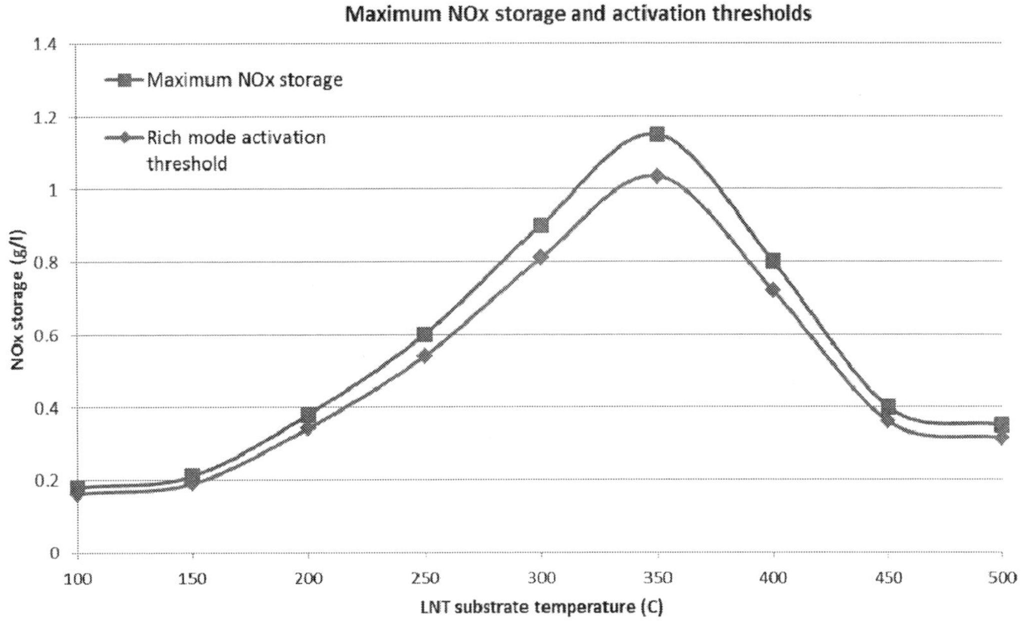

Figure 5: Dynamic NOx storage curve for LNT purge activation

A similar approach is also developed for the E-heater logic. Instead of a manual switch the E-heater would be activated under the following conditions:

- The LNT NOx fill ratio is slightly below the activation threshold.
- The LNT substrate temperature is below the minimum temperature threshold
- All the other conditions required to trigger a purge are met (see 2.4.1)

The E-heater activation logic is now a function of the NOx storage (or the filling ratio R-theta). The E-heater will be switched ON when R-theta is 90% of the activation threshold as shown in Figure 6:

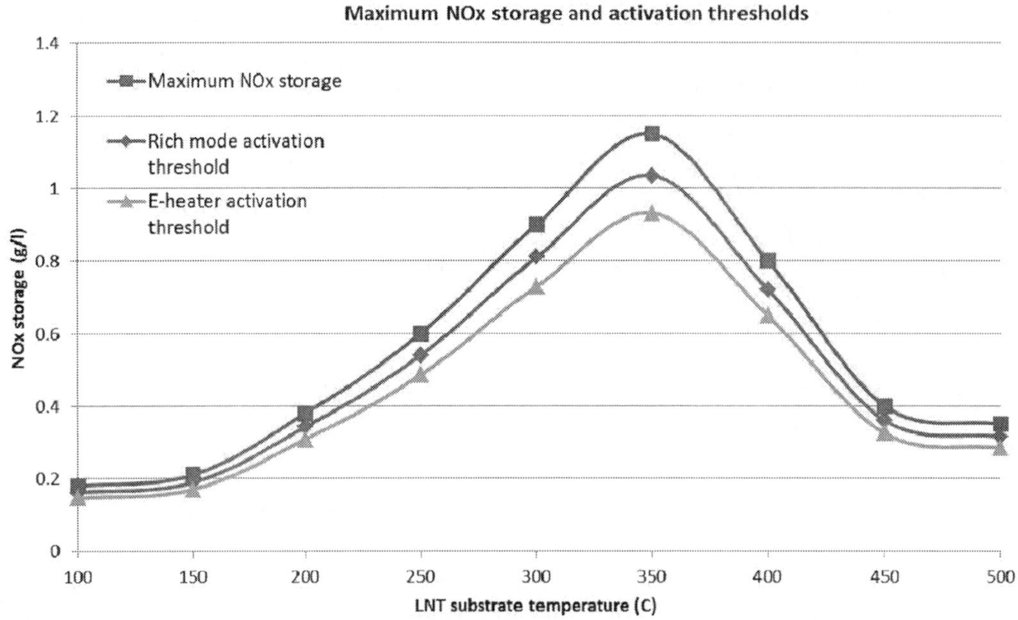

Figure 6: Dynamic NOx storage curve for E-heater activation

This combined E-LNT heating and purging strategy has been implemented in the ECU model with the following inputs and outputs:

- Inputs: R-theta, actual LNT substrate temperature, demanded LNT substrate temperature, engine operating conditions.
- Outputs: E-heater activation, E-heater target temperature, LNT purge activation.

Figure 7 shows an overview of the new functionality developed.

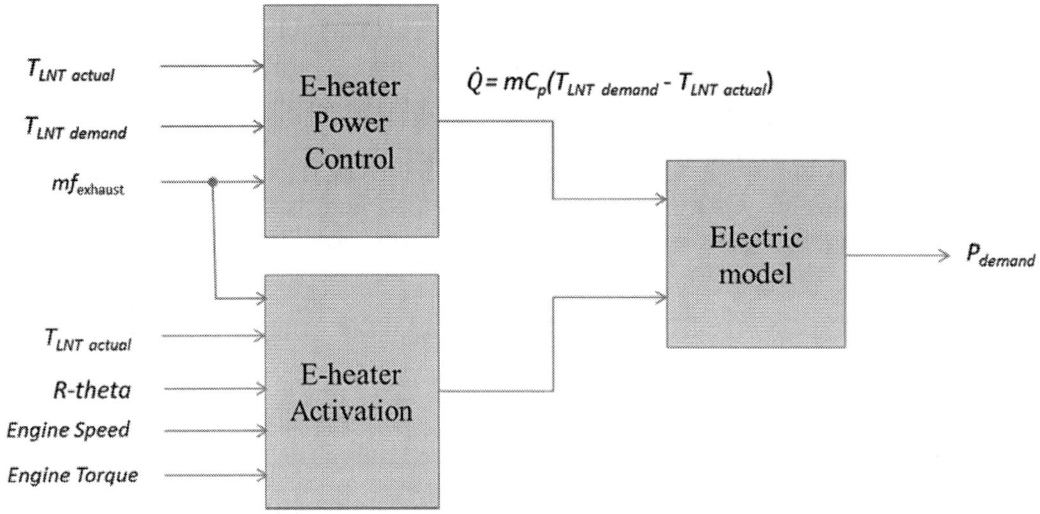

Figure 7: Integrated E-LNT control logic overview

The improved E-LNT system control logic has been tested for three different low load cycles. The simulation results and discussion are presented in the following section.

5 SIMULATION RESULTS AND DISCUSSION

5.1 Test Cycles

Following a basic study to understand the impact of each control variable of the improved E-LNT control strategy, a simulation plan was developed to assess the benefit of using the E-LNT system over three different city cycles.

5.1.1 Paris City Cycle

Paris driving cycle is a common test cycle used by most OEMs to validate calibration functions such as emissions or EAS in real world city driving conditions. Initially developed by PSA, this cycle is a 5.6 km long transient low speed cycle with an average speed of 21 km/h.

5.1.2 Artemis Urban Cycle

This cycle is a part of the Common Artemis Driving Cycles (CADC) devised by European Artemis Project based on statistical analysis of a large database of European real world driving patterns. The urban part is a 4.5 km long driving cycle with an average speed of 17.5 km/h.

5.1.3 NASR Cycle

NASR is a custom dynamometer cycle specifically developed during this study and is built from different low speed sections from EPA, Artemis and Paris cycles with extended engine idling periods. It can be representative of city stop-start driving conditions with very low exhaust gas temperatures.

5.2 Simulation Results

To analyse the impact of using the improved E-heater and purge logic at various configurations, 5 different configurations of E-heater power and desired LNT substrate temperature have been simulated and are given in Table 5 below:

Table 5: Simulation test matrix

Variables\Test Cases	Case 1	Case 2	Case 3	Case 4	Case 5
Max Heater Power (kW)	0	1.5	1.5	2	2
Target LNT Temperature (C)	0	225	250	225	250

In each case, the simulation is run for duration of one hour for all 3 test cycles.

Table 6 gives a summary of the simulations results obtained and the optimal case for each test cycle:

Table 6: Summary of simulation results showing improvement due to the E-LNT control strategy

	Paris	Artemis	NASR
Optimal Case	Case 2	Case 3	Case 2
Max Heater Power (kW)	1.5kW	1.5kW	1.5kW
Target LNT Temperature (C)	225 C	250 C	225 C
% Change in **CO2**	**+ 0.3%**	**+3.6%**	**+3.9%**
% Change in **NOx**	**0%**	**-19.3%**	**-22%**
% Change in **CO**	**-9%**	**-22%**	**-10%**
% Change in **HC**	**-18%**	**-18%**	**-11%**
Avg LNT temperature No Heater/ With heater	253/**254** °C	198/**215** °C	146/**168** °C

We can observe that a significant reduction in tailpipe NO_X emissions can be achieved with an acceptable CO_2 penalty when compared to the reference case without any E-LNT heating activation strategy. For example, in the case of the custom NASR cycle, 22% NOx reduction is obtained for 3.9% CO_2 penalty.

In the case of Paris cycle, the E-LNT system does not show a significant benefit in NOx emissions as the LNT substrate temperature is already within the boundary conditions for purge activation in the reference case.

For all cycles we can also note a reduction trend in tailpipe emissions of HC and CO.

Figure 8 shows the overall LNT conversion efficiency and the total duration of rich purge over each cycle. It is notable that in the case of the NASR cycle, where the LNT substrate temperature is below the minimum threshold, the use of the E-heater acts as a purge enabler and helps to improve the overall LNT conversion efficiency.

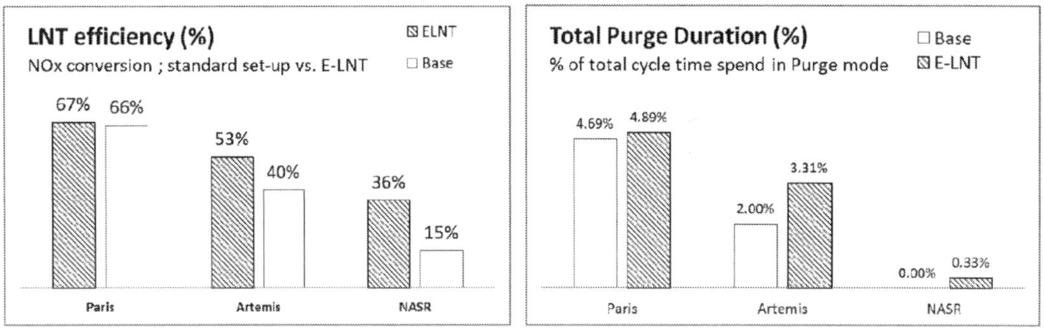

Figure 8: (left) Comparison of LNT conversion efficiency; (right) Total time spent in LNT purge mode

With the improved E-LNT control strategy, we can also see in Figure 9 that the fraction of energy consumed by the E-heater over the whole cycle represents a small percentage of the total energy produced by the engine. The impact of the E-heater usage in the total fuel consumption remains minimal.

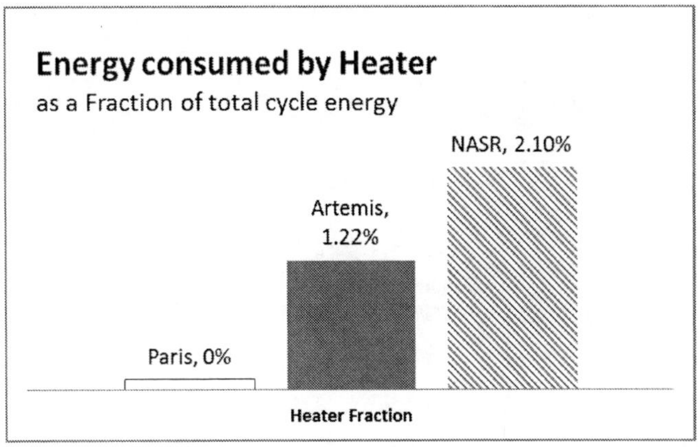

Figure 9: Comparison of E-heater's energy consumption

5.3 E-heater performance over the tests cycles

5.3.1 Paris Urban cycle

In the Paris cycle, with the control strategy developed, the E-heater activation is only required during the cold-start phase as shown in Figure 10. For the 2.4 L I5 engine, the exhaust temperatures are already favourable in this cycle, and the use of the E-LNT system does not provide an additional benefit for NOx conversion. However the cold start activation can help improving the overall HC/CO conversion of the LNT.

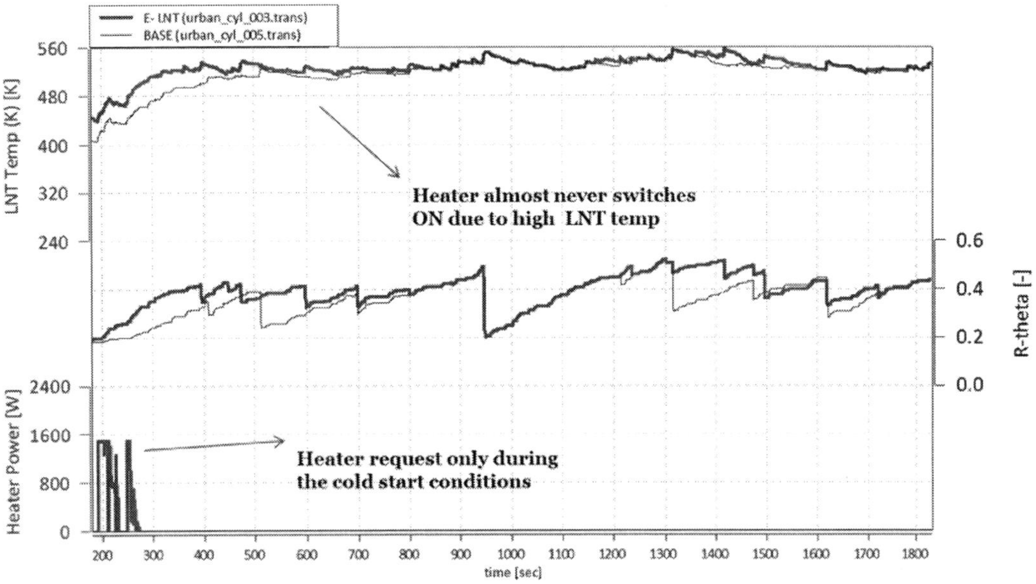

Figure 10: E-LNT system performance over Paris cycle

5.3.2 Artemis Urban cycle

Figure 11 shows the behaviour of E-LNT system over a section of Artemis cycle. In this case we can observe that the E-heater activation by increasing the LNT substrate temperature improves the NOx storage (R-theta). This results in an earlier purge trigger in the cycle, when compared to the reference case, and an improved overall LNT efficiency.

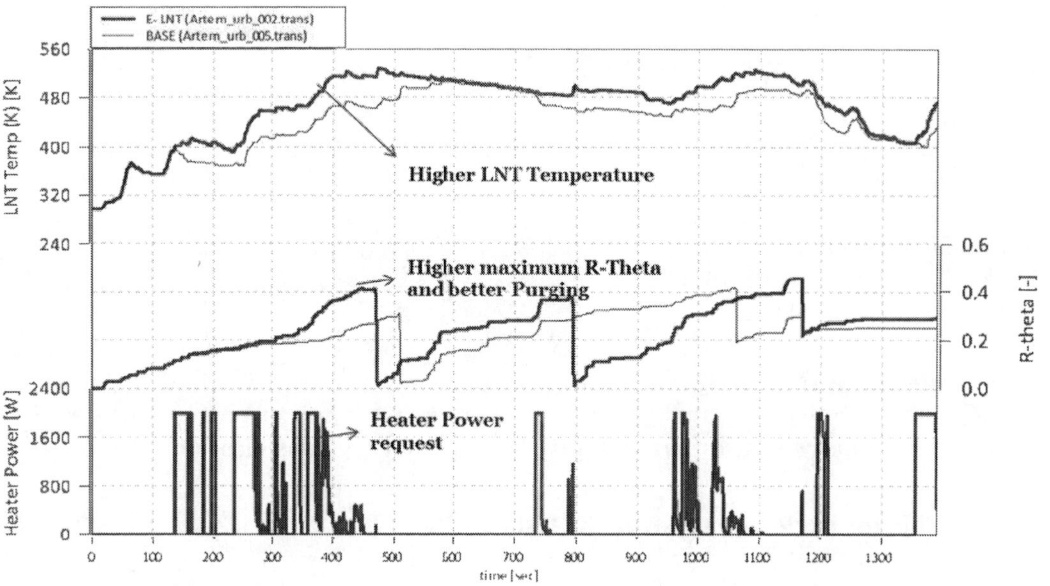

Figure 11: E-LNT system performance over Artemis urban cycle

5.3.3 NASR cycle

For the NASR cycle simulation run, the LNT substrate temperature in the reference case is always below the minimum purge temperature activation threshold of 200 °C (or 473 K) as we can see in Figure 12. The NOx filling ratio trace shows a saturation effect after a certain duration. With the purge being inhibited, this means all the engine out NOx will slip through the LNT.

The E-heater enables the rich conversion operation and improves the overall conversion efficiency by increasing the LNT substrate temperature.

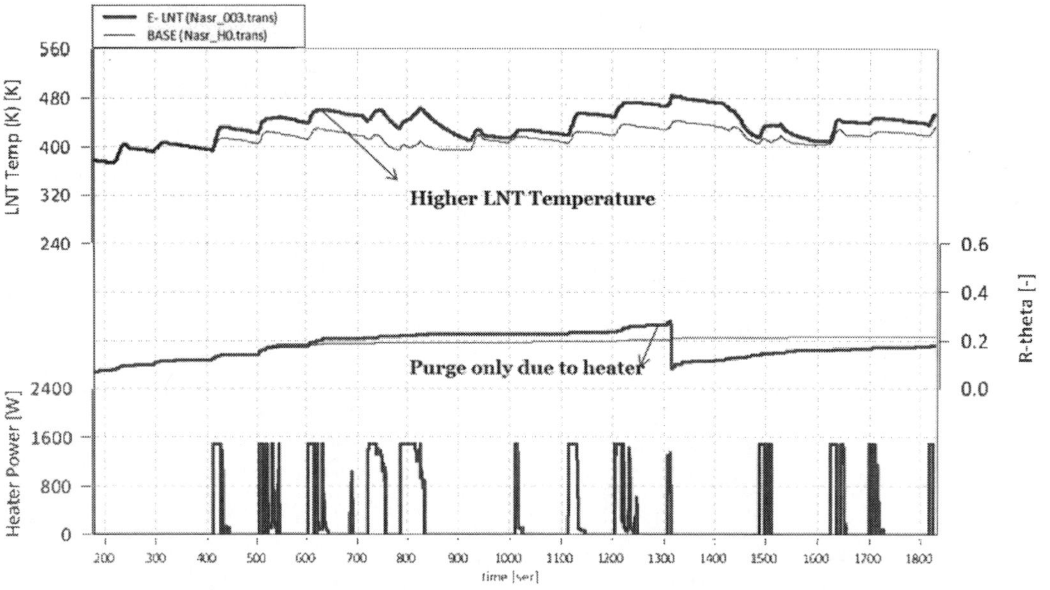

Figure 12: E-LNT system performance over NASR cycle

6 CONCLUSIONS

In this study we have demonstrated that the use of an E-LNT system combined with an optimized heater and purge strategy, offers the potential of significant NOx emissions reduction with acceptable fuel consumption penalties in low load or city driving cycle conditions.

In the specific conditions where the LNT substrate temperature is too low to request an efficient purge, the E-heater can be used to periodically heat up the LNT into the purge temperature boundary conditions and thus improve the tailpipe NOx emissions by up to 20% in some city cycles with a CO_2 penalty up to 4%. In this case the E-heater is used as a rich purge enabler and offers the best trade-off between NOx reduction and CO_2 penalty.

The increase in LNT temperature due to the E-heater also provides an additional benefit in overall HC/CO emissions for most of the cycles studied.

This study has been realised using a simulation model for a relatively large passenger car engine size and power output (I5D, 2.4 L, 150 kW), and a LNT formulation with improved capacity at low exhaust temperature. Further investigations with the combination of a lighter vehicle, a lower displacement engine and a less efficient LNT formulation constitute a possible next step. In such a case the resulting lower exhaust gas temperature will increase the need to address the LNT operation under city driving conditions. This would further demonstrate the effectiveness of the E-LNT system and the strategy developed in this project.

The system and the strategies developed in this study can potentially offer a viable NOx reduction technology utilising LNT, for example, for the upcoming RDE emissions standards for Euro 6c and beyond.

Finally, the E-LNT with the E-cat can be an attractive solution to be integrated in a Diesel hybrid powertrain with 48 V systems.

7 REFERENCES

[1] www.DieselNet.com, NOx Adsorbers
[2] SAE 2014-01-1491: Vehicular Emissions in Review, T. Johnson, 04/01/2014
[3] SAE 2014-01-2325: Development Trends for Commercial and Industrial Engines, Franke et al.v, 09/30/2014
[4] SAE 2008-01-0449: Development of a Diesel Emission Catalyst System for Meeting US SULEV Standards, Onodera et al., 17/04/2008
[5] SAE 2007-01-1241: Ceria Doped Ba-Alumina Oxides for Durable NOx-Storage Catalysts Efficient at Low Temperature, Belliere-Baca et al., 19/04/2007
[6] SAE 940465: Ultra-Low Power Electrically-Heated Catalyst System, Küper, Maus et al., 03/03/1994
[7] SAE 2011-24-0178: An alternative way to reduce fuel consumption during cold start: the electrically heated catalyst, Presti et al., 2011
[8] SAE 2012-01-1092: Electrically Heated Catalysts for Cold-Start Emissions in Diesel Aftertreatment, Chang et al., 16/04/2012

A novel working fluid for Organic Rankine Cycles (ORC)

A S PANESAR, M R HEIKAL, M MARENGO and **R E MORGAN**
School of Computing, Engineering and Mathematics, University of Brighton, UK

ABSTRACT

The use of ORCs on long-haul Heavy Duty Diesel Engines (HDDE) is a possible way to improve the specific fuel consumption. One of the key considerations in the research and development efforts for ORCs is to investigate and identify technical paths that may improve the practicality of such a heat to power conversion concept. For this, simple solutions are vital for a timely deployment of the technology to meet the anticipated CO_2 regulations.

To provide a potential solution, this paper presents the simulation results (using Aspen HYSYS) of a novel organic working fluid which has been especially formulated for the HDDE sector. Due to the unique heat transfer and expansion characteristics, the primary advantage of the novel working fluid included an equivalent performance to ethanol despite a 25% reduction in the total heat transfer area and a 25% reduction in the size of the expansion machine. The secondary advantages over ethanol included a higher molecular weight (66 vs. 46 g/mol) and a lower expansion enthalpy drop (90 vs 118 kJ/kg). Additionally, the novel fluid is expected to offer low global warming potential (< 20) and improved thermal stability relative to ethanol.

1 INTRODUCTION

In a typical internal combustion engine about 1/3rd of the total fuel energy is wasted in the form of exhaust gas heat. Converting this waste heat into usable mechanical or electrical power is seen as a key area in the development of low carbon powertrains [1, 2]. This is an ongoing area of intensified research, where numerous methods have been demonstrated. These methods include, but are not limited to, turbocompounding (mechanical, electrical), thermoelectric generators and fluid bottoming cycles [3-6].

Amongst the fluid bottoming cycle options, ORCs are shown to be better adapted to an output capacity of 5-25 kW and heat source quality < 400°C [7, 8]. ORCs are in fact being adopted as a premier technology for long-haul HDDEs when considering conversion efficiencies, technology readiness level, impending CO_2 legislations, absolute fuel consumption, base vehicle cost, space availability and weight penalty. Key ORC components like heat exchangers (HEX) and expansion machines (piston expanders, radial turbines) are becoming more viable due to a series of recent technological advancements and synergies with the current automotive components

[9-11]. The current market niche for ORCs is dependent on simplicity and affordability, with initial technology deployment on commercial vehicles expected in the 2020-2022 timeframe.

The two widely proposed fluids for HDDE applications are R245fa and ethanol. These fluids are not without their distinctive challenges. R245fa has a relatively high global warming potential (1030, relative to CO_2 for an integration time horizon of 100 years) and an unfavourable thermal stability temperature limit (< 250°C), while ethanol has a relatively lower molecular weight (46 g/mol) and requires a larger expansion machine size [12, 13]. Although, the molecular make-up of organic fluids fundamentally precludes the possibility of an ideal fluid, ethanol is generally considered to offer a more suitable trade-off [14].

2 BASELINE ETHANOL ORGANIC RANKINE CYCLE

This section presents the simulation results for an ethanol ORC with an Internal Heat Exchanger (IHE) to act as a reference for comparison. The simulations were conducted in an advanced chemical process modelling tool, Aspen HYSYS [15]. HYSYS contains physical, chemical and thermodynamic data for a wide variety of chemical compounds, as well as a selection of thermodynamic models required for accurate simulation of thermal and chemical systems. Only exhaust heat recovery, downstream of the aftertreatment devices, was considered at mid-speed mid-load (B50) from a 12.8 litre engine. This engine also utilised Exhaust Gas Recirculation (EGR). However, due to the lower EGR rates (< 15%) and further expected improvements in selective catalytic reduction technology, it is likely that by 2020-2022, HDDEs would offer exhaust as the only source of high quality and high quantity waste heat [16, 17]. Excluding EGR will also reduce the diesel particulate filter regeneration frequency. Although, regenerations can raise exhaust temperature above 600°C, ORC operation has to be stopped to avoid thermal degradation of the fluid.

Table 1 Assumptions and boundary conditions used in ORC simulations

Base engine performance (B50) ▪ BSFC = 195 g/kWh ▪ Crankshaft power = 158 kW *Non-insulated exhaust pipe* ▪ $T_{post-turbocharger\ turbine}$ = 420°C ▪ $T_{post-SCR}$ = 410°C ▪ $c_{p\ exhaust}$ = 1.14 kJ/kg°C ▪ $m_{exhaust}$ = 0.22 kg/s *Heat exchanger (counter-current)* ▪ $T_{HEX\ inlet}$ = 400°C ▪ $T_{max\ organic\ fluid}$ = 250°C ▪ $\Delta P_{organic\ fluid}$ = 0.5 bar ▪ $T_{pinch\ point\ HEX}$ = 15°C	*Air condenser (counter-current)* ▪ $T_{condensing\ organic\ fluid}$ = 90°C ▪ $T_{sub-cooling}$ = 5°C ▪ $\Delta P_{organic\ fluid}$ = 0.2 bar ▪ $T_{cooling\ air\ inlet}$ = 45°C ▪ $T_{cooling\ air\ outlet}$ = 65°C ▪ $\Delta P_{cooling\ fan}$ = 250 Pa ▪ $\eta_{fan\ and\ motor}$ = 65% *Internal heat exchanger (counter-current)* ▪ $T_{pinch\ point\ IHE}$ = 15°C ▪ ΔP_{IHE} = 0.2 bar (each side) *Expansion machine and pump* ▪ $\eta_{expansion}$ = 60% ▪ η_{pump} = 60% ▪ $\eta_{transmission}$ = 93%

Table 1 summarises the ORC modelling assumptions and boundary conditions used corresponding to realistic temperatures, component efficiencies, performances and pressure losses [9, 18, 19]. To limit the fan power requirement and the total engine cooling-module size, a high-temperature direct air-cooled ORC condenser was used and positioned after the low temperature engine radiator. Furthermore, the maximum temperature of ethanol was restricted to 250°C [20-22].

Fig. 1a shows the layout of the simulated ORC. The purpose of using an IHE was to internally utilise the exergy exiting the expansion machine, and hence avoid the potential loss in the condenser. The IHE increases the cycle thermal efficiency, decreases the heat recovery efficiency and has no impact on the net power. The improvement in the thermal efficiency results in reduced load on the condenser. The exclusion of low temperature exhaust heat recovery avoids corrosion and fouling in the HEX. Since the pressure difference across the expansion remains nearly equal, the specific work is unchanged.

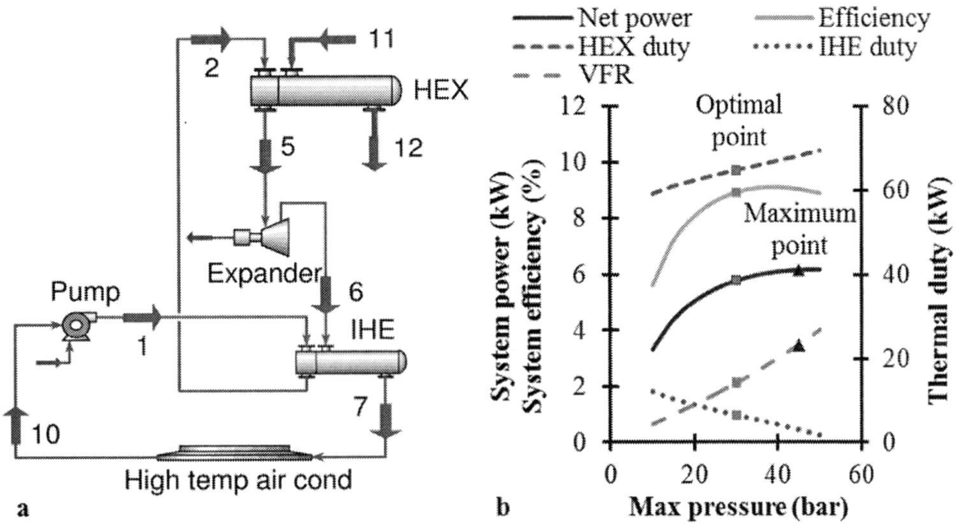

Figure 1 (a) ORC setup with IHE used in simulations
(b) Optimisation results for ethanol

The ORC system power was calculated according to Equation 1. This included a fixed 0.2 kW requirement for controls. Additionally, the increased fan power requirement at B50 was 1 kW.

$$ORC_{system} = \eta_{transmission}(W_{expansion} - W_{pump}) - W_{control} - W_{fan\ power} \quad \ldots(1)$$

Fig. 1b shows the pressure optimisation results for the ethanol ORC with an IHE. A maximum cycle pressure of 30 bar was considered optimal. This was since the rate of system power improvement was rather inadequate with higher pressures. The system power and thermal efficiency at 30 bar was ≈95% of the maximum, but with 2/3rd the size of expansion machine.

Table 2 summarises the optimal system performance and property values for ethanol. The system required a high expansion Volume Flow Ratio (VFR 14:1) and Pressure Ratio (PR 14:1), making it better suited to piston expanders. The requirement of a large expansion machine size with ethanol is fundamentally related to a higher $T_{boil\ low\ pressure} / T_{boil\ high\ pressure}$ value.

Table 2 Performance and property values for ethanol and E40

	Ethanol	E40
P_{max} bar	30	22
P_{min} bar	1.7	1.4
T_{max} $°C$	250	250
T_{min} $°C$	90	87
$T_{condensation}$ $°C$ (avg. for E40)	92	94
$T_{evaporation}$ $°C$ (avg. for E40)	200	204
m_{fluid} kg/s	0.067	0.087
$UA_{HEX} W/°C$	1640	1230
$UA_{IHE} W/°C$	252	190
$UA_{Cond} W/°C$	1361	1022
m_{air} kg/s	2.82	2.82
Q_{IHE} kW	6.6	10.9
Q_{HEX} kW	64.6	61
$PR_{expansion}$	14:1	11.7:1
$VFR_{expansion}$	14:1	10.6:1
W_{pump} kW	0.4	0.4
$W_{expansion}$ kW	7.9	7.9
$W_{fan\ power}$ kW	1	1
W_{system} kW	5.8	5.8
$\eta_{system\ thermal}$ %	8.9	9.4

Fig. 2 presents the component irreversibility contributions in the 30 bar optimal ethanol cycle. It is seen that the HEX made the highest contribution (38%), closely followed by the condenser (34%). The high heat transfer irreversibilities were principally due to the temperature differences during near isothermal evaporation/condensation and the high condensing temperature. The expansion (23%) and the pumping (1%) entropy generation rates which were related to the isentropic efficiencies were relatively medium and insignificant, respectively.

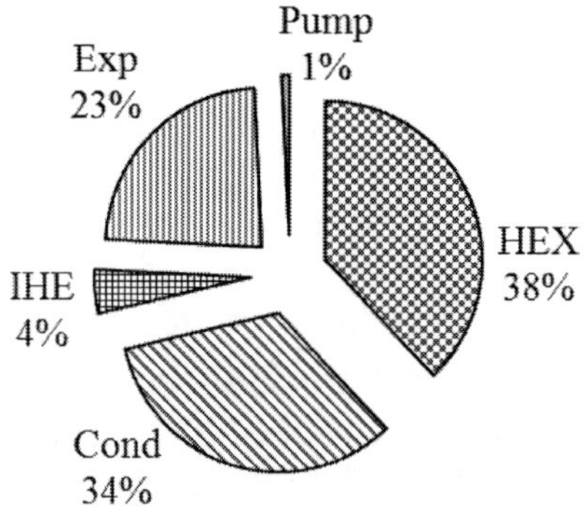

Figure 2 Irreversibility contributions in the fundamental ethanol ORC processes

The high $T_{boil\ low\ pressure} / T_{boil\ high\ pressure}$ value and heat transfer irreversibility drawbacks are addressed in the published literature by using high VFR expansion machines, more efficient HEXs/condensers and supercritical cycle operation [9, 11, 21]. The supercritical cycle offers lower HEX irreversibilities (avoids near isothermal evaporation) and may allow the exclusion of the IHE (IHE duty decreases with increasing pressure for a fixed temperature). However, supercritical ethanol cycle requires noticeably higher pressures (> 65 bar) and PR's (> 30:1). Although, efficient HEXs/condensers and high VFR expansion machines will be vital in the deployment of ORC systems. These approaches only indirectly address the above drawbacks. Hence, innovative pathways to improve ORC performance and cost-effectiveness remain a major challenge facing the research community.

3 NOVEL FLUID E40

In view of the above understanding, a study was undertaken to identify a single method which for the smallest change to the ethanol ORC with an IHE would translate to a noticeable benefit. The identified path was to formulate a novel organic fluid blend with unique heat transfer and expansion characteristics, while retaining ethanol as a noticeable blend component. Ethanol as a blend constituent was selected since this may allow for the carryover of the developed ethanol components by vehicle manufacturers, original equipment manufacturers, research centres and Tier 1,2 suppliers. An evolved screening methodology from previous works was developed and applied to examine over 200 well documented ethanol blends [23-26]. The description of this methodology is beyond the scope of this paper, nonetheless an attempt is made to itemise the relevant fluid properties.

This section presents the simulation results of a particularly useful blend with varying ethanol concentrations. The resulting different concentration blends were termed based on ethanol percentage by mass, e.g. E80 corresponding to 80% ethanol. (Note: the second blend constituent is undisclosed). Prior to a detailed heat transfer equipment calculation and design, the following was considered for the absolute size comparison. It was assumed that the overall heat transfer coefficient (U, $W/m^2 °C$) was similar for ethanol and the blends. Therefore, UA (W/°C), i.e. overall heat transfer coefficient multiplied by the heat transfer area (A, m^2), was considered as an indicator for the absolute heat transfer size comparison for HEX, IHE and condenser. Similarly, VFR defined as the ratio between the volumetric flow rates at the expansion outlet to inlet was considered as an indicator of the absolute size of the expansion machine. The blends were simulated for the same ORC setup (Fig. 1a), but with 25% reduced heat transfer equipment size (i.e. 25% lower UA for HEX, IHE, condenser) and 25% reduced expansion machine size (i.e. 25% lower VFR).

Fig. 3a presents the primary results of interest. Despite the system size reduction, the blends (with ≥40% ethanol) gave nearly equal system power to ethanol. This corresponded to Δ7 g/kWh of BSFC reduction. The improvements in system thermal efficiency (≈5%) and reduction in the expansion pressure ratio (≈15%) were also noticeable. The blend properties were calculated using the Wilson property package

[27]. The difference in expansion power was compared against an alternative property package. A low discrepancy of 2% was calculated.

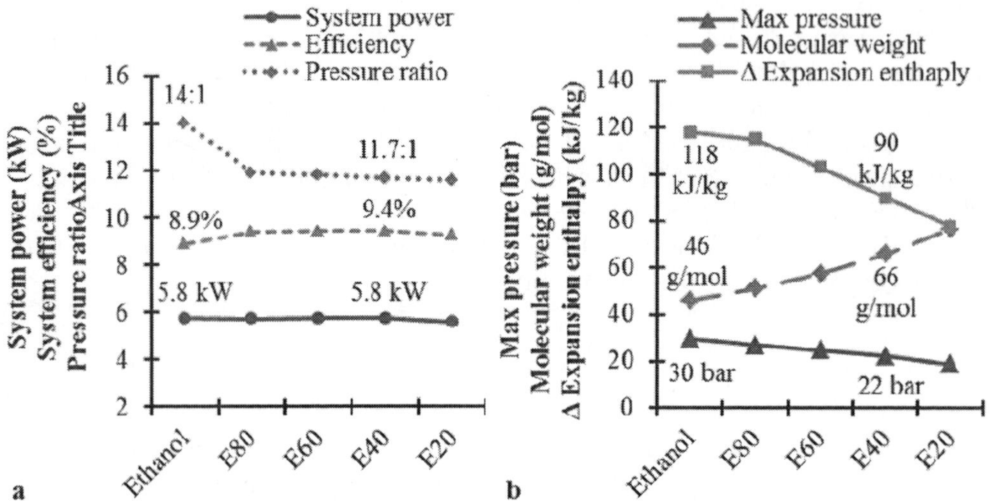

Figure 3 E40 parameters in comparison to ethanol of (a) primary interest and (b) secondary interest

Fig. 3b presents the secondary results of interest. Amongst the blends, E40 was considered as the most appropriate fluid giving equal system power to ethanol. This was because it offered a suitable trade-off between ethanol concentration, maximum system pressure and molecular weight. Compared to ethanol, E40 required a lower maximum system pressure (22 vs. 30 bar) and offered a much higher molecular weight (66 vs. 46 g/mol).

The lower maximum pressures are advantageous not only for turbines, but also for other positive displacement expanders. This is since under higher pressure operation not only pressure drops and leakages account for a significant portion of performance loss but the machine wear also increases [28]. As a consequence of the higher molecular weight of E40, the expansion enthalpy difference was 28 kJ/kg lower. The relatively lower pressure ratio and higher molecular weight bodes particularly well for single stage turbines as expansion machines. When using a turbine, a 20 g/mol increase in the molecular weight would translate to slightly increased turbine efficiency due to design considerations in < 100 kW output capacity [7]. The resulting system power improvements due to this were ignored. Since, a single method to reintroduce the recovered energy into the powertrain has not yet been identified by the HDDE sector. The use of E40 allows continued examination of both mechanical and electrical methods. Table 2 also summarises the system performance and property values for E40.

To understand the key characteristics of E40, which collectively resulted in nearly equal power to ethanol despite the reduced size, firstly compare the T-S diagrams (Fig. 4a vs. b). The unique heat transfer characteristics of E40 can be summarised as:

- *High ΔT variation during evaporation:* E40 offered a 30°C temperature rise (pt. 3 to pt. 4) compared to the near constant temperature of ethanol. This increases the mean heat addition temperature, and hence, the thermal efficiency. A high ΔT variation during evaporation mirrors the high heat source temperature drop (250°C) and reduces the heat transfer irreversibilities.

- *Higher internal heat recuperation:* The IHE duty for E40 corresponded to 20% of the total heat input, compared to 10% in ethanol. Higher internal heat recuperation decreases the condenser load. This offers lower condensation temperatures for the same cooling air flow rate, resulting in a higher thermal efficiency.
- *Low ΔT variation during condensation:* E40 offered a 15°C temperature rise (pt. 9 to pt. 8) compared to the near constant temperature of ethanol. A low ΔT variation during condensation mirrors the low heat sink temperature rise (20°C) and reduces the heat transfer irreversibilities.

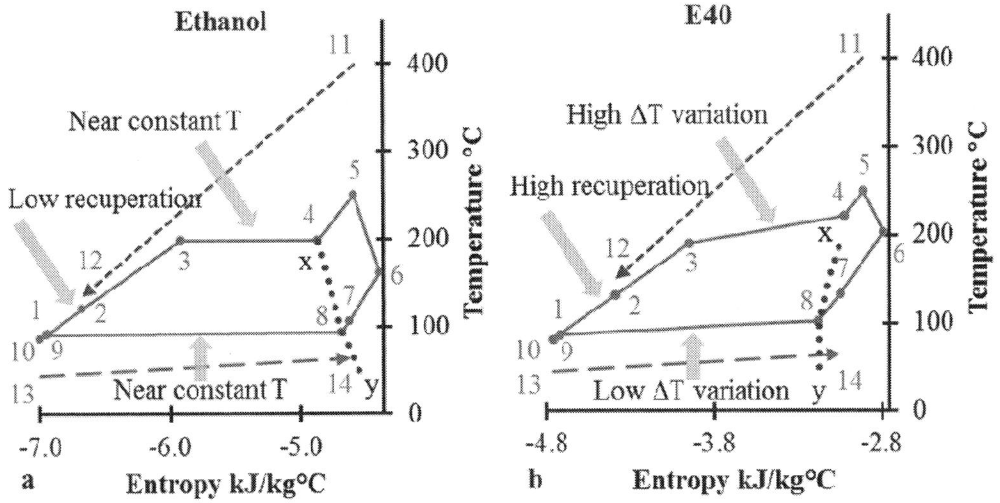

Figure 4 Heat transfer characteristic comparison between (a) ethanol and (b) E40

Secondly, the unique expansion characteristics of E40 can be summarised as:

- *Low $T_{boil\ low\ pressure}$ / $T_{boil\ high\ pressure}$ value:* For this illustration consider ethanol and E40 for phase-change at an assumed high and low pressure of 30 and 1 bar and with a fixed duty of 40 kW (Fig. 5). Under these conditions, ethanol gave a value of 0.74. Although E40 offers variable temperature, nonetheless, assuming the average temperature for 10 equal enthalpy intervals during phase-change gave a value of 0.70. This in effect means that under equal high and low system pressures, the maximum and minimum temperature difference for E40 is higher, giving greater area and hence improved efficiency.
- *Suitable for turbines and piston expanders:* E40 is a drying fluid (slope x-y, Fig. 4b), contrary to ethanol. This avoids droplets during expansion when expansion starts at the saturated vapour line. This along with a relatively higher molecular weight and lower pressure ratio makes it more suitable for turbines [10, 29].

It is important to highlight that the fluid characteristics in Fig. 5 are different to the typical approach of using a zeotropic blend. This is since, a zeotropic fluid offers similar temperature glides in the evaporation and condensation processes, and usually do not include the considerations of improving the pressure differential in the system.

Finally, Table 3 presents some key properties for ethanol and E40 for comparison. Both fluids contain carbon, hydrogen and oxygen molecules, and are expected to offer similar boiling points, freezing temperatures, thermal decomposition temperatures, auto-ignition temperatures, NFPA rating and environmental impact [30, 31]. The variation in liquid density, specific heat, thermal conductivity and viscosity is expected to be low (< 20%). Since ethanol and the undisclosed blend constituent both have been used for lubricating rotating shafts, E40 is also expected to offer this functionality [32]. Additionally, both fluids have virtually zero electric conductivity. Hence, both fluids could be used for cooling the electrical components of the ORC system. A preliminary survey result of metals and alloys suited for both ethanol and E40 include aluminium, brass, carbon steel, stainless steel and copper [33-37]. Furthermore, suitable O-ring materials and thermoplastics include ChemRaz, Kalrez, Kel-F, Polyether ether ketone and Polytetrafluoroethylene.

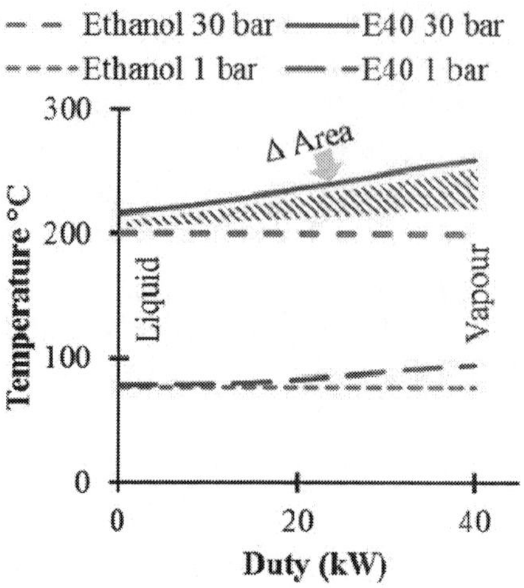

Figure 5 Expansion characteristic comparison between ethanol and E40

Table 3 Key fluid properties of comparison between ethanol and E40

	Ethanol	E40
Molecular make-up	Carbon Hydrogen Oxygen	Carbon Hydrogen Oxygen
Boiling point °C	77	78
Molecular weight g/mol	46	66
Liquid density kg/m³	736	795
Specific heat kJ/kg°C	3.31	2.61
Thermal conductivity W/mK	0.15	0.13
Viscosity cP	0.45	0.39
Auto-ignition temperature °C	363	≈390
Freezing temperature °C	-114	≈-100
NFPA Health	2	2
NFPA Flammability	3	3
NFPA Reactivity	0	0
Ozone Depletion Potential	0	0
Global warming potential	<20	<20
Electrical conductivity	Nil	Nil
Lubrication properties	+	++
Cost ($/L for 200L)	36	18
Turbine suitability	+ +	+ + +
Piston expander suitability	+ + + +	+ + + +

4 ECONOMIC ANALYSIS

A detailed economic review and an original cost analysis study conducted for small-scale (5-25 kW), high-production (>1000 units), ORC systems and its associated components resulted in the cost distribution shown in Fig. 6 [38-44]. Three relevant points in the present case were that:

- Approximately quarter of the total system costs were fixed and non-scalable.
- Scalable fluid expansion, fluid compression and heat transfer costs were closely related to VFR, pressure difference and UA, respectively.
- A similar distribution trend also existed for the size and weight of ORC components (Note: fixed and non-scalable percentage was around 30%).

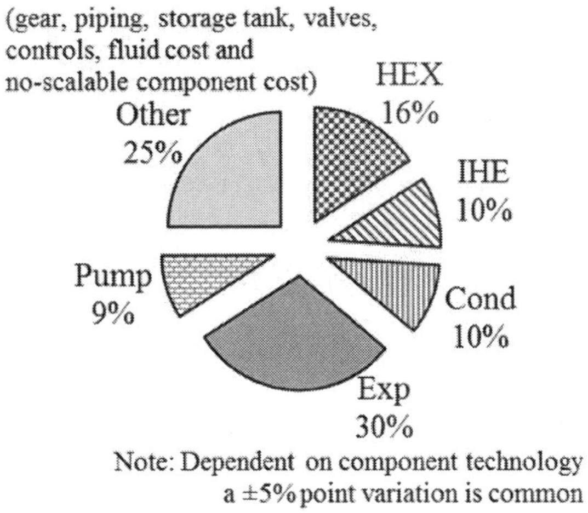

Figure 6 Small scale ORC component cost distribution

To estimate a techno-economical trade-off between ethanol and E40, the following was considered. According to Equation 2, it was assumed that the ratio of Cost/kW of the 30 bar ethanol cycle was 1.

$$\begin{aligned} Cost / kW = {} & 0.16UA_{HEX} + 0.10UA_{IHE} + 0.10UA_{condenser} \\ & + 0.30VFR_{expansion} + 0.09\Delta P_{pump} + 0.25 \end{aligned} \quad \ldots(2)$$

Therefore, when considering 25% lower UA, VFR and ΔP for E40, the Cost/kW reduced to 0.81. It is important to emphasise that marginal improvements expected in turbine efficiency and reduction in HEX cost (due to lower pressures) were ignored for E40 as a first approximation. A further benefit of E40 not considered was its lower fluid cost. Quotations from two different suppliers indicated that E40 fluid cost was expected to be around half of ethanol (Table 3) [45, 46].

5 CONCLUSIONS

A novel approach has been presented in this paper to facilitate the introduction of ORC systems for HDDE applications. The novelty of the concept developed here consisted of adapting the working fluid make-up to efficiently match the source/sink streams and reduce the cycle pressure difference for a fixed temperature difference. This was achieved by reducing the irreversibilities encountered during the phase-change process (despite subcritical pressures) and lowering the $T_{boil\ low\ pressure} / T_{boil\ high\ pressure}$ value (despite noticeable ethanol concentration). As a result, for an equal system power to ethanol ORC, the 40% ethanol blend ORC presented the advantages of 15-20% lower volume, weight and cost. This blend offered a higher energy density solution with a favourable system pressure, improving commercialisation potential. The decreased VFR requirement (10.9:1 vs. 14:1) and the increased molecular weight calculated (66 vs. 46 g/mol) allows the possibility of using both single stage radial turbines and piston expanders. This blend is expected to offer the carryover of the currently developed ethanol components, and continued use of both mechanical and electrical power generation methods.

6 REFERENCES

1. US DOE, Advanced Combustion Engine Research and Development, Annual Progress Report. 2011.
2. Baker, H., et al., Review of Low Carbon Technologies for Heavy Goods Vehicles, RD.09/182601.7. 2010.
3. Millikin, M. Daimler Trucks Introduces New Heavy-Duty Engine Platform for North America; DD15 Improves Fuel Economy by 5%. 2007; Available from: www.greencarcongress.com.
4. Kruiswyk, R., J. Fairbanks, and C. Maronde, An Engine System Approach to Exhaust Waste Heat Recovery, in Advanced Combustion Engine Technologies, Annual Progress Report. 2009, 2008.
5. LaGrandeur, J., et al., High-Efficiency Thermoelectric Waste Energy Recovery System for Passenger Vehicle Applications, in Advanced Combustion Engine Research and Development, Annual Progress Report. 2011, 2010, 2009.
6. Stanton, D. The Role of Waste Heat Recovery in Meeting Phase 2 US EPA Greenhouse Gas Regulations. in Diesel Emissions Conference & AdBlue Forum Europe. 2013.
7. Stine, W.B. and M. Geyer, Power Cycles for Electricity Generation, in Power From The Sun. 2001, www.powerfromthesun.net.
8. Crook, A.W., Profiting from Low-grade Heat: Thermodynamic Cycles for Low-temperature Heat Sources. 1994: Institution of Electrical Engineers, ISBN 9780852968352.
9. Seher, D., et al., Waste Heat Recovery for Commercial Vehicles with a Rankine Process in 21st Aachen Colloquium Automobile and Engine Technology, October 8-12, Aachen, Germany. 2012.
10. Halliwell, E., Testing Waste Heat Expanders for New Generation Diesels, in 10th International Conference on Turbochargers and Turbocharging, May 15-16, London, UK. 2012.
11. Edwards, S., et al., Waste Heat Recovery: The Next Challenge for Commercial Vehicle Thermomanagement. SAE Int. J. Commer. Veh., 2012. 5(1): p. 395-406.
12. Calm, J.M. and G.C. Hourahan, Physical, Safety, and Environmental Data Summary for Current and Alternative Refrigerants, in International Congress of Refrigeration, August 21-26, Prague, Czech Republic. 2011.

13. Zyhowski, G.J. Honeywell Refrigerants Improving the Uptake of Heat Recovery Technologies. 2013; Available from: www.honeywell-orc.com.
14. Teng, H., G. Regner, and C. Cowland, Waste Heat Recovery of Heavy-Duty Diesel Engines by Organic Rankine Cycle Part II: Working Fluids for WHR-ORC. 2007, SAE International, 10.4271/2007-01-0543.
15. Aspen Technology, HYSYS V 7.3. 2011.
16. Johnson, T., Diesel Emission Control in Review. SAE International Journal of Fuels and Lubricants, 2009. 2(1): p. 1-12.
17. Cloudt, R., F. Willems, and P. van der Heijden, Cost and Fuel Efficient SCR-only Solution for Post-2010 HD Emission Standards. SAE Int. J. Fuels Lubr. 2(1):399-406, 2009.
18. Fu, J., et al., An open steam power cycle used for IC engine exhaust gas energy recovery. Energy, 2012. 44(1): p. 544-554.
19. Badami, M., et al., Design and performance evaluation of an innovative small scale combined cycle cogeneration system. Energy, 2008. 33(8): p. 1264-1276.
20. Arunachalam, P.N., et al., Waste Heat Recovery from Multiple Heat Sources in a HD Truck Diesel Engine Using a Rankine Cycle - A Theoretical Evaluation. 2012, SAE International, 10.4271/2012-01-1602.
21. Teng, H. and G. Regner, Improving Fuel Economy for HD Diesel Engines with WHR Rankine Cycle Driven by EGR Cooler Heat Rejection. 2009, SAE International, 10.4271/2009-01-2913.
22. Abdurashidova, A.A., et al., The thermal properties of water-ethanol system in the near-critical and supercritical states. High Temperature, 2007. 45(2): p. 178-186.
23. DDBST. Online Dortmund Data Bank Search. accessed on 10.10.2014; Available from: ddbonline.ddbst.com.
24. Ponton, J.W. Homogeneous azeotrope databank. Web Tools for Process Engineers. 2007; Available from: eweb.chemeng.ed.ac.uk.
25. Lide, D.R. and H.V. Kehiaian, CRC Handbook of Thermophysical and Thermochemical Data, Volume 1. 1994: Taylor & Francis, ISBN 9780849301971.
26. Horsley, L.H., Azeotropic Data-III, Volume 3. 1973: American Chemical Society, ISBN 9780841205338.
27. Aspen HYSYS V 7.3, Physical Property Methods. 2011.
28. Vincent, L. and Q. Sylvain. Advances in ORC expander design. in International Symposium On Advanced Waste Heat Valorisation Technologies, September 13-14, Kortrijk, Belgium. 2012.
29. Cogswell, F.J., et al., Design Of An Organic Rankine Cycle For Waste Heat Recovery From A Portable Diesel Generator, in International Mechanical Engineering Congress & Exposition, November 11-17, Denver, USA. 2011.
30. NFPA, NFPA 704: Standard System for the Identification of the Hazards of Materials for Emergency Response. 2012.
31. US EPA. Ozone Layer Protection. 2014; Available from: www.epa.gov.
32. Park, T., et al., A Rankine Cycle System for Recovering Waste Heat from HD Diesel Engines - Experimental Results. 2011, SAE Technical Paper, 10.4271/2011-01-1337.
33. eFunda. O-Ring Compatibilities. accessed on 10.10.2014; Available from: www.efunda.com.
34. Ingersoll Rand, Chemical Compatibility Guide, FORM 8677-P 0508. 2008.

35. Cole-Parmer. Chemical Compatibility Database. accessed on 10.10.2014; Available from: www.coleparmer.co.uk.
36. allorings.com. O-Ring Fluid Compatibility Guide. accessed on 10.10.2014; Available from: www.allorings.com.
37. McQuay International. Application Guide AG 31-007: Refrigerants. 2002; Available from: www.daikinmcquay.com.
38. Thekdi, A.C. Waste Heat to Power Economic Tradeoffs and Considerations. in 3rd Annual Waste Heat to Power Workshop, September 25, Houston, USA. 2007.
39. Roos, C.J., An Overview of Industrial Waste Heat Recovery Technologies for Moderate Temperatures Less Than 1000F. 2009, Northwest Clean Heat and Power Application Center. WSUEEP09-26.
40. Dickey, H.K. Low Temperature Geothermal Power Generation with HVAC Hardware. in Geothermal Resource Council, Annual meeting, September 30-October 3, Sparks, USA. 2007.
41. Matches. Process Equipment Cost Estimates. 2014; Available from: www.matche.com.
42. Peters, M., K. Timmerhaus, and R. West, Plant Design and Economics for Chemical Engineers. 2003: McGraw-Hill Education, ISBN 9780072392661.
43. Quoilin, S. and V. Lemort, Technological and Economical Survey of Organic Rankine Cycle Systems, in European Conference on Economics and Management of Energy in Industry, April 14-17, Algarve, Portugal. 2009.
44. Lopes, J., et al., Review of Rankine Cycle Systems Components for Hybrid Engines Waste Heat Recovery. 2012, SAE International, 10.4271/2012-01-1942.
45. Spectrum Chemical Manufacturing Corp. 2014; Available from: www.spectrumchemical.com.
46. ScienceLab. 2014; Available from: www.sciencelab.com.

Heat, AC & Energy Management

Effects of heated seats on thermal comfort and heater energy consumption in vehicles

H OI
Nissan Motor Co., Ltd, Japan

Y TOCHIHARA
Kyushu University, Japan

ABSTRACT

Subjective experiments were conducted to investigate the effects of heated seats. Heated seats are effective in improving occupant thermal comfort in cool environments both in the initial warm-up period and under steady-state conditions. The optimal thermal sensation and skin temperature of the areas in contact with a heated seat were higher when the air temperature was lower. This result indicates that a higher seat temperature is suitable at lower air temperatures. The room operative temperature at which the occupants felt a "neutral" overall thermal sensation was reduced by at least 3°C by using the heated seat. Heated seats can reduce the total heater energy consumption of vehicles.

1 INTRODUCTION

In recent years, eco-friendly vehicles such as electric vehicles (EVs) have gained greater public attention due to the increasing concern about global warming. One issue of EVs is the deterioration of energy efficiency when the heater system is operated. EVs have no thermal source like the engine in a conventional vehicle. The energy for operating the heater system in an EV is thus supplied by the main battery. Therefore, the operation of the heater system has a major impact on the distance that an EV can travel before needing a battery charge. The cruising range of an EV with a fully PTC (positive temperature coefficient) electric heater system operating is reduced to about one-third of that without heating [1]. Even for conventional vehicles which have engines, improvement of fuel efficiency can result in an insufficient thermal source. Some vehicles have an auxiliary electric heater or a control to increase the engine speed for operating the heater, which affects fuel efficiency. Therefore, the vehicle cabin environment needs to be designed so that it can be made thermally comfortable while using less heater energy.

Most vehicles today use convective heaters (HVAC) that heats the air in the whole vehicle cabin to indirectly warm the occupants. However, the large amount of heat loss from the vehicle severely limits the heating capability of the system. To reduce heater energy consumption, the use of local heating devices that directly heat the occupants may be more effective. The present study focused on heated seats and investigated their effects on thermal comfort and heater energy consumption in vehicles.

2 METHODS

Two experiments (Experiment A and B) were conducted. Eight male university students in good health were selected as experimental subjects in each experiment. All subjects provided written informed consent. All experimental protocols were approved by the Institutional Review Board of Kyushu University. The clothing worn during the experiments comprised an undershirt, long-sleeved shirt, jacket, underpants, trousers, socks and shoes. All clothing was provided for the subjects. The clo value measured with a thermal manikin was 1.0 clo (=0.155 m^2°C/W).

The experimental equipment is shown in Figure 1. A vehicle seat with a leather surface was used in the two experiments. Silicone rubber heaters were installed between the leather and polyurethane layers of the seat, where the subject came into contact with the seat. The areas of the heaters in the seat back and cushion were 0.080 m^2 and 0.144 m^2, respectively.

2.1 Experiment A

In experiment A, the effects of the heated seat on thermal comfort during the initial warm-up period were investigated. The subjects' mean (standard deviation) characteristics were age of 22.5 years (1.1 years), height of 175.0 cm (1.3 cm), weight of 57.2 kg (3.2 kg) and body fat percentage of 14.3% (3.0%).

A subject was put in a pre-conditioning room (air temperature = 22°C; relative humidity = 50%) where the person rested in a sedentary position for 40 min. The subject then moved to the test room (air temperature = 5, 10, 15, or 20°C; relative humidity = 50%) and sat on the heated seat, which was switched on at that time if the test condition included use of the heated seat. Subjects provided responses at 1-min. intervals for 20 min. while seated on the heated seat.

A total of eight test conditions were used, consisting of a combination of the air temperature in the test room (5, 10, 15, or 20°C) and the on/off status of the heated seat. Power supply to the heated seat was kept constant during the test at 60 W.

Skin temperature was measured at 16 points using epoxy-coated thermistor probes. The mean skin temperature (Tsk) was calculated by modifying the weighting by the Hardy and DuBois 7-point method to add the skin temperature of the areas in contact with the seat.

$$Tsk = (7 \cdot T_1 + 17.5 \cdot T_2 + 14 \cdot T_3 + 5 \cdot T_4 + 9.5 \cdot T_5 + 13 \cdot T_6 + 7 \cdot T_7 + 3.375 \cdot T_8 + 3.375 \cdot T_9 + 3.375 \cdot T_{10} + 3.375 \cdot T_{11} + 3.375 \cdot T_{12} + 3.375 \cdot T_{13} + 3.375 \cdot T_{14} + 3.375 \cdot T_{15}) / 100$$

where component temperatures were: T_1, forehead; T_2, chest; T_3, forearm (left); T_4, hand (left); T_5, anterior thigh (left); T_6, leg (left); T_7, foot (left); T_8, upper back (right); T_9, upper back (left); T_{10}, lower back (right); T_{11}, lower back (left); T_{12}, buttock (right); T_{13}, buttock (left); T_{14}, posterior thigh (right); and T_{15}, posterior thigh (left).

In this paper, the mean skin temperature calculated using the temperatures of the right upper back, left upper back, right lower back and left lower back is defined as the "seat back skin temperature". The mean skin temperature calculated using the temperatures of the right buttock, left buttock, right posterior thigh and left posterior thigh is defined as the "seat cushion skin temperature".

Thermal sensations (overall, area in contact with seat back and area in contact with seat cushion) were reported using a 7-point scale: -3, cold; -2, cool; -1, slightly cool; 0, neutral; +1, slightly warm; +2, warm; and +3, hot. Thermal comfort (overall, area in contact with seat back and area in contact with seat cushion) were reported on a 7-point scale: -3, very uncomfortable; -2, uncomfortable; -1, slightly uncomfortable; 0, neutral; +1, slightly comfortable; +2, comfortable; and +3, very comfortable. In addition, desirable seat temperatures (seat cushion and seat back) were measured by using 3 selected options: -1, "a higher temperature would be better"; 0, "the current temperature is fine"; and 1, "a lower temperature would be better".

2.2 Experiment B

In experiment B, the effect of the heated seat on heater energy consumption in the vehicle was investigated. The subjects' mean (standard deviation) characteristics were age of 22.1 years (1.0 years), height of 168.9 cm (1.9 cm), weight of 61.7 kg (8.7 kg) and body fat percentage of 19.0% (4.7%).

A subject was put in the pre-conditioning room (air temperature = 25°C; relative humidity = 50%) where the person rested in a sedentary position for 30 min. The subject then moved to the test room (air temperature =10 or 20°C; relative humidity = 50%) and sat on the heated seat. The heated seat was pre-heated before the subject sat on it if the test condition included use of the heated seat. Subjects provided responses at 5-min. intervals for 30 min. while seated on the heated seat. In this experiment, the mean of the last five minutes (from 25 to 30 min.) was analyzed for discussing the results in a steady-state condition.

A total of four test conditions were used, consisting of a combination of the air temperature in the test room (10 or 20°C) and the on/off status of the heated seat. The seat surface temperature was controlled at 37°C when the heated seat was used. The average power supply to the heated seat from 25 to 30 min. was 10 W.

Thermal sensations and thermal comfort were measured in the same way as in experiment A.

2.3 Data analysis

Measured data were analyzed by a repeated measures analysis of variance (ANOVA). If Mauchly's sphericity test showed significant results, the Greenhouse-Geisser epsilon correction was used. To test differences, the t-test and multiple comparison by the Bonferroni method were used. Differences were considered statistically significant at the level of $p<0.05$.

2.4 Calculation of heater energy consumption by the HVAC system

Heater energy consumption by the HVAC system in steady-state conditions was estimated using a simple model that is illustrated in Figure 2. Heater energy consumption by the HVAC system was calculated under a condition without any solar heat load.

In real vehicle cabins, the effect of radiation, which is created by the temperature difference between the outdoor and indoor environment, can not be ignored. Therefore, in the present study, an operative temperature that included the effect of radiation was used. When the air velocity is low, the operative temperature can be calculated as an average of the air temperature and the radiant temperature.

The heater energy consumption required by the HVAC system Q (W) to keep the vehicle cabin at a certain operative temperature was calculated with the following equation, *COP* parameter was added to the equation we previously proposed [2].

$$Q = 2\,(T_o - T_{out})\,(R_1 + R_2)\,[1 / (R_1 + R_2) + V \cdot \rho \cdot C_p] / (2 \cdot R_1 + R_2) / COP$$

where T_o (°C) is the operative temperature in the vehicle cabin, T_{out} (°C) is the air temperature in the outdoor environment, R_1 (°C/W) is the thermal resistance between the outdoor environment and the interior surface, R_2 (°C/W) is the thermal resistance between the interior surface and the cabin environment, V (m^3/s) is blower air volume, ρ (kg/m^3) is the air density (= 1.2 kg/m^3), C_p (J/kg°C) is the specific heat capacity of air (= 1006 J/kg°C), and *COP* (N.D.) is the coefficient of HVAC performance.

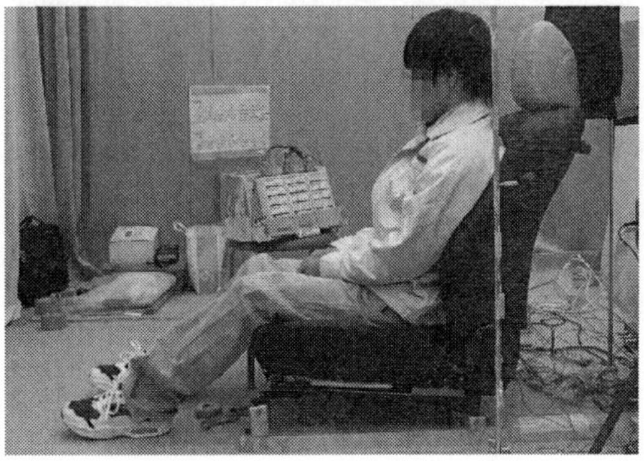

Fig. 1. Experimental equipment.

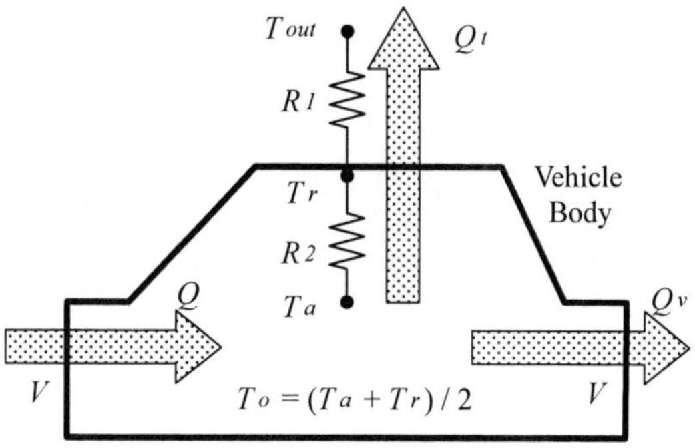

Fig. 2. Model for calculating HVAC system energy consumption.

3 RESULTS

3.1 Thermal sensations and thermal comfort in initial warm-up period (Experiment A)

Use of the heated seat increased the mean skin temperature under all air temperatures. Mean skin temperatures over time, comparing heated and unheated seats at 5, 10, 15 and 20°C, are shown in Figure 3(a). Under all air temperature conditions, significant main effects of the heated seat, time and their interaction were observed by ANOVA. Faster significant increases were observed when the air temperature was higher. Mean skin temperatures remained elevated during testing when the heated seat was used and stayed largely stable when it was not used. Mean skin temperatures when the heated seat was not used were about 31.0, 31.5, 32.5, and 33.0°C at air temperatures of 5, 10, 15, and 20°C, respectively.

Overall thermal sensations over time with heated and unheated seats at 5, 10, 15 and 20°C are shown in Figure 3(b). Under all air temperature conditions, significant main effects of the heated seat, time and their interaction were observed by ANOVA. Use of the heated seat increased the overall thermal sensation under all air temperatures. When the unheated seat was used, overall thermal sensations remained around -2.5 (between "cold" and "cool") at 5°C and at around 0 ("neutral") at 20°C. When the heated seat was used, overall thermal sensations reached about -0.5 (between "slightly cool" and "neutral") at 5°C and +2.0 ("warm") at 20°C.

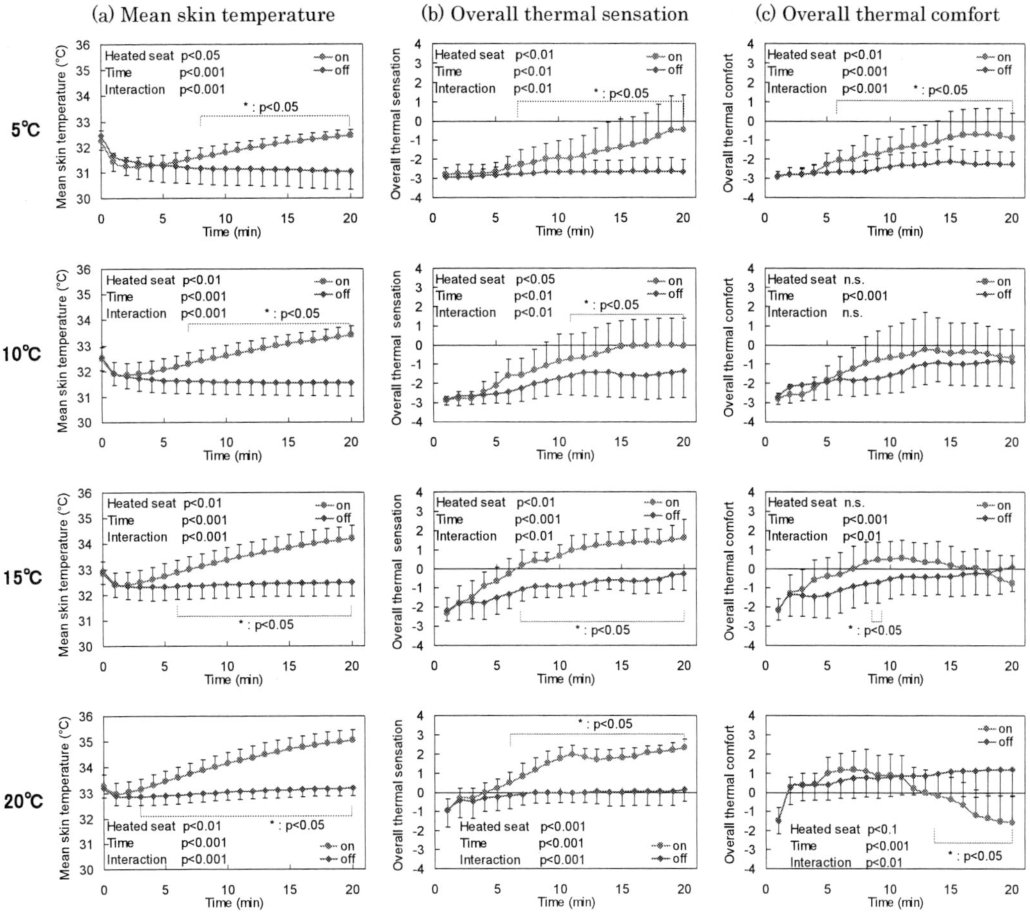

Fig. 3. (a) Mean skin temperature, (b) overall thermal sensation, and (c) overall thermal comfort relative to time with the heated seat on/off at 5, 10, 15 and 20°C. Values are mean with S.D. The horizontal bars indicate the ranges, a significant difference ($p<0.05$) was observed between the heated seat on and off.

Results for overall thermal comfort are presented in Figure 3(c). At 5 and 20°C, significant main effects of the heated seat, time and their interaction were observed by ANOVA. At 10°C, a significant main effect of time was observed. At 15°C, significant main effects of time and the interaction of the heated seat and time were observed. Use of the heated seat improved overall thermal comfort at 5 and 15°C, whereas it reduced overall thermal comfort at 20°C. Overall thermal comfort was around +1 ("slightly comfortable") when the unheated seat was used at 20°C. However, in the latter part of the test when the heated seat was used at 20°C, overall thermal comfort continued to decrease and reached -1.5 (between "slightly uncomfortable" and "uncomfortable") by the end of the test. Overall thermal comfort didn't reach 0 ("neutral") when the heated seat was used at 5 and 10°C.

Use of the heated seat also improved seat back and seat cushion thermal comfort significantly at 5 and 10°C, whereas it reduced these indices at 20°C.

3.2 Optimal conditions during initial warm-up (Experiment A)

When the heated seat was used, subjects changed the desirable seat temperature response from "a higher temperature would be better" to "the current temperature is fine" at around 12, 10, 6, 2 min. at air temperatures of 5, 10, 15, 20°C, respectively.

The seat back and seat cushion skin temperatures at which the subjects changed the desirable seat temperature response to "the current temperature is fine" when the heated seat was used are shown in Figure 4(a). A significant main effect of the region (seat back vs. seat cushion) was observed by ANOVA. The optimal seat back skin temperature was higher than the optimal seat cushion skin temperature. Moreover, a tendency toward a difference in the air temperature ($p<0.1$) was also observed. The optimal skin temperature of the areas in contact with the seat tended to be higher when the air temperature was lower.

The optimal thermal sensation of the areas in contact with the seat was affected by the air temperature. The seat back and seat cushion thermal sensations at which the subject changed the desirable seat temperature response to "the current temperature is fine" when the heated seat was used are shown in Figure 4(b). A significant main effect of the air temperature was observed by ANOVA. Optimal thermal sensations of the areas in contact with the seat were higher when the air temperature was lower. On the other hand, no significant difference was observed by region. The optimal thermal sensation of the areas in contact with the seat during the initial warm-up period was about +1.5 (between "warm" and "slightly warm") at 5 and 10°C, +0.5 (between "slightly warm" and "neutral") at 15°C and 0 ("neutral") at 20°C.

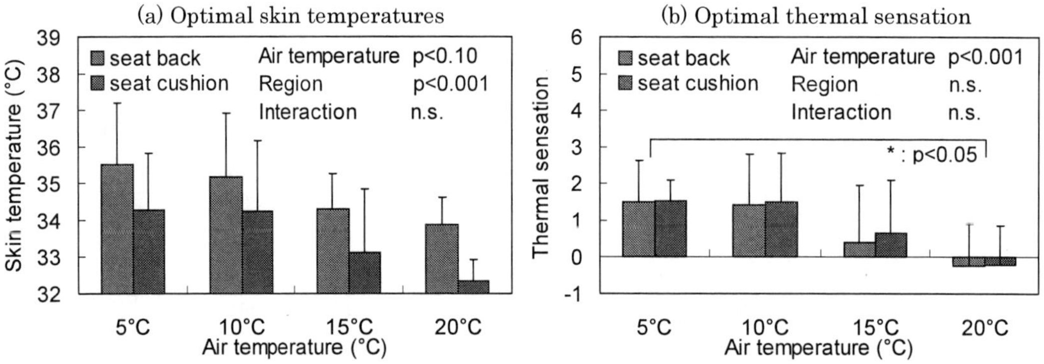

Fig. 4. (a) Seat back and seat cushion skin temperatures and (b) seat back and seat cushion thermal sensation at which subjects changed the desirable seat temperature response from "a higher temperature would be better" to "the current temperature is fine" when using a heated seat. Values are mean with S.D.

3.3 Thermal sensations and thermal comfort in steady-state conditions (Experiment B)

The results for the overall thermal sensation are shown in Figure 5(a). Significant main effects of the heated seat were observed by ANOVA. The use of the heated seat increased the overall thermal sensation. When the heated seat was not used at 10°C, the overall thermal sensation was about -2, "cool" and it improved to about -1.3, between "cool" and "slightly cool", with the use of the heated seat. At 20°C, the overall thermal sensation was higher than 0 ("neutral") without the heated seat.

The results for overall thermal comfort are presented in Figure 5(b). Although no significant main effect of the heated seat was observed, the interaction of the heated seat and temperature was observed by ANOVA. Significant difference was observed between the heated seat on and off at 10°C. However, no difference was observed at 20°C. At 20°C, the overall thermal comfort was higher than 0 ("neutral") and a higher score was reported when the heated seat was not used.

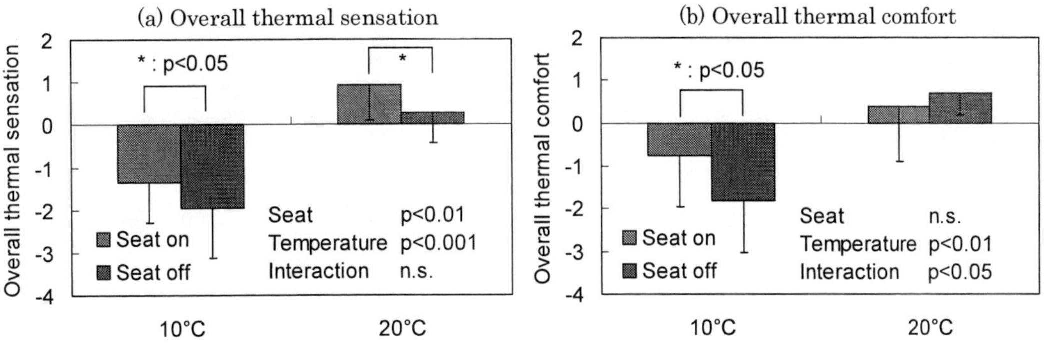

Fig. 5. (a) Overall thermal sensation and (b) overall thermal comfort in each condition. Values are mean with S.D.

3.4 Optimal operative temperature in steady-state conditions (Experiment B)

The optimal *To*, i.e., the operative temperature at which the subjects felt a "neutral" overall thermal sensation, was estimated from the experimental data by linear interpolation. In the climate chamber, the air temperature and the radiant temperature were almost the same. Therefore, the operative temperature was equal to the air temperature.

The results of this calculation are shown in Figure 6. Each line crossed the thermal sensation 0 line at approximately 19°C without the heated seat and at 16°C with it. The optimal *To* was reduced by 3°C by using the heated seat.

3.5 Heater energy required in vehicles in steady-state conditions

The heater energy required in a vehicle to maintain a "neutral" thermal sensation in steady-state conditions was calculated by adding the energy consumption for the heated seat to the HVAC energy consumption. The energy consumption by the HVAC system was calculated using the equation shown in the Methods section. It was calculated under the assumption that two heated seats were used. The energy consumption per one heated seat was assumed 10 W, which was determined from measured energy consumption in the experiment B; 9-11 W in a steady-state condition regardless of air temperature. Thus the total required heater energy was calculated as follows:

without heated seats: energy required by the HVAC system (To = 19°C) (W)
with heated seats: energy required by the HVAC system (To = 16°C) + 2•10 (W)

The values of vehicle heat resistance were determined from the data reported by Fujita [3] and Jung et al. [4]: during vehicle driving R_1 = 0.011°C/W and R_2 = 0.008W°C /W. The total amounts of heater energy required, as calculated for conditions involving an outdoor air temperature of $Tout$ = 0°C, a blower air volume of V = 0.05 m^3/s and COP =

1.0 or 2.0, are shown in Figure 7. The energy required for the HVAC system to maintain the operative temperature of the vehicle cabin at 19 and 16°C was 2719 and 2290 W when *COP*=1.0, and 1360 and 1145 W when *COP*=2.0, respectively. The total required energy was 2719 W (without heated seats) and 2310 W (with heated seats) when *COP* = 1.0, and 1360 W (without heated seats) and 1165 W (with heated seats) when *COP* = 2.0. The use of heated seats reduced the total amount of heater energy required by 409 W when *COP* = 1.0 and by 195 W when *COP* = 2.0.

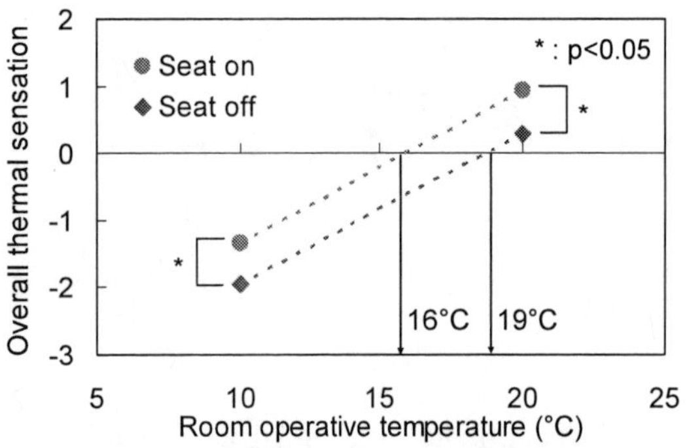

Fig. 6. Calculated optimal room operative temperature.

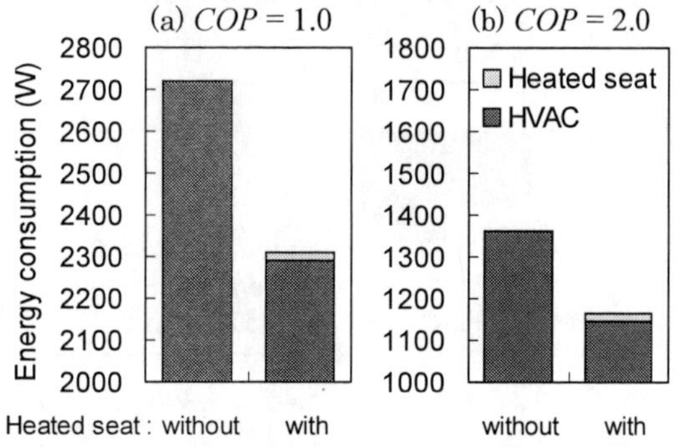

Fig. 7. Calculated required heater energy without/with heated seat at (a) *COP* = 1.0 and (b) *COP* = 2.0, respectively. (T_{out} = 0°C, V = 0.05 m^3/s)

4 DISCUSSION

4.1 Effects of heated seat on thermal sensations and thermal comfort

In Experiment A, a significant increase in overall thermal sensation was observed at 5, 10, 15 and 20°C during the initial warm-up period (Figure 3(b)). Significant improvements in overall thermal comfort were observed at 5°C and 15°C (Figure 3(c)). The overall thermal comfort with the heated seat at 10°C was also higher than that without the heated seat and significant increases in seat back and seat cushion thermal comfort were observed at 10°C. Although heated seat alone was not enough to make occupant's thermal comfort "neutral" at 5 and 10°C (Figure 3(c)), these results indicated that heated seats are effective to improve thermal comfort at 5, 10 and 15°C. On the other hand, no improvement in overall thermal comfort was observed when the heated seat was used at 20°C, but rather a significant decrease was observed. Subjects reported being between "slightly uncomfortable" and "uncomfortable" in the latter half of the test at 20°C (Figure 3(a)). When overall thermal comfort was negative, the overall thermal sensation was around "warm" (Fig. 3). Zhang et al. [5] reported no need for heating or cooling vehicle seats at an air temperature of 21.7°C, very close to 20°C. In the present study, the overall thermal sensation was around "neutral" (Figure 3(b)) and overall comfort was around "slightly comfortable" (Figure 3(c)) when the heated seat was not used at 20°C. In addition, the mean skin temperature under the same conditions was about 33°C (Figure 3(a)), within the range of 33-34°C that is generally regarded as a comfortable skin temperature [6]. These results suggest that the environment without the heated seat was suitable at 20°C, whereas the environment at this temperature with the heated seat was uncomfortable because the seat was too warm.

In Experiment B, a significant increase in overall thermal sensation was observed when the heated seat was used at 10 and 20°C in a steady-state condition (Figure 5(a)). A significant increase in overall thermal comfort was observed when the heated seat was used at 10°C (Figure 5(b)). When the heated seat was not used at 20°C, the overall thermal sensation was higher than 0 "neutral", and overall comfort was between "neutral" and "slightly comfortable". Presumably, these results were produced by the same mechanism as was described for the results in Experiment A; i.e., when the overall thermal sensation was higher than 0 "neutral", the heated seat did not improve thermal comfort.

In summary, the use of heated seats appears effective in improving thermal comfort both in the initial warm-up period and under steady-state conditions in a cool environment, but it is not effective when the air temperature is 20°C or more.

4.2 Optimal skin temperature and thermal sensation of areas in contact with the seat during the initial warm-up period

The optimal skin temperature of the areas in contact with the seat differed by region. The optimal seat back skin temperature was higher than the optimal seat cushion skin temperature (Figure 4(a)). However, the optimal thermal sensation of the areas in contact with the seat did not differ by region (Figure 4(b)). This suggests that the skin temperature for achieving the same thermal sensation differs by region; the required skin temperature of the areas in contact with the seat back is higher than that of the areas in contact with the seat cushion.

The optimal skin temperature of the areas in contact with the seat was also affected by the air temperature (Figure 4(a)), as the optimal thermal sensation of the areas in contact with the seat was higher when the air temperature was lower (Figure 4(b)). The optimal thermal sensation was around "neutral" at 20°C and between "slightly warm" and "warm" at 5 and 10°C. Zhang et al. [7] conducted subjective experiments in which they varied subjects' local skin temperatures individually. They concluded that the local sensation at which maximum comfort is felt shifts with the overall sensation of the body. When the overall sensation is cold, warmer local sensations are desired. The optimal thermal sensation of the areas in contact with the seat is thus higher when the air temperature is lower. The optimal skin temperature of the areas in contact with the seat is therefore higher when the air temperature is lower. These results indicate that a higher seat temperature is suitable at lower air temperatures.

4.3 Optimal operative temperature in steady-state conditions

Brooks and Parsons [8] reported that a heated seat reduced neutral air temperatures by nearly 5°C. Zhang et al. [5] also pointed out that a heated seat expanded the "90% satisfied range" of room temperatures by 5.5°C in a downwards direction. These studies showed that heated seats had a larger effect than the 3°C (Figure 6) seen in our study. In the study by Brooks and Parsons, the seat temperatures were adjusted by the subjects themselves. In the study by Zhang et al., the heater power of the heated seat was larger when the room temperature was lower. Thus, the seat temperatures in these two studies might have been higher than the level of 37°C in the present study, when the room temperature was low. This difference might have caused the observed differences in the effects of heated seats. These results suggest that heated seats can reduce the optimal room temperature by at least 3°C and maybe even more depending on the seat temperature.

4.4 Heater energy required in vehicles in steady-state conditions

The present study indicates that heated seats are effective in reducing total heater energy consumption in vehicles. The total amount of heater energy required can be reduced by 409 W by using heated seats when *COP* = 1.0. Even if *COP* = 2.0, which means a more efficient heater such as a heat pump system is used, it can be reduced by 195 W. As mentioned above, Brooks and Parsons [8] and Zhang et al. [5] reported even larger effects of heated seats. According to their results, an even greater energy reduction can be achieved.

5 CONCLUSION

The effects of heated seats on thermal comfort and heater energy consumption in vehicles were investigated. Our main findings are as follows:

(1) Heated seats are effective in improving occupant thermal comfort in a cool environment both in the initial warm-up period and under steady-state conditions, but they are not effective when the air temperature is 20°C or higher.
(2) The optimal thermal sensation of the areas in contact with the seat is higher when the air temperature is lower, at around "neutral" at 20°C and between "slightly warm" and "warm" below 10°C. The optimal skin temperature of the areas in contact with the seat is also higher when the air temperature is lower. These results indicate that a higher seat temperature is suitable when the air temperature is lower. In addition, the optimal skin temperature of the areas in contact with the seat back is higher than that of the areas in contact with the seat cushion.
(3) The room operative temperature at which the occupants felt a "neutral" overall thermal sensation was reduced by at least 3°C by using a heated seat.
(4) Heated seats can reduce total heater energy consumption in vehicles.

ACKNOWLEDGEMENTS

This paper includes our previous research published by Taylor & Francis in Ergonomics [2] and Elsevier in Applied Ergonomics [9].

REFERENCES

[1] Nakane, S., Kadoi, M., Seto, H. and Umezu, K. (2010) Air conditioning system for electric vehicle. Journal of society of automotive engineers of Japan, 64(4), 35-40. (in Japanese with English abstract)
[2] Oi, H., Yanagi, K., Tabata, K. and Tochihara, Y. (2011) Effects of heated seat and foot heater on thermal comfort and heater energy consumption in vehicle. Ergonomics, 54(8), 690-699. doi:10.1080/00140139.2011.595513
[3] Fujita, A. (2005) A study of air conditioning heat load in cabin. JSAE annual congress (spring), May 2005 Yokohama. Tokyo: Society of automotive engineers of Japan, Proceedings no. 68-05, 1-4. (in Japanese with English abstract)

[4] Jung, M., Kemle, A., Strauss, T. and Wawzyniak, M. (2011) Interior Heating for Hybrid and Electric Vehicles. ATZ (Automobiltechnische Zeitschrift), 113, 36-40.
[5] Zhang, Y., Wyon, D., Fang, L. and Melikov, A. (2007) The influence of heated or cooled seats on the acceptable ambient temperature range. Ergonomics, 50(4), 586-600. doi:10.1080/00140130601154921
[6] Olesen B.W. and Fanger P.O. (1973) The skin temperature distribution for resting man in comfort. Archives des sciences physiologiques, 27(4), 385-393.
[7] Zhang, H., Arens E., Huizenga, C. and Han, T. (2010) Thermal sensation and comfort models for non-uniform and transient environments. Part II: Local comfort of individual body parts. Building and Environment, 45, 389-398. doi:10.1016/j.buildenv.2009.06.015
[8] Brooks, J.E. and Parsons, K.C. (1999) An ergonomics investigation into human thermal comfort using an automobile seat heated with encapsulated carbonized fabric. Ergonomics, 42(5), 661-673. doi:10.1080/001401399185379
[9] Oi, H., Tabata, K., Naka, Y., Takeda, A. and Tochihara, Y. (2012) Effects of heated seats in vehicles on thermal comfort during the initial warm-up period. Applied Ergonomics 43(2), 360-367. doi:10.1016/j.apergo.2011.05.013

Optimisation of thermal comfort and fuel consumption by using coupled numerical tools

E ANDRÈS, V MAZET and **S BRUCK**
Renault SAS, Alliance Systems Engineering Division, France

ABSTRACT

A numerical simulation tool has been developed to couple 1) a model of thermal systems including the cabin in the AMESim environment with 2) a model of the hybrid drivetrain and vehicle in the Matlab/Simulink environment. This numerical platform enabled us to assess the benefits from the use of an electric coolant pump instead of a mechanical pump for the high temperature engine cooling circuit of a plug-in hybrid concept (PHEV). Results show that using an electric coolant pump significantly reduces the energy consumption and strongly improves the thermal comfort at low ambient temperature.

DEFINITIONS / ABBREVIATIONS

BSFC: Brake Specific Fuel Consumption
HVAC: Heating Ventilating Air Conditioning
ICE: Internal Combustion Engine
NEDC: New European Driving Cycle
OEM: Original Equipment Manufacturer
PTC: Positive Temperature Coefficient
SOC: State Of Charge

1 INTRODUCTION

Over the last decade, electrification of vehicles has increased due to the needs of fuel economy and greenhouse effect gases emissions reduction. Vehicle electrification helps improving the global fuel efficiency but has a negative impact on the cabin comfort as it reduces heat availability. As a consequence, OEM have to obtain a trade-off between thermal comfort, fuel consumption, and vehicle performance requirements. That is in this context that Renault has recently presented its ultra-low fuel consumption vehicle concept EOLAB [8]. Being able to predict the impact on the vehicle global energetic consumption (fuel or electricity) of any modification made on vehicle is then a subject of great importance for OEM and suppliers.

A numerical simulation tool is developed, which couples 1) a model of the thermal systems (including a high temperature and a low temperature cooling circuit, a HVAC, a cabin) in the AMESim environment with 2) a model of the hybrid drivetrain and vehicle in the Matlab/Simulink environment. The electric coolant pump (already known for its utility in saving energy [1], [6]) is used as a demonstrator of the numerical platform potential. Its influence on the thermal comfort regulation is clearly shown for the cooling circuit of a gasoline plug-in hybrid (PHEV).

The advantages of the electric coolant pump have already been described for thermal engine [1], [4], [6]. The latter shows fuel consumption reduction and radiator size reduction. The use of a global numerical platform such as the one presented in [7] has also been described in the case of a dual mode coolant pump for a Diesel engine [3]. From [5] is already seen a numerical platform coupling thermal and vehicle model this time for a Diesel hybrid electric vehicle. A rapid description of all technical solutions for fuel economy and cabin heating improvements can be seen. What is presented in this paper are the detailed results on fuel economy and thermal comfort improvement obtained thanks to an electric coolant pump on gasoline plug-in hybrid vehicle using our coupled numerical platform.

This article is divided into three parts: description of the hybrid drivetrain, of the numerical platform and the application to value the use of an electric coolant pump.

2 PLUG IN HYBRID DRIVETRAIN

The study has been conducted on the EOLAB vehicle concept [8]. The hybridation concept studied is a parallel architecture, Figure 1. The electrical motor and the internal combustion engine (ICE) can be mechanically linked to the wheels, simultaneously or not, thanks to an innovative transmission. This transmission uses fixed gear ratios. Two main driving modes can be distinguished:

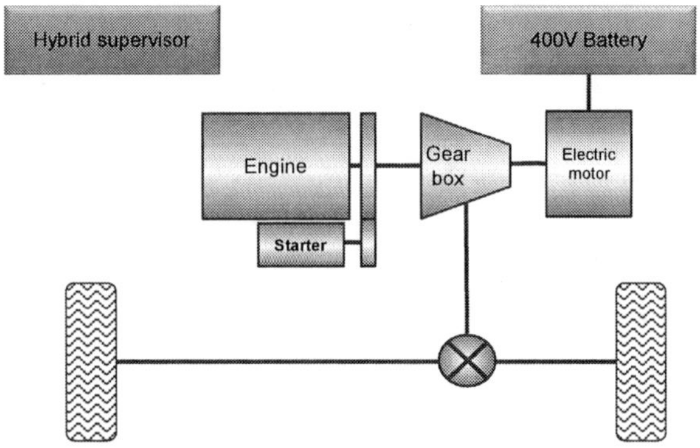

Figure 1 PHEV architecture

- battery charge depleting: the electric motor drives the vehicle. The battery contains enough energy to allow full electric drives, up to 120km/h, with a maximal electrical range of approximately 60km. The internal combustion engine (ICE) can be fired for short periods to assist the electric motor when the torque requirement is too high for the electric motor alone.
- battery charge sustaining: the electric motor and the ICE drive the vehicle simultaneously. The battery state of charge (SOC) is regulated at its low level. In this case the electric motor is used either as a generator to charge the battery and to reach optimal brake specific fuel consumption (BSFC) of the ICE at high load, or as motor to avoid the high BSFC of the ICE at very low load.

3 NUMERICAL PLATFORM

3-1 Thermal model (AMESim)

The complete system is modelled using LMS Imagine.Lab AMESim. The model contains the ICE coolant loop, the HVAC and the cabin, all modelled in a transient 0D formalism.

Each branch of the cooling circuit is represented by a coolant volume and a pressure drop. The engine thermal behaviour is modelled by a lumped-mass model, [2], using two thermal inertias for the metallic masses of the engine: one for engine oil, one for coolant contained inside the engine, one for air in the under hood. Heat exchanges between these thermal inertias are identified from dyno-test bench measurements made on NEDC. The HVAC model contains the blower, the mixing flap and the PTC heater (Positive Temperature Coefficient).

The cabin model contains four isothermal moist air zones to represent the whole cabin: one for the front feet, one for the front heads, one for the rear feet, one for the rear heads. These four air zones can receive air mass flow rate from the HVAC and from surrounding air zones. The moist air volumes also exchange heat with the cabin walls by convection and radiation and with the surrounding air zones by mass transfer. Finally, the solar load is also modelled. The cabin walls are modelled as a stack of layers to represent the different materials involved. This model requires CFD calculations to know the aeraulic inside the cabin regarding the total air mass flow rate, the air distribution and the recycling rate.

3-2 HVAC Control

The control of the HVAC is based on already in use strategies for conventional vehicle. Blower command, recycling, mixing, distribution flaps positions, and PTC heater command are controlled in the model. The control value is the air temperature at driver's head, which has to be maintained in the comfort range. In this study, the conventional regulation has been adapted:

- firstly, to compensate the lack of power at the heater, when the ICE is off.
- secondly, to reach the optimum of energy consumption for thermal comfort.

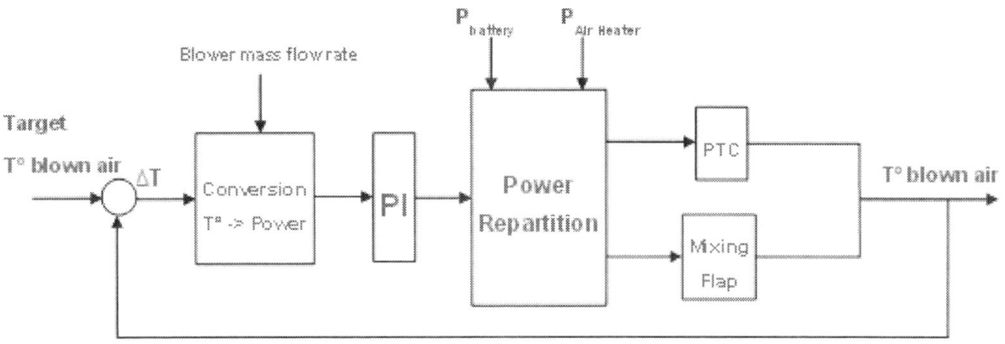

Figure 2 Strategy for thermal control optimization

3-3 Plug In Hybrid Drivetrain (SIMULINK)

This hybrid drivetrain is simulated using Matlab Simulink. Each component (transmission, electric motor, ICE), is modeled using measured operating maps (maximal torque, efficiency, torque losses, energetic consumption map). The remarkable point in this model is its hybrid energy supervisor. It optimizes at each time step the operating point of the powertrain (ICE and electric motor) to minimize the global energetic consumption of the vehicle. It determines the best torque distribution between electric motor and ICE and the gear ratio to use. As the transmission gear ratios are fixed, the engine speed is ruled by the driving cycle.

3-4 Coupling

Coupling consists in the inter-connection of the two models described above, Figure 3.

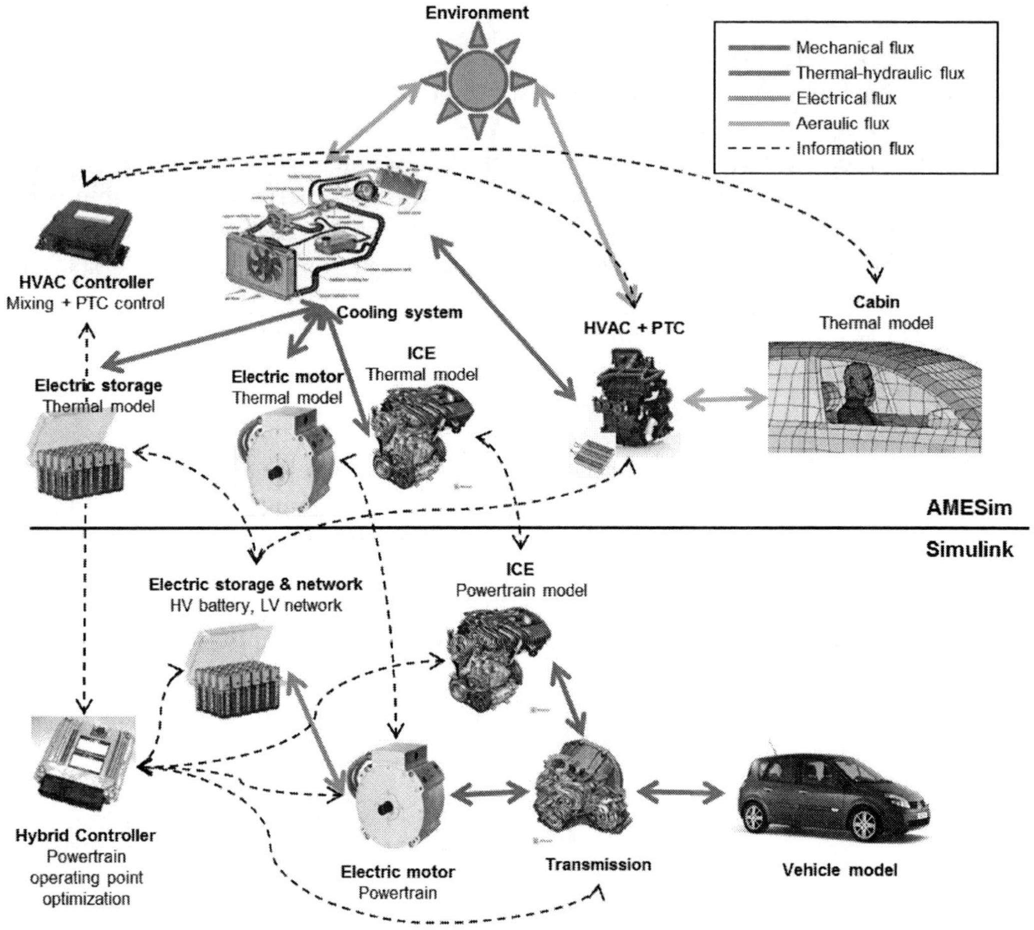

Figure 3 Simulink and AMESim coupling

Coupling models is the only way to evaluate the retroactive effects on each model. Increasing the PTC heater electrical consumption increases the battery depletion. It induces the rise of energy need from the ICE to maintain the battery charge level (in battery charge sustaining). Retroactively, it leads to more heat rejection from ICE to the coolant, and then decreases the load on the PTC heater. Fuel consumption or the EV range can then be evaluated, with only one simulation run, in regard to the thermal comfort in the car cabin. Following variables are exchanged during the co-simulation:

Table 1 Main informations exchanged between Simulink and AMESim models

Power from PTC heater	AMESim → Simulink
Driver's head cabin temperature	
ICE coolant temperature	
Engine rotational speed	Simulink → AMESim
Engine effective torque	
Engine state	
Vehicle speed	

4 APPLICATION TO THE E- COOLANT PUMP ENHANCED VALUE

On drives where the ICE is fired-up, the coolant, recovering heat from the thermal engine, is used to heat up the passenger cabin by the way of a standard cabin heater. On pure electric drive, PTC heater is used to supplant heat recovered from the ICE. The latter causes a reduction of the vehicle electric range and then a fuel consumption increase. The solution presented here uses the electric coolant pump during pure electric drives in order to value heat stored in the volume of coolant and in the metallic masses of the engine. Numerical simulations are performed using the platform described above to evaluate its benefits on thermal comfort and fuel consumption.

4-1 Driving conditions

Simulations are performed with a vehicle speed profile based on a real driving cycle representative of a mean customer use. Cycle length is 18,7km, driven in 2490s, which leads to an average speed of 27km/h, Figure 4. This cycle is realized in battery charge sustaining mode: ICE status (engine on = 1, engine off= 0) is also plotted, Figure 4.

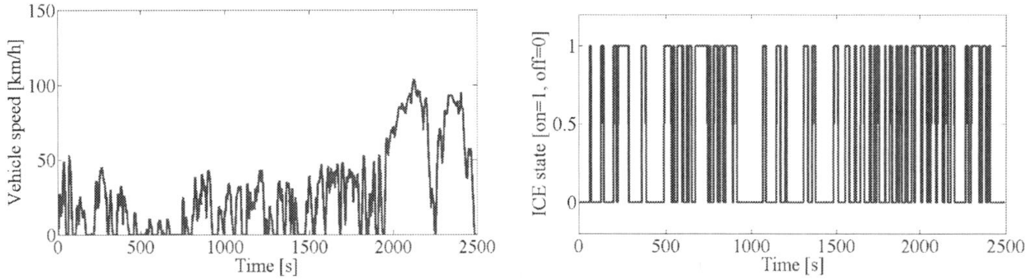

Figure 4 Vehicle speed and ICE status on the considered driving cycle

4-2 Simulation conditions

The study uses the numerical platform presented in chapter 3. This part presents the benefits for thermal comfort and fuel consumption of a continuous coolant circulation kept thanks to the electric coolant pump, even when ICE is turned off. The study is conducted with an ideal pump (efficiency = 1). This assumption has a minor effect on the global energy balance due to relative small energy consumption of the electric coolant pump. The simulations were run in a -5°C ambiance. At this cold temperature, thermal comfort is difficult to maintain with a mechanical coolant pump when the ICE is off, Figure 5. When ICE is off, air temperature in the cabin at the driver's head quickly drops below 25°C. As the coolant flow is stopped in the cabin heater, the only heat source for cabin heating is the PTC heater. In this case it is clearly insufficient to maintain comfort.

The following strategy is implemented in the AMESim thermal model:

- coolant flow rate is regulated by a proportional integrator (PI) filter, aiming to keep the coolant temperature difference between engine inlet and outlet at a predetermined value. This value may vary regarding ICE operating point, ambient temperature, coolant temperature, …
- when the coolant temperature in the engine is below 40°C, the electric coolant pump is forced rotating at 900 rpm (close to the mechanical coolant pump rotational speed at idle).
- if the cabin power need is below 3kW, and if the coolant temperature is below 40°C, the coolant flow rate is increased,
- if the cabin power need is above 3kW and the coolant temperature below 40°C, the coolant flow rate remains at its low level and PTC heater is switched on.

4-3 Results

In this part, benefits of the electric coolant pump for thermal comfort and energetic consumption are precisely evaluated. Results of the co-simulation AMESim – SIMULINK are presented Figure 5 to Figure 8. On each figure results for electric coolant pump is the green continuous line; results for mechanical coolant pump is the blue dotted line.

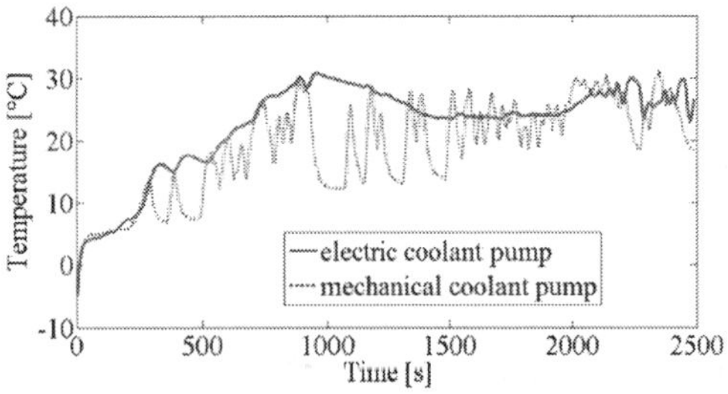

Figure 5 Air temperature at driver's head in the cabin front

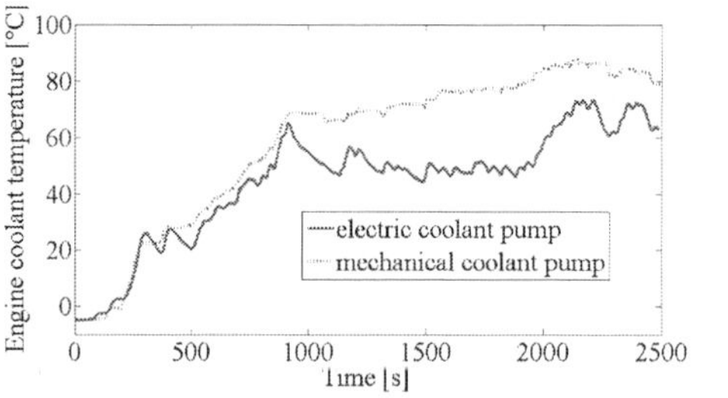

Figure 6 Coolant temperature at ICE outlet

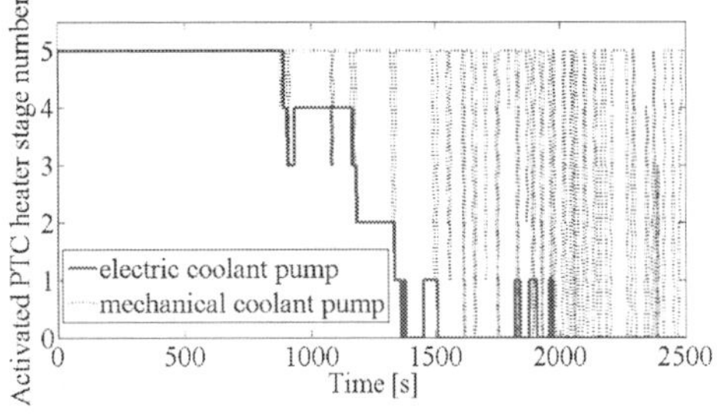

Figure 7 Activated PTC heater stages number

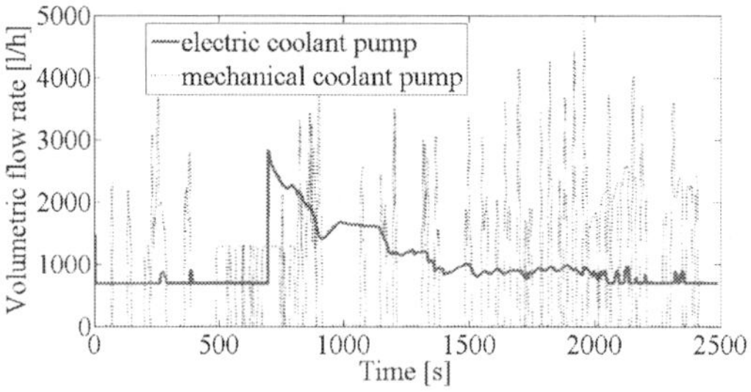

Figure 8 Coolant flow rate at coolant pump outlet [in l/h]

Figure 5 shows the air temperature at driver's head. Air temperature obtained is higher with the electrical coolant pump. Keeping a continuous coolant flow rate in the cabin heater completely erases the large temperature fluctuations and averages the temperature level. Thermal comfort can now be achieved by -5°C, whereas it was not with a mechanical pump.

Figure 6 presents the coolant temperature at ICE outlet. With the electrical coolant pump, the value stabilizes around 50°C. This value is consecutive to the calibration of the control of the electric coolant pump and can be modified according to engine needs.

Figure 7 shows the activated PTC heater stage number. With a mechanical coolant pump, activated PTC heater stage number oscillates between 0 and 5 (full power) at high frequency. With the electric coolant pump, the PTC heater level stays at its maximum during the engine coolant warm up, then decreases continually until the end of the driving cycle. This clearly shows the use of the heat stored in the coolant in the second part of the driving cycle.

Figure 8 shows coolant flow rate across the coolant pump. During warm up and while the coolant temperature stays below 40°C, flow rate remains at its minimal value. Above 40°C, the flow rate reaches a maximum value, and then decreases slowly with the progressive end of PTC heater use.

As the use of the electric coolant pump leads to colder ICE temperatures, Figure 6, the engine oil temperature also loses about 10°C at the end of the driving cycle, Figure 9. Therefore, friction losses are increased and then generate a rise of fuel consumption. The coupling is thus absolutely essential to confirm the benefits of the electric coolant pump.

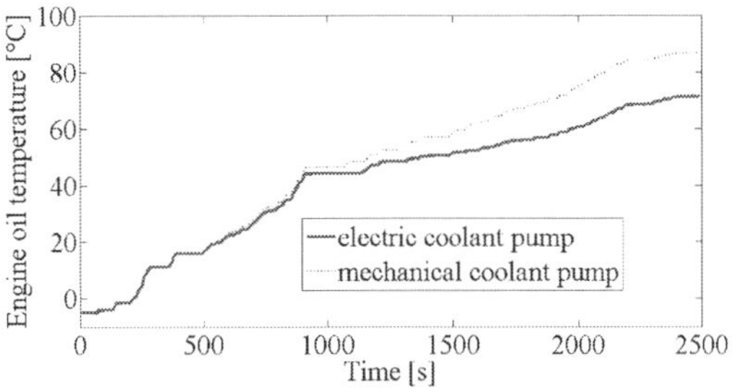

Figure 9 Engine oil temperature

Figure 10 shows the energy consumption during the driving cycle, normalized by its value obtained with mechanical coolant pump. Compared with the mechanical coolant pump, the energy consumed by the electric coolant pump and by the PTC heater is almost cut down by half with the electric coolant pump. The energy consumption of the electric coolant pump is negligible compared to the PTC heater one. The electric coolant pump brings huge benefits on PTC heater activation. This result can be visible on the fuel consumption, Figure 11. It allows fuel saving up to 8.6% on the driving cycle. This value is quite close to the one presented in [6], which was measured on a physical support.

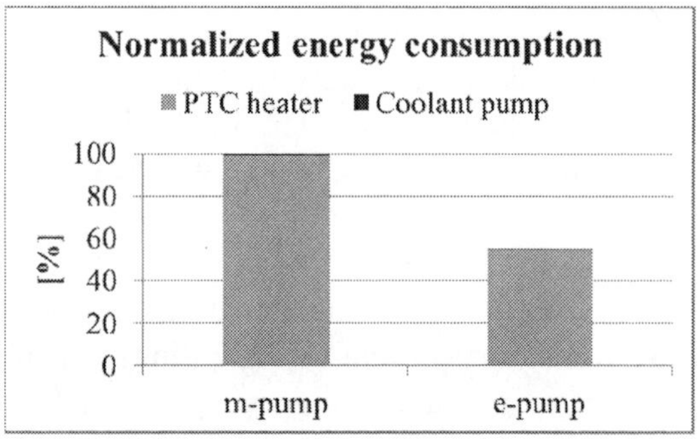

Figure 10 Comparison of coolant pump and PTC heater power consumption

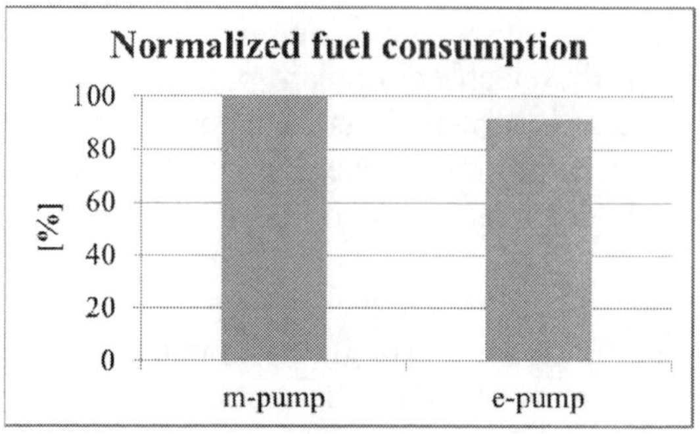

Figure 11 Comparison of fuel consumption

Simulations have also been performed for other ambient temperatures: 0°C, 5°C and 10°C. The benefit of the electric coolant pump reaches its maximum at 0°C: fuel consumption is reduced by more than 10%, Figure 12. When ambient temperature increases, benefits decrease, linked with PTC heater activation. Figure 13 shows fuel consumption normalized by its value at 10°C for the mechanical coolant pump. With an electric coolant pump, the fuel consumption at 0°C is equivalent to the one with the mechanical pump at 10°C.

This numerical platform has enabled us to assess the benefits of the use of an electric coolant pump instead of a mechanical pump for a PHEV, at low ambient temperatures. The results of numerical simulations show that the use of an electric coolant pump reduces the energy consumption significantly. Moreover, the thermal comfort is enhanced at low outside temperature levels compared to a solution using a mechanical coolant pump

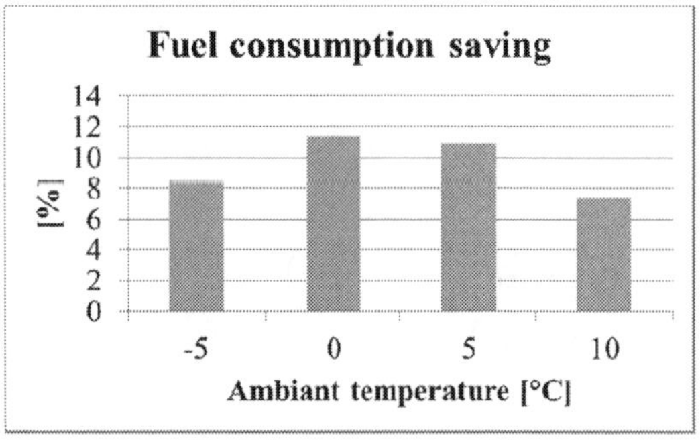

Figure 12 Fuel consumption saving

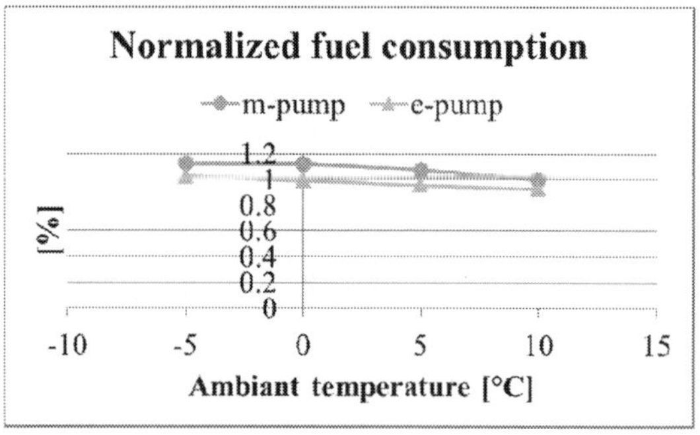

Figure 13 Normalized at T=10°C fuel consumption

5 CONCLUSION – OUTLOOK

Numerical simulations performed with the exposed platform confirm the benefits of using an electric coolant pump instead of the mechanical one. In a -5°C ambiance, the cabin thermal comfort is enhanced. In the same time, fuel consumption is decreased by 8.6% thanks to a better use of the PTC heater. The benefit on fuel consumption has been confirmed at other ambient temperatures: 0°C, 5°C and 10°C. The biggest fuel saving (10%) is obtained at 0°C.

Such results are very helpful to quantify the benefits of one solution to another. Vehicle future thermal systems are getting more and more complex, especially for highly electrified ones. Exhaustive exploration and validation in the whole operating range on test bench become too expensive and time consuming. Numerical tools which can couple thermal models with hybrid drivetrain models are now mandatory to optimize the system and insure its reliability.

REFERENCES

[1] Cho, H., Jung, D., Filipi, Z. S., Assanis, D. N., Vanderslice, J. & Bryzik, W. (2004) *Application of controllable electric coolant pump for fuel economy and cooling performance improvement*. IMECE 2004-61056.

[2] Minami, K., Otsubo & H., Nakai, I. (2011) *Powertrain thermal modelling for vehicle simulation*. VTMS10 Conference paper, May 15-19 2011, Birmingham UK, 201-210, Institution of Mechanical Engineers.

[3] Negandhi, V., Jung, D. & Shutty, J. (2013) *Active thermal management with a dual mode coolant pump*. SAE n° 2013-01-0849.

[4] Gubbiotti Ribeiro, E., Pinto de Andrade Filho, A. & Luiz de Carvalho Meira, J. (2007) *Electric water pump for engine cooling*. SAE n° 2007-01-2785.

[5] Revereault, P., Rouaud, C. & Marchl, A. (2010) *Fuel economy and cabin heating improvements thanks to thermal management solutions installed in a diesel hybrid electric vehicle*. SAE n° 2010-01-0800.

[6] Kawamoto, N., Naiki, K., Kawai, T., Shikida, T. & Tomatsuri, M. (2009) *Development of new 1.8-liter engine for hybrid vehicle*. SAE n° 2009-01-1061.

[7] Steinberg, P. & al. Wärmemanagement des Kraftfahrzeugs VI. (2008) Expert Verlag, Haus des Technik, p203-241.

[8] http://media.renault.com/download/media/specialfile/61732_1_5.aspx

Underhood & Simulation

Cooling airflow virtual component characterisation

D WELLMAN[1] and **S WAKELAM**[2]
[1] Jaguar Land Rover Thermal and Aerodynamic Systems Engineering (TASE) UK
[2] EXA UK

ABSTRACT

To be able to simulate cooling airflow components in the virtual world within a full vehicle model the subtleties of modelling these components must be understood. To further our understanding, this paper will cover the virtual methodology used to validate independent testing of heat exchangers in isolation and in combination. As the challenges of packaging the modern vehicle increase, space is becoming more and more of a premium. In addition to this the challenge of cooling the vehicle is increasing, with more heat being rejected to various fluid circuits to manage economy and emission benefits. This can drive multiple heat exchanger stack ups with very close together core locations. To be able to have better grasp of vehicle correlation data these stack ups are isolated in the physical world then validated using virtual methods. This will give insight into understanding any virtual compromises caused in the full vehicle development. Ultimately this will lead to better selections of heat exchangers early on in the virtual vehicle development.

1 INTRODUCTION

At Jaguar Land Rover (JLR) cooling airflow simulations are conducted over a full vehicle virtual model over a number of cycles within the vehicle development timeline. The desire at JLR is to reduce the amount of physical signoff we conduct by moving to a more virtual process. To enable this we must be confident in our correlation between our virtual dataset and that produced in the wind tunnel. This is a stepping stone to full real world condition validation. Correlation is highly important, providing a robust dataset and strong standing for any future vehicle programmes.

The airflow around a full vehicle model is very complex and transient and this can also be said about underhood flows. The accurate prediction of airflow underhood is fundamental in understanding the cooling performance of the vehicle, helping to predict the amount of heat rejected to ambient and ultimately the vehicle performance impact. Better understanding of the flow through cooling airflow components like heat exchangers will enable us to move closer to a reduced physical validation programme.

To achieve this, the problem is isolated to allow more of a focus on the components in question i.e. the heat exchangers. Cooling airflow components were tested in isolation at a third party test facility to characterise them and then virtually modelled to verify the computational fluid dynamic (CFD) assumptions. By isolating the component, a controlled environment is set up to enable detailed study of the governing parameters that affect vehicle performance. Accurate pressure drop for example is paramount in understanding the resistance of the heat exchanger, hence allowing attributes like vehicle drag to be more accurately predicted and ultimately not over-engineering a vehicle. Accurate heat rejection is also important as this affects the air temperature leaving the heat exchanger and impacting on either the fan blades or downstream heat exchanger if part of a stack up. The scope of this initial study included some stack configurations to be tested and virtually assessed to determine whether any relationships or design rules could be derived from the proximity of one heat exchanger to another. This was also important in understanding the virtual aspect, to investigate if there were any limitations on heat exchanger proximity to still maintain accuracy.

2 PREVIOUS STUDIES

Numerous works have been carried out on the effect of installing heat exchangers and within full vehicle models to validate full vehicle testing [2, 3, 4, and 5]. The heat exchangers used in this work are typical of heat exchangers found in the full vehicle models at Jaguar Land Rover. This paper aims to eliminate the complexities of a full vehicle model to over check the approximations being used for the heat exchanger in an isolated environment. This then will allow greater confidence in full vehicle analysis.

3 PHYSICAL TESTING

Components from the JLR vehicle range were selected to give a good spread of the product portfolio. These were then sent to Young Calibration who acted as an accredited third party testing facility to produce the experimental results [1]. To enable sufficient data to be able to virtually assess several tests were devised.

3.1 Empty tunnel testing

To enable the true aerodynamic pressure loss of a cooling airflow component to be obtained the empty tunnel was first tested to assess the pressure contribution from the experimental setup. This then could be subtracted from any testing to give a true aerodynamic pressure taking into account any entry losses

$$\Delta P_{True} = \Delta P_{Component} - \Delta P_{Rig} \quad (1)$$

3.2 Heat exchanger testing

Components were tested with and without coolant flows to obtain the pressure difference over the core when tested adiabatically and non-adiabatically. Pressures were measured along with upstream temperature and velocity to enable airside calculations. Also thermal images were taken to study the internal coolant flow pattern. On the coolant side coolant mass flow and inlet and outlet temperatures and pressures were also recorded to enable heat rejection calculations. Coolant used in all experiments was Texaco XLC and Water at a 50:50 Volume Mix. Heat exchangers were also assembled into stacks and core spacing was varied to assess whether this gap change affected any of the measured variables. These entire tests were conducted over a range of airflows at a specified temperature.

3.3 Tunnel Setup

Two separate experimental tunnels were used to verify the results in the external test house. A closed section tunnel and open equivalent. For the purposes of this paper only the open tunnel section results are considered. The two tunnels results were found to be the same when several over checks were completed between various heat exchangers. Open section tunnel testing is more flexible and allows the study of heat exchanger stacks to be conducted without the influence of any downstream walls affecting the clamping and mounting of these units. The open tunnel can be represented as a schematic shown by Figure 1. Please note this diagram is not to scale.

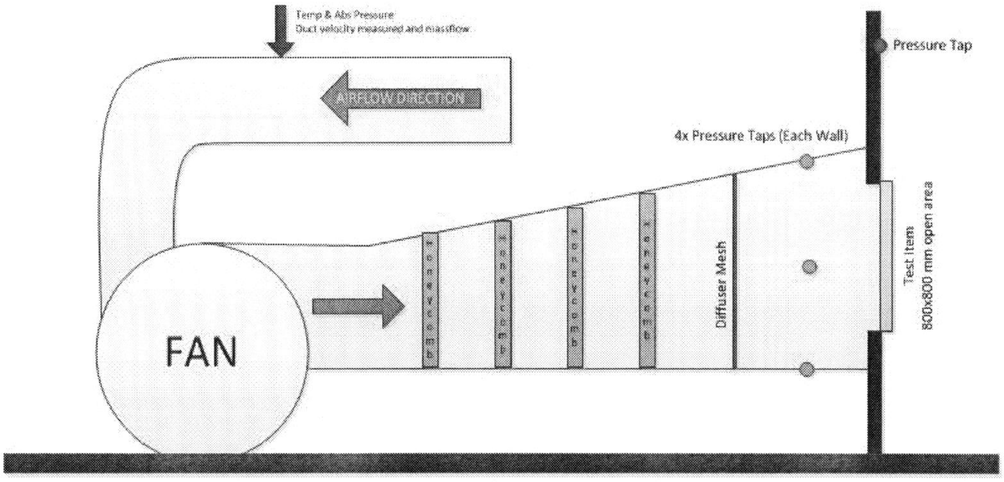

Figure 1 – Open section tunnel schematic

Pressure drop measurements for the open section tunnel were conducted by averaging the four upstream pressure tapings into a single measurement and subtracting this from an ambient pressure probe located on the downstream side of the test item. This provides a pressure delta across the test item. Heat exchangers were mounted in the place of the test item and any airflow gaps sealed to minimise leakage. Figure 2 depicts a typical mounting of the heat exchanger in the test environment.

Figure 2 – Typical heat exchanger mounting strategy

4 VIRTUAL TESTING

Virtual testing was conducted using EXA PowerFLOW®, a Lattice Boltzmann Equation (LBE) based solver. An overview of the mathematical model used in the formulation of this method is presented in [3]. The main advantage of a LBE method is that it does not need any special iterative procedures to conserve mass, momentum and energy so the equations become numerically very efficient and robust [2]. This numerical efficiency allows for larger computational lattices, lending itself to full scale vehicle modelling nicely. To validate the cooling airflow components we must use the same methods employed by the full scale vehicle models at JLR initially to ensure congruency in the modelling.

The experimental setup was represented virtually by modelling the open section tunnel geometrically as close to the physical world as possible in the expanding section. Some assumptions were made up and downstream to simplify the model.

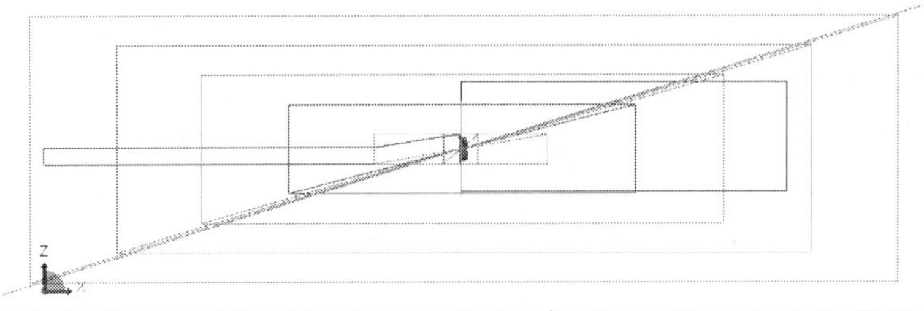

Figure 3 – Virtual test rig

To make sure there were no spurious velocities or pressures coming from the inlet the location was extended to dampen out these effects. To minimise extra computation time caused by this extension, resolution is reduced. The lattice is sized according to that of a typical full vehicle model at JLR according to the best practice developed as part of previous correlation exercises. This served as a good starting point for the virtual rig.

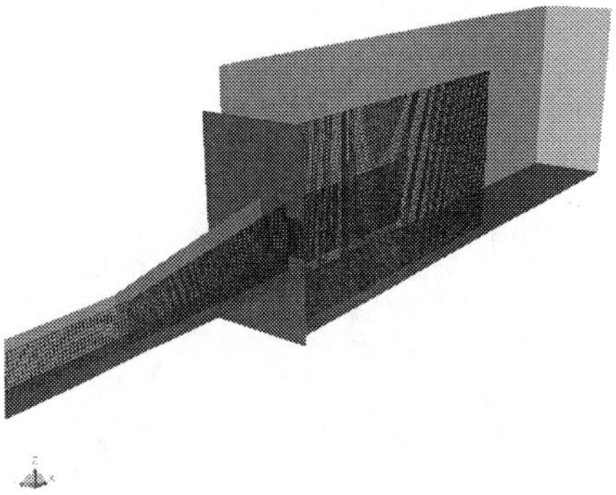

Figure 4 – Airflow lattice

The virtual simulations were also coupled with a one dimensional solver to capture the heat transfer between the internal coolant of the heat exchanger and cooling airflow [6]. This method approximates the heat exchanger to be that of a variable heat rejection porous medium. In other words fins and tubes of the heat exchanger are not modelled, but represented by a porous medium. The heat transfer coefficients are represented by an interpolation of the experimental measured coolant and airflow data using a sandwich function given by (2)

$$HTC = \frac{1}{K_C \left(\frac{1}{\dot{m}_a}\right)^{\alpha} + D + K_H \left(\frac{1}{\dot{m}_c}\right)^{\beta}} \tag{2}$$

Where $\dot{m}_a$ is the air mass flow rate, $\dot{m}_c$ is the coolant mass flow rate and K_H, K_C, D, α and β are interpolation coefficients calculated by the 1D tool representing the thermal resistivity of air, coolant and the metal of the heat exchanger unit respectively [6]. These coefficients are interpolated using a Monte Carlo curve fit approximation to provide a map of heat transfer coefficient. In a PowerFLOW® simulation heat exchangers are working at a variety of operating points. To account for these differences between the lab tests and simulations a correction is used to change the local heat transfer coefficient based on the thermal diffusivity of air. This is estimated using Sutherland's law in [6].

5 RESULTS AND DISCUSSION

5.1 Adiabatic Heat Exchangers

Figure 6 shows the experimental results along with the virtual testing for one heat exchanger. The testing included a blanking plate only run to verify the tunnel setup without a test item present and to allow the actual pressure loss of an isolated heat exchanger to be calculated according to (1). The isolated adiabatic heat exchanger results are presented below in Figure 7.

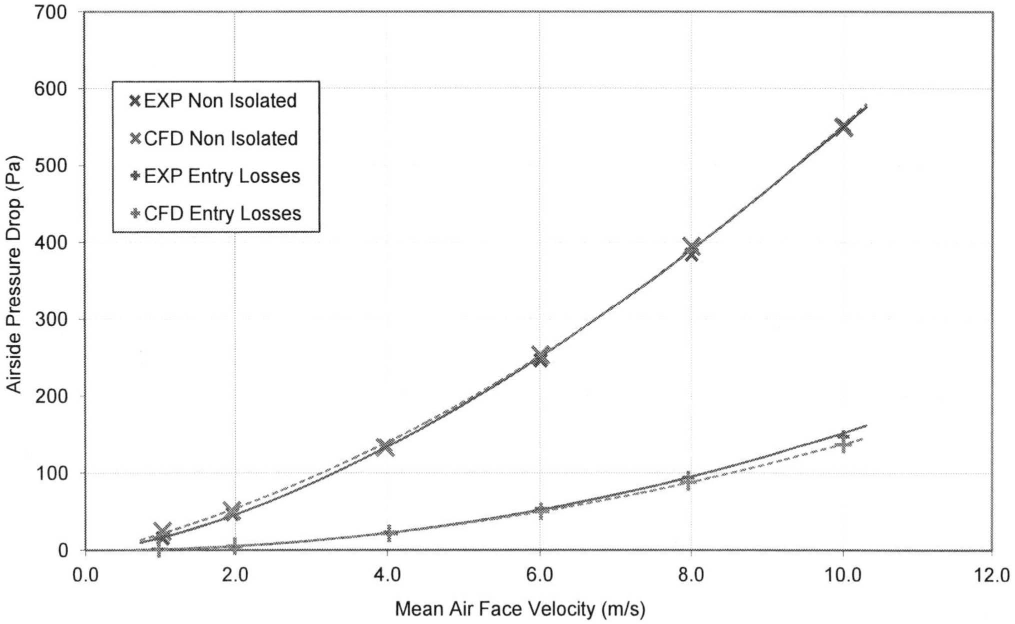

Figure 6 – Adiabatic results HX 4

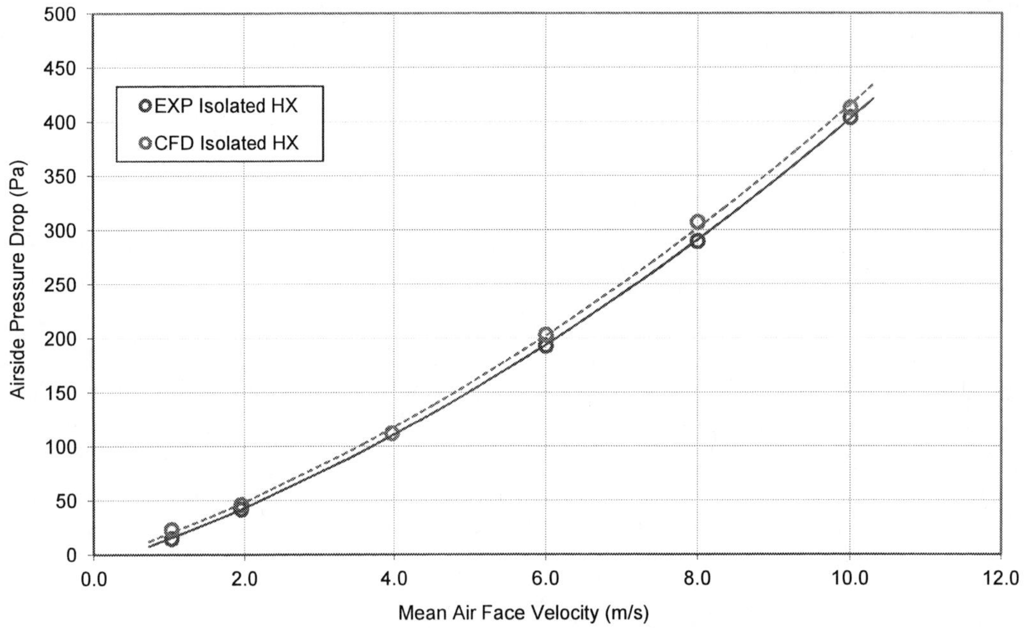

Figure 7 – Isolated adiabatic results HX 4

The agreement between the virtual rig and the experimental rig for the isolated heat exchanger is very good. The very good alignment between the CFD and experimental data for the entry loss test verifies the assumptions made in modelling the wind tunnel virtually. Further improvements could be made to this prediction by incorporating friction into the walls. Typically the CFD is within 5% of the experimental results. The lower airflow range has a larger error but this can be attributed to measurement inaccuracies at these low airflow rates. Isolating the heat exchanger from the entry losses is important for the implementation into full vehicle models as the inlet losses (particularly for small heat exchanger core areas) can be significant and lead to over engineering of designs and unnecessary increases in cooling drag contributions.

5.2 Working Heat Exchanger

After the verification of the adiabatic heat exchanger the coolant flow rate was introduced and a working heat exchanger was assessed. Results for the heat exchanger are presented in Figure 8. This set of data can also be compared to that of its cold pressure drop also shown in Figure 8.

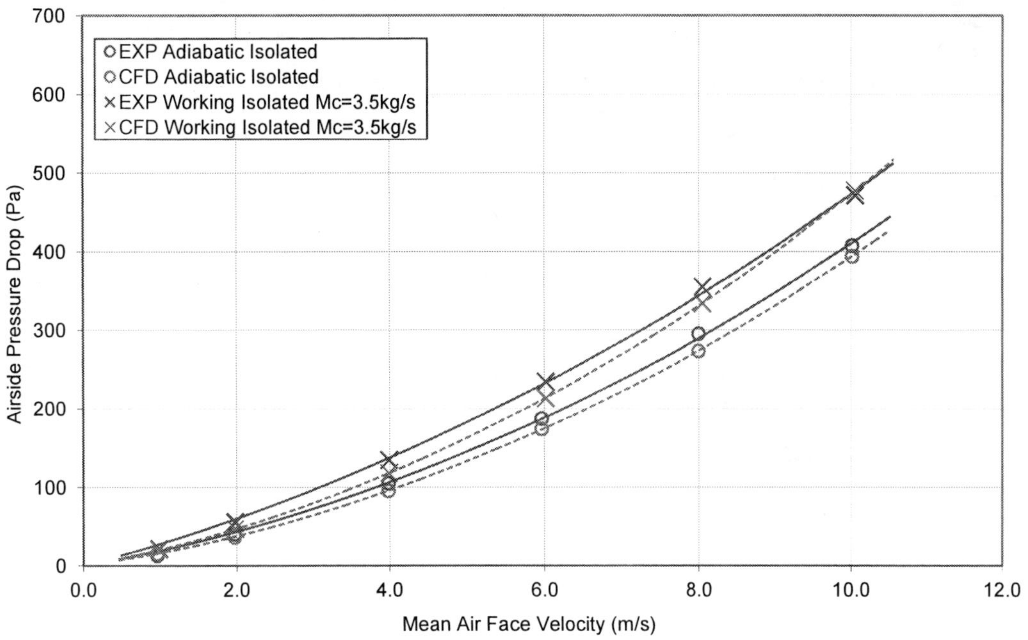

Figure 8 – Working vs. adiabatic airside pressure drop for HX 7

The working heat exchanger produces a higher pressure drop due to the elevated temperature of the exit airflow, thus changing the density and affecting the pressure. With the addition of heat the simulated pressure drop is not a constant offset from the experimental results. This could be due to the effective thermal diffusivity correction being applied. The local heat transfer coefficient is changed to account for the change in air on temperature. More work is required to further understand this correction and the implications it has on in vehicle installations.

This pressure difference can also be seen in Figure 9. This figure shows the working heat exchanger heat rejection characteristics for a variety of coolant flow rates. The uppermost plot is akin to that of Figure 8. The reader can observe that as the pressure correction gets better at a higher airflow rate. Also the heat rejection prediction is closer to that of the experiment. For lowest coolant flow rate the heat rejection is following the same trend as the experiment but is at a constant offset. This can be attributed partly to the error in prediction of the heat exchanger resistance presented in Appendix A. Mainly this error in heat rejection is due to the errors in the interpolation conducted by PowerCOOL® when translating the bench data into local heat transfer coefficients. In all cases where the heat exchanger resistance is not computed properly the heat rejection is not matching. These are only small discrepancies, but can make a difference when validating a full vehicle setup. The measurement variation in the experimental data is typically around 50 Pascal over all the working heat exchanger tests. This is within the CFD delta to the experimental results. These results show very good correlation with the experimental data.

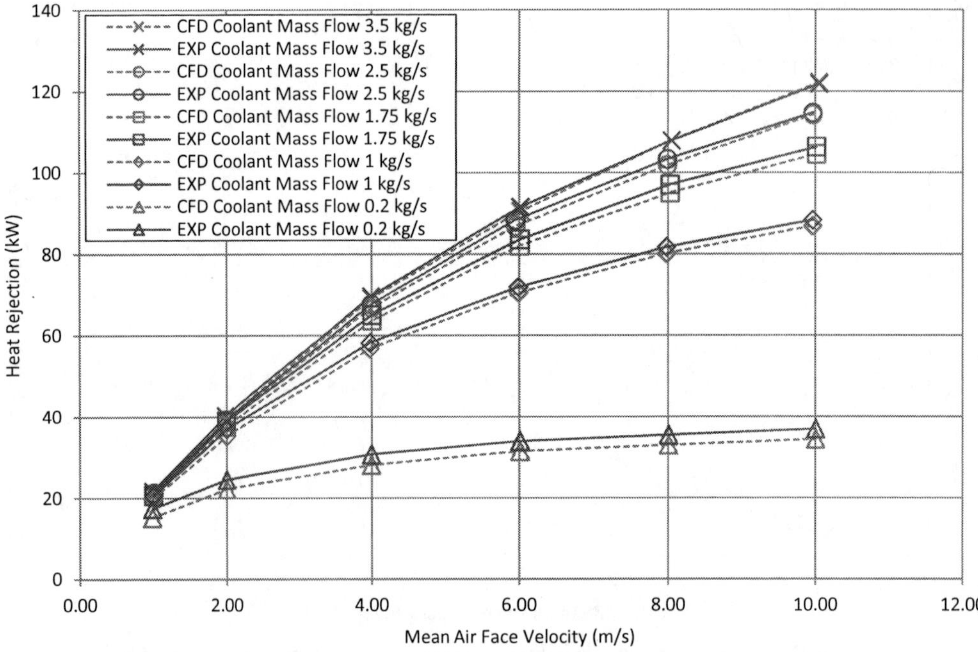

Figure 9 – Heat rejection results

5.3 Full Heat Exchanger Stack

To verify these findings heat exchanger stacks were analysed in the wind tunnel rig. Results for the working heat exchanger pressure drop are presented in Figure 10. These experimental results suggest that as the heat exchangers are spaced further apart that their overall pressure drop reduces as the pressure recovers slightly. This indicates that the stagnation caused by the downstream heat exchanger is compounded by other upstream heat exchangers if placed too close together. The same phenomenon is observed for an adiabatic heat exchanger but the absolute pressure drop is lower due to the absence of heat transfer. The results for this are omitted from this paper. The CFD method used to solve seems to struggle with the smaller offsets. The CFD is not resolving the spacing between the heat exchangers well enough and as Figure 11 depicts the "pinch" points of close heat exchanger proximity are not being calculated correctly. In addition to this the lack of mixing with airflow passing around the coolers elevates the air on temperature to the last heat exchanger creating a large pressure correction from the bench tested single heat exchanger. This geometrical spacing requires further study to determine what is required in a CFD model to be able to resolve these small core gaps. Since core spacing is constantly being squeezed, this study is important in predicting overall system resistance. As the spacing increases between the cores the resolution becomes less of an issue. This can also be seen in Figure 10 where the large core offset is closer to that of the experimental data.

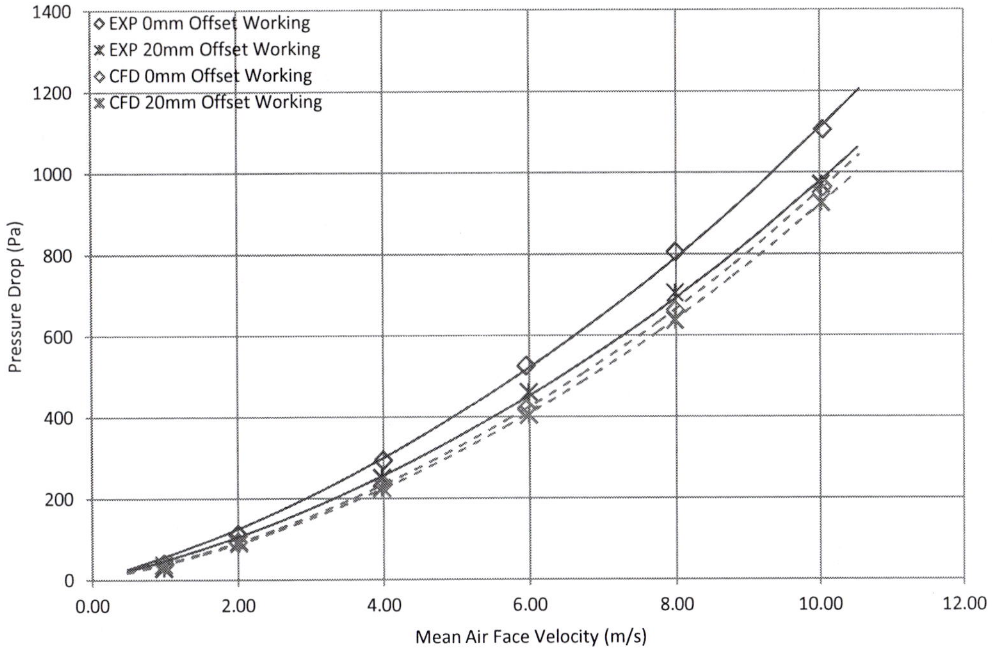

Figure 10 – Stack 2 working pressure drops at various core offsets

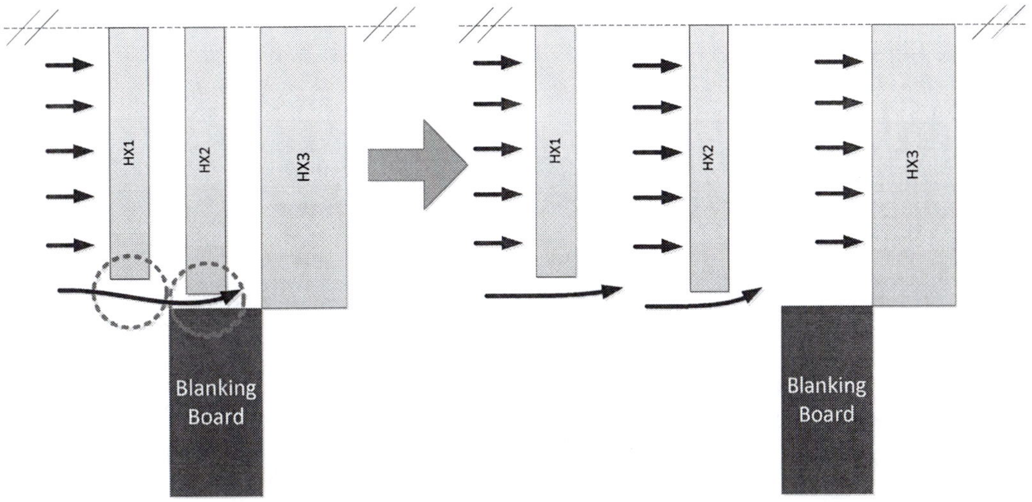

Figure 11 – Stack up schematic for no and max core offset

Figure 12 presents the heat rejection data for the stack compared to that of the single heat exchanger (as in Figure 9). Introduction of a working extra heat exchanger upstream of the main radiator significantly affects its ability to reject heat due to the reduced temperature delta. Also the decrease in resistance produced by offsetting allows the first heat exchanger exit air to be cooled by airflow not passing through the cooler. Hence the main radiator performance is very slightly improved at this offset core position. The CFD agreement for heat rejection is good, within a 5% error. However due to the lack of pressure resolution with the narrow heat exchanger spacing, the heat rejection is more elevated than in the experiment since system resistance is lower. The accurate prediction of the pressure drop of a system affects its ability to reject heat. This can lead to errors when correlating a full vehicle model to a test vehicle. The upstream heat exchanger agrees excellently with the experimental results as there is no interpolation being carried out here to correct the heat transfer coefficients. Moreover the heat rejection is almost identical to that of a single heat exchanger as expected since this is the most upstream heat exchanger rejecting heat into the airflow. The upstream condenser was an adiabatic loss in this stack simulation.

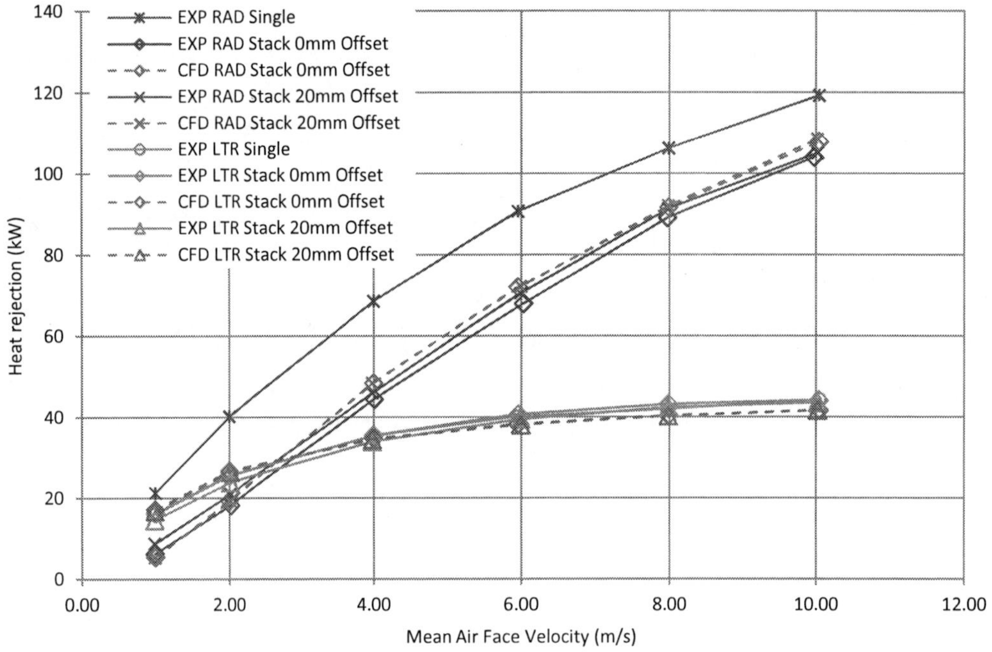

Figure 12 – Stack vs. single heat rejection

CONCLUSION

Overall the CFD agreement with experimental data is very good. PowerFLOW® is able to predict adiabatic pressure drops over porous media with a high degree of accuracy. The simulations seem to be slightly more inaccurate when temperature is introduced. The thermal diffusivity correction and interpolation applied to the heat transfer coefficient seems to slightly under predict the resistance of the heat exchanger resulting in a variation in performance. For all future simulations the adiabatic pressure loss should be input into the simulation as this is the true aerodynamic pressure drop of the core. If a working pressure drop is input for an aerodynamic simulation this will lead to an over prediction of cooling drag contribution. The interpolation carried out requires further work to understand the limitations of the method as this seems to affect the heat rejection at low coolant flow rates. More coolant flow rates here would help to achieve better results. However all results presented here are within the experimental errors quoted so the validation is satisfactory.

When simulating a stack the CFD method is very reliant on the resolution in the model to properly resolve narrow gaps around heat exchangers and provide accurate system resistance. The interpolations and corrections to HTC, and ultimately pressure are stressed particularly when core spacing becomes very close. This needs to be the subject of further work to identify the requirements for accurately predicting the pressure of a multi heat exchanger stack. This will enable better correlation of full test vehicles and ultimately a shift towards virtual signoffs. As a general design rule to increase cooling affect and reduce resistance, cores should be spaced with maximum allowable offset to encourage mixing and maximise heat rejection. It is also important to match the geometrical relationships between radiator cores as unnecessary areas of restriction can be created.

In summary the CFD method is very reliant on good input data to the simulation to enable the most accurate prediction of flow phenomena. Particularly this is important when stack ups of heat exchangers are simulated. It is also important to understand the relative position of the heat exchangers in the stack up compared to the walls of ducting or the test section. Also achieving the best interpolation of bench data to minimise heat transfer correction errors and therefore by proxy system pressure drop will help to produce the correct heat rejection and therefore fan performance resulting in accurate mass flow predictions through the heat exchanger cores.

REFERENCE LIST

[1] Young Calibration Website. 2014. *ISO 17025 Accreditation.* [ONLINE] Available at:http://www.youngcalibration.co.uk/about/iso-17025-accreditation. [Accessed 27 November 14].

[2] Alajbegovic A. Sengupta R. Jansen W. "Cooling Airflow Simulation for Passenger Cars using Detailed Underhood Geometry" *SAE World Congress 2006, Detroit,* SAE Paper 2006-01-3748.

[3] Alajbegovic A. Konstantinov A. Amodeo J. Jansen W. "Simulation of Cooling Airflow under Different Driving Conditions" *SAE World Congress 2007, Detroit,* SAE Paper 2007-01-0766.

[4] Lafferty C. Alajbegovic A. Horrigan K. "Under-hood Thermal Simulation of a Class 8 Truck" *SAE World Congress 2007, Detroit,* SAE Paper 2007-01-4280.

[5] Amodeo J. Alajbegovic A. Jansen W. "Thermal Management Simulation for Passenger Cars – Towards Total Vehicle Analysis" *Mira Conference 2006.*

[6] PowerCOOL User's Guide, Release 4.4 Revision 2, Exa Corporation, Boston, Massachusetts, 2011.

APPENDIX

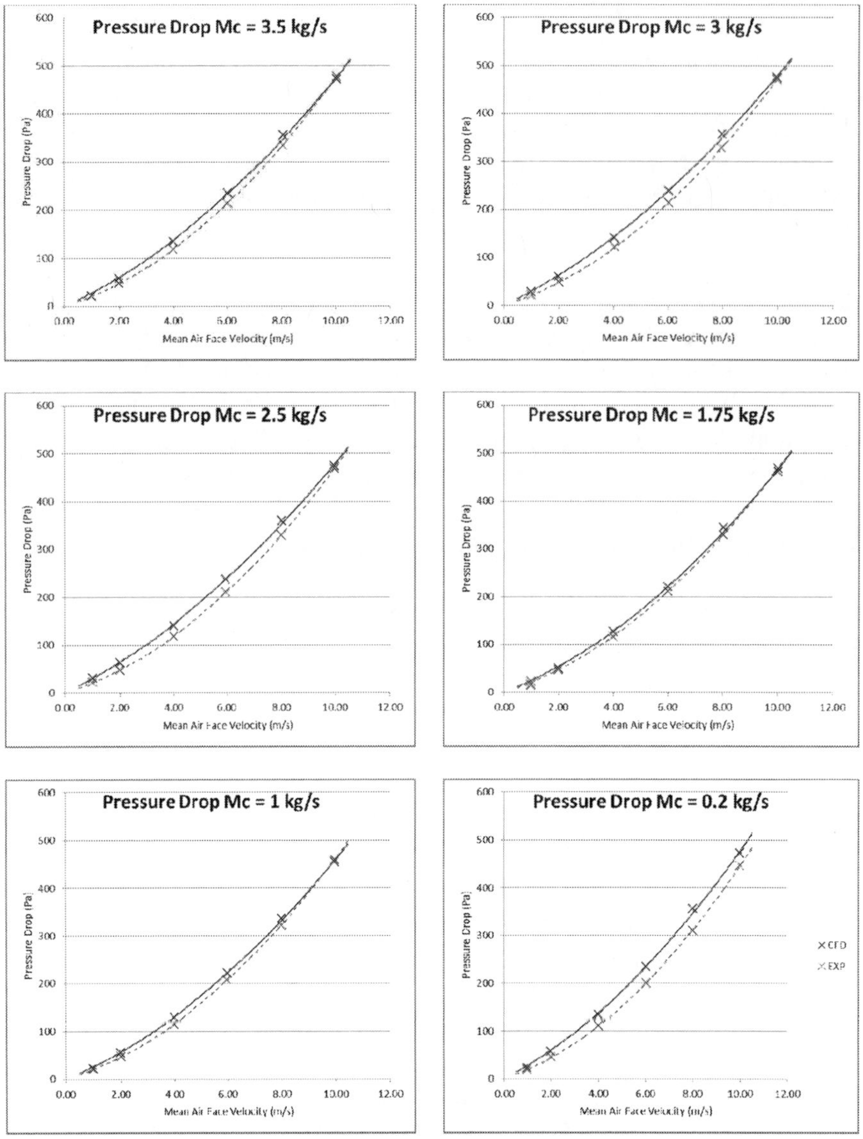

Appendix A – Working (hot) pressure drop data

Prediction of transient cooling package behaviour with multiple air paths

A PRICE[1], E TATE[2], Z YANG[2] and V STAELENS[3]
[1] Bentley Motors Ltd, UK
[2] Exa Corporation, USA
[3] Exa Corporation, UK

ABSTRACT

A combination of 3D CFD and system models can accurately predict coolant temperatures in a vehicle with a single cooling air path on simple drive cycles. However, some vehicles have complex cooling packages with multiple airflow paths and undergo drive cycles in extreme ambient conditions and vehicle load points. Predicting the transient temperatures of multiple vehicle fluids over multiple air paths introduces new challenges.

In this work, a full vehicle 3D CFD simulation of multiple stabilized operating points is used to characterize the non-uniform airflow boundary conditions at the entrance and exit to each air path. The method used to select a minimal number of operating points is discussed. These operating point characterizations are combined to generate time-varying boundary conditions which are applied to the system model of the heat exchangers in each air path. This system model is further extended to include the powertrain heat generation and fluid flow rates.

To validate this methodology, it is applied to a vehicle with three cooling airflow paths. In the central air path, the cooling package has three heat exchangers in addition to a radiator. On either side of the radiator there are two charge air coolers located in separate air paths. The driver's side charge air cooler also has a differential cooler. The heat exchangers for each air path are oriented at different angles with respect to the vehicle centreline. This vehicle is tested and simulated on a drive cycle consisting of 300kph+ thermal stabilization followed by a full load ride and handling course which requires significant heat rejection, followed by a key off soak.

GLOSSARY OF TERMS

CAC – Charge Air Cooler
PAS – Power Assist Steering
HLHA – High Load, High Ambient
RMS – Root Mean Squared
ATF – Automatic Transmission Fluid
Diff – Differential
HVAC – Heating, Ventilation, and Air Conditioning
CFD – Computational Fluid Dynamics
ETK – ECU Monitoring Device

1 INTRODUCTION

The cooling package on a vehicle rejects heat from the powertrain and HVAC system into the ambient. This requires airflow through the cooling package. This airflow requires power. The power is either obtained directly from the powertrain to drive a fan or indirectly through increased vehicle drag. The size and shape of the cooling apertures are a distinctive part of a vehicle and satisfy functionality and design. An aperture which is too small may require an excess of fan power and noise to cool the vehicle. If an aperture is too large, the vehicle drag will be unnecessarily large. This increase in drag will have a direct impact on both the performance and efficiency of the vehicle experienced by the customer. The best cooling package minimizes the power used on a fuel economy test cycle, relies on an aperture which integrates with the vehicle styling, has satisfactory fan noise and is able to reject the necessary waste heat from the vehicle. This rounded approach to design will enable the production of a vehicle to be at its most efficient not only in terms of performance but also in cost and complexity.

For vehicles with high power density and low drag, multiple stacks of heat exchangers utilizing separate airflow paths and different orientations may be used to increase flexibility for under hood packaging and vehicle grill design. This flexibility in under hood packaging is paramount in the larger engine GT segment due to the reduction in space, but the requirement for large engine performance.

Previous analysis work at Bentley has consisted of component characterisation and linear extrapolation of data with averaged and lumped inlet conditions from bench testing and CFD simulations to predict steady state sections of a transient drive cycle. This produces errors in the results due to the inherently transient nature of all drive cycles. This often leads to incorrect component selection and can incur unforeseen project cost late in the life cycle. The benefit of developing a transient analysis process for thermal component selection early in the vehicle life cycle can therefore be easily realised.

Prior work demonstrated accuracy and process benefits from using CFD results to establish boundary conditions on a system model of a cooling system with one stack of heat exchangers [2]. This work builds on the established accuracy of CFD simulations to generate results that include recirculation under the hood and external to the vehicle body [3]. The use of CFD allows accurate incorporation of both under hood heat recirculation and non-uniform boundary conditions on the cooling package. Many cooling systems have multiple stacks of heat exchangers. This work demonstrates that CFD results can be used in cooling systems with multiple airflow paths. To demonstrate the benefits, The Bentley Continental GT W12 Speed was modelled in PowerFLOW and GT Suite. The cooling system on this vehicle is challenging because of its high power density and power to weight ratio.

The Bentley Continental GT W12 Speed has a cooling package with three distinct cooling air paths. The large quantity of airflow is needed because of the capability of the powertrain and the expectation for uncompromised operation in high ambient temperatures. The rated power of the engine is 467 kW and is able to accelerate the vehicle from 0 to 100 kph in 4.2 seconds with a top speed of 331 kph [1]. This power is achieved from a 12 cylinder, 6.0 litre engine with twin-turbochargers and twin air-to-air coolers for the charge air. Because of the high power level for the powertrain, the engine coolant, transmission oil, and differential oil each have dedicated heat exchangers. Additionally, the power steering and HVAC system reject heat through dedicated heat exchangers. At a peak, in excess of 270 kW of heat may be rejected into the cooling air stream. This demonstrates a 58% relationship between engine power and vehicle heat rejection requirements which further emphasises the challenges with the efficiencies of the W12 engine.

2 MULTIPATH COOLING PACKAGE OVERVIEW

The cooling package in this study consisted of three cooling stacks as illustrated in Figure 1. The cooling airflow is split between the three stacks. The mounting for these three stacks are not planar as illustrated in the top view in Figure 2. They are mounted with offsets and at angles to each other. The central cooling stack contains the majority of the heat exchangers. To the left of the central cooling stack, a charge air cooler is located in front of the differential oil cooler. To the right of the central cooling stack, a charge air cooler is located. The central cooling stack consists of a multiple heat exchangers of varying dimensions shown in Figure 2. The cooling airflow passes over the power steering oil cooler, condenser, transmission oil cooler, and then the radiator. Figure 3 illustrates how the air at the front of the vehicle is split among the different cooling packages.

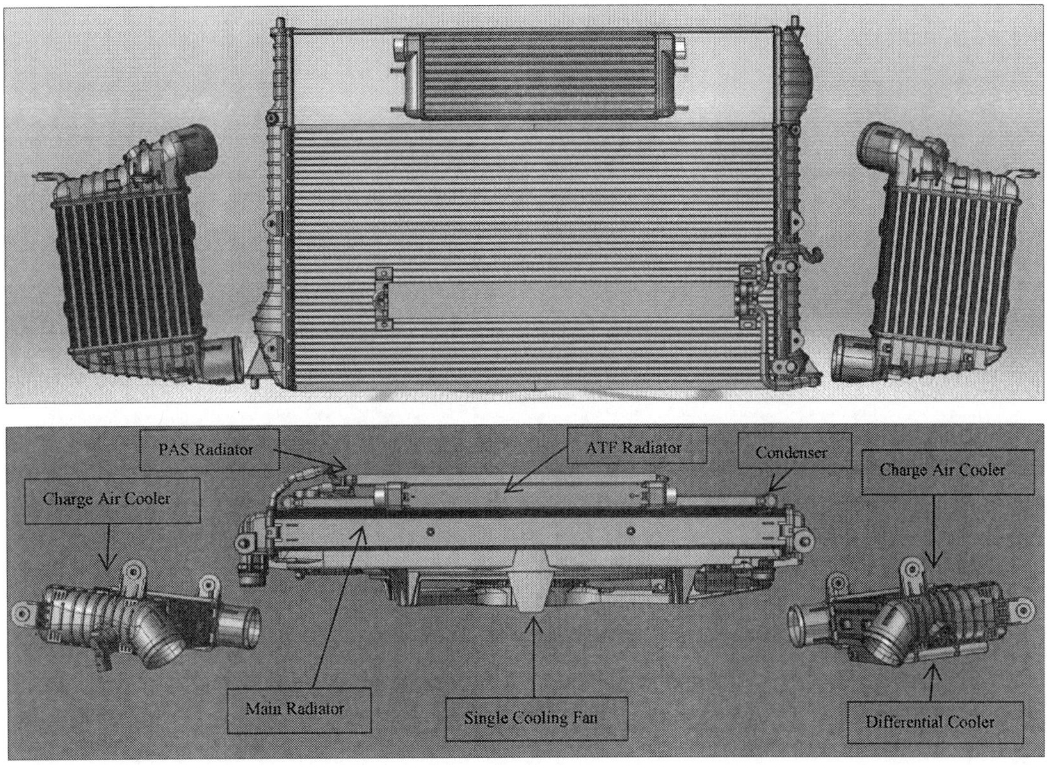

Figure 1 - Cooling System Geometry & Top View of Cooling Package Orientation

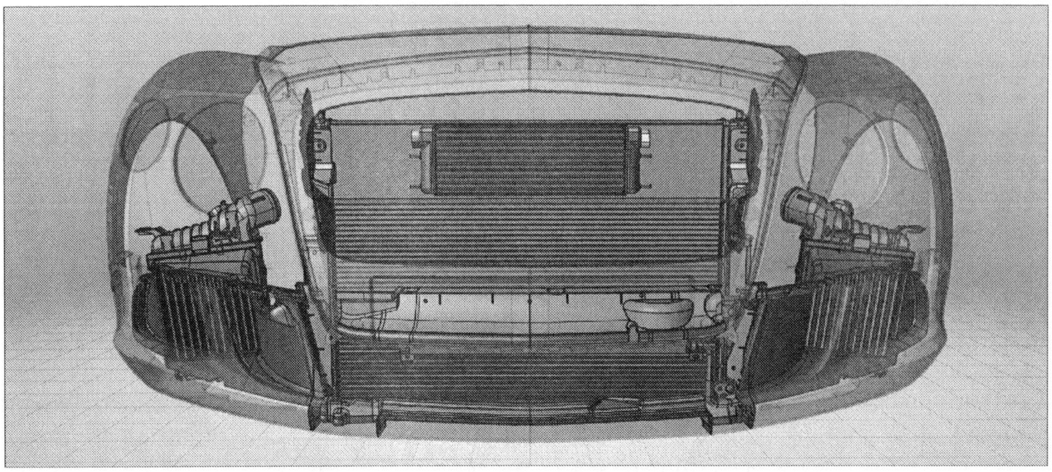

Figure 2 - Cooling Package Orientation

Figure 3 - Airflow through Cooling System

3 SYSTEM OPERATION

In this cooling system, the fluid flow rates in each heat exchanger vary with different operating conditions determined by vehicle speed, engine speed, and engine load. The engine coolant, transmission fluid, and power steering fluid are driven through their respective circuits by pumps which are driven by the engine. The flow for each of these circuits flows through the heat exchangers once the fluid temperature exceeds a design threshold and a thermostat opens. The differential fluid flow is driven by a pump whose speed varies with vehicle speed. The air flow in the right and left charge air coolers is related to engine power. The condenser's refrigerant flow is related to the cooling demand in the cabin.

In a drive cycle, the heat rejection into the different fluids in the powertrain varies with the temperature of the fluid, the rate of heat generation, and the stored thermal energy in the different components. The ability of different powertrain components to store and release thermal energy leads to one of the major complications in designing a cooling package for this class of vehicle. This problem is the generation of heat at high power levels, with the need to dissipate the heat at lower vehicle speeds. This effect is illustrated in Figure 4, in this example, the vehicle is operated at high speeds and high road loads. Once the powertrain is thermally stabilized at a higher speed, the average speed is reduced, but the powertrain is operated at a high average power level to provide repeated accelerations starting at about 1500 seconds. When this high average engine power level occurs, high average engine heat generation occurs. Since the vehicle is operating at lower speeds, the amount of cooling airflow is reduced. The combination of high engine power levels and reduced airflow causes rising coolant temperatures.

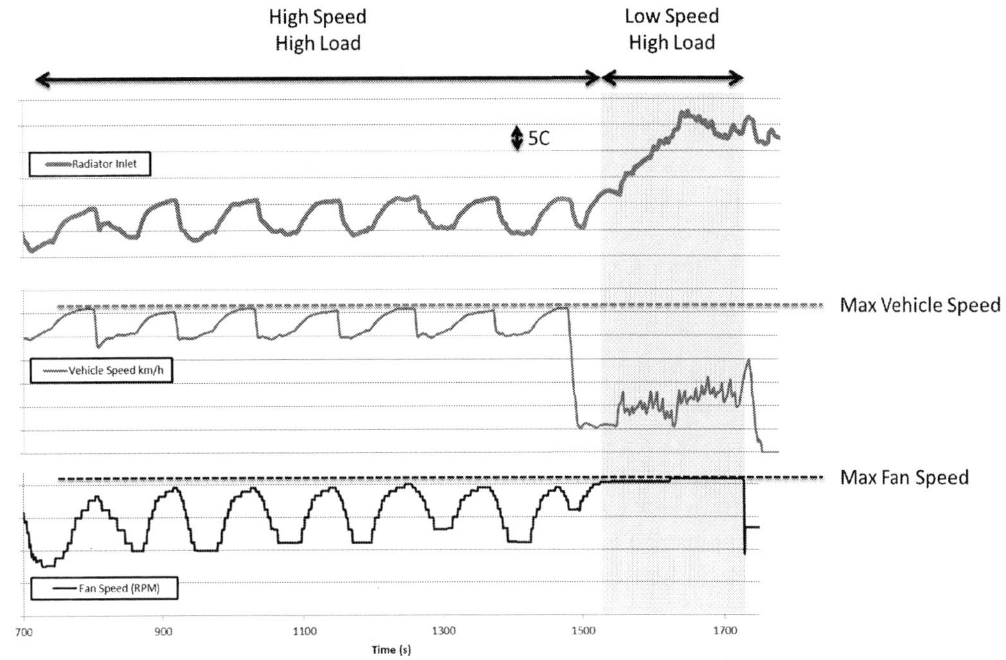

Figure 4 - Increasing Coolant Temperature at Lower Speeds

This illustrates one of the key challenges in design of the cooling system for a passenger vehicle. A minimal capability cooling system must ensure that at the peak ambient temperature and at the maximum rated speed, the cooling system rejects the powertrain and HVAC heat. However, this minimal system will not ensure that the vehicle does not reduce available torque or overheat when subjected to higher average heat rejection than the road load. This deration of torque and overheating of the engine cooling system is not acceptable for a Bentley customer and hence is a frequent focal point in the design.

A maximal capability cooling system would ensure that the vehicle can reject the heat from the engine operating at full power while the vehicle is stationary. This kind of cooling system is unrealistic. The heat exchanger size and fan size would be both prohibitive and needless in a passenger vehicle. The design would carry excess cost and weight for the average and spirited customer drive cycle. This increased weight would have a direct impact on the performance and the economy experienced by the customer. Therefore, an ideal cooling system is found by designing for sufficient heat rejection at various steady state operating conditions and transient drive cycles.

To understand this trade-off, consider the graph in Figure 5. This illustrates an exemplary drive cycle where the vehicle starts at a constant speed, accelerates, cruises, brakes and returns to constant speed driving. Between the events marked by 1 and 2, the vehicle is accelerated at full engine power from 30% of max vehicle speed to 50% of max speed. Upon reaching 50% of max speed, the vehicle cruises for the period marked between 2 and 3. Between 3 and 4, the vehicle decelerates to 30% max speed again. Figure 6 shows the cooling system capability and the engine heat rejection versus vehicle speed. A key concept is that during a transient, the engine may be delivering its maximum power, and generating the maximum amount of waste heat independent of vehicle speed. However, the cooling system's ability to reject heat is a function of vehicle speed, fan speed, and coolant flow rates. So the cooling system cannot reject its maximum amount of heat and maintain coolant temperatures at lower vehicle speeds. This variation of cooling capability is illustrated in Figure 6. During an aggressive drive cycle, as illustrated in Figure 5, the engine is able to generate more heat than the cooling system can reject. This occurs at the point marked by 2 in Figure 5 and Figure 6. This kind of transient can be handled with acceptable coolant temperatures, because the coolant and engine block store thermal energy during part of the transient, and release it during other parts of the drive cycle. Therefore, a cooling system which balances performance and economy requires consideration of operation on both steady state conditions and defined drive cycles.

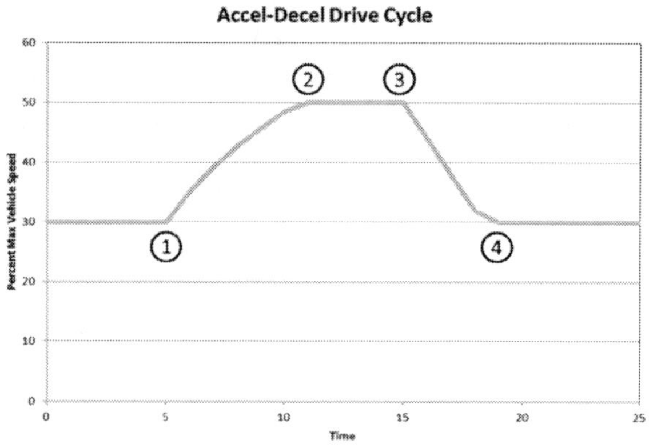

Figure 5 - A Transient Drive Cycle

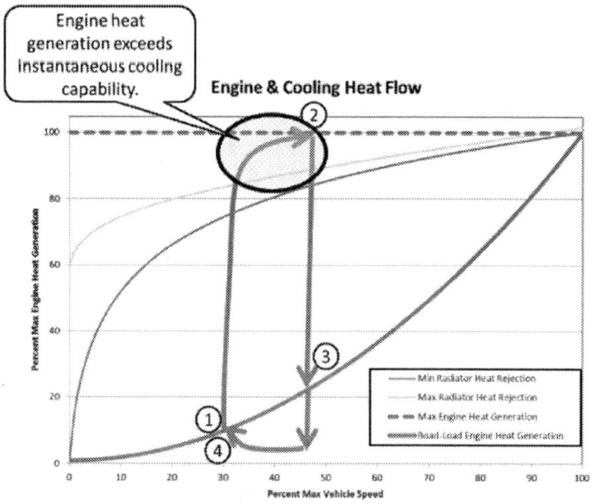

Figure 6 - Heat Generation and Rejection on a Drive Cycle

To illustrate this effect in testing, consider the plot shown in Figure 7. This plot shows engine power and radiator heat rejection rescaled so the plots overlay. As an approximation, the heat rejection of the engine into the coolant will be proportional to the engine brake power. When the engine power is rescaled so the steady state engine power is proportional to the steady state heat rejection in testing, the pattern of generating engine heat in excess of cooling power, followed by cooling in excess of engine heat is seen. For example, when the Continental GT is driven at max speed around a track, during the straight-aways, the engine will be rejecting a maximum amount of heat almost continuously. However, the heat rejection by the radiator lags the engine heat. This is illustrated in Figure 7 between about seconds 1180 and 1240. When the vehicle is moving through the corners, the speed and engine heat rejection are lowered. This occurs between seconds 1240 and 1290. During this period, the radiator rejects more heat than the engine generates. At high speeds, this cycling results in stable coolant temperatures because the high air velocity. However, during the low speed, high load driving from about 1500 seconds to 1700 seconds, the lower average speeds result in lower average cooling capacity, causing the coolant temperature to rise.

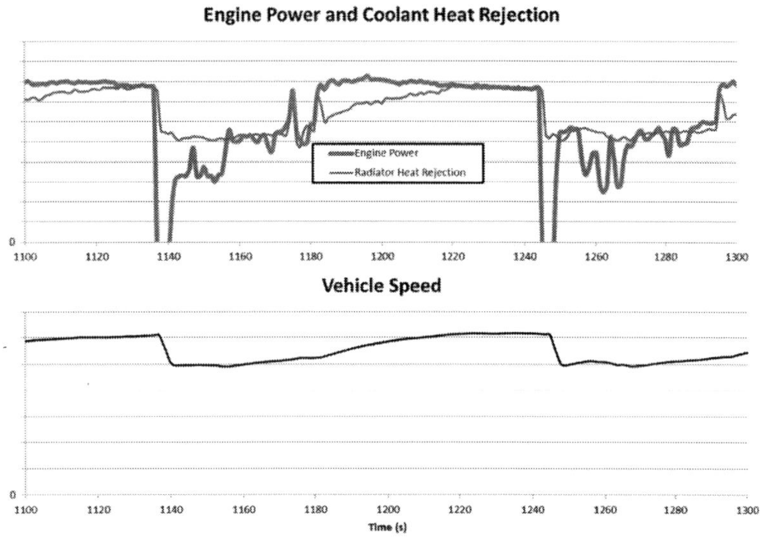

Figure 7 - Engine Power and Radiator Heat Rejection at High Speed and High Load

4 ANALYSIS AND DRIVE CYCLE

The goal of this work was to predict the heat rejection and temperatures at the outlets of the heat exchangers on the High Load High Ambient cycle. This cycle is illustrated in Figure 8. It consists of an extended idle to warm up the powertrain, followed by a high speed operation, then high power, low speed operation followed by an extended idle. The focus of the simulation was on the high and low speed portions the test cycle. This occurs between about 1000 seconds and 1800 seconds.

The cycle represents the worst case thermal driving cycle. The high speed and high load section represents an autobahn drive into and out of high traffic areas. This is then followed by a low speed high load section which represents a customer exiting the autobahn and driving up a mountain road in a spirited fashion followed by a key off event at the top.

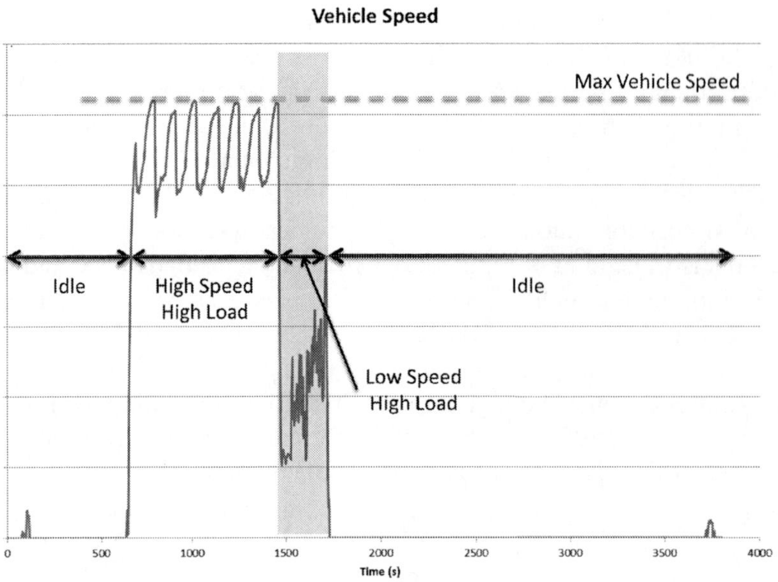

Figure 8 - High Load High Ambient Cycle Vehicle and Fan Speeds

To model the cooling package, the boundary conditions were established by a combination of PowerFLOW, system models, and test measurements. PowerFLOW was used to establish airflow boundary conditions at different operating conditions. A combination of component data, system models and test measurements were used to establish the time varying inlet conditions at the heat exchangers. Figure 9, Figure 10, and Figure 11 show how these models were used to simulate the boundary conditions on the cooling package. When simulated, the PowerFLOW simulations were performed at the vehicle level. A single GT Suite model was used to model the entire cooling package: right, centre, and left cooling stacks.

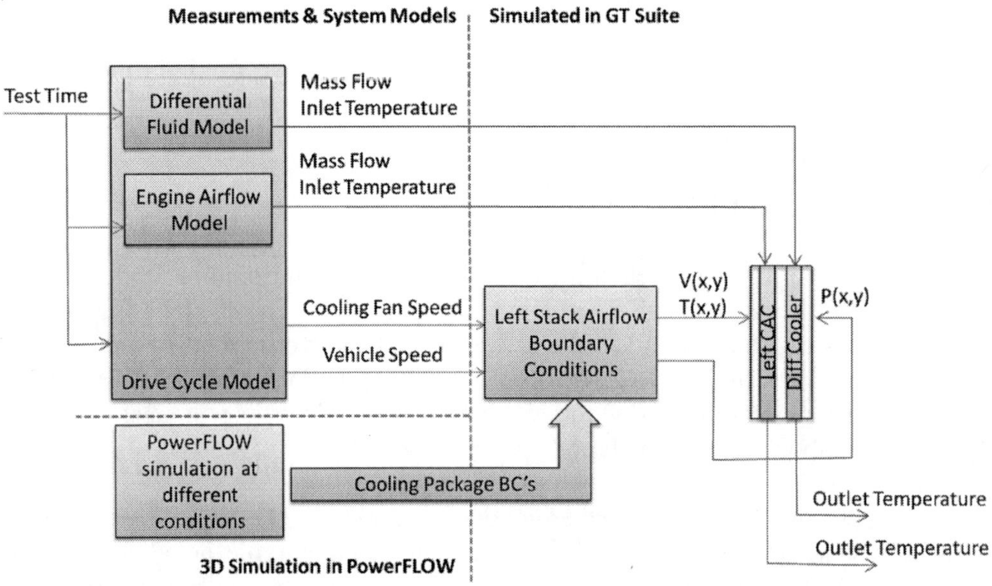

Figure 9 - Simulation Configuration for Left Cooling Stack

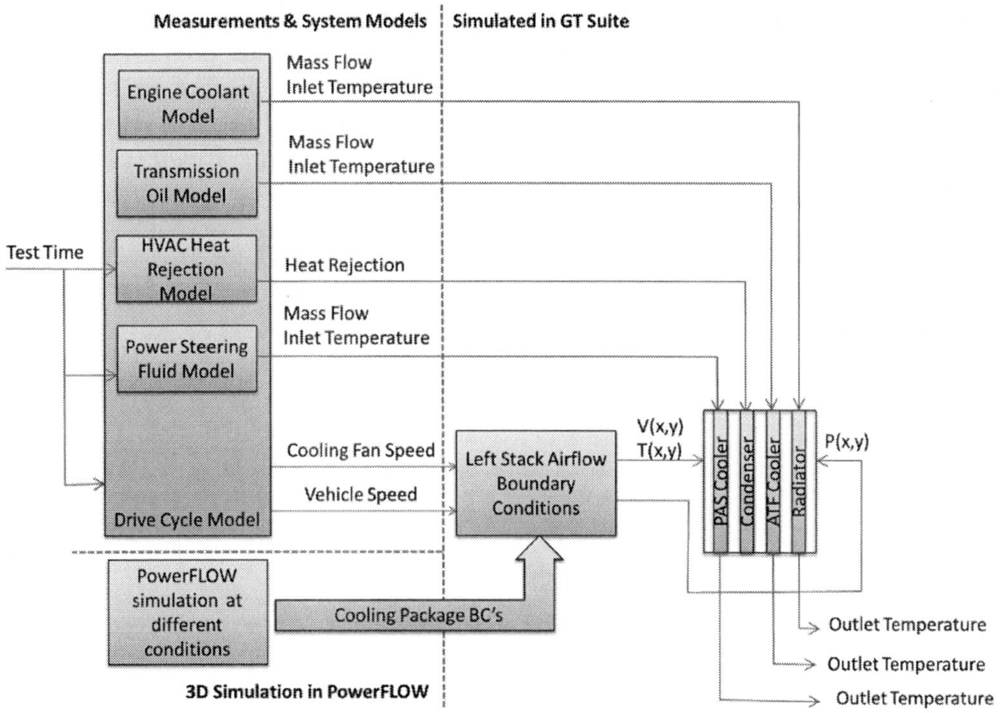

Figure 10 - Simulation Configuration for Central Cooling Stack

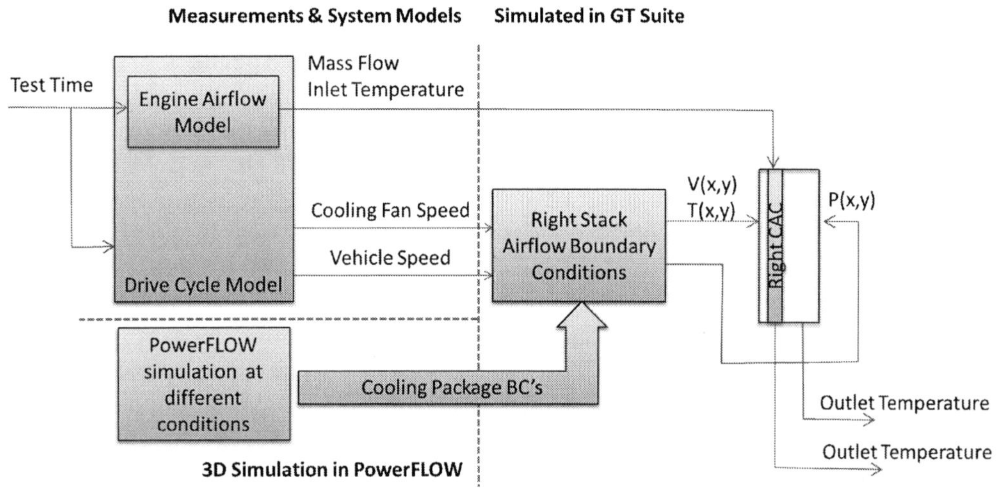

Figure 11 - Simulation Configuration for Right Cooling Stack

In Figure 9, Figure 10, and Figure 11, the models which generate the fluid mass flow were created from component data and bench testing of subsystems. The fluid temperatures were directly measured on the vehicle. The heat rejection by the condenser was obtained from supplier data. The vehicle speed was measured during the test. The cooling fan speed was modelled from the electrical voltage applied during the test.

The inlet conditions for the heat exchangers in the cooling package could quickly be generated for the full duration of the cycle using system models. However, full 3D airflow simulation to generate the airflow boundary conditions for the duration of the

High Load High Ambient cycle is prohibitive. Therefore a sampling and response surface strategy were used. This approach is feasible because the airflow at the boundary of the cooling package stabilizes quickly and the vehicle repeatedly visits the same operating conditions. Figure 12 shows the distribution of vehicle speed and fan speed over the test cycles. The values in Figure 8 were sampled every second and plotted with transparency so that darker regions indicate areas where the vehicle spends more time. To minimize extrapolation of operating conditions, six points along the convex hull of the test conditions were chosen. Additional points were chosen to provide additional data in regions where the vehicle spent significant time. These points are illustrated on Figure 12 by the filled red circles. Each of these operation points is identified by the number one through eight.

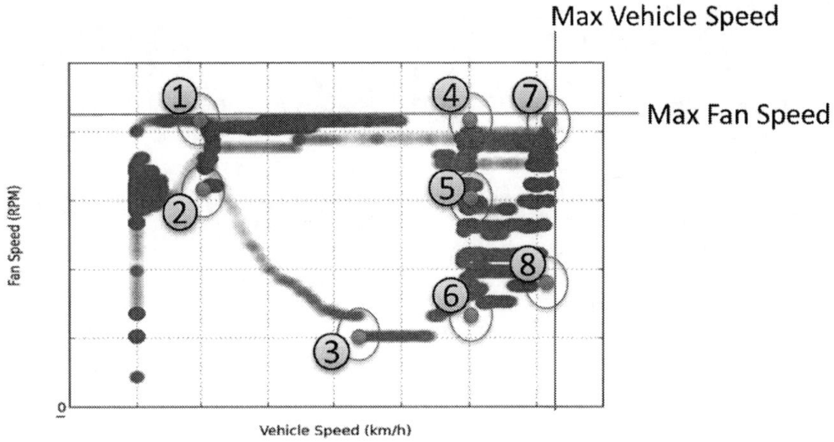

Figure 12 - Distribution of Vehicle and Fan Speeds

After the cooling airflow conditions were determined, the temperature and mass flow or heat rejection needed to be established for each heat exchanger. This is done to properly characterize under hood recirculation of hot gas from the engine compartment to the face of the cooling package. Characterization of the temperatures due to recirculation requires a reasonable model of the heat rejection from the cooling package. This meant that for each heat exchanger, an inlet temperature and mass flow for the PowerFLOW simulation needed to be specified. This was challenging because the temperature and mass flow occurred in a distribution about the velocity and fan speed operating points. This is illustrated in Figure 13. For example, when the vehicle operates at about 20% of top speed and the fan runs near max fan speed, several different flow rates occur because the engine is in different gears and under different throttle positions. This is illustrated on Figure 13 by the black dots imposed on the red traces showing coolant temperature, coolant flow, fan speed, and vehicle speed. For the operating point of vehicle speed and fans speed chosen, the coolant flow varied over 20% of the max flow rate and the temperature varied by more than 10 degrees C. Because of this variation, selection of operating points for each heat exchanger was done by calculating the average inlet temperatures and flow rates that occurs for operating points near a given vehicle and fan speed.

For each of the selected operating points, a full vehicle PowerFLOW model was simulated. From the plane in front of each cooling stack, the velocity and temperature distribution were recorded. The pressure was recorded from the backside of each cooling stack. It was necessary to measure each cooling stack independently since the three cooling stacks were non-planar.

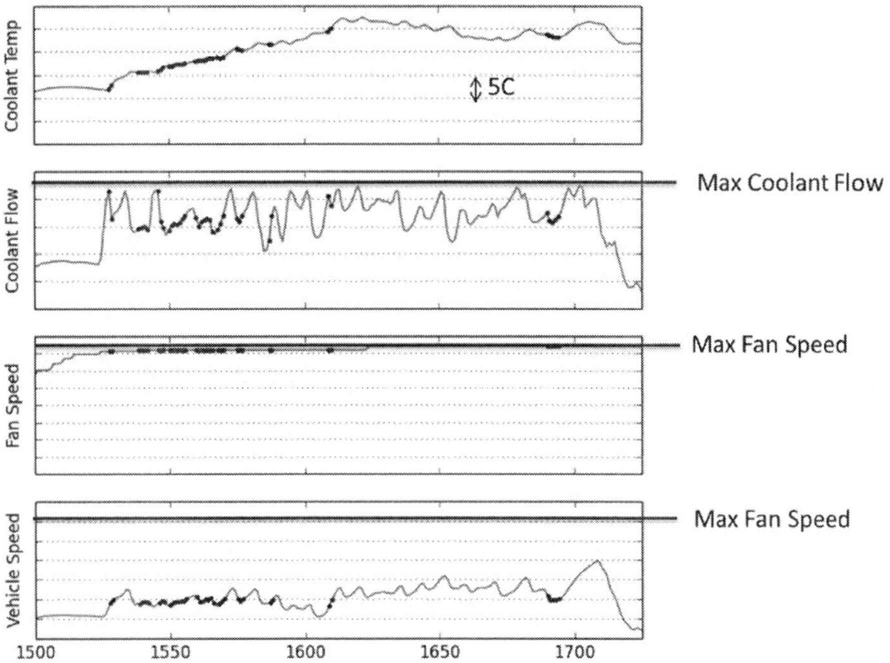

Figure 13 - Coolant Flow versus Vehicle Speed and Fan Speed Operating Point

Figure 14 illustrates the variation in airflow at the entrance to each air path in the cooling package. The image on the left corresponds to the inlet to the left cooling stack. The image on the right corresponds to the right cooling stack. In the centre cooling stack, the large region with low air flow corresponds to the location of the PAS cooler. The grid variation is related to grill geometry. These velocity variations were obtained from the PowerFLOW simulation at high vehicle speed. However, this non-uniformity persists even at lower speeds. Figure 15 shows the non-uniformity of the air temperature at the cooling package inlet. For the centre stack, the maximum temperature non-uniformity occurs at lower speeds. For the right and left CAC, the temperature non-uniformity increases with speed. At higher speeds, the centre stack inlet temperatures are nearly uniform.

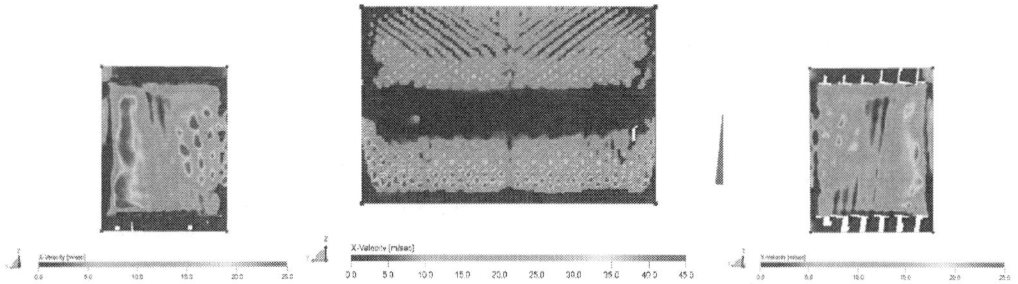

Figure 14 - Airflow Velocity on Cooling Package Inlets

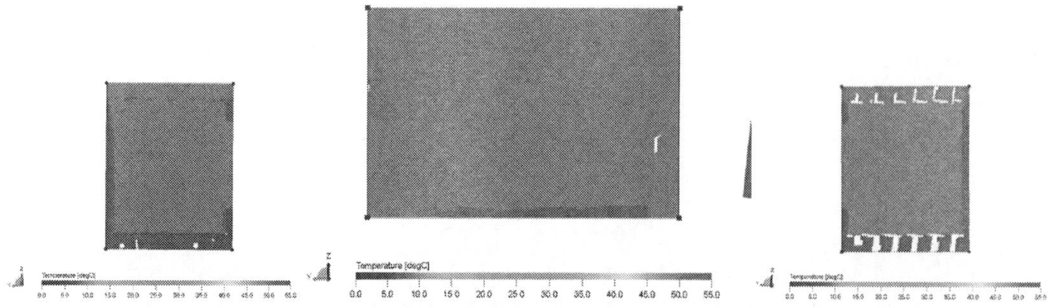

Figure 15 - Air Temperature at Cooling Package Inlets

To quantify this non-uniformity of the airflow, the method suggested in [4] is used. This non-uniformity index measures average absolute deviation from a uniform velocity. The equation for calculating this index based on samples from a regular grid of air velocity is shown Figure 16. If the velocity is uniform, the index will be zero. For the case where half of a radiator is blocked, the velocity non-uniformity index will have a value of one. The non-uniformity index of the air velocity at the different operating points is shown in shown in Table 1. This table shows that the centre stack has significantly more variation in airflow velocity than the right and left hand stacks. The case number refers to the operating points in identified in Figure 11.

$$i_{vel} = \frac{1}{n} \cdot \sum_{k=0}^{n} \frac{|V_k - \bar{V}|}{\bar{V}}$$

Figure 16 - Velocity Non-Uniformity Index Equation

Table 1 - Air Velocity Non-Uniformity Index

	Case 1	Case 2	Case 3	Case 4	Case 5	Case 6	Case 7	Case 8
Right Stack	0.324	0.346	0.311	0.324	0.331	0.325	0.328	0.330
Centre Stack	0.719	0.674	0.798	0.789	0.795	0.783	0.787	0.794
Left Stack	0.299	0.317	0.327	0.333	0.335	0.336	0.339	0.340

To characterize the temperature non-uniformity, a modified form of the equation shown in Figure 16 is used. This equation is shown in Figure 17. It is calculated from a regularly gridded sampling of the temperature at the inlet to the cooling package. This index provides insight into the deviation of temperatures at the inlet to the cooling package. A value of zero indicates that the inlet temperatures are perfectly uniform. A value of one occurs if half of the inlet has a temperature deviation one degree above average and half has a temperature deviation one degree below the average value. The temperature indexes are shown in Table 2. These indices show that at high speeds, the centre stack has nearly uniform temperature boundary conditions. However, at low vehicle and fan speeds, the non-uniformity in temperature is larger. The right and left stacks have about the same amount of temperature non-uniformity at all operating conditions.

$$i_{temp} = \frac{1}{n} \cdot \sum_{k=0}^{n} |T_k - \bar{T}|$$

Figure 17 - Temperature Non-Uniformity Index Equation

Table 2 - Temperature Non-Uniformity Index

	Case 1	Case 2	Case 3	Case 4	Case 5	Case 6	Case 7	Case 8
Right Stack	0.198	0.430	0.439	0.284	0.334	0.283	0.317	0.330
Centre Stack	0.220	0.688	0.066	0.048	0.060	0.057	0.057	0.057
Left Stack	0.213	0.289	0.481	0.419	0.436	0.480	0.467	0.420

The air flow and air temperature data from PowerFLOW was converted into boundary conditions for use in GT Suite. Two different types of boundary conditions were generated. In the first case, the average velocity and temperature were generated for GT Suite. In the second case, the PowerFLOW results were sampled on a regular grid to generate a 2D boundary condition for the GT model. Experimentation was used to determine that a grid size of 45 elements by 45 elements is a good balance between execution speed and accuracy. Since a limited number of operating points were simulated in 3D, a response surface method was used to generate boundary conditions for other operating points.

To generate the GT suite model, the geometry used in PowerFLOW was discretized. This was done by creating a 3D parametric model in GT Suite which defines the detailed internal and external geometry of the heat exchangers. Additionally, the heat transfer characteristics from bench tests and the mass of heat exchanger are used to create the system model. The visual representation of this model is shown in Figure 18. After the definition of the stack was completed, it was converted into the system model illustrated in Figure 19.

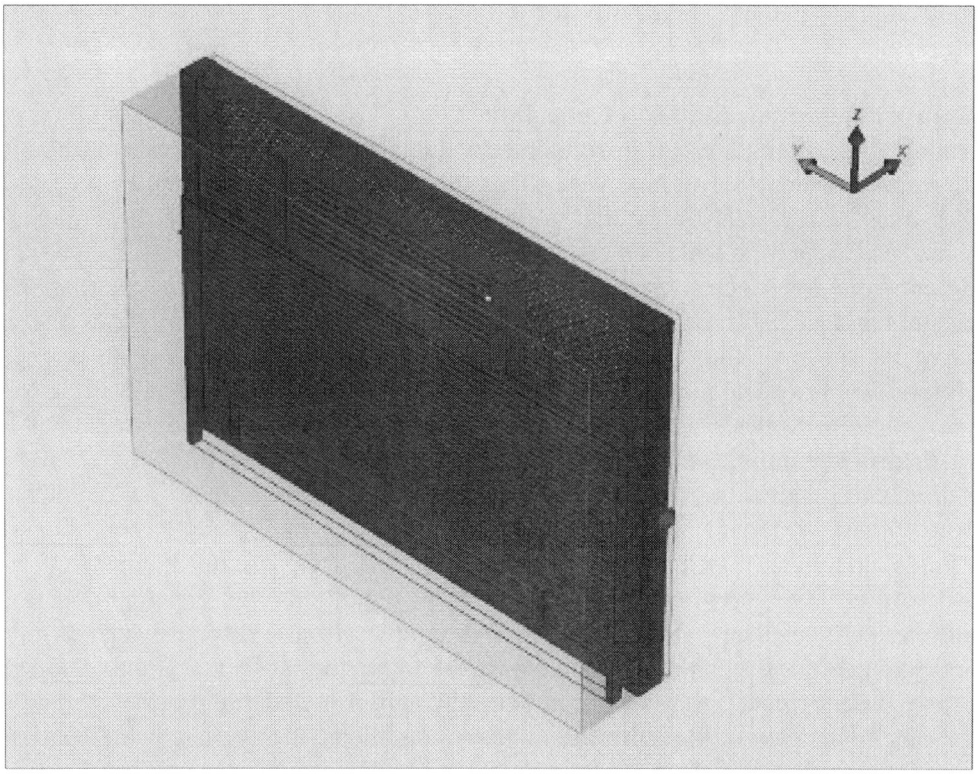

Figure 18 - Cool3D Representation of Centre Stack in Cooling Package

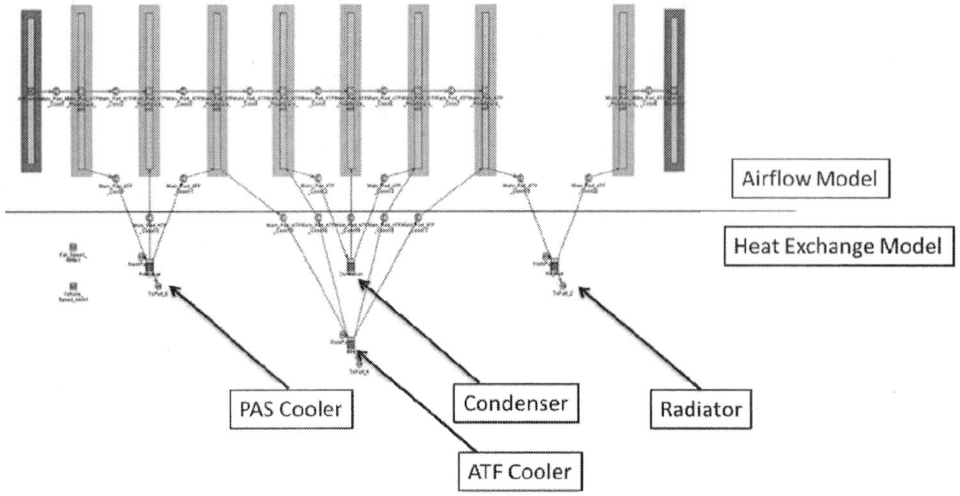

Figure 19 - Discretized System Model of Centre Stack in Cooling Package

5 VALIDATION

The validation was performed for two system modelling techniques. In the first technique, the uniform boundary conditions were applied to the inlet of the three stacks. This approach provides faster execution, but sacrifices some accuracy. The second technique applies 2D, non-uniform boundary conditions to each stack. This approach provides higher accuracy, especially for the heat exchangers which partially cover the stack.

5.1 Instrumentation and Data Collection

To validate this simulation, an instrumented vehicle was built and driven on the High Load High Ambient (HLHA) cycle. This vehicle used thermocouples to collect temperatures at the inlet and outlet for each heat exchanger except the condenser. Pressure responses were collected from the inlet and outlet of each heat exchanger using inline pressure transducers. Data was collected from the powertrain via diagnostic connectors and an ETK. Components were bench tested to characterize flow rates for different vehicle operating conditions. This information was used to create boundary conditions for the system model for all boundary conditions except airflow.

5.2 Transient Simulation Setup

The drive cycle simulations of the cooling package were performed in GT Suite. As shown in Figure 9, Figure 10, and Figure 11, the time varying boundary conditions for the fluids and heat rejection were establish using other system models and test measurements. The cooling airflow boundary conditions at different operating conditions were calculated from PowerFLOW, sampled from a response surface model to provide data on a regular grid to support the input format in GT Suite. The same response surface model was used for both the uniform and non-uniform boundary conditions. In the case of the uniform boundary conditions, the average value was used. The differences at the highest load case between uniform and non-uniform boundary conditions can be seen below in figure 20 and 21. It is clear to see the flow velocity peaks and non-uniform behaviour of the cooling air flow.

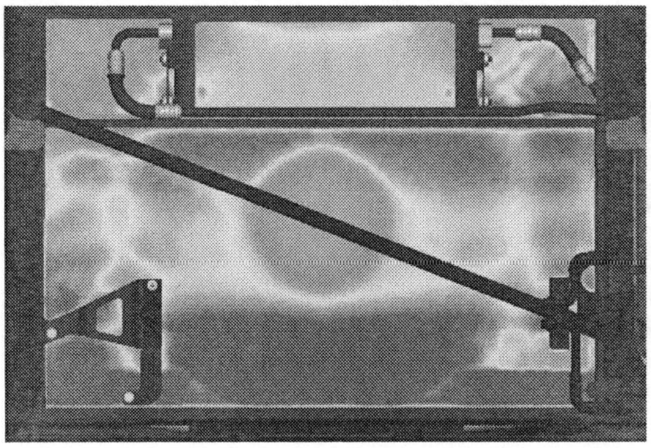

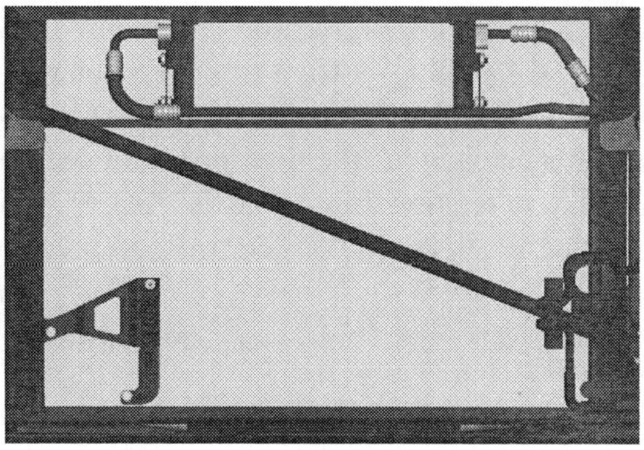

Figure 20 - Velocity Magnitude Heat Exchanger Boundary Conditions Comparison

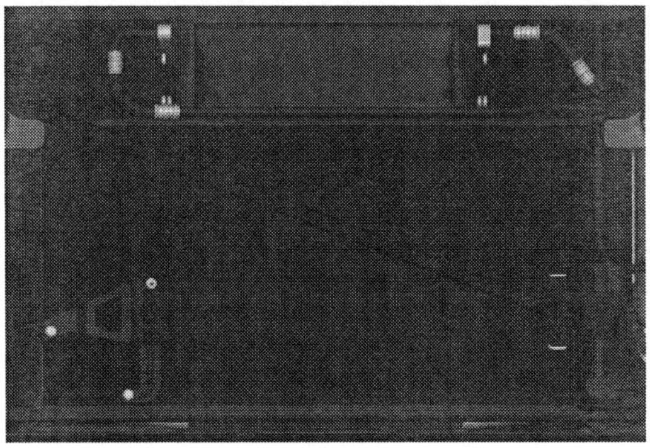

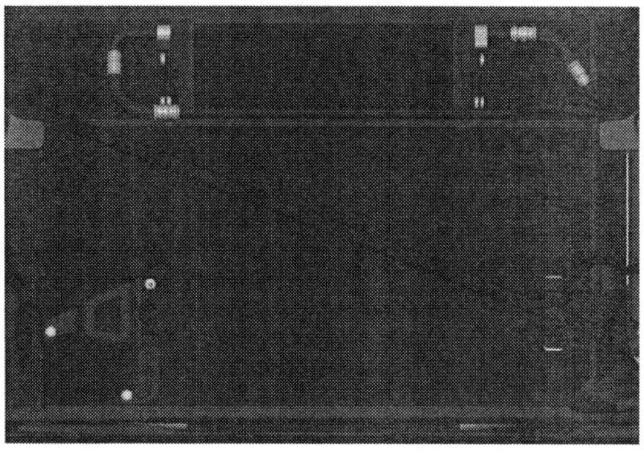

Figure 21 - Temperature Heat Exchanger Boundary Conditions Comparison

5.3 Simulation Results

In the centre stack, four heat exchangers interact. The PAS cooler rejects heat which flows into the condenser, then into the radiator. The ATF cooler rejects heat which flow into the radiator. The heat rejection by the condenser was simulated as a uniform heat rejection into the air stream. Transient models of these heat sources upstream of the radiator were included in the system model of this cooling stack. The CFD results which were used to generate the boundary conditions for the stack included effects such as under hood recirculation, which warmed the air above ambient conditions.

Figure 22 shows the PAS fluid temperatures at the outlet of the PAS cooler from both measurement and simulation. During the majority of the simulation, this error is less than 1C. The simulation shows good tracking to measured values even when large temperature excursions occur. For this heat exchanger the air boundary conditions were the average temperature and velocity at the face of the cooling package. The fluid flow rate was determined using a model map of flow rate versus engine speed.

Figure 23 shows the ATF cooler outlet temperature. One of the modelling challenges with this cooler was the prediction of the fluid flow through the cooler. The flow model used only considered transmission input speed to calculate the flow. However, this simplistic model is only correct when the transmission is in gear. Inspection of the measured temperature in Figure 23 shows significant drops in the outlet temperature when the vehicle changes speed and shifts. During shifts, ATF flow stays internal to the transmission, and does not flow through the cooler. The lower flow rates result in lower outlet temperatures. Additionally from about 1500 to 1700 seconds, the vehicle is in the ride and handling portion of the test undergoing almost continuous shifting. During this continuous shifting period the oil pressure in the ATF circuit changes very frequently and hence changes the pressure field seen by the pressure pump, changing the flow rate the fixed relationship pump is able to deliver. The error in the flow rates due to the shifting results in the larger errors in predicted temperature.

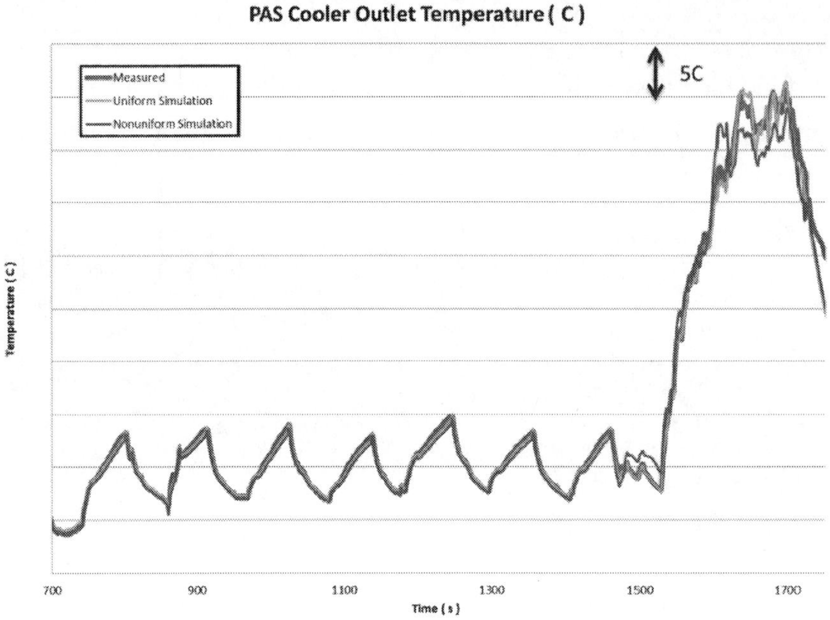

Figure 22 - PAS Cooler Outlet Measured and Simulated Temperature

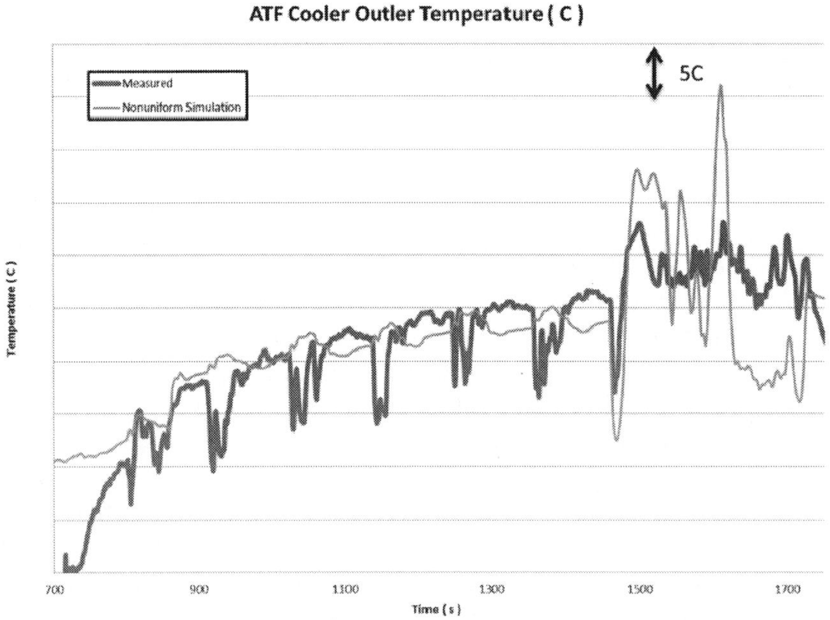

Figure 23 - ATF Cooler Outlet Measured and Simulated Temperature

As noted earlier, the engine coolant flow rates were simulated based on a characterization of the coolant flow from a test bench. The simulated current flow is shown in Figure 24. The flow rate is a function of engine speed. When accelerating relatively constantly, these flows follow a regular pattern which tracks the engine speed in the gear during the acceleration. However, during this cycle, when the vehicle speed changes rapidly, the engine speed varies more, causing the coolant flow rate to change dramatically as seen between 1550 and 1700 seconds in the cycle.

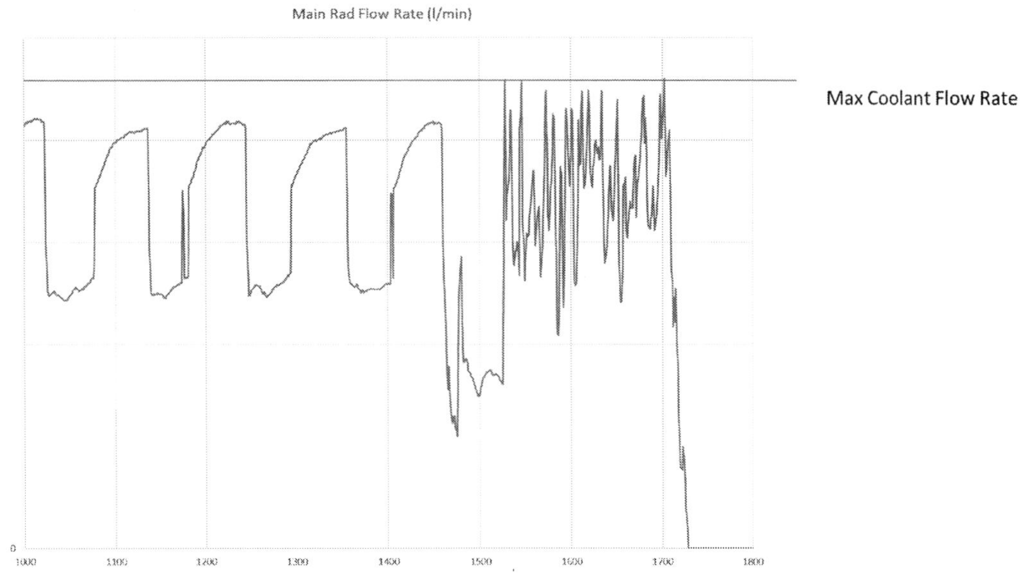

Figure 24 - Simulated Coolant Flow Rates

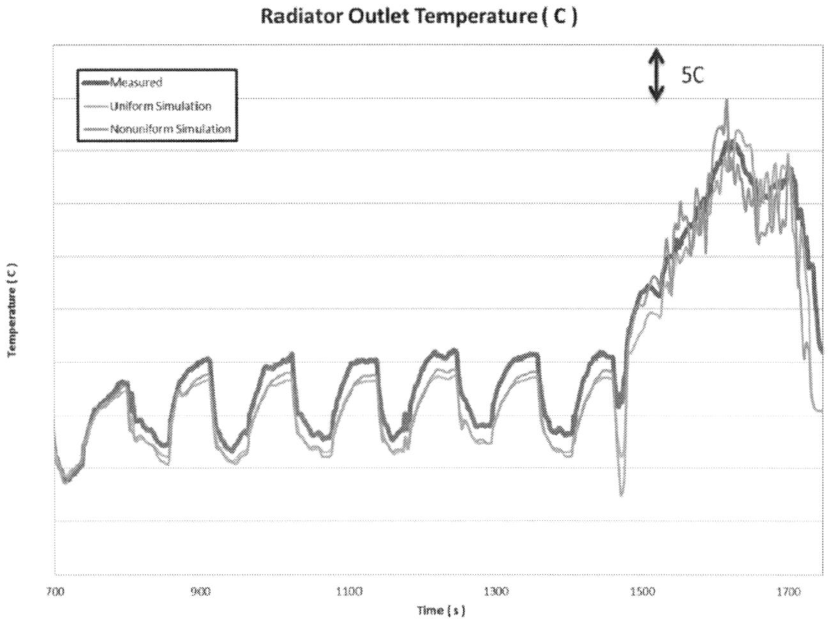

Figure 25 - Comparison of Radiator Measured and Simulated Temperature

The results for the radiator simulation are shown in Figure 25. Using the uniform boundary conditions, the radiator outlet temperatures were predicted within about 2C of the measured data during the drive cycle. This accuracy is achieved with the time varying changes in the airflow, coolant flow and heat rejection from heat exchangers upstream in the airflow. The use of non-uniform boundary conditions on the cooling stack reduced the error in the output temperature prediction by about 0.2C during the high speed portions of the test. Examining the test results, the coolant temperature is predicted within 0.5C around 800 seconds. However, between about 800 seconds and 1500 seconds, there is about a 1.5C error. This bias appears to be due to the use of a simplistic coolant flow model which assumes flow is a function of engine speed only. During the ride and handling part of the test, seconds 1500 through 1750, the simulation based on uniform airflow provides the result which best tracks the test results. During this interval, the non-uniform boundary conditions results in larger swings in error. This is believed to be caused by the interpolation between non-uniform airflow patterns where few PowerFLOW characterizations were performed.

Figure 26 and Figure 27 show the measured and predicted outlet temperatures for the charge air coolers (CAC). Two different modelling approaches were used with the CAC cooling airflow boundary conditions. For the non-uniform case, the same system model as used for the centre stack was evaluated. For the uniform case, an alternative heat exchanger model which applied average mass flow was used. Recall that the airflow through the left CAC flows through the differential oil cooler before exiting that stack, while the right CAC is alone. In the case of the left CAC, this average mass flow approach provided a good fit to measured outlet temperatures during high speed operation. For the right CAC, the non-uniform conditions provided a slight improvement on using uniform conditions during high speed operation. For both sets of boundary conditions, the transients during the low speed, high load operations showed increased errors.

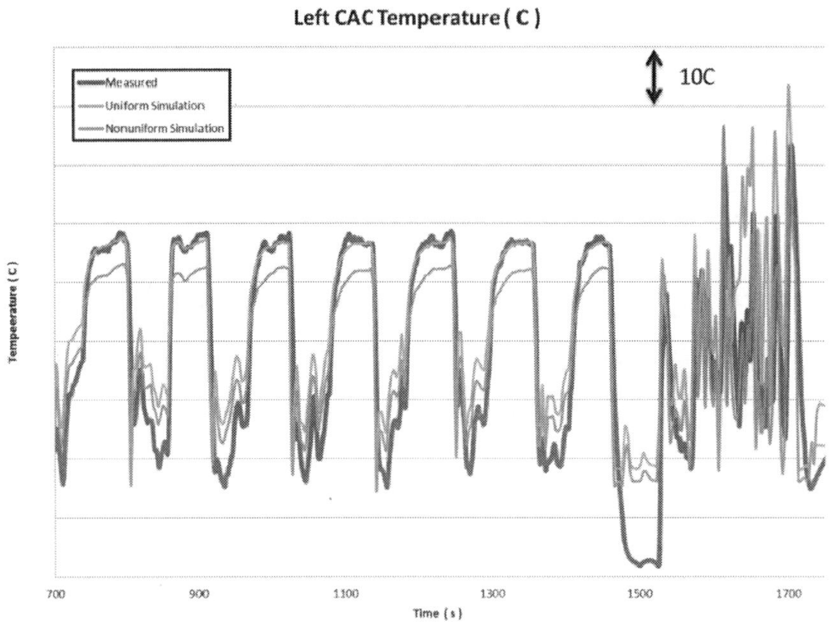

Figure 26 - Left CAC Measured and Simulated Temperature

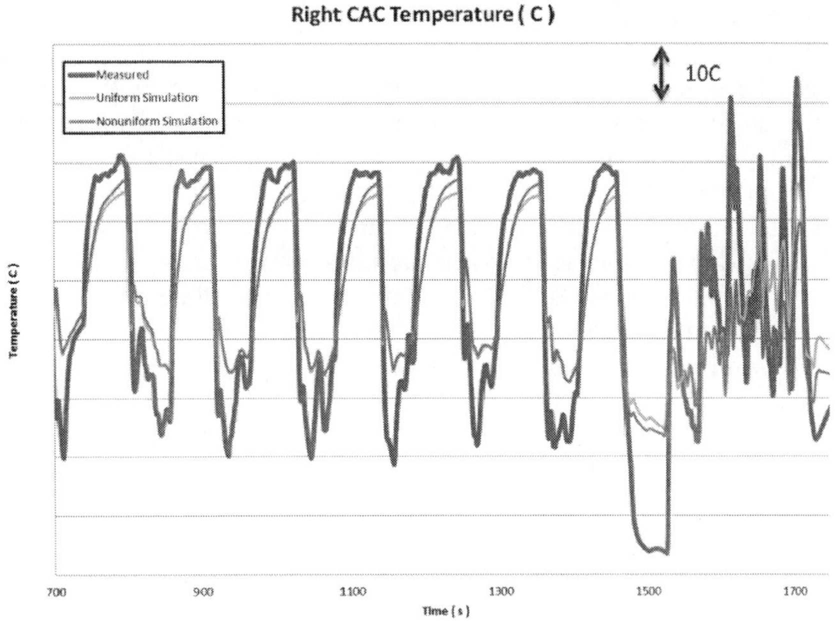

Figure 27 - Right CAC Measured and Simulated Temperature

The temperatures for the differential oil cooler are shown in Figure 28. Of the systems modelled, this fluid had the largest error and largest differences between the use of uniform and non-uniform boundary conditions. The non-uniform boundary conditions reduced the error in predicted outlet temperatures. However, the general shape of the outlet temperatures deviated appreciably. The combination of heat rejection from the preceding CAC and the simple flow model used for the differential fluid flow are suspected sources of this behaviour.

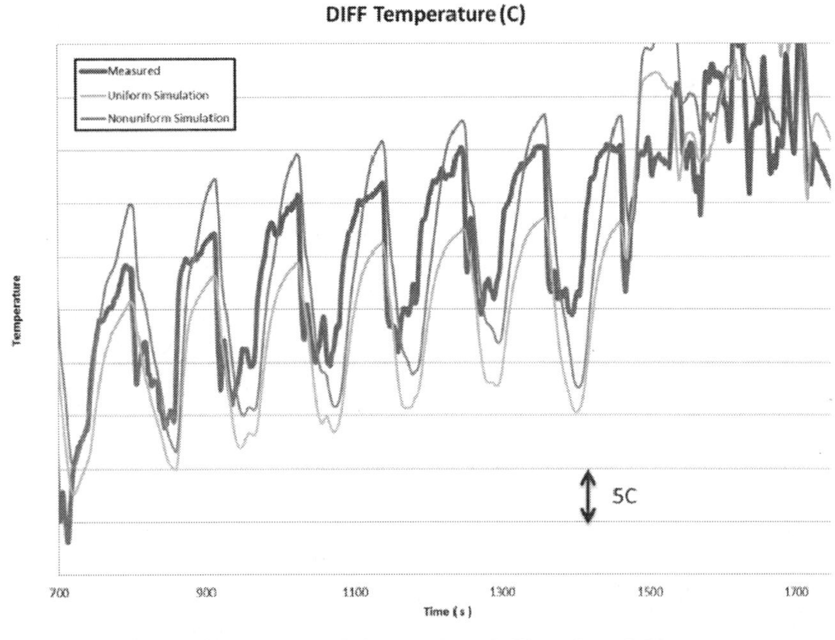

Figure 28 - Differential Measured and Simulated Temperatures

5.4 Results Summary and Discussion

These results show that combining CFD with system models can accurately predict temperature where multiple heat exchangers are interacting with the airstream. Furthermore, multiple stacks of heat exchanges can be combined in a single model, allowing all of the heat rejection in a complex cooling package to be simultaneously evaluated. When the CFD boundary conditions are simplified to a uniform value, this process was able to predict coolant temperatures with a typical error of about 1.5 degrees C during the HLHA cycle. When the non-uniform boundary conditions were applied to the boundary conditions of the cooling package, the error was reduced, in periods of stabilized operating, by about 25%. However, during transients, this advantage was significantly reduced.

To summarize the simulation results for each of the different heat exchangers, the mean and RMS errors were calculated. Since the accuracy of the simulations varied with the operating conditions, these values are computed for the high and low speed portions of the HLHA test cycle. The mean value is used to help understand if the simulation will accurately predict average heat rejection. The RMS error is used to quantify the magnitude of the instantaneous errors. These results are summarized in Table 3 through Table 6. To simplify comparison between the methods the best result are in bold print.

The similarity of the results for both uniform and non-uniform boundary conditions indicate that for this level of fidelity in modelling this vehicle, the benefits of a non-uniform boundary conditions marginally contribute to improved results. However, with resolution of issues in modelling the coolant and oil flow rates along with removal of other error sources, these results should be examined again.

Table 3 - Summary of Temperature Errors at High Speed for Uniform Boundary Conditions

	PAS Cooler	ATF Cooler	Radiator	Left CAC	Right CAC	Diff Cooler
Mean Error [C]	**0.003**	**1.34**	1.60	-3.26	**0.19**	5.80
RMS Error [C]	**0.27**	5.56	1.76	6.63	8.49	7.31

Table 4 - Summary of Temperature Errors at High Speed for Non-Uniform Boundary Conditions

	PAS Cooler	ATF Cooler	Radiator	Left CAC	Right CAC	Diff Cooler
Mean Error [C]	0.37	-1.63	**1.59**	**1.02**	-0.42	**0.32**
RMS Error [C]	0.46	**4.25**	**1.73**	**6.24**	**8.44**	**5.76**

Table 5 - Summary of Temperature Errors at Low Speed for Uniform Boundary Conditions

	PAS Cooler	ATF Cooler	Radiator	Left CAC	Right CAC	Diff Cooler
Mean Error [C]	0.71	-5.67	**1.80**	-7.73	**-1.32**	-3.94
RMS Error [C]	**1.44**	**7.07**	**3.39**	14.31	13.45	7.68

Table 6 - Summary of Temperature Errors at Low Speed for Non-Uniform Boundary Conditions

	PAS Cooler	ATF Cooler	Radiator	Left CAC	Right CAC	Diff Cooler
Mean Error [C]	**0.65**	**1.95**	1.95	**0.83**	2.44	**-3.17**
RMS Error [C]	2.38	7.58	4.21	**10.27**	**13.29**	**6.12**

During the course of this work, several sources of error were identified. Future work focused on improving these results should address two key areas. The first area is improvement of the internal flow models. Each of the flow models rely on using engine and vehicle operating conditions to specify the instantaneous flow rate. However, these fluid systems are complex networks of pipe, restrictions, valves, and pumps.

A starting point for improved system modelling should include an in-vehicle validated flow model for each cooling fluid. Since fan speed is critical for accurate simulation of under-hood airflow, direct measurement of fan speed should be performed to validate the fan speed models. The second area for improvement is in geometry and external environmental modelling. The HLHA cycle includes high speed operation and significant body forces. Parts of the cycle involve aggressive manoeuvring which moves the front wheel away from a neutral location, altering airflow. The presence of ride height control mechanisms can significantly alter airflow. High speed travel through corners can also alter airflow because of the effective vehicle yaw. Future validation efforts should measure body position and ride height during the track testing. The resulting postures changes should be incorporated into the CFD characterization of cooling airflow. One potential source of error in this model is fluid structure interaction. Appreciable aerodynamic forces are applied to the front fascia and duct work. A check for deformation at high speeds affecting leakage, airflow distribution, and recirculation should be performed. Finally, because of the position of the apertures near the ground, multiple ambient air temperature sensors should be used to detect any temperature gradients in drawing hot air off the track surface.

6 CONCLUSIONS

In previous projects at Bentley the HLHA cycle would be the first time that the hardware is tested as a full system outside of the 1D analysis stage. This provides no ability for error and hence many of the cooling systems are unknowingly over specified at this stage in the vehicle life cycle. There is also a large impact on the project cost model in order to frequently visit the test sites all over the world to gain the correct ambient conditions and the ability to perform this high load drive cycle. Performing further 1D analysis coupled with 3D CFD capability allows for a reduction in this risk and a real insight into the performance over this HLHA cycle. This approach ultimately impacts positively not only on the company business model but also on the customer in terms of vehicle performance and efficiency.

In the pursuit of this work the importance of high quality and complete input data has been emphasised. For future vehicle development tests a catalogue of required measurement points has been created and implemented into test vehicle build plans.

This will ensure not only good quality input data but also a very good basis to validate the model itself as a sign off activity for each vehicle project.

This validated analysis will be utilised on the next generation Continental GT to drive design decisions early in the product life cycle, to avoid the incurrence of unnecessary cost and to provide the customer with the best driving experience on offer.

7 REFERENCES

[1] Bentley Motors, Detailed Product Specification, http://www.bentleymotors.com/en/models/continental/continental-gt-speed/detailed-specification.html, Jan 21, 2013.

[2] Jansen, W., Amodeo, J., Tate, E., Yang, Z., "Drive Cycle Simulation of a Tiered Cooling Pack Using Non-Uniform Boundary Conditions," SAE Technical Paper 2014-01-0654, 2014, doi:10.4271/2014-01-0654.

[3] Alajbegovic, A., Xu, B., Konstantinov, A., Amodeo, J. et al., "Simulation of Cooling Airflow under Different Driving Conditions," SAE Technical Paper 2007-01-0766, 2007, doi:10.4271/2007-01-0766.

[4] Jama, H., Watkins, S., Dixon, C., Ng, E., "Airflow Distribution through the Radiator of a Typical Australia Passenger Car", 15th Australasian Fluid Dynamics Conference, The University of Sydney, Sydney, Australia, 13-17 December 2004.

Dual use heater core – increase in trailer tow capacity and analysis of control set-points with active grill shutters

S UPPULURI and **A NAIKNAWARE**
Computational Sciences Experts Group, USA

ABSTRACT

The concept evaluated in this paper includes a modification to the HVAC airbox to enable heating the cabin, cooling the engine or both on demand regardless of the passenger's cabin heating and cooling requirements without affecting the cabin comfort. This is achieved with a cooling door that can be selectively positioned such that at least some of the air moving through the heater core is directed outside the passenger cabin to the underhood area of a vehicle, thereby providing supplemental engine cooling on demand regardless of the passenger's cabin heating and cooling requirements. In addition, this technology enhances the operation of Active Grill Shutters by keeping it closed for a longer period of time for a given drive cycle.

1 INTRODUCTION

As a standard design practice, the front end cooling pack (radiator and cooling fan) of cars on the road are sized for a worst case thermal scenario such as a vehicle towing a 1300 kilogram trailer driving up a grade on a 43C day. This scenario describes less than 1% of driving time, but 100% of vehicles have cooling packs designed around a worst case contingency. This over design means an oversized cooling pack and fan leading to higher drag and lower fuel economy for the everyday driver.

This paper explores alternatives to this approach by leveraging the heater core. The heater core is capable of removing 10% - 20% of total engine waste heat depending on the size of the heater core and the engine design. The heater core is already plumbed into the engine with plenty of coolant flow during all operating conditions. The heater core could be used to supplement engine cooling, on-demand, and perhaps provide enough cooling of the engine during peak cooling requirement times, so that the front-end cooling pack can be optimized for regular driving, which would reduce the front end airflow requirements, reducing the drag of the vehicle and reducing the size of the fan required. This will improve fuel economy by reducing the parasitic loads, or alternative provide higher trailer towing capacity to existing vehicle by providing more heat rejection capacity. The trouble is, when it is hot outside and the engine needs extra cooling, the driver is also hot, and unlikely to turn on the heater.

2 DUAL USE HEATER CORE CONCEPT

This concept is illustrated in Figure 1. The figure shows a simplified illustration of the HVAC airbox for ease of understanding. The Dual Use Heater Core (DUHC) technology introduces a modification to the existing HVAC airbox by include a door (called the "cooling door" for reference purposes in the rest of this paper) that will vent the hot air blowing across the heater core back outside to the underhood therefore providing engine cooling without affecting cabin comfort. The cooling door is positioned automatically by the Engine Control Unit ("ECU") dependent on a combination of parameters of the engine such as engine coolant temperature, engine oil temperature, cabin temperature requirement etc. The blower speed and the position of the cooling door are adjusted by the ECU depending on how much supplemental engine cooling is required.

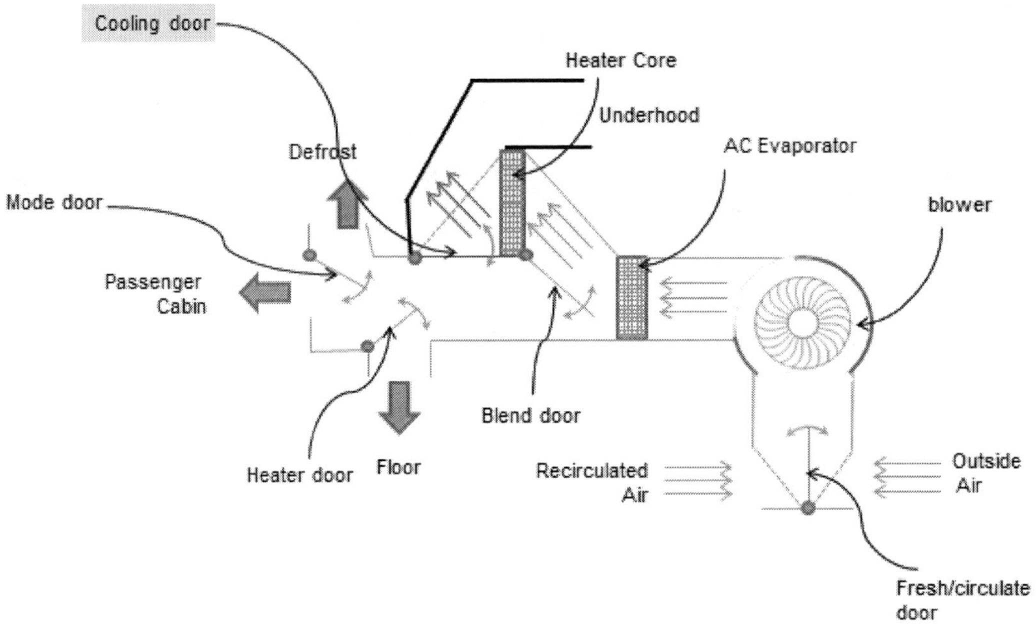

Figure 1 The HVAC airbox modification with an additional hot air vent door, "Cooling door"

During extreme operating conditions and when the driver has AC on, the blend door allows some air to go to the heater core while maintaining the cool airflow to the passenger cabin. Figure 2 shows positioning of the blend door and cooling door in such a scenario. This diverted airflow passes through the heater core to provide supplemental engine cooling. The cooling door opens, and the hot air from the heater core is vented back into the underhood. The blend door is positioned such that cabin comfort is not affected. The amount of airflow through the heater core will determine the amount of supplemental engine cooling that the heater core will provide. The blower is operated at a higher speed to deliver enough airflow to the passenger cabin and to the heater core. The cooling door position and the blower speed is controlled automatically by the ECU depending on the engine cooling and cabin comfort requirements.

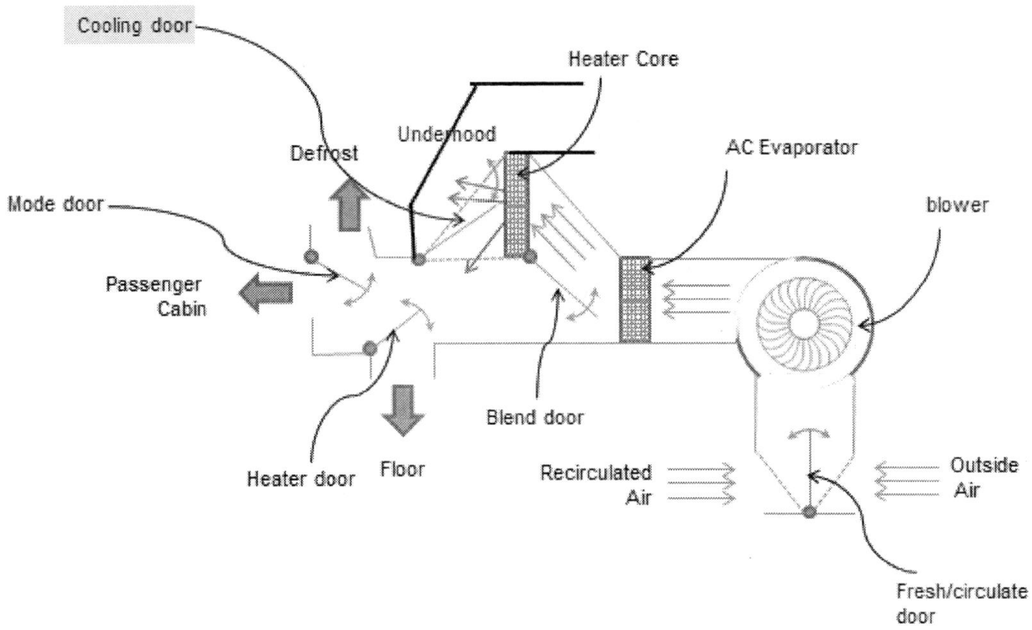

Figure 2 Operating mode with part of hot air being vented and part being directed into the passenger cabin

When passenger cabin heating is also desired in addition to supplemental engine cooling, the engine cooling door would be positioned such that some of the hot air from the heater core is sent into the passenger cabin and the rest is vented into the underhood providing both passenger cabin heating and supplemental engine cooling.

When supplemental engine cooling is not desired, the engine cooling door would remain closed, mimicking the HVAC air box operation of today (Figure 3).

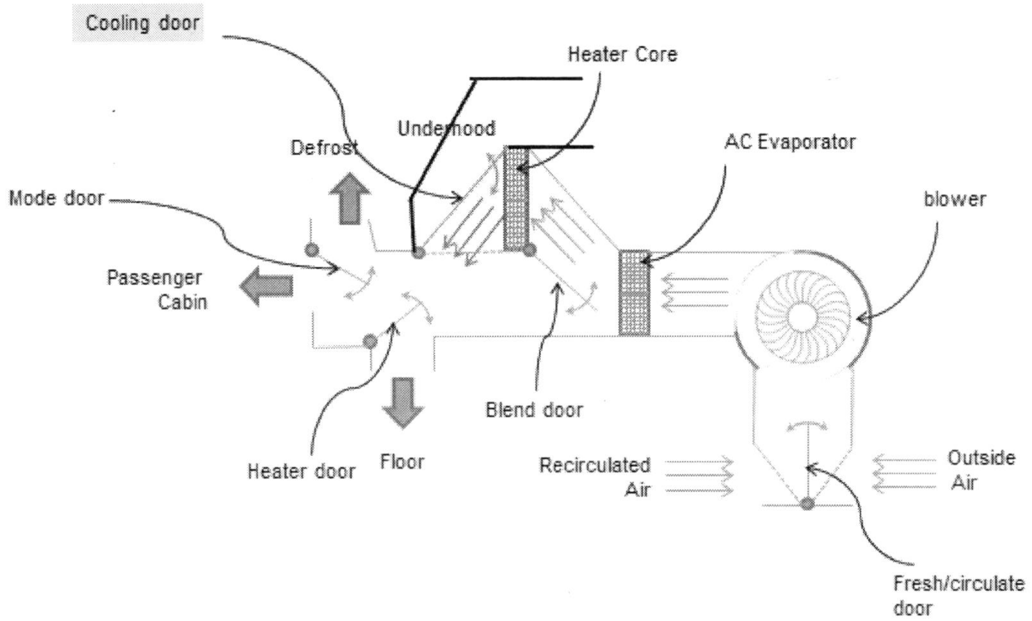

Figure 3 Operating mode with all hot air being directed into the passenger cabin

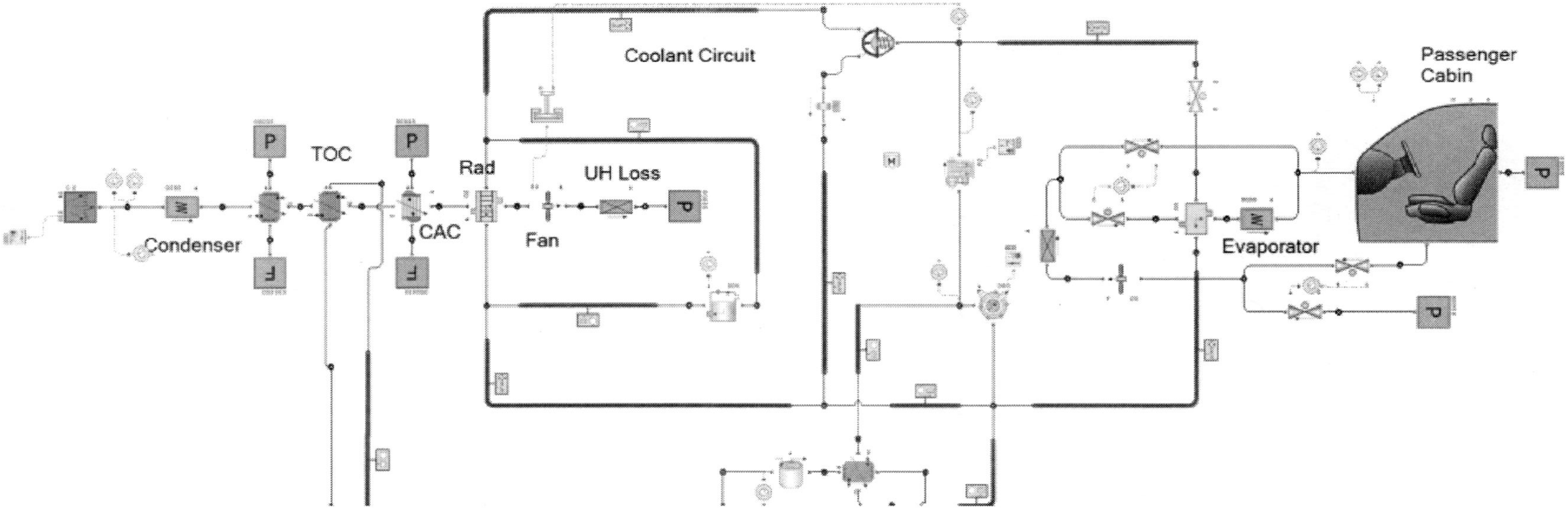

Figure 4 1D cooling system model to evaluate the DUHC concept

3 ANALYSIS

Before hardware testing, the efficacy of this design was computationally evaluated in Flowmaster using a system model capturing the trade-offs between the systems. The schematic of the powertrain cooling system developed for the analysis is shown in figure 4.

The coolant, engine oil and transmission oil flow sub-systems were modelled using standard representations of the passages, bends, and other restrictions. The thermal and pressure loss data for the major components such as pumps and heat exchangers represented that of sedan with a 2.2L naturally aspirated gasoline engine. The coolant sub-system consisted of thermostat valve opening system, cabin heater interaction system and pump which controls the coolant flow in system. Engine thermal sub-system consists of engine thermal inertia and combustion heat interaction between various engine parts. Engine oil system consist the heat interaction between various engine parts and engine oil. Transmission oil sub-system is a simplified circuit and consists of thermal interaction between various transmission parts and transmission oil. The interactions between the fluids and the engine structure were modelled using thermal bridges which implements equations for convection and conduction heat transfer based on the coefficients provided. The front-end airflow sub-system of the vehicle was modelled with heat exchangers ordered and stacked as in the vehicle, and the fan and stagnation pressure at the front the vehicle drove the airflow through the heat exchangers. The exact amount of airflow was determined by where the fan curve intersected the total airflow resistance through the cooling pack. The details of this type of 1D system model has been described in earlier work. [8]

The 1D cooling model was used to evaluate the DUHC concept at a few steady-state operating conditions first. The steady state condition was the thermal design point of the vehicle where the ambient temperature was 43C, and the vehicle was going up a grade with a maximum trailer weight of 1300kg. The airflow across the heater core was applied and the required front end airflow was calculated to maintain a fixed top tank coolant temperature of 110C. The summary of the results is shown in table 1 below. It was found that for every unit of airflow that was sent through the heater core, there was about 1.5X units of airflow reduction in required front-end airflow. This was because the heater core operates at a higher effectiveness than the radiator. As an example, when 0.071 m^3/s airflow was sent across the heater core, the front end airflow reduced from 1.18 m^3/s to 1.071 m^3/s. At this operating condition, the heater core has an effectiveness of 0.95 and the radiator had an effectiveness of 0.78. The scaling effect of airflow reduction would be a direct function of difference in effectiveness of the heater core and the radiator for a given operating condition.

Table 1 Summary of required airflow at front end with and without DUHC

		Without Dual Use Heater Core	With Dual Use Heater Core
Top Tank temperature	**deg C**	110	110
Airflow through Heater Core	**m^3/s**	0	0.071
Required front-end airflow	**m^3/s**	1.180	1.071

There is another variable to consider. When the AC is on, and additional air is sent through the evaporator, the heat load of the condenser goes up. This, in turn, puts additional heat load on the cooling pack. The effect of the condenser load was evaluated computationally. In one scenario, when additional 0.071 m^3/s airflow was sent through the evaporator, the condenser heat load went up by 40%. When additional 0.094 m^3/s was sent through the evaporator, the condenser heat load went up by 80%. However, when complete system simulation was performed to account for this added condenser heat load, the benefit was still net positive. For every unit of airflow sent through the heater core, there was a 1.2X units of airflow reduction in required front-end airflow.

The key benefit of this technology is an optimized front-end cooling module for everyday driving leading to better fuel economy while still having the heat rejection capacity for the worst case scenario. The size of the front-end cooling module and the size of the cooling fan could be reduced. If the front-end cooling module size and architecture were fixed for, let's say, legacy reasons, then adding this technology would give the vehicle a higher trailer-tow rating as the engine cooling capability has been increased. In addition, this technology enhances the operation and benefits of other fuel-saving technologies such as AGS by enabling it to remain closed for a longer period of time. AGS are shutters in front of the cooling pack that partially block the airflow. The primary purpose of the active grill shutters is to reduce the drag when the full cooling capacity of the cooling is not desired and therefore improve fuel economy. The impact on AGS operation is important as a majority of OEMs are evaluating AGS for future vehicles, not just in the luxury segments but also in the lower cost segments. [1,2,3,4,5,6]. The EPA expects that the aerodynamic improvement with AGS will be significant. The EPA states on their website that "EPA will provide a credit for active aerodynamic technologies that can demonstrate a reduction in aerodynamic drag of 3% or more. The credit will be 0.6 g/mile for cars and 1.0 g/mile for trucks when the reduction in aerodynamic drag is around 3%" [7].

3.1 Control Strategy Evaluation

Since the DUHC will be operated directly by the ECU, set-points need to be determined to evaluate the optimum operating zone. Figure 5 below shows a typical range of operation and the set points for each of these technology controls being studied vis-a-via, the Thermostat, DUHC and AGS. The typical order for set-points is the Thermostat, DUHC and the AGS in that order of increasing coolant or engine oil temperature. The thermostat has the least impact on the fuel economy where as the activation of the AGS has the greatest impact on fuel economy of the three. The drive cycle considered in this analysis was non-standard road load drive cycle with increasing speeds from 80 kph to 120 kph with ambient temperature of 43C.

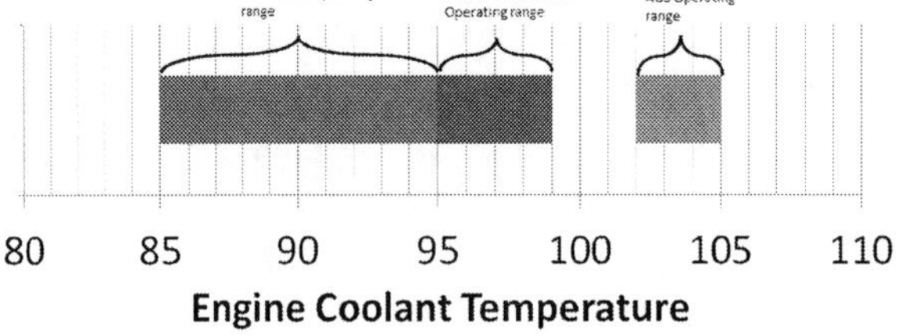

Figure 5 Typical range of operation for thermostat, dual-use heater core and AGS

3.2 Effect of set-point overlap

The first analysis was performed was comparing a scenario where the set-points of the thermostat, DUHC and AGS overlapped to a scenario where the operating ranges were completely separated. In the baseline scenario where the set-points are overlapped, the thermostat is set to open at 89C and be fully open at 110C, The DUHC is set to start operating when the coolant temperature reaches 97C and shut-down when the coolant temperature drops below 89C. The AGS is set to open when the coolant temperature reaches 105C and shut off when the coolant temperature reaches 102C. The overlap of the set points in the baseline scenario is shown in the figure 6 below. This baseline scenario with overlapping set-points is compared with an alternate scenario (case 1) where there is no overlap of the set points.

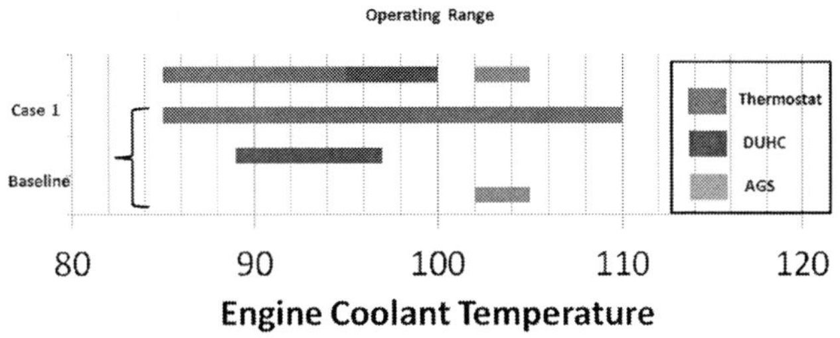

Figure 6 Set points for the Thermostat, DUHC and the AGS

One would expect that when there is overlap of the set points and the thermostat does not fully open until much after the DUHC and AGS are active, then the system is not taking full advantage of all the available coolant flow, and therefore would lead to more frequent use of DUHC and AGS. Computational results indicate that this is only partly true.

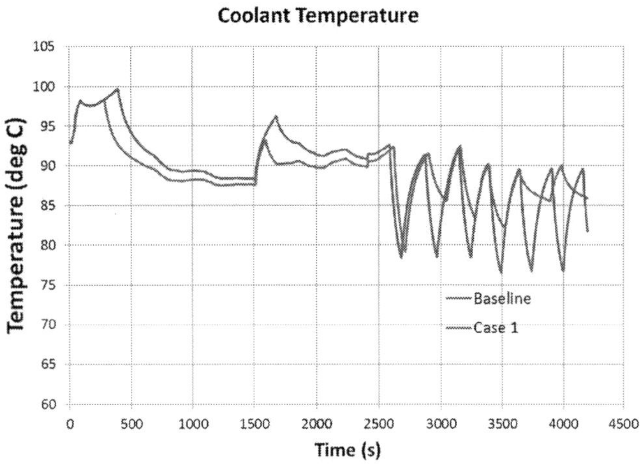

Figure 7 Coolant temperature comparison between Baseline and Case 1 runs

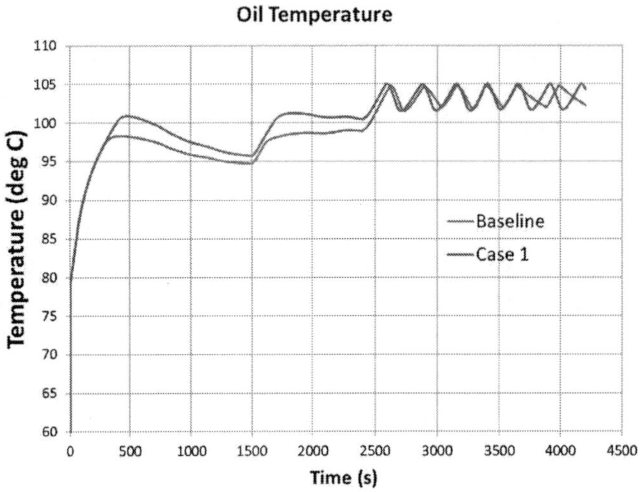

Figure 8 Engine oil temperature comparison between Baseline and Case 1 runs

Figure 7 and 8 show that, for the baseline case, the coolant temperature and the oil temperature in the initial part of the drive cycle (at lower speeds) is maintained at the lower temperature. This is because the DUHC is triggered at lower temperature in the baseline case. As also expected, the thermostat for the Case 1 is much more open leading to more coolant flow through the radiator and the heater core (see figure 9).

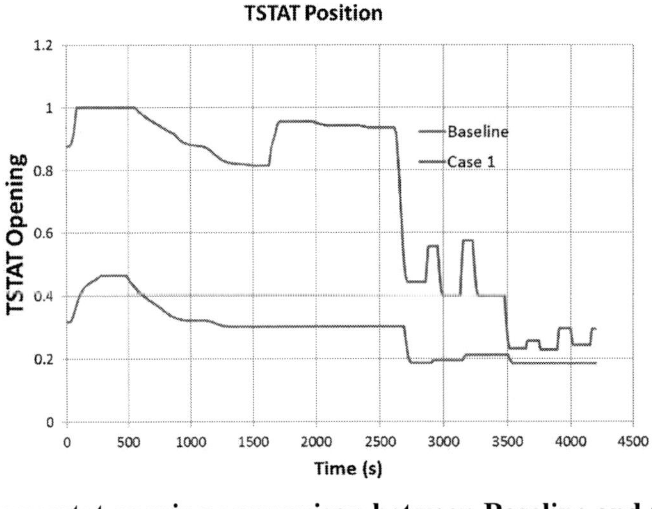

Figure 9 Thermostat opening comparison between Baseline and Case 1 runs

One would assume that the effectiveness of the DUHC will be diminished when the thermostat is open less, but as one can observe in figures 10 and 11 that the radiator heat duty and heater core heat duty in the early part of the drive cycle are nearly identical. This is because, at the operating conditions of 50 and 60 mph, the heater core and the radiator are airflow limited and additional coolant flow does not make a significant difference.

For the later part of the drive cycle (when the vehicle is moving at 70 and 80 mph), it is observed that when significantly higher airflow is available during the AGS open conditions, there is a notable difference in radiator heat duty, coolant and engine oil temperatures.

The take away from this analysis is that at lower road load speeds such as 40 and 50 mph, the heat exchangers are airflow limited and therefore a loss in coolant flow due to a partially opened thermostat does not have a significant impact on the performance of the technologies such as DUHC and AGS. At higher speeds such as 70mph and 80mph, one would want as much coolant flow available as possible to maximize the cooling effect of the available airflow. Therefore, it is desirable to have a fully open thermostat before other technologies come into effect.

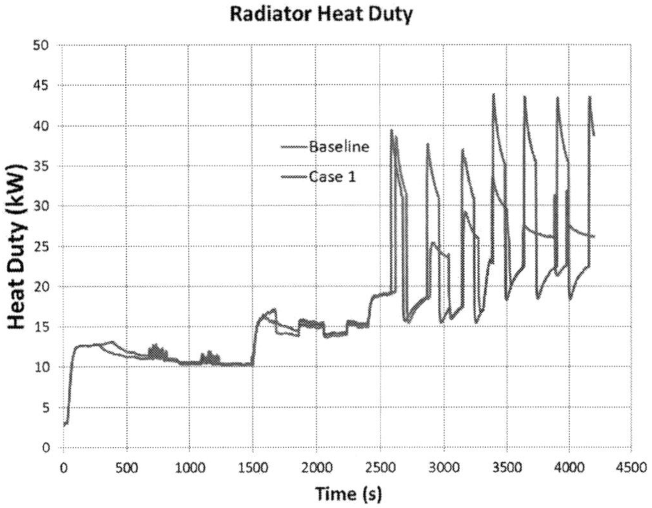

Figure 10 Radiator heat duty comparison between Baseline and Case 1 runs

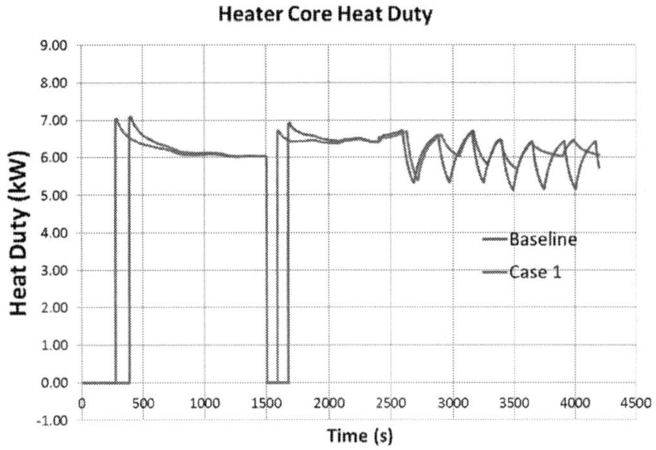

Figure 11 Heater core heat duty comparison between Baseline and Case 1 runs

3.3 Effect of DUHC on AGS closure

The second part of the analysis was to evaluate the effect of the DUHC on the AGS closure rate. As shown in case 2, the DUHC set-points were set below the AGS to enable the DUHC to provide cooling before the AGS opens. This is because the fuel economy penalty to run the blower is lower than that of opening the AGS.

Figures 12 and 13 below show the comparison of the engine oil temperature and the AGS positions. It can be noted that when the DUHC is activated, the oil temperature is stable at lower speeds (50 and 60 mph) as the DUHC provides enough cooling to maintain the oil temperature at a steady level. The AGS remains closed during these speeds. At higher speeds, such as 70 and 80 mph, the AGS does open, but with a lesser frequency than when there was no DUHC in operation.

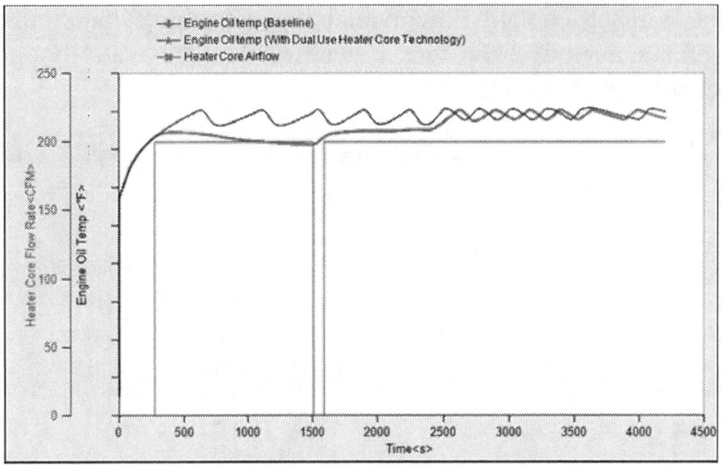

Figure 12 Effect of DUHC- Comparison of Engine oil temperatures

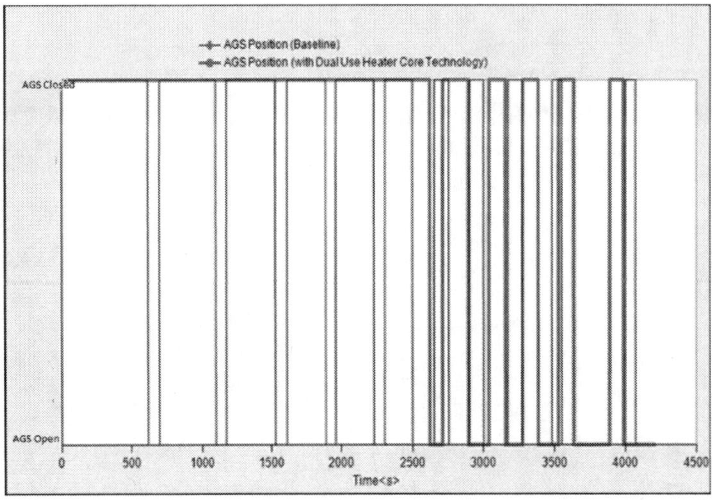

Figure 13 Effect of DUHC, comparison of AGS closure and cycling

The main takeaway from this is that DUHC technology helps maintain the AGS closed for longer compared to a vehicle without a DUHC technology. In addition, the precise control of additional cooling that the DUHC offers a much finer control of the fluid temperatures.

3.4 Effect of gap between set-points

The third part of the analysis investigates the effect of the temperature gap between the set points of the thermostat, DUHC and the AGS. Figure 14 below shows the set points of the three cases evaluated.

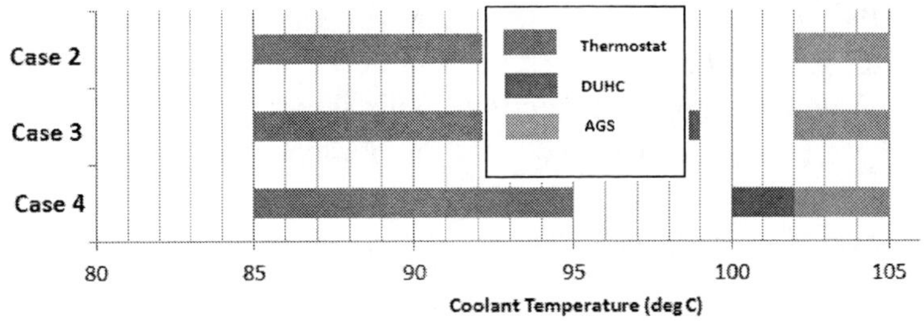

Figure 14 Set points for the Thermostat, DUHC and the AGS to compare the gap between set points

The set points of the thermostat and the AGS are kept constant through the three cases, and the open/close set-points of the DUHC are varied such that there is 2 deg C, 4 deg C and 7 deg C gaps between the TSTAT fully open temperature and when the DUHC initializes. Figures 15 and 16 below show the oil temperature, DUHC operation and AGS operation for these three scenarios.

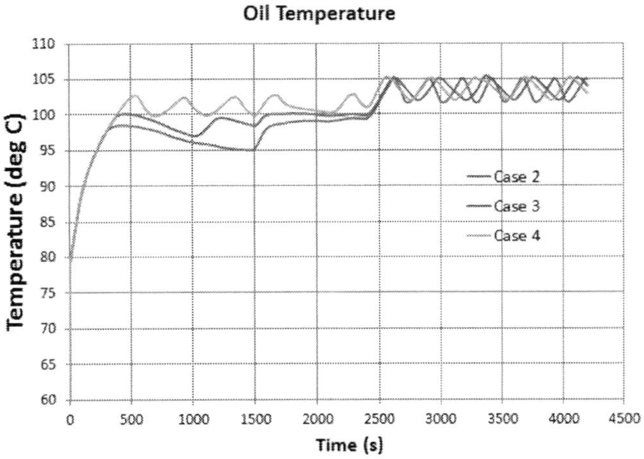

Figure 15 Comparison of engine oil temperature for changing set point gaps

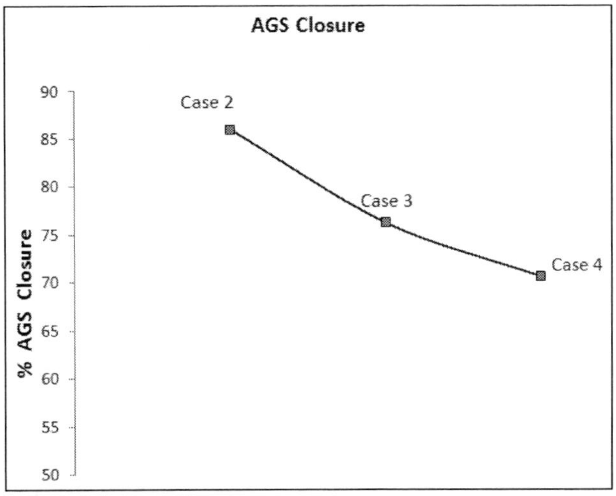

Figure 16 Change in AGS closure rate for changing set point gaps

As noted from the cycling of the oil temperatures and AGS closure rate, case 2, for which the DUHC is activated the earliest has the highest AGS closure rate which is desirable from an aerodynamic perspective. This is because, the earlier the DUHC is turned on, the more heat it can remove from the coolant (and therefore the oil), leading to a lower AGS open rate. The take away from this analysis, is that it is desirable to have DUHC activated as early as possible, to enable it to have a maximum impact on the heat removal.

3.5 Increase in AGS activation time

The final part of this analysis evaluates the effect of increasing the AGS activation time keeping the thermostat and DUHC activation periods and set points the same. Figure 17 below show the set point for the three cases evaluated.

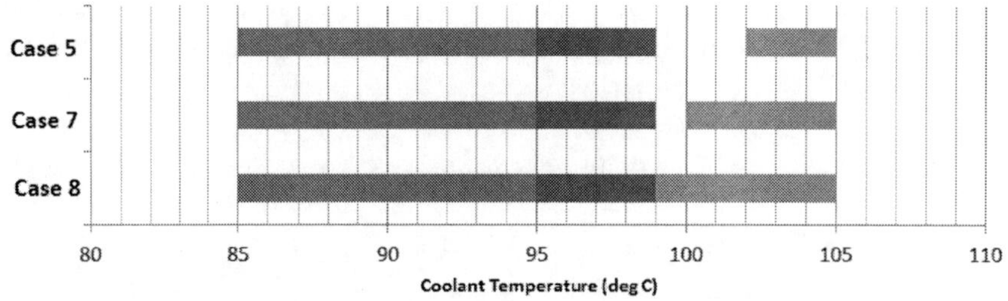

Figure 17 Set points for the Thermostat, DUHC and the AGS to evaluate increase in AGS activation time

The figures 18 and 19 show the engine oil temperature and the percentage of AGS closure for three cases with increasing AGS closure windows. It can be noted from the analysis data that increasing the AGS closure rate by advancing the AGS shut off temperature increases the magnitude of fluctuations in temperatures. As expected, advancing the AGS shut off temperature decreases AGS cycling but at the expense of increasing the AGS open time.

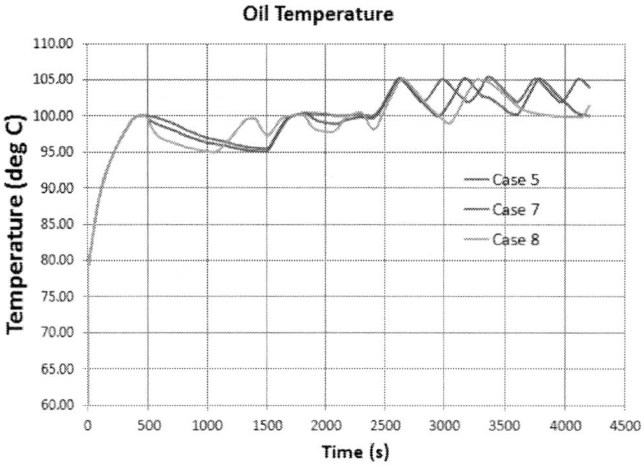

Figure 18 Comparison of engine oil temperature with increasing AGS activation time

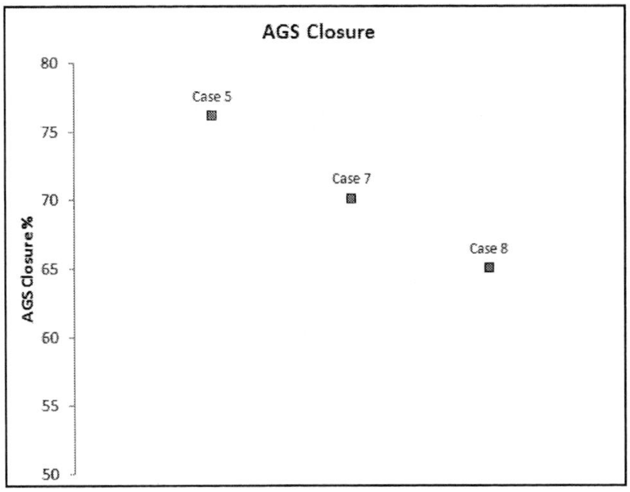

Figure 19 Change in AGS closure rate for increasing AGS activation time

4 FUTURE WORK

As the development of this concept progresses with a partner OEM, various trade-off's will be studied to assess all potential impacts. On the NVH side, the noise in the cabin will be measured to assess the impact of having a higher capacity blower. On the climate control and energy management side, the reduction in the potential benefit because of the increased power consumption of the blower will be studied. Currently, the design of the vent paths are being investigated to leverage the aerodynamics of the specific vehicle to the drive more airflow across the heater core without additional blower power.

5 SUMMARY/CONCLUSIONS

The paper shows the benefits of the DUHC technology and the effects of its set-points in relation to the thermostat and the AGS. Specifically the paper concludes that

a) The DUHC technology provides supplemental cooling to the engine and enables the AGS to remain closed for longer periods during road load driving conditions. The DUHC also increases the trailer tow rating of the vehicle by providing additional engine cooling.

b) For partial driving conditions, the overlap of the thermostat set point with the DUHC and the AGS have minimal impact on their efficacy as the system is airflow limited and therefore not sensitive to small changes in coolant flow rate.

c) Increasing the AGS duration by moving the AGS shut of temperature to an earlier temperature has undesirable effects both from a fluid temperature and AGS cycling perspective – specifically advancing the AGS shut off temperature decreases AGS cycling but at the expense of increasing the AGS open time and larger fluctuations in fluid temperature.

6 REFERENCES

1. El-Sharkawy, A., Kamrad, J., Lounsberry, T., Baker, G. et al., "Evaluation of Impact of Active Grille Shutter on Vehicle Thermal Management," SAE Int. J. Mater. Manuf. 4(1):1244-1254, 2011, doi:10.4271/2011-01-1172.
2. The New Mercedes-Benz E-Class Blue EFFICIENCY" Press Release, March 2009, http://www.emercedesbenz.com/Mar09/11_001616_The_2010Mercedes_Benz_E_Class_BlueEFFICIENCY_Technology.html
3. Air Vent Control on multiple BMW Efficient Dynamics Vehicles, http://www.bmw.com/com/en/newvehicles/7series/sedan/2008/allfacts/effdyn/air_vent_control.html
4. Next-Gen Ford Focus unveiled! U.S. and Euro models finally united", Sam Abuelsamid, www.autoblog.com, January 11, 2010 (info from Press Release)
5. "New York debuts for Sonata Hybrid and Turbo", www.sae.org, April 8, 2010 (info from Press Release)
6. "Chevy Cruze Eco 'feels' wind change to better fuel economy", Suzanne Ashe, www.cnet.com, August 12, 2010 (info from Press Release)
7. http://www.epa.gov/otaq/climate/documents/420d11004.pdf
8. Scott, T., Chang, F., Khandaker, M., and Uppuluri, S., "Transient Thermal Modeling of Power Train Components," SAE Int. J. Passeng. Cars - Mech. Syst. 5(2):962-973, 2012, doi:10.4271/2012-01-0956.
9. Flowmaster v 7.8.4 Reference help

Heat & AC

Simulative comparison of conventional and secondary loop automotive refrigeration systems

J C MENKEN, J E KOERNER, T A WEUSTENFELD and **K STRASSER**
AUDI AG, Germany

J KOEHLER
University of Braunschweig, Germany

Recent attempts to find energy-efficient thermal management systems for electric and plug-in hybrid electric vehicles have led to secondary loop systems. As an alternative approach to meet dynamic heating and cooling demands, these systems reduce the refrigerant charge. Besides advantages such as hermetic encapsulation and a compact, modular design of the refrigeration cycle, secondary loop systems contain an additional layer of thermal resistance. This might influence the system's transient behavior as well as passenger compartment comfort during cool-down or heat-up. Instead of direct refrigerant-to-air heat exchangers, these novel systems are equipped with refrigerant-to-water plate heat exchangers used as indirect condensers and evaporators. The conventional condenser and evaporator in the vehicle front end and in the air handling unit are replaced by water-to-air heat exchangers. This work presents Modelica-based models to examine the transient behaviour of four possible R134a system setups: direct condenser + direct evaporator (conventional refrigeration cycle), direct condenser + indirect evaporator, indirect condenser + direct evaporator, indirect condenser + indirect evaporator (fully integrated secondary loop system). Under defined use cases and environmental conditions, simulations are conducted to compare these four system designs while taking into account the possibility of using the secondary loop system in an efficient heat pump mode for cabin heating.

NOTATION

P	Power	EVAP	Evaporator
E	Energy	COND	Condenser
T	Temperature	ICE	Internal Combustion Engine
$\dot{Q}$	Heat Flow Rate	CHX	Cabin Heat Exchanger
φ	Relative Humidity	pwel	Power Electronics
$\dot{q}_{sol}$	Solar Radiation	emch	Electric Machine
Pr	Prandtl Number	col	coolant
Re	Reynolds Number	ref	refrigerant
Nu	Nusselt Number	amb	ambient
ρ	Density	dEdC	Direct Evaporation/Direct Condensation
p	Pressure	dEiC	Direct Evaporation/Indirect Condensation
d_h	Hydraulic Diameter	iEdC	Indirect Evaporation/Direct Condensation
v	Velocity	iEiC	Indirect Evaporation/Indirect Condensation
t	Time		

INTRODUCTION

For over a century vehicles have been powered by an internal combustion engine (ICE). Growing public awareness of resource shortages and climate change have driven efficiency increases and the development of new propulsion concepts. Customer demands for increasingly efficient vehicles and a comfortable passenger compartment environment create new challenges in heat and energy distribution within the vehicle. Additionally, the move towards greater electrification of the vehicle through hybrid electric vehicles (HEV), plug-in hybrid vehicles (PHEV) and battery electric vehicles (BEV) leads to new thermal management requirements for new components such as energy storage, electric machines and power electronics. These components experience significant transient thermal loading; therefore, much has been invested in order to ensure appropriate and efficient operating conditions [1]. As the number of components requiring active thermal management increases, so do weight, size, cost and system complexity. One method for meeting these thermal management demands while reducing the cost of electric drive systems is through an integrated thermal management system. In this system, the various components (e.g. power elctronic, electric motor, energy storage etc.) are connected via secondary coolant loops to a compact refrigeration cycle as well as to coolant-to-air heat exchangers in the vehicle front end and passenger compartment.

Various applications of secondary loop refrigeration systems can be found in the literature. Wang et al. [2] offer a vaste review on secondary loop refrigeration systems including performance and risk assessment of flammable refrigerants. Applications in the fields of commercial refrigeration, residential air conditioning/heat pumping and mobile air conditioning are presented. The focus in the following will be on vehicle thermal management systems. During the past few years, research in vehicle thermal management via secondary loops has significantly increased. Ghodbane et al. [3] describe a secondary loop system with the HFC refrigerant R152a and demonstrate equal to less cooling performance at a comparable energy consumption compared to the baseline R134a-system. Kowsky et al. [4] present a Central Thermal Management Unit called HPAC (Heat Pump Air Conditioning) for the use in electric and hybrid cars using the battery waste heat as the energy source during the heat pump mode. Downsizing the compact refrigerant cycle to approximately 450mm x 275mm x 225mm reduces the amount of refrigerant and offers packaging flexibility. It is stated that a reduction of refrigerant emissions by at least 50% and an improved electric drive range can be reached. In this presented design, the charge of the tested refrigerant R134a can be reduced by 24% compared to conventional mobile air conditioning systems. The occurring effects in such systems can even be increased by the addition of phase change materials in the secondary loops [5] or using an ice storage [6]. This can be beneficial in order to meet CO_2 emission standards for passenger cars.

This work focuses on the comparison of conventional and secondary loop coolant systems using a simple automotive refrigerant loop containing a compressor, a condenser (refrigerant to air or refrigerant to coolant), an expansion valve, an evaporator (refrigerant to air or refrigerant to coolant) and a refrigerant reservoir (accumulator).

METHODS

Boundary Conditions and Operation Modes

As the requirements of the vehicle thermal management system change over the year depending on the driving cycles and environmental conditions, system evaluation is not a trivial task. In order to use representative data describing operating conditions of the evaluated system, the boundary conditions have to be specified. One system configuration in this work uses the waste heat of electric drivetrain components (battery, electric machine, power electronics) in an integrated secondary loop system. Because of the transient thermal behaviour of the electric drivetrain components and the velocity dependency of the air mass flow rate through the vehicle front end, real world driving cycles must be considered. Additionally, the environmental conditions influence the vehicle's transient behaviour. Therefore, the ambient temperature $T_{air,amb}$, relative humidity $\varphi_{air,amb}$ and solar radiation $\dot{q}_{sol}$ need to be specified.

To describe driving behaviour, the three Artemis driving cycles (urban, road and motorway limited to v=130km/h) are used. Figure 1 shows the velocity profiles over time [7] as well as the calculated drivetrain losses [8]. Taking into account the occurring distribution of urban, sub-urban and motorway profiles, the assumed weighting factors of the three Artemis cycles are shown in Table 1. These factors are based on the general distribution of traveled distances in Germany (0.33 urban, 0.36 road and 0.31 motorway) according to Rumbolz et al. [9] combined with the respective durations of the three given Artemis driving cycles.

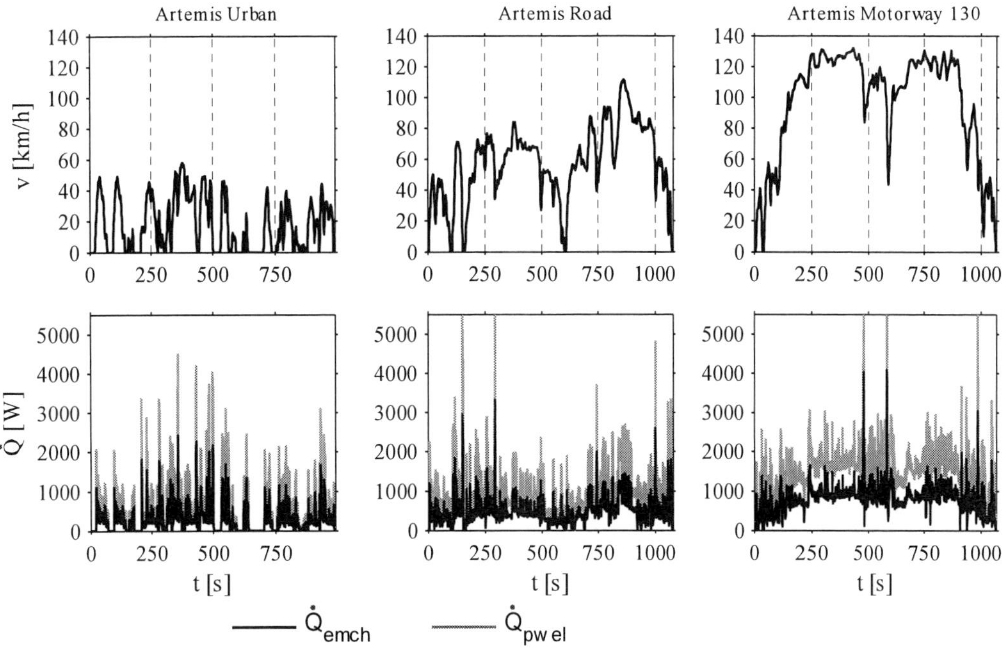

Figure 1 – Velocity profiles of evaluated driving cycles Artemis urban, road and motorway [7]. Drivetrain heat losses of electrical machine and power electronics [8].

Table 1 – Weighting factors for driving cycles Artemis urban, road and motorway [9].

Driving cycle	Weighting factor
Artemis Urban	0.68
Artemis Road	0.21
Artemis Motorway 130	0.11

Regarding the ambient conditions, Table 2 displays the values for $T_{air,amb}$, $\varphi_{air,amb}$ and $\dot{q}_{sol}$ including weighting factors [8] used in this work.

Table 2 – Ambient conditions based on European weather data for system evaluation used in this work [8].

Use case	Weighting factor	$T_{air,amb}$ [°C]	$\varphi_{air,amb}$ [%]	$\dot{q}_{sol}$ [W/m²]
Heat-up (winter)	0.44	5	75	170
Re-heat (autumn)	0.45	15	64	318
Pull-down (summer)	0.12	30	42	538

For the evaluation of the system configurations, a battery powered electric vehicle is assumed. Because of the lack of an internal combustion engine, the missing heat load during winter and reheat operation mode is either provided by an electrical heater within the supply flow of a coolant-to-air cabin heat exchanger or by a primary refrigerant cycle operated in a heat pump mode. In the following, the operation modes for cabin heat-up, re-heat and cool-down are listed and described in detail. The re-heat operation mode is distinguished by dehumidification of the instream air and subsequent re-heating before entering the passenger compartment.

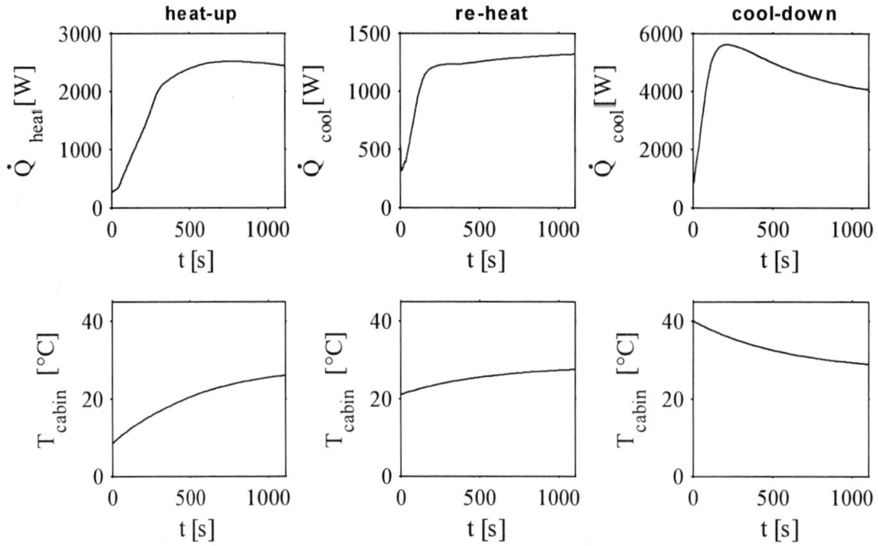

Figure 2 – Air side cooling or heating load at water-to-air or refrigerant-to-air heat exchanger within the air handling unit and respective cabin reference temperatures for heat-up, re-heat and pull-down based on [8].

Based on experiments and preliminary simulation, the cooling or heating load at the refrigerant-to-air or coolant-to-air heat exchangers is predefined as shown in Figure 2. $\dot{Q}_{heat}$ indicates a heat-up of the instream air, whereas $\dot{Q}_{cool}$ indicates the air side cooling load. It should be noted that during re-heat operation, only the cooling load at the first heat exchanger is displayed. It is assumed that the heating load at the second cabin heat exchanger is either supplied by an electric heater or it is assumed that the heat load from the indirect condenser is sufficient. Based on simulational and experimental results, this can be a justified assumption.

System Setup

In order to compare the performance of conventional direct and secondary loop refrigeration systems, four system configurations are examined. Figure 3 shows these different setups containing direct or indirect condensers and evaporators.

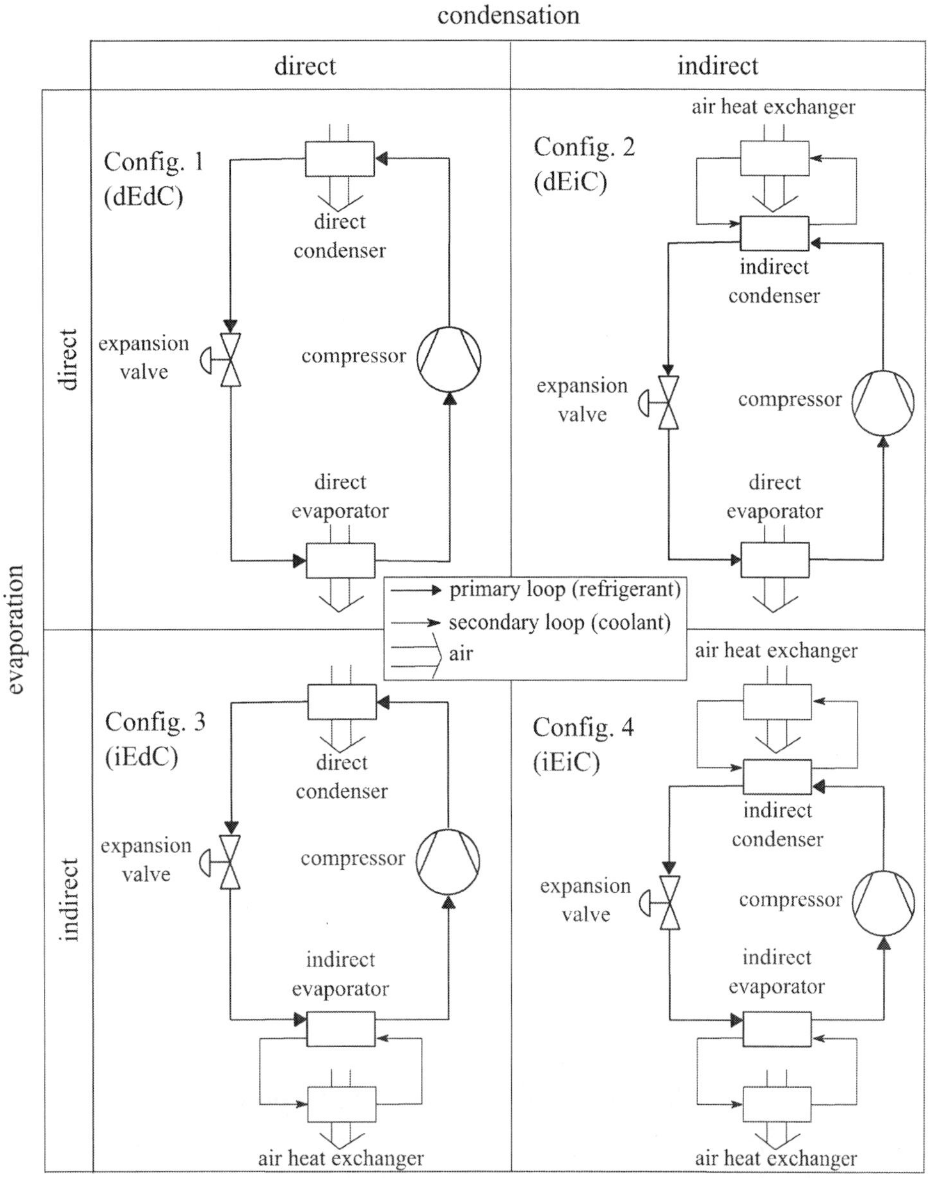

Figure 3 – Overview on examined system configurations.

Configuration 1 (dEdC) represents a conventional system with a refrigerant-to-air evaporator in the air handling unit and a refrigerant-to-air condenser in the front end. Implemented in a vehicle, this configuration would contain the highest refrigerant charge and highest refrigerant side heat losses and pressure drops due to the long hoses and tubes connecting the cycle components within the engine compartment. However, the transient behaviour during cabin cool-down is benchmark due to no additional layer of heat transfer. However, during heat-up and re-heat operation, the entire required heat load must be provided by an electric water heater. Figure 4 shows the simulation setup regarding the air handling unit for all use cases of this configuration with the cabin heat exchanger (CHX) and/or the evaporator (EVAP).

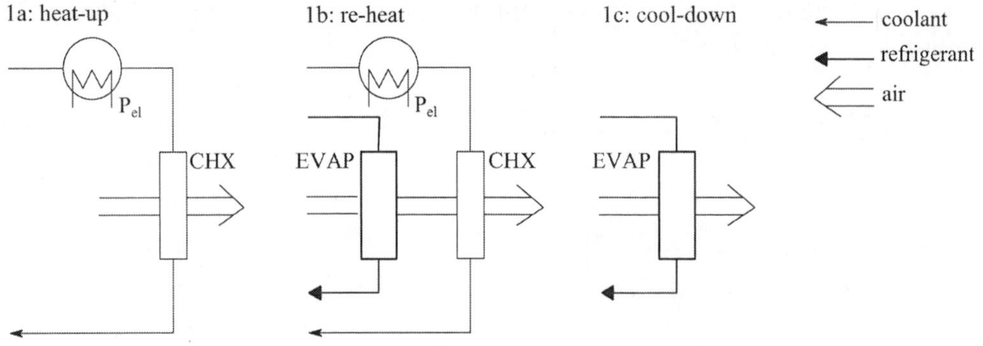

Figure 4 – Air handling unit setup for configuration 1 (dEdC) during heat-up, re-heat and cool-down operation mode.

Configuration 2 (dEiC) consists of a refrigeration cycle with a secondary loop on the high pressure side. With this configuration it is possible to operate the refrigeration cycle in a heat pump mode during re-heat, meaning that the air is cooled down and dehumidified at the direct evaporator (EVAP), and then re-heated with the help of the condensation heat by the cabin heat exchanger (CHX). At first, electric drivetrain components can generally not be used as a heat source at low ambient temperatures (i.e. if the vehicle has been standing overnight.). Figure 5 shows the simulation setup regarding the air handling unit for all use cases of configuration 2 (dEiC).

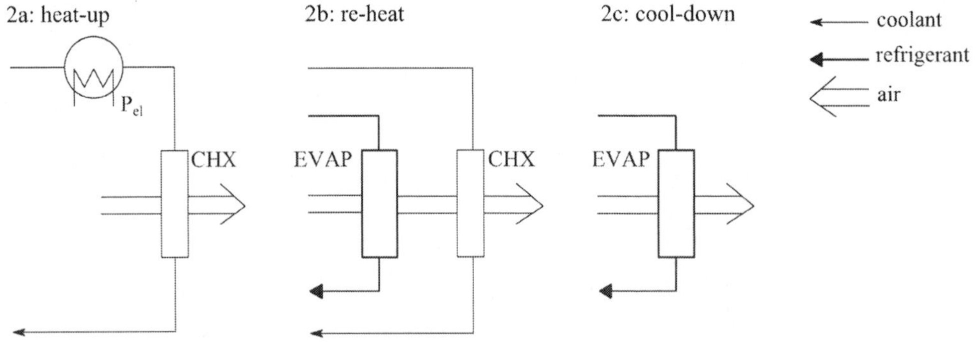

Figure 5 – Air handling unit setup for configuration 2 (dEiC) during heat-up, re-heat and cool-down operation mode.

Configuration 3 (iEdC) consists of a refrigeration cycle with a secondary loop on the low pressure side. With this configuration it is not possible to use the condensation heat during heat-up or re-heat. To cool down the passenger compartment in summer, a cabin heat exchanger is supplied with cold water from the indirect evaporator. Figure 6 shows the simulation setup regarding the air handling unit for all use cases of this configuration.

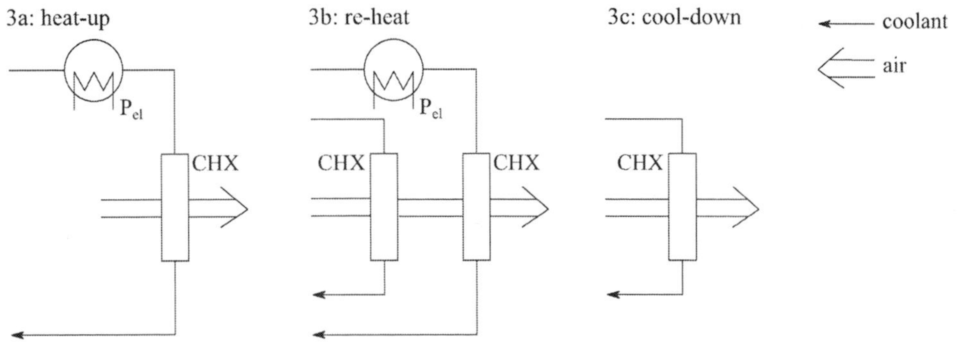

Figure 6 – Air handling unit setup for configuration 3 (iEdC) during heat-up, re-heat and cool-down operation mode.

Configuration 4 (iEiC) consists of a refrigeration cycle with a secondary loop on the high and low pressure side (fully integrated secondary loop system). Due to the fact that all control and energy flow distribution happens on the coolant side, a range of new options like a heat pump mode and the incorporation of electric drivetrain components to the thermal management system emerge. Additionally, the ambient air can be used as heat source during this heat pump mode. Figure 7 shows the simulation setup regarding the air handling unit for all use cases of configuration 4 (iEiC).

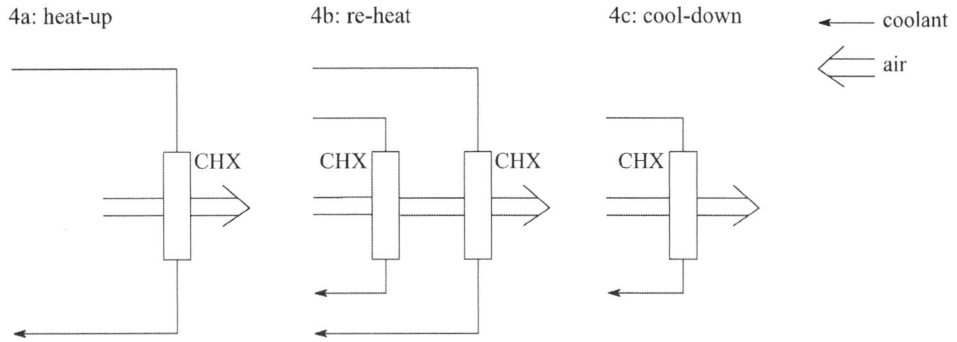

Figure 7 – Air handling unit setup for configuration 4 (iEiC) during heat-up, re-heat and cool-down operation mode.

Based on experimental results and to ensure a high performance of the indirect condenser, all configurations using a secondary coolant loop have a common coolant mass flow rate of $\dot{m}_{col} = 0.2\frac{kg}{s}$, except during heat up on the high pressure side. To force the pressure to rise on the high pressure side, the coolant mass flow rate is set to $\dot{m}_{col} = 0.03\frac{kg}{s}$. It would also be possible to use the temperature air flap instead of reducing the coolant mass flow rate.

Modeling and Simulation

All system setups presented in this work are modeled with the TIL Library [10]. They consist of a 34cm³ electrical scroll compressor, a condenser, an electrical expansion valve, an evaporator and a refrigerant reservoir on the low pressure side (accumulator). The systems containing a secondary loop either on the condensing or evaporating side are equipped with coolant-to-air heat exchangers within the air handling unit or the vehicle front end, respectively. Figure 8 exemplarily shows the model of the refrigerant cycle of configuration 4 (iEiC).

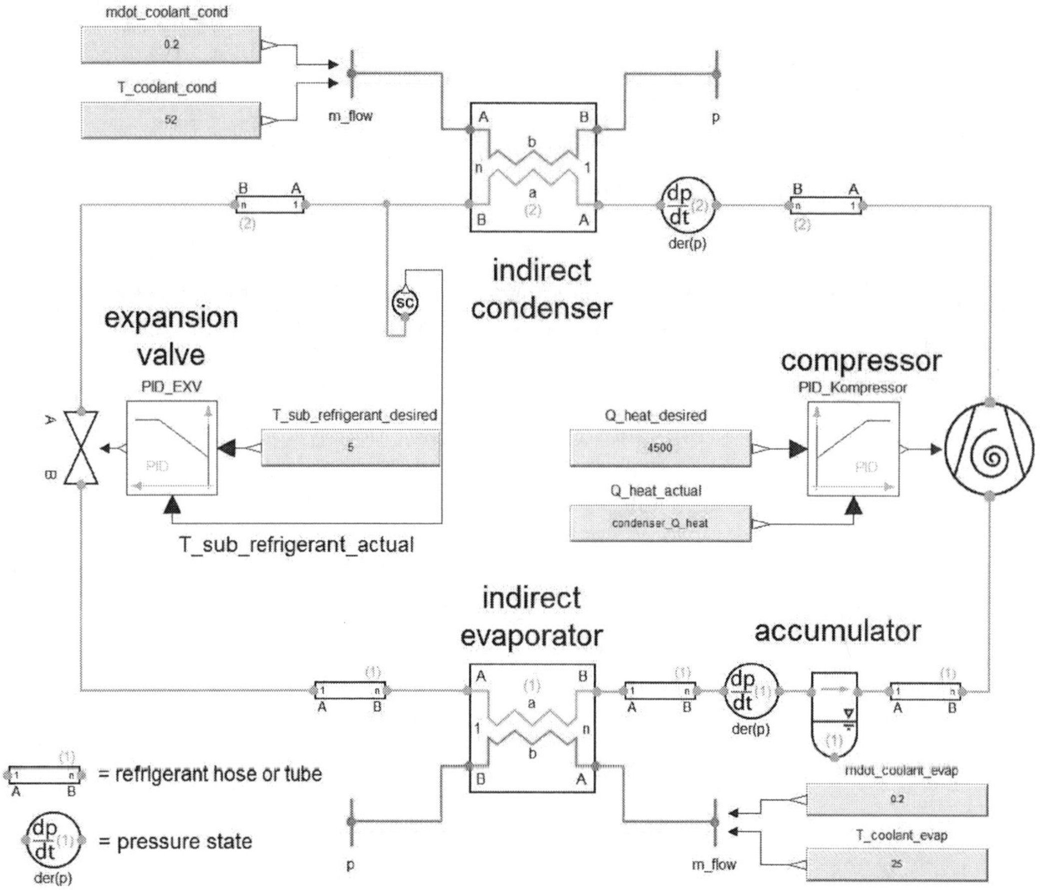

Figure 8 – Refrigerant cycle modeled with the TIL library. Representation of configuration 4 (iEiC).

The compressor is described with a loss-based model according to Schedel et al. [11]. For heat transfer within the evaporator and condenser on the refrigerant side, a plate heat exchanger model is used. The area expansion factor of the plates is obtained by the respective chevron-plate-pattern [12]. Using empirical correction factors of the Dittius-Boelter-Correlation according to Akers/Deans/Crosser [13], the indirect condenser's heat transfer coefficient for Re>50,000 can be calculated as follows (C=5.03, k=0.33):

$$\mathrm{Nu} = \mathrm{C} \cdot \mathrm{Re}^{\mathrm{k}} \cdot \mathrm{Pr}^{1/3}$$

The electric expansion valve is used to control the subcooling of the refrigerant at the condenser outlet. It is modeled with a variable valve opening area according to the Bernoulli-equation:

$$\dot{m}_{ref} = A_{eff} \cdot \sqrt{2 \cdot \rho_{ref,in} \cdot (p_{ref,in} - p_{ref,out})}$$

The refrigerant side heat transfer during evaporation is described with the Lévèque [12] correlation. The Nusselt number is defined as follows:

$$Nu = 1{,}615 \cdot [(\xi \cdot \frac{Re}{64}) \cdot Re \cdot Pr \cdot d_h/l]^{1/3}$$

Consequently, the heat transfer coefficient can then be obtained:

$$\alpha = Nu \cdot \lambda/d_h$$

Due to the design of the accumulator and the oil bleed hole at the bottom of the inner J-tube, the vapour quality of the refrigerant at the accumulator outlet is unsaturated gas [14]. Therefore, a constant outlet vapour quality of $x_{ref,accu,out} = 0.99$ is assumed.

The geometry of the refrigerant to air condenser and evaporator are based on typical automotive components from a compact vehicle. The heat transfer of these components is derived from experimental data.

In order to guarantee comparable conditions, all heat exchangers are modeled with typical pressure loss based on measurements.

All dynamic calculations of the physical properties of the components as well as coolant and refrigerant within the heat exchangers and connections elements are cell-based. This also means that the refrigerant charge has to be chosen appropriately.

A test stand of this primary cycle and the appropriate model validation are presented in previous publications [15], [16].

RESULTS

Figure 9 shows the results of the overall energy consumption for each transient simulation for cabin heat-up, re-heat and cool-down during the respective Artemis driving cycle.

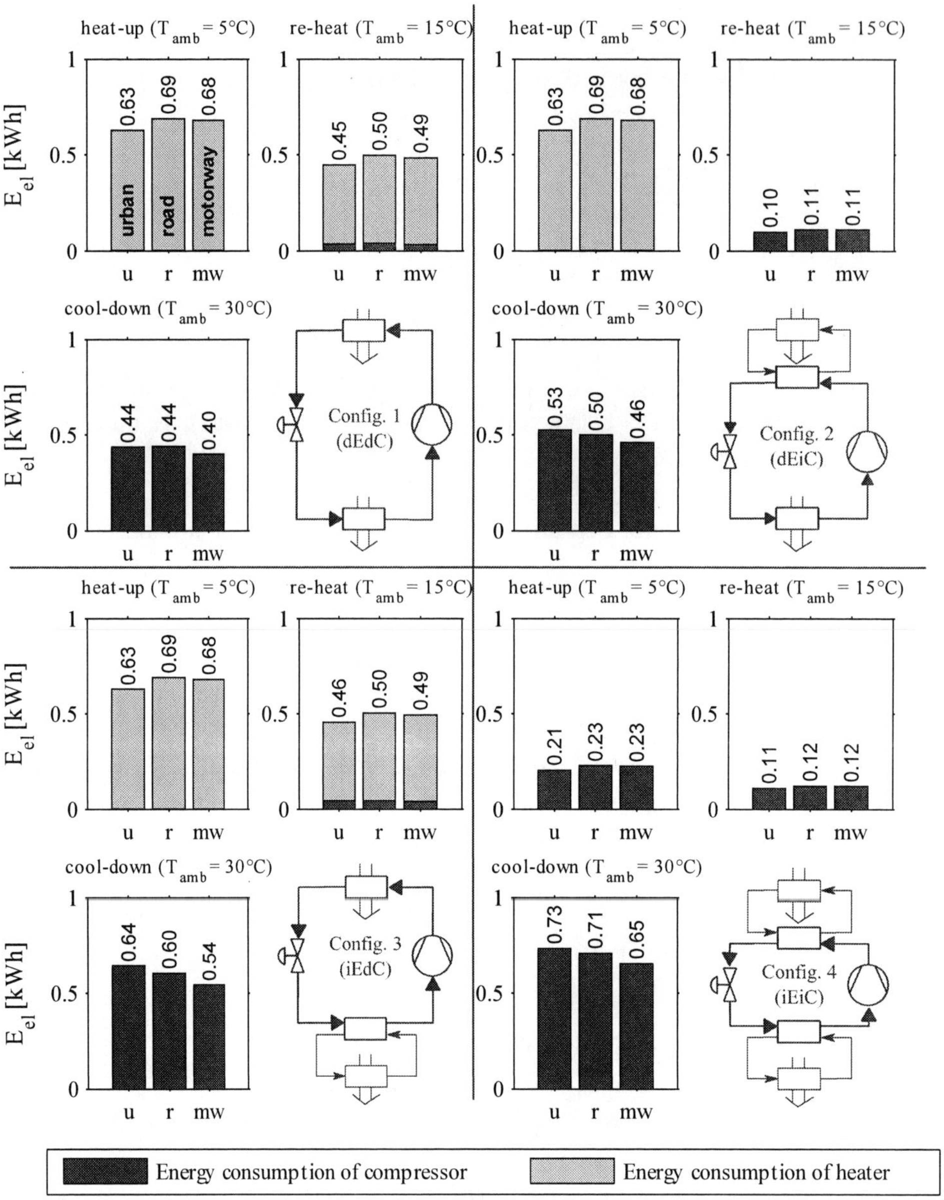

Figure 9 – Results of the energy consumption for all evaluated transient simulations.

Multiplied with the weighting factors for the distribution of driving behaviour shown in Table 1 and the weather date in Table 2, the average energy consumption for an Artemis driving cycle as well as the average energy consumption for each configuration can be obtained from Figure 10.

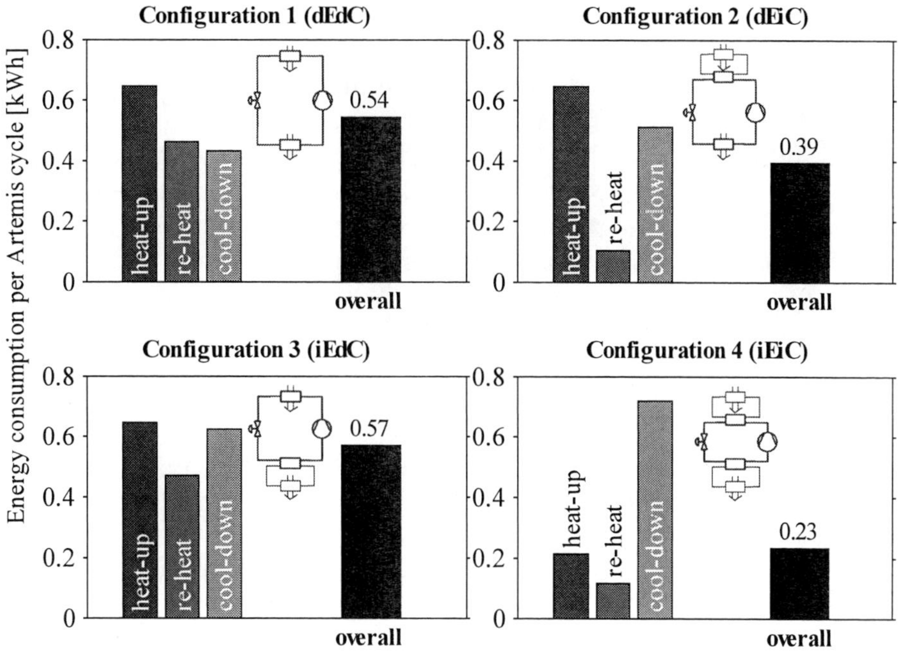

Figure 10 – Weighted energy consumption for an average Artemis driving cycle.

DISCUSSION

During cabin heat-up with configuration 1 (dEdC), 2 (dEiC) and 3 (iEdC), a significantly higher energy consumption can be seen due to the fact that all the heat load has to be brought to the system with an electric heater. In contrast to this, it is possible to operate configuration 4 (iEiC) in an efficient heat pump mode.

During re-heat mode, configurations 2 (dEiC) and 4 (iEiC) show an efficient performance using the condensation heat load to re-heat the dehumidified air.

For cabin cool-down during hot summer conditions, configuration 1 (dEdC) and 2 (dEiC) show the best performance in terms of average energy consumption. This is because of the additional layer of thermal resistance between coolant and refrigerant on the low pressure side. Due to this coolant's thermal mass, a higher amount of energy has to be put up in configuration 3 (iEdC) and 4 (iEiC). It can be seen, that the heat transfer on the low pressure side has a significant influence on the overall system behaviour and can limit the transient performance of the system.

When considering the prevalence (i.e. weighting factors) of the considered environmental conditions, efficiency is crucial especially during heat-up and re-heat mode; however, performance (i.e. time to comfort within the passenger compartment) should not be neglected.

SUMMARY

This work presents a simulative comparison of four different configurations of a refrigerant system using the refrigerant R134a: direct condenser + direct evaporator (conventional refrigeration cycle), direct condenser + indirect evaporator, indirect condenser + direct evaporator, and indirect condenser + indirect evaporator (fully integrated secondary loop system). Under defined use cases and environmental conditions (heat-up, re-heat and cool-down), simulations are conducted to compare the four system designs while taking into account the possibility of using the secondary loop system in an efficient heat pump mode for cabin heating.

It has been shown that a fully integrated secondary loop system can be used to reduce the energy consumption of the thermal management system significantly. This is due to the fact that a heat pump using the waste heat of the electric drive train components can be utilized during winter operation for efficient cabin heat-up at cold start conditions or during re-heat operation.

The next step in the evaluation of secondary loop systems is the inclusion of daily driving patterns, including parking and standing times between driving phases. It is thereby possible to account for storage effects within the secondary coolant loops due to their thermal masses.

REFERENCES

[1] K. Bennion and M. Thornton, "Integrated Vehicle Thermal Management for Advanced Vehicle Propulsion Technologies", in *SAE 2010 World Congress*, Detroit, Michigan, 2010.

[2] K. Wang, M. Eisele, Y. Hwang and R. Radermacher, "Review of Secondary Loop Refrigeration Systems", *International Journal of Refrigeration,* Nr. 33(2), 2010.

[3] M. Ghodbane, T. D. Craig and J. A. Baker, "Demonstration of an energy-efficient secondary loop HFC-152a mobile air conditioning system", in *Final Report for the U.S. Environmental Protection Agency*, 2007.

[4] C. Kowsky, E. Wolfe, L. Leitzel and F. Oddi, "Unitary HPAC system", *SAE International Journal of Passenger Cars - Mechanical Systems,* Nr. 5(2), 2012.

[5] N. C. Lemke, J. Lemke and J. Koehler, "Secondary Loop Systems for Automotive HVAC Units under different Climatic Conditions", in *International Refrigeration and Air Conditioning Conference*, Purdue, 2012.

[6] M. Eisele, Y. Hwang and R. Radermacher, "Utilization of Ice Storage in Secondary Loop Automotive Air-Conditioning Systems", in *SAE 2013 World Congress & Exhibition*, Detroit, MI, USA, 2013.

[7] M. Andre, "The ARTEMIS European driving cycles for measuering car pollutant emissions", *Science of the Total Environment,* pp. 79-84, 2004.

[8] T. A. Weustenfeld, J. C. Menken, K. Strasser and J. Koehler, "Central Thermal Management for Electric Vehicles using a fully integrated Primary Loop Refrigeration System", in *Thermal Management System Symposium*, Denver, 2014.

[9] P. Rumbolz, A. Piegsa and H.-C. Reuss, "Messung der Fahrzeug-internen Leistungsflüsse und der diese beeinflussenden Größen im "real-life" Fahrbetrieb", in *VDI-Tagung Innovative Fahrzeugantriebe*, Dresden, 2010.

[10] TLK Thermo GmbH and University of Braunschweig, *TIL Library, version 3.2.3.* 2014.

[11] F. Schedel, G. Suck, S. Foersterling, W. Tegethoff and J. Koehler, "Effizienzbewertungen von Wärmepumpen in Hybridfahrzeugen mit Hilfe der verlustbasierten Modellierung von Scrollverdichtern", in *DKV-Tagung*, Hannover, 2013.

[12] H. Martin, "Pressure Drop and Heat Transfer in Plate Heat Exchangers", in *VDI Heat Atlas*, Heidelberg, Springer Verlag, 2010, pp. Mm1-Mm7.

[13] W. W. Akers, H. A. Deans and O. K. Crosser, "Condensation Heat Transfer within Horizontal Tubes", *Chemical Engineering Progress Symposium,* Nr. 55, pp. 171-176, 1959.

[14] H. Raiser, T. Heckenberger, W. Tegethoff, J. Koehler and S. Foesterling, "Transient behaviour of R744 vehicle refrigeration cycles and the influence of the suction side accumulator design", *SAE Technical Paper,* no. 2006-01-0162, 2006.

[15] J. C. Menken, T. A. Weustenfeld and J. Koehler, "Experimental Comparison of the Refrigerant Reservoir Position in a Primary Loop Refrigerant Cycle with Optimal Operation", in *International Refrigeration and Air Conditioning Conference*, Purdue, 2014.

[16] J. C. Menken, T. A. Weustenfeld and J. Koehler, "Reduktion der Kältemittelfüllmenge im Pkw durch den Einsatz von kompakten Kältemittelkreisläufen", in *DKV-Tagung 2014*, Düsseldorf, 2014.

Next generation header materials for corrosion resistant heat exchangers

V SASS, G BERMIG, H JANSSEN and **S SCHLUETER**
Hydro Aluminium Rolled Products GmbH, Germany

ABSTRACT

Today's automotive condensers are typically based on a design with extruded microchannel tubes combined with corrugated fins and header tubes made from rolled aluminium strip. Increasing demands for corrosion lifetime relentlessly expose any inconsistency in the galvanic compatibility of the individual components, which concerns especially the match between rolled and extruded aluminium alloys.

Consequently a new header material has been developed which was for the first time optimized for galvanic compatibility with extruded microchannel tubes. Thus the hazard of galvanic corrosion in the header – tube joint region is eliminated and the use of zinc containing filler alloys on the header is no longer necessary. The lower total zinc load also reduces zinc enrichment in brazing joints and helps to prevent premature fin loss.

1 INTRODUCTION

Brazed aluminium condensers have become a standard for automotive air conditioning systems [1]. The most common designs are parallel flow condensers based on microchannel tubes (also known as MultiPort Extruded tubes (MPEs)), which offer high efficiency and allow for high operating pressures. Recently the corrosion lifetime of condensers has come into focus and the requirements for future projects are expected to go beyond the current industry standard of 40 days component lifetime in the SWAAT corrosion test acc. to ASTM G85-A3. In order to meet these expectations a thorough understanding of the corrosion mechanisms in heat exchangers and the interactions between the individual components is required. One of the weak points in the corrosion design of typical condensers is the mismatch in the electrochemical corrosion potential between header materials and the microchannel tube alloys which often results in premature leakages in the header – tube joint region. A newly developed header alloy with matched electrochemical corrosion potential (ECP) is expected to address this problem effectively.

2 CORROSION DESIGN FOR BRAZED ALUMINIUM HEAT EXCHANGERS

Resistance against environmental corrosion is a fundamental requirement for any automotive heat exchanger. In a typical brazed aluminium heat exchanger different tailored alloys are used for the different heat exchanger components, e.g. fins, tubes or manifolds, with slightly different chemical compositions and accordingly different electrochemical corrosion potential values. The different components of a heat exchanger form a galvanically coupled system and corrosion attack in an aggressive environment will be focused on the components with the least noble corrosion potential, whereas the remaining parts will be cathodically protected. This principle is commonly used to achieve long corrosion lifetimes by a so-called "corrosion design" of heat exchangers. This is based on the use of aluminium alloys with a matched electrochemical corrosion potential (ECP) for the different heat exchanger components. As schematically shown in Fig. 1 an alloy with the most noble corrosion potential is selected for the thin walled fluid containing elements, i.e. flat tubes, a less noble alloy is used for thick walled fluid containing elements (header) and the least noble alloy acts as fin material. Thus any corrosion attack will start at the fins, which act as a sacrificial anode for the fluid containing tubes and long component lifetimes can be achieved before the first tube leakage will occur.

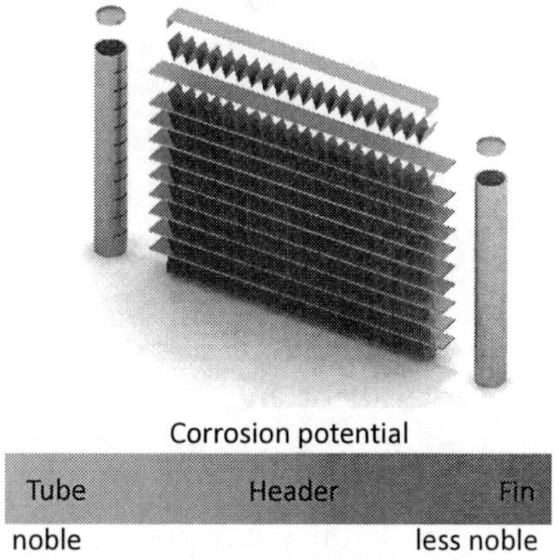

Fig. 1: Schematic illustration of material selection by corrosion potential for corrosion design of brazed aluminium heat exchangers.

The manufacturers of heat exchangers sheet materials have developed material packages with matched corrosion potentials for all types of automotive exchangers will allow for an optimum corrosion design.

The material selection becomes more complicated if rolled sheet material is combined with extruded components. This situation is typically for parallel flow condensers (PFC), where often extruded microchannel tubes are used. The aluminium alloys commonly used for microchannel tubes typically contain less alloying elements like manganese or copper than rolled heat exchangers alloys in order to allow for a productive extrusion process. Since these alloying elements significantly shift the ECP of an alloy in a more noble direction while kept in solid solution as shown in Fig. 2, typical extrusion alloys like EN-AW 3102 have a less noble ECP than common header alloys based on the EN-AW 3003 or EN-AW 3017 alloy compositions and will thus not be galvanically protected by these header alloys. In order to achieve an efficient corrosion protection of the microchannel tubes in many cases zinc containing coatings

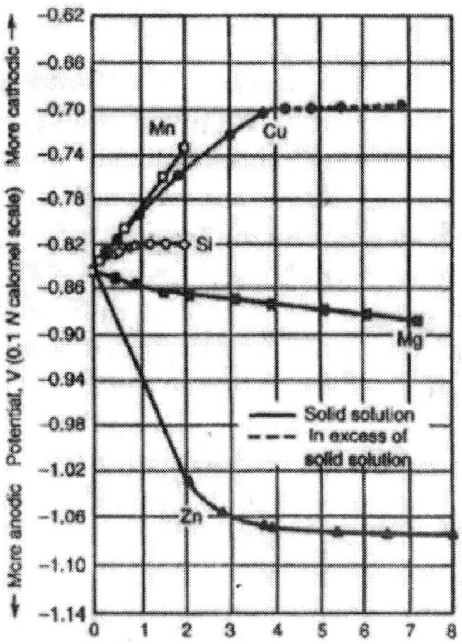

Fig. 2: Influence of different alloying elements in solid solution on the electrochemical corrosion potential of aluminium alloys [2].

are employed. Such coatings shift the ECP of the tube surface in a less noble direction and create sacrificial layer, which provide cathodic protection and enhance lateral instead of pitting corrosion.

Common zinc containing coatings for microchannel tubes are formed by zinc arc spray (ZAS) or zinc containing flux coatings like (Zn Flux or ZnSilFlux) [3]. Careful control of the zinc amount and the resulting diffusion profile after brazing is necessary in order to avoid excessive zinc enrichment at the tube surface and in the tube – fin braze joints. The typical fin material for condensers is a braze clad EN-AW 3003 type core alloy with an addition of 1,2% or more zinc. The superficial zinc content of the microchannel tubes after brazing should be kept below the zinc content of the fins in order to avoid zinc enrichment and preferential corrosion of the braze joints between tubes and fins, which result in a premature fin loss and a reduction of heat exchanger efficiency [4]. Using a combination of optimized zinc coatings and advanced longlife microchannel tubes, however, corrosion of the heat exchanger core can be controlled and lifetimes of well above the current standard of 40 days in SWAAT can be achieved. This development shifts the focus on another critical corrosion issue, namely the header – tube joint and the adjacent fin-free tube region. As already pointed out above, the typical header alloys have a nobler ECP than the microchannel tube alloys. Thus galvanic coupling will induce corrosion at the header - tube braze joint. An example is given in Fig. 3 which shows a header – tube joint in a brazed heat exchanger which failed in a SWAAT corrosion test after only 17 days. The header alloy is a standard material with an ECP in as brazed condition of approx. -725mV, the uncoated microchannel tube is made of EN-AW 3102 alloy with a measured ECP of -742mV. This mismatch in the galvanic potential induced severe corrosion of the tube.

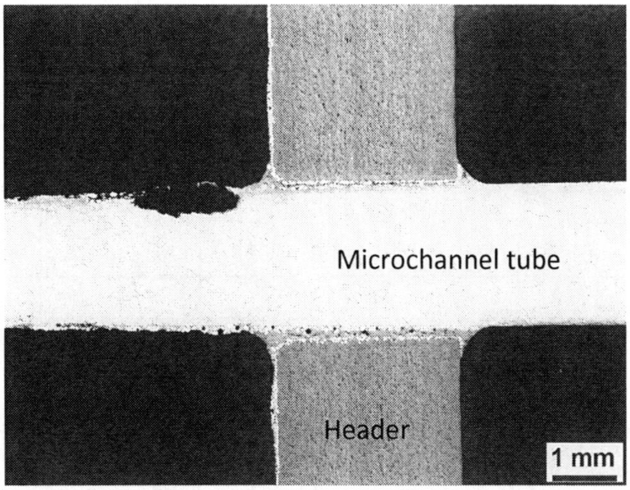

Fig. 3: Cross section of a header – microchannel tube joint in a heat exchanger which failed after 17 days of SWAAT corrosion testing. A mismatch of the ECP between the standard header material (-725mV) and the EN-AW 3102 tube alloy (-742mV) induced severe corrosion of the tube by galvanic coupling.

A zinc containing coating on the tubes will delay the corrosion progress, but once the superficial zinc layer is consumed rapid dissolution of the tube material will occur. This effect was highlighted in a recent study [4], in which the effect of different tube and header alloys as well as several zinc containing tube coatings on the lifetime of brazed heat exchangers in a SWAAT corrosion test was investigated. Since in a first test round most leakages were observed at the header - tube braze joints, a second test round was carried out with the header – tube region shielded by plastic foil wrapping. Depending on the material and coating combinations improvement in SWAAT lifetime of more than 40 days were observed. This result clearly shows the need for header materials with a matched ECP and the drastic potential gain in condenser corrosion lifetime by the use of such header materials.

3 DEVELOPMENT OF A NEW HEADER ALLOY WITH A MATCHED CORROSION POTENTIAL FOR MICROCHANNEL TUBES

The key to adjust the ECP of aluminium alloys is the control of the content of alloying elements in solid solution. In common heat exchanger alloys for rolled products the main alloying element is manganese, which shifts the ECP significantly in a more noble direction (see Fig. 2). The same effect is caused by copper, which is added to many alloys in order to increase mechanical strength and to deliberately adjust the ECP. Alloying elements which shift the ECP in a less noble direction are mainly magnesium and zinc. Both elements should be avoided in a header alloy due to a negative effect on brazing properties or in the latter case due to the high diffusion rate, which makes control of the zinc distribution after brazing unpredictable.

Thus the control of the ECP has to be achieved by adjusting the chemical composition and the thermo-mechanical processing simultaneously in order to minimize the content of alloying elements in solid solution after brazing. The new header alloy developed by Hydro with the designation HA 3903-R achieves this goal with a balanced Cu-free composition shown in Tab. 1.

Tab. 1: Chemical composition range of header alloy HA 3903-R. All values are given in wt.-%.

	Chemical elements								others	
	Si	Fe	Cu	Mn	Mg	Cr	Zn	Ti	each	total
min.				0.9						
max.	0.7	0.7	0.05	1.5	0.10	0.20	0.10		0.05	0.15

The HA 3903-R alloy is available with a wide variety of cladding alloys on one or both sides depending on customer requirements. The most common tempers for header applications are the soft state O-Temper and the half hard temper H14, but also temper H24 is available.

The mechanical properties of the new alloy in as delivered condition and after a simulated brazing process are shown in Tab. 2. The as brazed values were determined after a simulated Controlled Atmosphere Brazing (CAB) process with a heating rate of 0,9K/sec and a 5min holding period at 600°C. The mechanical properties are equal to typical header alloys and superior to EN-AW 3003 standard material.

Tab. 2: Typical mechanical properties of HA 3903-R material in as delivered condition and after a simulated brazing process. The as brazed values depend on the brazing conditions.

Alloy	Temper*	Typical Thickness	Typical strength As delivered		Typical strength As brazed	
		[mm]	$R_{p0.2}$	R_m	$R_{p0.2}$	R_m
HA 3903-R	O	0.5 – 3.0	min. 40	110 - 150	min. 45	120 - 150
	H14	0.5 – 3.0	min. 140	150 - 190	min. 45	120 - 150

4 CORROSION PROPERTIES OF THE NEW HA 3903-R HEADER ALLOY

4.1 Electrochemical corrosion potential

The ECP of the new header alloy HA 3903-R measured according to ASTM G69 standard against a saturated calomel electrode is in the range between -740mV and -760mV both for as delivered and as brazed condition. In Fig. 4 the ECP of HA 3903-R in as brazed condition is shown together with the values of typical standard header alloys and microchannel tubes measured in the Hydro corrosion labs.

As intended the corrosion potential of the HA 3903-R header alloy is significantly more negative than that of comparable alternative header alloys and is also more negative than the ECP of most common microchannel tube alloys.

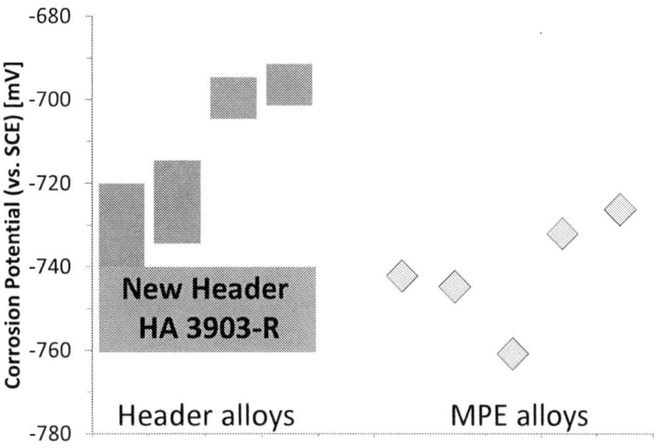

Fig. 4: Illustration of the electrochemical corrosion potential range of HA 3903-R compared to typical standard header and microchannel tube alloys.

4.2 Galvanic compatibility tests

Galvanic compatibility between two materials can be easily checked by a galvanic corrosion test as defined in the DIN 50919 standard. In this test two dissimilar coupon samples are immersed in an electrolyte (acidified (pH 2.8 to 3.0) synthetic seawater solution as defined in ASTM G85-A3) and the galvanic current between the samples is measured over a period of 48h. The direction of the galvanic current indicates which of the samples is corroding (anode) and which is protected (cathode). A further measurement of the mass loss of the samples reveals the extent of the corrosion attack.

Tests were carried out with the new header material HA 3903-R and a standard header alloy (EN-AW 3005 mod.) combined with microchannel tube alloy samples EN-AW 3102 and 9108 provided by Sapa Precision Tubing.

The measured ECP values of the samples in as brazed condition were -759mV for HA 3903-R, -721mV for the standard header alloy, -742mV for EN-AW 3102 and -744mV for 9108 tube alloy. The results of the galvanic corrosion tests are given in Tab. 3.

Tab. 3: Results of galvanic corrosion tests acc. to DIN 50919. Colours indicate galvanic compatibility between header and microchannel tube alloys (green: galvanic current flows from header to tube, tube is protected). Numbers represent the mass loss of tube samples after 48h test duration given in mg/m^2.

	Microchannel tube alloy	
Header alloy	EN-AW 3102	9108
HA 3903-R	16	62
Standard Header	344	250

The results demonstrate clearly the positive effect of using galvanically compatible alloys for header and tubes in brazed aluminium condensers. Changing from a standard header alloy to the newly developed HA 3903-R effectively suppresses galvanic corrosion of microchannel tubes and reduces corrosion mass loss by a factor of 4 to 20 depending on the exact alloy combination.

4.3 Corrosion tests with minicore samples

For a further test of the new header material minicore samples were brazed as shown in Fig. 5. The material combination used was:

- Header material: HA 3903-R with one sided braze clad 7,5% EN-AW 4045 1,5mm thickness, O-Temper
- Microchannel tubes: EN-AW 3102 uncoated, 16mm x 1,3mm cross section
- Fins: EN-AW 3003 mod. + 1,5% Zn with 10% EN-AW 4343 braze clad in 0,10mm thickness, H14

At one side of the minicores a fin free area with a width of about 20mm was left in order to deprive the tubes deliberately of the galvanic protection from the fins. Due to the simple design of the minicores the time until the first leakage could not be determined. A visual inspection after 40 days SWAAT corrosion testing showed some corrosion of the fins, but no severe corrosion attack of the tubes was observed. This result is confirmed by metallographic investigations of tubes in the most exposed fin free areas. As shown in Fig. 5b no severe corrosion attack is observed.

Further corrosion tests with automotive condensers were conducted by a major heat exchanger manufacturer which clearly confirmed the positive effect on corrosion lifetime of using the new header alloy HA 3903-R instead of conventional Cu-containing header alloys. The results of these investigations will be published in due time.

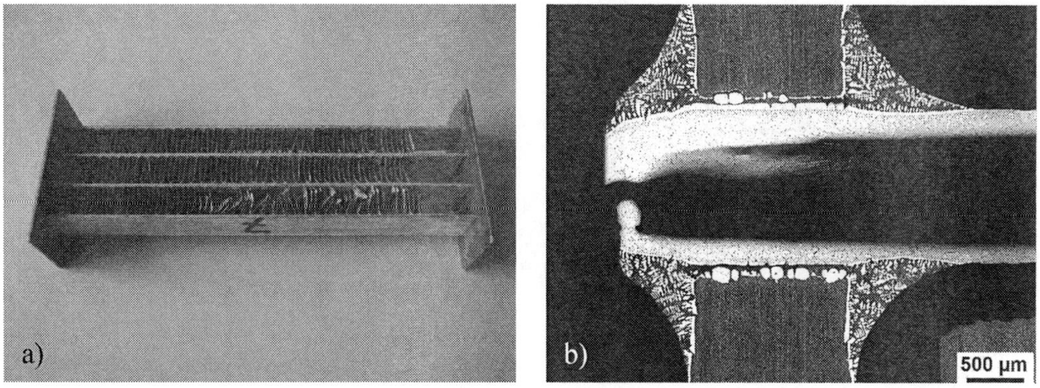

Fig. 5: Brazed minicore sample with HA 3903-R header material after 40 days of SWAAT corrosion testing. a) Complete minicore. b) Cross section through header – tube braze joint.

5 CONCLUSIONS

In order to further improve corrosion lifetime of brazed automotive condensers beyond the current standards a careful match of the electrochemical potential of all materials is required, including both rolled and extruded particles. An important step towards this goal is the introduction of a new Cu-free header alloy HA 3903-R, which was specifically designed for galvanic compatibility with extruded microchannel tubes. Apart from preventing corrosion issues in the header – tube – joint section the use of the new header material also offers the potential to reduce the total zinc load in the corrosion system, which will offer further benefits like prevention of premature fin loss due to excessive zinc enrichment in fin – tube brazing joints.

REFERENCE LIST

[1] Sugihara, A. and Lukas, H., "Performance of Parallel Flow Condensers in Vehicular Applications," SAE Technical Paper 900597, 1990, doi:10.4271/900597.

[2] J. Davis: "Corrosion of aluminium and aluminium alloys", ASM International, 2011.

[3] D. Steiner et al., "Pre-coated tubes for brazed heat exchanger – experience with Zn containing fluxes", Proc. of 8th Int. Congress Aluminium Brazing, Düsseldorf, 3rd – 5th June 2014.

[4] J. H. Nordlien et al., "Recent advances in corrosion resistant alloy-coating combinations for Al microchannel tubing", Proc. of 8th Int. Congress Aluminium Brazing, Düsseldorf, 3rd – 5th June 2014.

Author Index

G

H

I

J

K

L

M

U

V

W

Y

Z